Basic Anatomy and Physiology of the Human Body

Basic Anatomy and Physiology of the Human Body

J. Robert McClintic, PhD
California State University, Fresno, California

John Wiley & Sons, Inc.   New York • London • Sydney • Toronto

Photos of models in Chapters 1 and 9 by George Roos.

Library of Congress Cataloging in Publication Data

McClintic, J   Robert, 1928-
    Basic anatomy and physiology of the human body.

    Includes bibliographies and index.
    1. Human physiology.   2. Anatomy, Human.   I. Title.
QP34.5.M3   612   74-32195

ISBN 0-471-58174-7

Printed in the United States of America

10-9 8 7 6 5 4 3 2 1

This book is dedicated
to my wife
Peggy
whose patience and assistance
aided its completion

and to my daughters
Cathleen, Colleen, Marlene
who make it all worthwhile

Preface

This book is designed to meet the needs of an introductory one-quarter course or a one-semester combined course in human anatomy and physiology. Persons who want to know how their body is made and works, and persons who are preparing for health-oriented careers will find the material interesting and stimulating. Instructors of introductory courses in human anatomy and human physiology will also find the text appropriate because of its outstanding artwork in anatomy and the conciseness and clarity of the physiological presentations. In each chapter there are summary tables and step-by-step "flow sheets" that enable the reader to check his grasp of basic concepts before proceeding. The summaries at the end of chapters provide a final review of the chapter content. There is a complete glossary, which not only gives the definitions of words but also provides an opportunity to understand their derivation and pronunciation.

The discussion of each system or organ is introduced by a primarily pictorial presentation of the system's or organ's development. This is followed by a well illustrated discussion of normal structure and function, which leads to an understanding of the changes described in the section on clinical considerations. This section discusses the clinical conditions that are most common and the ones that are presented daily in the mass communications media. The discussion emphasizes the disruption of normal processes and the inherent ability of the body to compensate for the abnormality. Thus the reader acquires a coordinated picture of the body's system from the embryo onward.

Accepting the ancient premise that "one picture is worth a thousand words," the chapters that are primarily anatomical in orientation are condensed to the excellent anatomical artwork, and to tables that direct attention to the essential concepts and features. Reference is provided to the living body by photographs and discussion where appropriate.

The text material is up to date and relevant as, for example, in the areas of immunity, transplantation, genetics, development, aging, contraception, and venereal disease. Heavy emphasis is not placed on chemical and biochemical processes or on cellular ultrastructure but, instead, on the concepts that should be learned and that will be a basis for further education.

The user of this book, in addition to acquiring *knowledge* of human anatomy and physiology, will gain an *appreciation* of the interrelationships of his own body parts and the relationship of his body to nature in general. The end result will be an appreciation of what is involved in the concept of "good health," and the reader will be able to make judgments about the validity and appropriateness of the statements that constantly reach us through the mass media.

I thank the staff of Wiley and especially Robert L. Rogers, biology editor, for encouragement and support during the preparation of the manuscript. I also am greatly indebted to my wife Peggy McClintic, MSN, whose patience, constructive criticism, typing, and proofreading contributed greatly to the production of this excellent textbook.

The responsibility for any errors or omissions is mine.

J. Robert McClintic, PhD

Contents

chapter 6 The Skeleton

chapter 7 Articulations

chapter 8 The Structure and Properties of Muscular Tissue, with Emphasis on Skeletal Muscle

chapter 9 The Skeletal Muscles

chapter 10 The Basic Organization and Properties of the Nervous System

chapter 11 The Spinal Cord and Spinal Nerves

chapter 12 The Brain and Cranial Nerves

chapter 13 The Autonomic Nervous System

chapter 14 Blood Supply of the Central Nervous System; Ventricles and Cerebrospinal Fluid

chapter 15 Sensation

chapter 16 Body Fluids and Acid-base Balance

chapter 24 The Urinary System

chapter 25 The Reproductive Systems

chapter 26 The Endocrines

Epilogue

Basic Anatomy and Physiology of the Human Body

chapter 1
An Introduction to the Structure and Function of the Human Body

chapter 1

The study of living organisms

The science of biology studies living things, and includes many subdivisions. ANATOMY is the subdivision that deals with the study of the structure of an organism. In the study of the human body, considerable knowledge of structure may be gained by looking with the naked eye; this acquaints us with the *gross anatomy* of the object being viewed. Dissection, which involves cutting and teasing of body parts, aids the viewing of items of interest that lie covered by other structures, and is, in fact, the basis of the word anatomy (G. *ana,* up + *temnein,* to cut). Viewing the smaller units of body organization may require the use of microscopes of various types. *Microscopic anatomy,* including cytology (study of cells), histology (study of tissues and organs), and developmental anatomy or embryology (study of how the body develops and grows), enables us to study the fine structure and origins of the body components.

PHYSIOLOGY studies how the body and its parts work or function. The word physiology means the study of the nature of things (G. *physis,* nature, + *logos,* study) and the discipline draws on many other areas of knowledge to explain body function. For example, physiology draws on *anatomy* for a structural basis of function; on *chemistry* and *physics* to aid in the definition of basic substances the body contains, and laws the body operations follow; on *biochemistry* to aid in the understanding of the complex chemical reactions that occur in the body; on *biophysics* to aid in the explanation of electrical and physical phenomena that occur in the body; and on *genetics* and *embryology* to explain the processes involved in determination of body function, growth, and development.

Anatomy and physiology thus combine to give one a broad and exciting view of the make-up and activity of the shell we inhabit during our days on earth (or in space).

Some generalizations about body structure and function

As we proceed with the study of the human body, it is well to keep in mind some basic ideas about the body. These ideas, or generalizations, will give a direction and purpose to our study. The reader is encouraged to add his own generalizations to those given below.

1. THE BODY HAS SEVERAL LEVELS OF ORGANIZATION. The basic units of structure and function of the human body are its CELLS. It has been estimated that there are 1×10^{14} (1 followed by 14 zeros, or 100 trillion) cells in the body. With this many individual units demanding nutrients and producing wastes, problems of supply and removal would seem to be too much to overcome, yet the organization of the body has solved these problems. Cells that are similar in structure and function, together with their associated intercellular material (the substance between the cells), form TISSUES. There are four primary tissue groups that compose the body.

Epithelial tissues cover and line internal and external body surfaces.

Connective tissues connect and support body parts.

Muscular tissues can shorten or contract to cause movement.

Nervous tissues conduct nerve impulses through the body.

Two or more tissues put together in a specific pattern to carry out a particular job, form an ORGAN. Several organs working together to carry out a larger body process, form a SYSTEM. Many systems are combined to form the human body (Fig. 1.1). Although there are obvious external differences between the male and female, "inside" we are all nearly the same. The only basic differences between the sexes lie in the organs of the reproductive systems. All other functions are carried out by nearly identical organs with similar functions. Table 1.1 presents an introduction to the systems and organs of the body and their functions.

2. THE FUNCTIONS CARRIED OUT BY THE LOWER LEVELS OF ORGANIZATION ARE ALSO CARRIED OUT BY THE BODY AS A WHOLE. Thus, understanding of the cellular level contributes to knowledge of tissue, organ, and system levels, and ultimately to knowledge of the whole organism. As examples of the dependence of whole body function on cells, we may cite the following comparisons between cellular and organism activity.

Cells are EXCITABLE or capable of responding to changes in the external and internal environ-

---

FIGURE 1.1.   Male and female adults.

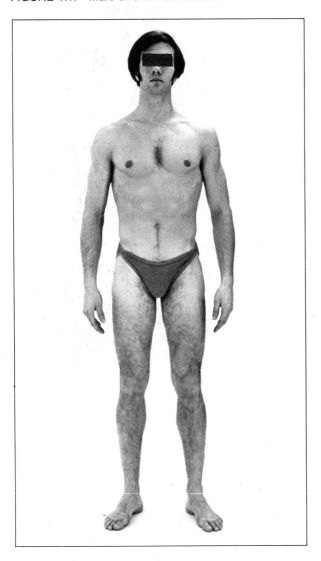
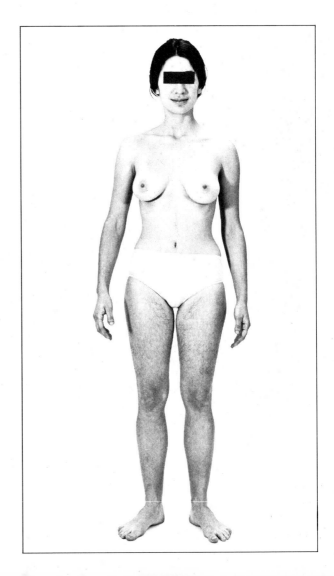

TABLE 1.1 Body Systems, Their Organs, and Functions

System	Major organs or tissues	Main function(s) of the system
Integumentary	Skin, hair, nails, skin glands	Protection; temperature control
Skeletal	Bones and joints, cartilages	Support; protection; storehouse of minerals; blood cell formation
Muscular	Skeletal muscles	Cause body movement
Circulatory	Blood, heart, blood vessels	Carry nutrients and wastes; circulate the blood; carry the blood
Lymphatic	Lymph, lymph organs (tonsils, nodes, spleen, thymus), lymph vessels	Return tissue fluid to blood vessels; create immunity; form blood cells
Respiratory	Nose, throat, larynx, trachea, lungs	Supply oxygen; eliminate carbon dioxide; regulate acid-base balance
Digestive	Mouth, esophagus, stomach, intestines, liver, pancreas	Digest and absorb nutrients; excrete wastes
Urinary	Kidney, ureters, bladder, urethra	Excrete wastes and regulate blood composition
Reproductive	Male: testes, ducts, accessory glands	Produce sperm and secretions of the semen
	Female: Ovaries, uterine tubes, uterus, vagina, mammary glands	Produce eggs; nourish offspring
Nervous	Brain, spinal cord, peripheral nerves, organs of special sense	Control many body activities, allow appreciation of changes internally and externally
Endocrine	All glands secreting hormones into the blood stream (e.g.: pituitary, thyroid, parathyroid, adrenals, pancreas, testes, ovaries)	Control metabolism, growth, and development of the body

ments of the body; so the body as a whole responds to change.

Cells INGEST or take in materials; the whole body eats food.

Cells DIGEST foods and metabolize them to release energy for body activities or formation of new materials; the body as a whole carries on digestion and metabolism.

Cells EXCRETE or rid themselves of wastes of their activity; the body rids itself of wastes in the urine and feces, and through the skin.

Cells produce, from body fluids, useful products

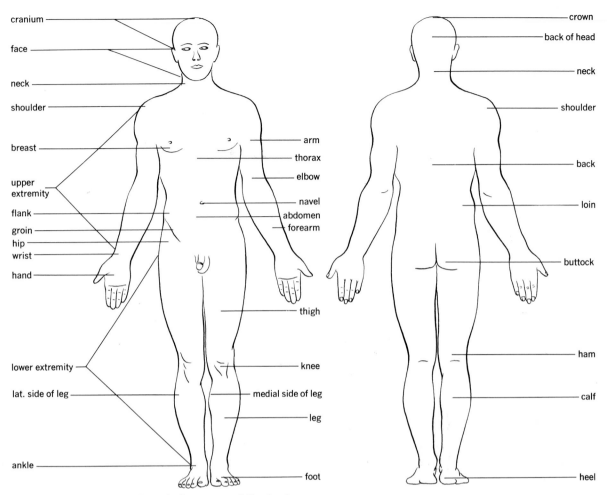

FIGURE 1.2. General descriptive areas of the body.

known as secretions; the process of SECRETION is widely utilized to produce digestive enzymes, hormones, and other materials used in the body.

Most cells REPRODUCE themselves for purposes of repair, growth, and continuance of a given line of cells; the body reproduces itself for continuance of the species.

MOVEMENT of materials occurs within cells, and the whole organism may be caused to move through its environment.

These functions are not merely lists of what the cell or organism does, but are CRITERIA OF LIFE as well. Something that exhibits these activities may properly be considered living.

3. THE BODY, PARTICULARLY ITS FUNCTIONS, IS ORGANIZED TO MAINTAIN HOMEOSTASIS AND INSURE SURVIVAL OF CELLS. A system of checks and balances operates to maintain body composition and function nearly constant within the very narrow limits necessary for survival of individual cells, and therefore the organism. The term HOMEO-STASIS describes this nearly constant internal state the body normally maintains in functions such as composition of body fluids, body temperature, and levels of acids and bases. Much of physiology deals with discovering and describing the controlling mechanisms that insure continued body homeostasis. Alterations of homeostasis form the basis for diagnosis of disease and abnormal function, and also the basis for instituting treatment intended to restore normal function.

4. MANY BODY FUNCTIONS ARE DETERMINED BY STRUCTURE, and knowledge of structure may enable prediction of function. A "common sense" relationship often exists between structure and function. For example: muscle cells must be elongate structures or they cannot shorten effectively; bone must be a hard strong tissue to protect and

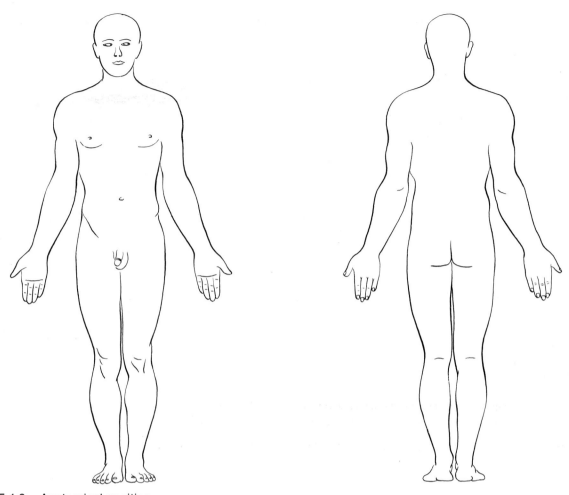

FIGURE 1.3.   Anatomical position.

to serve as the support for the body. Many examples of the relationship between structure and function will be cited in the chapters that follow.

5. BOTH STRUCTURE AND FUNCTION CHANGE AS THE INDIVIDUAL AGES. A newborn is mostly water and fat, and many body functions are not mature. An adult is different chemically from an infant, containing more solids and relatively less water. Old age is associated with lowered levels of body function and a "return to infant state" in terms of teeth and certain other body structures. Each age group has certain attributes that characterize it, and an infant should not be considered a "small adult" any more than a young or middle aged adult should be considered to have the same structure and level of function as an octogenarian (80 years of age).

Application of these generalizations to the study of the human body will be made in later chapters.

The following sections introduce some basic terminology that will be encountered many times in later chapters. These terms will increase understanding and appreciation of body structure and function.

Gross body areas

An aid to the naming of gross body areas is presented in Figure 1.2. These commonly used terms are often referred to in identifying an area in which organs may be located, or in which there is discomfort or pain in the individual.

Terms of direction

The location and relationships of many body parts may be described by words that are used when the body is in ANATOMICAL POSITION (Fig. 1.3). In this position, the body is standing erect, the eyes are level and directed forward, the arms are at the sides with the palms forward, and the feet are parallel, with the heels close together. The terms are:

Anterior or ventral. The front or belly side of the body, or something in front of the original point of reference; for example, the sternum (breastbone) is located on the anterior or ventral part of the chest.

Posterior or *dorsal.* The back side of the body, or something in back of the original point of reference; for example, the spine is located on the posterior or dorsal part of the body.

Superior. Above, toward the head, or something higher on the body than the original point of reference; for example, the head is superior to the neck.

Inferior. Below, toward the feet, or something lower on the body than the original point of reference; for example, the feet are inferior to the knees.

Medial. A line running from the center of the forehead to between the feet defines the midline of the body. Medial implies a position closer to the midline; for example, the nose is medial to the eyes.

Lateral. A position away from, or to the side, from the midline; for example, the ear is on the lateral part of the skull.

External. This term is most frequently used to refer to something that is toward the surface of the body or a hollow organ. The term superficial is often used to mean the same thing; for example, the skin covers the external surface of the body.

Internal. This term refers to a position beneath the body surface, or toward the interior of a hollow organ. The term deep is used in the same fashion; for example, the internal organs of the abdomen are seen when the abdominal wall is opened.

Proximal. This term refers to a position closer to the point of attachment of a body part to the midline; for example, the shoulder is proximal to the elbow; the knee is proximal to the ankle.

Distal. The opposite of proximal, distal implies further from the point of attachment or midline; for example, the wrist is distal to the elbow, or, the knee is distal to the hip. The use of the terms proximal and distal is illustrated in Figure 1.4.

FIGURE 1.4. The use of the terms proximal and distal with reference to the limbs.

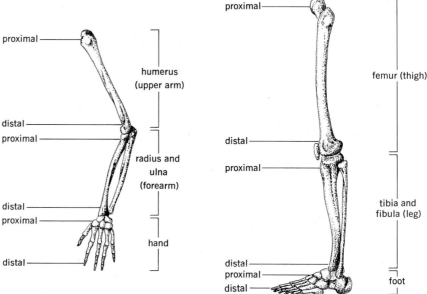

Planes of section

Further information about the location of body parts can be provided by the study of sections cut in various directions or planes through the body. These are described by the following terms.

Midsagittal (sagittal) section. The body is divided equally into right and left portions.

Parasagittal plane. Any plane parallel to the midsagittal section, but not in the midline.

Coronal plane. The body is divided into front and back portions.

Transverse (horizontal) plane or cross section. The body is divided into superior (upper) and inferior (lower) portions.

These planes are shown in Figure 1.5.

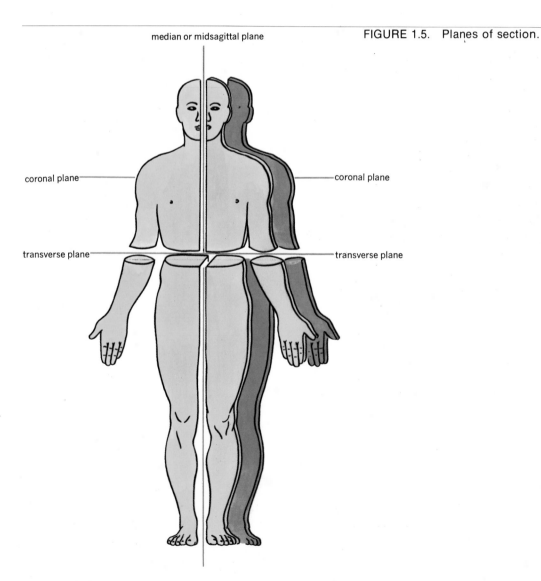

median or midsagittal plane

coronal plane

coronal plane

transverse plane

transverse plane

FIGURE 1.5.   Planes of section.

Location of organs

Body cavities

Most body organs and systems are found within the body cavities (Fig. 1.6), and location of a given organ does not vary greatly between individuals. A DORSAL CAVITY, located on the posterior or back aspect of the body, contains the brain, spinal cord, and parts of the nerves attaching to these organs. A VENTRAL CAVITY, located on the anterior or belly side of the body is subdivided by the muscular diaphragm into an upper *thoracic* or chest cavity, and a lower *abdominopelvic* cavity.

The thoracic cavity contains the thymus, esophagus, trachea, bronchi, lungs, heart, and great vessels entering and leaving the heart. The abdominopelvic cavity contains the organs of digestion, excretion, reproduction, several endocrines, and many blood vessels to and from these organs. These organs together are usually called the VISCERAL ORGANS, or the VISCERA. Some of these body organs are shown projected on the external body surface in Figure 1.7. Many of the organs of the abdominopelvic cavity (e.g., spleen, liver, kidneys, and uterus) may be *pal-*

FIGURE 1.6. The true body cavities. *(A)* Lateral view showing ventral and dorsal cavities. *(B)* Anterior view showing divisions and subdivisions of the ventral cavity.

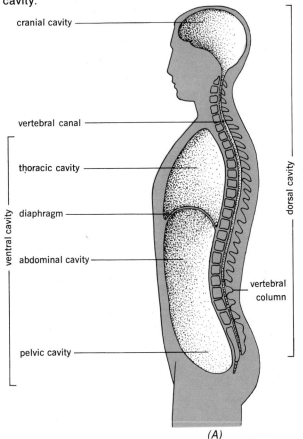

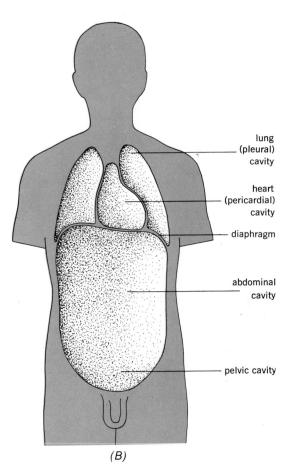

(A)

(B)

pated (felt) to determine their size, shape, and location, thus, knowledge of their position is important.

Reference lines

The positions of many body organs may be described by locating them with reference to imaginary lines drawn on the chest and abdomen. The MIDSTERNAL AND MIDCLAVICULAR LINES lie on the anterior chest (Fig. 1.8); the ANTERIOR AND POSTERIOR AXILLARY LINES lie on the side of the chest (Fig. 1.9); the VERTEBRAL AND SCAPULAR LINES lie on the back (Fig. 1.10).

On the abdomen, several unnamed lines may be drawn that subdivide the abdomen into QUADRANTS known as the right upper (RUQ), left upper (LUQ), right lower (RLQ), and left lower (LLQ). Other lines divide the abdomen into NINE SMALLER AREAS. The quadrants and smaller areas are shown on Figure 1.11.

The reader should be able to describe what organs lie in each of these abdominal regions.

FIGURE 1.7.   Surface projections of some visceral organs. *(A)* Anterior view. *(B)* Posterior view.

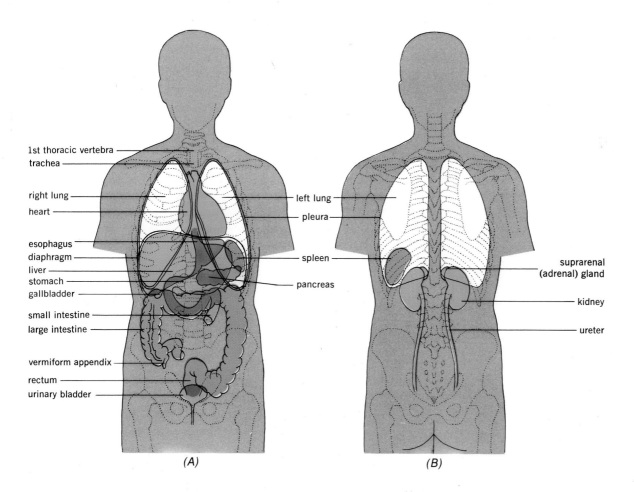

1st thoracic vertebra
trachea
right lung
heart
esophagus
diaphragm
liver
stomach
gallbladder
small intestine
large intestine
vermiform appendix
rectum
urinary bladder

left lung
pleura
spleen
pancreas

suprarenal
(adrenal) gland
kidney
ureter

(A)   *(B)*

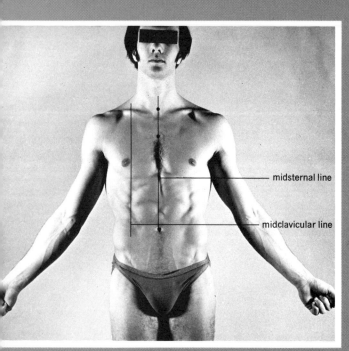

FIGURE 1.8.   Anterior thoracic reference lines.

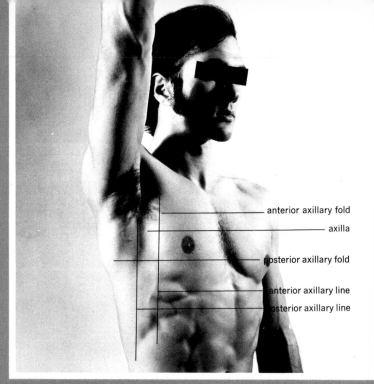

FIGURE 1.9.   Lateral thoracic reference lines.

midsternal line

midclavicular line

anterior axillary fold

axilla

posterior axillary fold

anterior axillary line

posterior axillary line

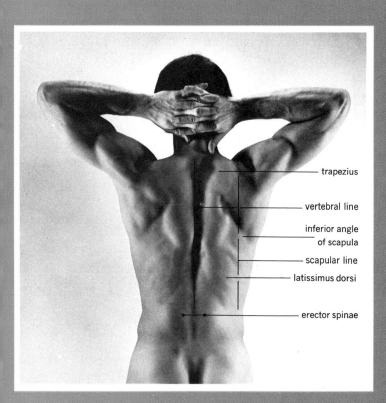

FIGURE 1.10.   Posterior thoracic reference lines.

trapezius

vertebral line

inferior angle
of scapula

scapular line

latissimus dorsi

erector spinae

FIGURE 1.11.   Abdominal reference lines.

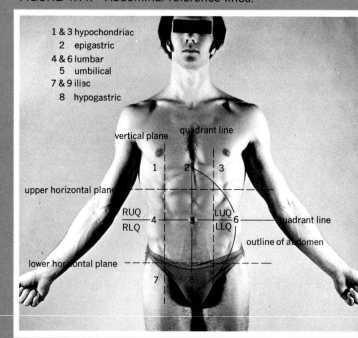

1 & 3 hypochondriac
2   epigastric
4 & 6 lumbar
5   umbilical
7 & 9 iliac
8   hypogastric

vertical plane

quadrant line

upper horizontal plane

RUQ
RLQ

LUQ
LLQ

quadrant line

outline of abdomen

lower horizontal plane

12

Summary

1. Anatomy.

   a. Studies the structure of living organisms.

   b. Is divided into several areas of study including gross anatomy (what may be seen with the naked eye), microscopic anatomy, and developmental anatomy.

2. Physiology.

   a. Studies the function of living organisms.

   b. Draws on chemistry, physics, anatomy, biochemistry, biophysics, genetics, and embryology to explain body activities.

3. Some generalizations to bear in mind about body structure and function include:

   a. There are several levels of body organization including cells, tissues, organs, systems, and the whole organism.

   b. Functions carried out by lower levels of organization are often repeated at higher levels.

   c. The body is organized to maintain structure and function within narrow limits to insure cell and organism survival. The maintenance of a nearly constant internal structure and function is called homeostasis.

   d. Structure often determines function.

   e. Structure and function are different according to age.

4. The body may be described by a series of terms that indicate direction and position.

   a. Anterior indicates front; posterior indicates back.

   b. Superior indicates higher; inferior indicates lower.

   c. Medial indicates toward the midline; lateral, away from the midline or to the side.

   d. External indicates toward the surface; internal indicates beneath the surface or inside.

   e. Proximal indicates closest to point of attachment; distal, away from the point of attachment.

5. Planes of section further indicate location of body parts.

   a. A midsagittal plane creates right and left portions.

   b. A coronal plane creates front and back portions.

   c. A transverse plane cuts across a structure and creates upper and lower portions.

6. Organs are located within cavities of the body.

   a. Dorsal cavities contain the brain and spinal cord.

   b. Ventral cavities contain the visceral organs.

7. Reference lines provide an additional way of locating organs within the body.

Questions

1. What does the study of each of the following disciplines contribute to our knowledge of anatomy and physiology?

    a. Gross anatomy.

    b. Histology.

    c. Chemistry.

    d. Embryology.

2. What are the various levels of organization in the body, and what is each composed of?

3. What is meant by homeostasis? Name several processes that must be maintained within narrow limits to insure survival.

4. Using terms of direction, describe the relative positions of the following:

    a. The elbow and the hand.

    b. The "belly button" and the hip.

    c. The thorax (chest) and the head.

    d. The skin and any organ in the abdominal cavity.

5. What is the meaning of each of the following?

    a. Midsagittal plane.

    b. Cross section.

    c. Loin.

    d. Flank.

    e. Cranium.

    f. Leg.

6. What are the body cavities and what does each contain?

7. What organs would be found in:

    a. The RUQ.

    b. The umbilical region.

    c. The right hypochondriac region.

Readings

Cannon, W. B. *The Wisdom of the Body*. Norton. New York, 1963.

Gray, Henry. *Anatomy of the Human Body*. Edited by C. M. Goss. 29th ed. Lea and Febiger. Philadelphia, 1973.

Langley, L. L. (ed). *Homeostasis. Origins of the Concept*. Benchmarks Books Publishing Program. Stroudsburg, Pa., 1973.

Morris' *Human Anatomy*. Edited by B. J. Anson. 12th ed. McGraw-Hill. New York, 1966.

chapter 2
The Cellular Level of Organization and Function

chapter 2

Cell structure and function

Chapter 1 indicates that the basic units of structure and function of the body are its cells. Individual body cells are of many different sizes* and shapes (Fig. 2.1). The smallest body cells are found in the bloodstream, while the largest (longest) are nerve cells. Shape and size may depend on the functions the cell serves. For example: muscle cells must be long if they are to shorten; nerve cells must be long to reach from the brain or spinal cord to organs in outlying areas; red blood cells have a flattened shape to permit rapid pas-

*Units commonly used to refer to sizes of cells or cell parts are:
μ = micron or 1/1000 mm (millimeter); about 1/25,000 inch.
$m\mu$ = millimicron or 1/1000 micron; about 1/25,000,000 inch.
$\overset{\circ}{A}$ = angstrom or 1/10 $m\mu$; about 1/250,000,000 inch.

sage of gases (O_2, CO_2) through their membranes.

Regardless of size or shape, most cells have three parts that are responsible for the basic structure and function of the cell. These parts are: a cell (plasma) membrane; the cytoplasm; and a nucleus (Fig. 2.2). Each part has subdivisions, or components, that enable the cells to carry out their many and varied functions.

The cell membrane (Fig. 2.3)

The cell membrane forms a delicate, self-repairing, selective barrier between one cell and its neighbors, and between a cell and its environ-

FIGURE 2.1 Diverse forms of mammalian cells. (Not to the same scale.)

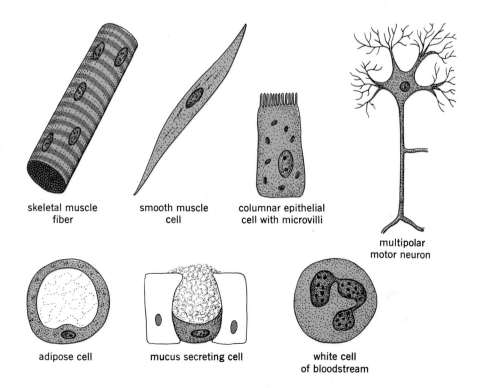

skeletal muscle fiber

smooth muscle cell

columnar epithelial cell with microvilli

multipolar motor neuron

adipose cell

mucus secreting cell

white cell of bloodstream

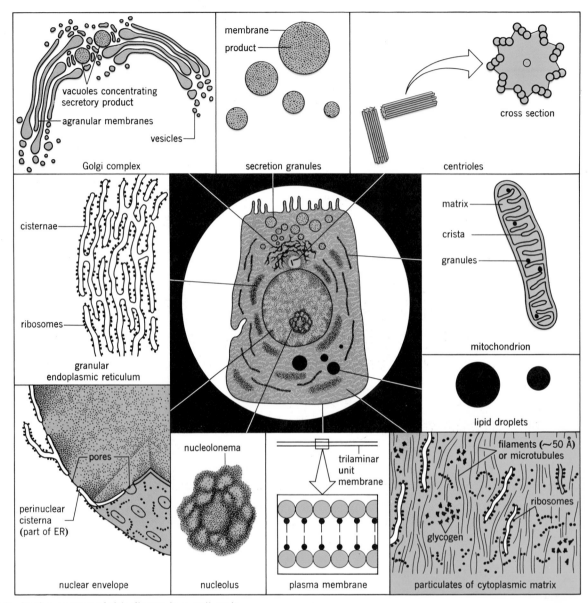

FIGURE 2.2   In the center of this figure is a cell as it would appear by light microscopy. The peripheral drawings represent the ultrastructure of the cellular components as seen in the electron microscope.

ment. The term *selective*, or semipermeable, implies that the membrane exerts control over what passes through it, and does not allow all substances to pass with equal ease. This selectivity protects the cell from the entry of many toxic or unwanted materials, and prevents loss of vital internal components. The structure of the membrane is not definitely known, but it contains lipids (fatty substances), and proteins, in a three-layered (trilaminar) "sandwich," or globular "micellar" arrangement (Fig. 2.4). (Micelles are small units of one substance in another.) "Pores," or openings in the membrane, seem to be present, and are areas of freer passage for small molecules (H_2O, inorganic salts) through the membrane.

The cytoplasm

The cytoplasm is all of the cellular substance within the cell membrane, excluding the nucleus. It may be characterized as the "factory area" of the cell, in which many of the basic chemical reactions of the cell (and body) occur. The cytoplasm RECEIVES raw materials, BREAKS DOWN (catabolizes) substances to usable compounds or forms of energy, MANUFACTURES (synthesizes) new substances (proteins, lipids, carbohydrates), PRODUCES and PACKAGES SECRETIONS for delivery elsewhere in the cell or body, DELIVERS materials to other parts of the cell, or to the circulation, and EXCRETES waste products. All these different cytoplasmic activities are prevented from interfering with one another because they occur on, or within, formed bodies in the cytoplasm, called

FIGURE 2.3.   The gross structure of the cell membrane as seen in the electron microscope. mv, microvillus; cm, cell membrane. × 148,000.

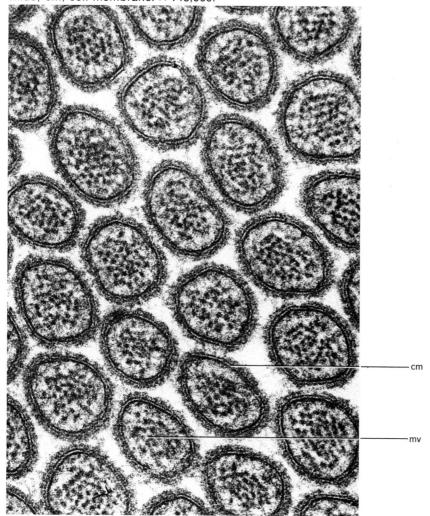

cm

mv

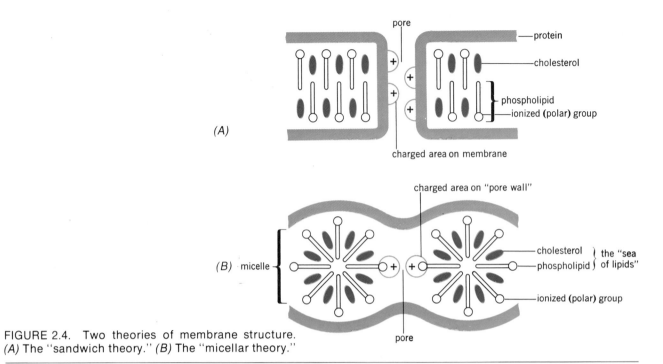

FIGURE 2.4.   Two theories of membrane structure.
(A) The "sandwich theory." (B) The "micellar theory."

FIGURE 2.5.   Nucleus of a tracheal cell.
NM, nuclear membrane; Nuc, nucleolus;
NP, nuclear pores; CH, chromatin; NF,
space containing nuclear fluid. (Courtesy
Norton B. Gilula, The Rockefeller Uni-
versity) × 15,000.

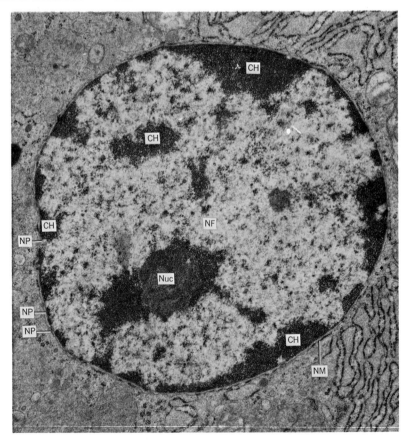

TABLE 2.1   The Cytoplasmic Organelles

Name of organelle	Where found in cell	Structure or appearance	Functions	Comments
Endoplasmic reticulum (ER)	Throughout cytoplasm	Branching, membrane-lined tubules; connect to plasma membrane and nucleus	Transport cell fluids and chemicals	Two types: smooth, with no granules; rough, with granules on outer surface
Ribosomes	On outer surface of ER, or free in cytoplasm	A small granule, composed of nucleic acid and protein	Synthesize proteins, such as, enzymes and blood proteins	May form chains called "polyribosomes" for synthesis of large proteins
Mitochondria	Randomly in cytoplasm	Round or oval bodies with 2-layered membrane; inner layer has folds called shelves or cristae	Makes ATP, an energy source for cell activity	"Powerhouse" of cell. Active cells have more mitochondria than less active cells
Golgi apparatus or body	Usually between nucleus and cell membrane in one part of cell	Membrane lined channels and expanded areas	Makes lipids; assembles connective tissue collagen molecules, packages secretions	Well developed in cells secreting enzymes and producing collagen
Lysosomes	Randomly in cytoplasm	Membrane surrounded "sacks" of enzymes	Break large molecules to smaller ones	Can destroy cell if sacks break ("suicide packets")
Central body	On one side of the nucleus	2 centrioles and an area called the centrosome	Aid in cell division	If lacking, cell division cannot occur

ORGANELLES (tiny organs). The appearance of these organelles is shown in Figure 2.2, and their structure and functions are summarized in Table 2.1.

The nucleus (Fig. 2.5)

The nucleus is a pliable structure that governs cellular activity. It is the "brains" of the cell, or the "president" of the corporation that forms the rest of the cell. In the resting (not dividing) cell, the nucleus shows a two-layered MEMBRANE, penetrated by pores that are many times larger than those of the cell membrane. Granular CHROMATIN MATERIAL, which forms the chromosomes at the time of cell division, and a large NUCLEOLUS, float in the NUCLEAR FLUID. The nucleus CONTROLS CELLULAR ACTIVITY through the production of nucleic acids that are passed from the nucleus to the cytoplasm through the nuclear pores. These nucleic acids are responsible for the synthesis of proteins that form enzymes, hormones, and other controlling chemicals. The details of this control are presented in Chapter 4.

Cellular composition

The basic substances that compose the cells, and therefore the body, are those that are most plentiful on earth. The uniqueness of the body lies in the way in which these materials are put together into the complex molecules forming the body.

Four elements form more than 95 percent of the body. They are:

Oxygen (O) 65 percent
Carbon (C) 18 percent
Hydrogen (H) 10 percent
Nitrogen (N) 3 percent

The remaining 5 percent of elements composing the body include six of major importance. These elements and their concentrations are:

Calcium (Ca) 2 percent
Phosphorus (P) 1 percent
Potassium (K) 0.35 percent
Sulfur (S) 0.25 percent
Sodium (Na) 0.15 percent
Magnesium (Mg) 0.05 percent

O, C, H, N, P, and S are typically combined into larger molecules of carbohydrate, lipids, proteins, nucleic acids, and other substances that form the basis of body structure and function. The list to follow presents some of the more important compounds these elements form.

WATER composes 55–60 percent of the cell substance. It is a good solvent, and causes many materials to ionize (assume electrical charges) in solution and thus become more reactive chemically. It also is not usually toxic to the cell, and is an excellent medium for heat transfer. It thus plays a great role in regulation of body composition and temperature.

The INORGANIC SUBSTANCES, many of which are electrolytes, usually consist of positively charged units (cations) and negatively charged units (anions). Some of the common inorganic substances are:

Substance	Chemical Symbol	
Hydrogen	H^+	
Sodium	Na^+	
Potassium	K^+	
Calcium	Ca^{++}	
Magnesium	Mg^{++}	
Chloride	Cl^-	
Phosphate	HPO_4 or $PO_4^=$	} radicals, con-
Sulfate	$SO_4^=$	} sisting of
Bicarbonate	HCO_3^-	} several elements that stay together in chemical reactions.

The inorganic substances are responsible for the establishment of forces that cause water movement between the cell and its environment, act as components of buffer systems which resist change in the body's acid-base balance, and aid in the establishment of excitability in all cells.

CARBOHYDRATES are starches and sugars, and are composed of C, H, and O, with H and O usually in the same ratio as in water; that is, 2:1. Their names commonly end with the suffix -ose. Glucose, fructose, and galactose are 6-carbon SIMPLE SUGARS (monosaccharides), the building units of more complex carbohydrates. They have the general formula $C_6H_{12}O_6$. Sucrose, lactose, and maltose are 12-carbon DOUBLE SUGARS (disaccharides). They have the general formula $C_{12}H_{22}O_{11}$, and are found in many carbohydrates eaten in the diet, and as breakdown products in the metabolism of more complex carbohydrates.

Glycogen or animal starch, is a storage form of glucose, and consists of hundreds or thousands of glucose units put together in a polymer to create a POLYSACCHARIDE. Carbohydrates are easily metabolized by most cells, and serve as the preferred source of energy for fueling most cellular activity. They may also form structural compounds in the body, such as the protein-polysaccharides of connective tissue.

LIPIDS, or fatty substances, are composed of carbon, hydrogen, and oxygen (CHO), with only small amounts of oxygen in the molecule. They are insoluble in water, and include the TRIGLYCERIDES (neutral fats), STEROLS (cholesterol and steroids), and combinations of lipids and non-lipid substances (phospholipids, lipoproteins). Lipids and their combinations are important sources of energy, are constituents of cell membranes, insulate against loss of body heat in the skin, give shape and form to the body, and form the basis of many important body hormones.

PROTEINS are complex molecules composed of C, H, O, and N in about an 8:1:3:2 ratio. They consist of smaller "building blocks" known as amino acids held together by peptide bonds. Fibrous proteins are elongated chains of amino acids and are found in muscle, tendons, and skin. They shorten and form strong supporting fibers. Globular proteins are folded proteins that form blood proteins and many enzymes.

TABLE 2.2   A Summary of the Chemicals of Living Material

Substance	Location in cell	Function
Water	Throughout	Dissolve, suspend, and regulate other materials; regulate temperature
Inorganic salts	Throughout	Establish forces to govern water movement, pH, buffer capacity
Carbohydrates	Inclusions (non-living cell parts)	Preferred fuel for activity
Lipids	Membranes, Golgi apparatus, inclusions	Reserve energy source; give form and shape; protection; insulation
Proteins	Membranes, cytoskeleton, ribosomes, enzymes	Give form, strength, contractility, catalysts, buffering
Nucleic acids		
DNA	Nucleus, in chromosomes and genes	Direct cell activity
RNA	Nucleolus, cytoplasm	Carry instructions; transport amino acids
Trace Materials		
Vitamins	Cytoplasm	Work with enzymes
Hormones	Cytoplasm	Work with enzymes to activate or deactivate enzymes
Metals	Cytoplasm, nucleus	Specific functions in various synthetic schemes, e.g.: synthesis of insulin (zinc), maturation of red cells (cobalt, copper, iron)

ENZYMES act as ORGANIC CATALYSTS to alter the rates of chemical reactions in the body. These chemical reactions proceed rapidly in the body under mild temperature and acid conditions. Without enzymes, these reactions would proceed too slowly to be of value to the body. An enzyme is a protein that may require metals, vitamins, or other materials as cofactors to exhibit full activity. Enzymes are generally named according to their actions and/or what they work on, and typically end with the suffix -ase.

NUCLEIC ACIDS are complex compounds of nitrogenous bases, sugar, and phosphoric acid in the form of long helices (coils). Two major types of acid are recognized: DNA (deoxyribo-nucleic acid) is a double helix found in chromatin, mitochondria, and cilia, and appears to direct the synthesis of proteins. RNA (ribonucleic acid) is a clover-leaf-shaped single helix, found primarily in or on the nucleolus, and in the cytoplasm. It conveys instructions for protein synthesis (messenger-RNA) and transports amino acids (transfer-RNA).

A variety of TRACE SUBSTANCES are also necessary for normal cellular function. Vitamins, certain metals (zinc, copper, cobalt) are necessary for normal enzyme function. A summary of the chemicals found in the living material is presented in Table 2.2.

Passage of substances through membranes or cell

The chemicals the cell uses for energy sources, buffering, synthesis of new materials, and other activities, are ultimately derived from the cell's environment. Similarly, wastes and products of the cell's activity may be passed from the cell to the cell's environment. Such substances must pass through, at least, one membrane to be utilized by, or to be passed out of the cell. Two general categories of processes are responsible for causing materials to cross cellular membranes.

PASSIVE PROCESSES rely on differences in concentration or pressure to move materials across a membrane, or through a solution. Movement of the substance is always from its area of higher concentration or pressure to the area of lower concentration or pressure. The *rate* of movement decreases as concentrations or pressures approach equality in the solution or on the two sides of a membrane, and net movement stops when equality of concentration or pressure is reached. The cell contributes nothing to the process in terms of energy, or to creation of the forces causing movement, and thus is "at the mercy" of the process. Other than the concentration or pressure differences, only the semipermeable nature of the membrane influences passage of materials across it. DIFFUSION (Fig. 2.6) may occur with or without the presence of a membrane, and depends on scattering of solutes (dissolved substances) in solvents (dissolving medium), due to collisions between the moving molecules, until the solute becomes evenly distributed throughout the solvent. If diffusion of solute occurs through a membrane, the passage of the solute is determined by: its ability to dissolve in the lipids of the membrane (fat solubility), its size, and whether

FIGURE 2.6. Diffusion. Molecules of dye from the area of higher dye concentration to the area of lower dye concentration.

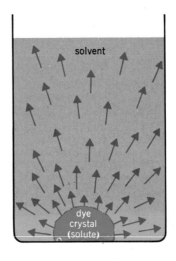

the solute carries an electrical charge. In general, small, highly lipid soluble molecules, with no charge, or a charge opposite to that of the membrane, will pass most rapidly through cell membranes. Diffusion occurs in the body in the transfer of respiratory gases through cell and capillary membranes, and in the movement of any solute through a solvent. If a spray of something odoriferous is released in a room, it soon becomes uniformly distributed in the room by diffusion. OSMOSIS is a specialized form of diffusion, referring to the passage of water molecules through a membrane. In order for water *only* to diffuse, there must be a membrane that restricts the passage of solutes, and a gradient for water, that is, a greater water concentration on one of the two sides of the membrane. Solute concentrations are inversely related to solvent concentrations, thus water movement occurs *from* the area of greater solvent concentration, or *to* the area of greater solute concentration. To illustrate the operation of the process, consider the placing of cells having selective membranes, in solutions containing different concentrations of solute (Fig. 2.7). If the cell is placed in a solution that contains a solute concentration equal to that of the cell, water concentration will be equal inside and outside of the cell, and there will be

equal rates of diffusion of water molecules into and out of the cell. The cell has been placed in an ISOTONIC (*iso*, equal) solution and will neither swell nor shrink. If the cell is placed in a solution that has less solute in it than in the cell, water concentration is greater outside the cell, and more water will flow into the cell than out of it. The cell will swell. In this situation, the cell has been placed in a HYPOTONIC (*hypo*, less) solution. If the cell is placed in a solution containing more solute than in the cell, water concentration is greater in the cell and more water will leave the cell than enter. The cell will shrink. In this situation, the cell has been placed in a HYPERTONIC (*hyper*, more) solution. Since the greatest number of body solutes are the inorganic substances, their concentrations within the cell and in its environment are very important in terms of regulating osmosis. Examples of osmosis as it occurs in the body include, water movement through the kidney tubules during urine formation, movement of tissue water into capillaries, and absorption of water through the walls of the alimentary tract. FILTRATION is a passive process that depends on a *pressure* difference on two sides of a membrane. In that case, the membrane acts like a sieve, and the pressure forces across the membrane anything small enough to pass through the

FIGURE 2.7.   Osmosis. Changes in cell size in *(A)* isotonic, *(B)* hypotonic, and *(C)* hypertonic solutions. Open circles represent water molecules, solid circles represent solute molecules. Arrows indicate flow of water, with length denoting greatest direction of flow.

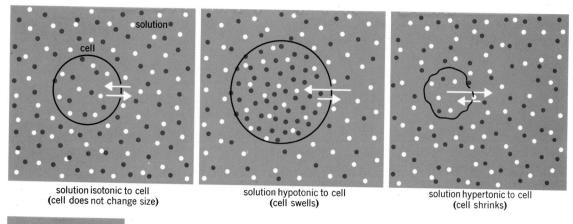

solution isotonic to cell
(cell does not change size)

solution hypotonic to cell
(cell swells)

solution hypertonic to cell
(cell shrinks)

• water molecules
• solute molecules

"water goes to where the solute concentration is greatest"

membrane pores. Filtration thus tends to separate large molecules from small ones. It occurs in the passage of materials out of capillaries under the pressure of the blood (created by the heart action). Filtration and osmosis are shown in Figure 2.8, which depicts what happens in a capillary.

ACTIVE PROCESSES rely on the cell to contribute materials or energy to the process of getting materials across the membrane. Active processes cause materials to cross membranes in spite of

concentration or other gradients, and they insure nearly complete passage of a substance across a membrane. In ACTIVE TRANSPORT, the cell provides a large molecule called a *carrier*, which is located in the membrane, and which can attach to and move a substance through the membrane. The cell also provides a *source of energy* (usually ATP) to activate the carrier, and *enzymes* to attach the activating molecule to the carrier. The scheme is shown in Figure 2.9. Active transport

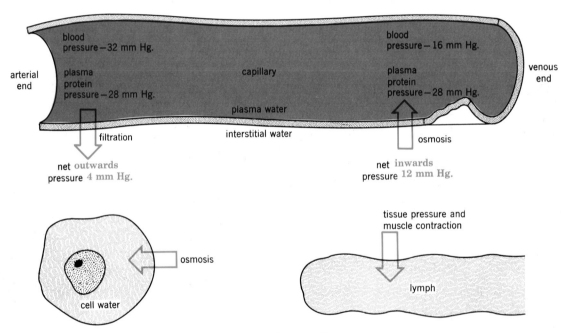

FIGURE 2.8.   Filtration and osmosis between a capillary and a body cell.

FIGURE 2.9.   A diagrammatic representation of active transport. CS, carrier system. The transport mechanism is energized by ATP.

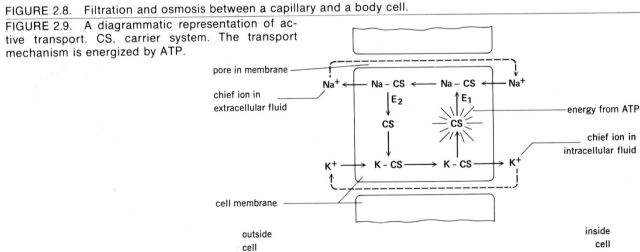

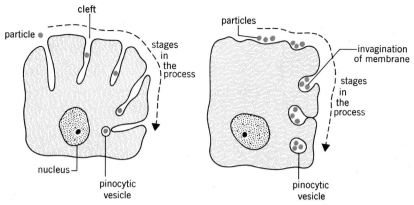

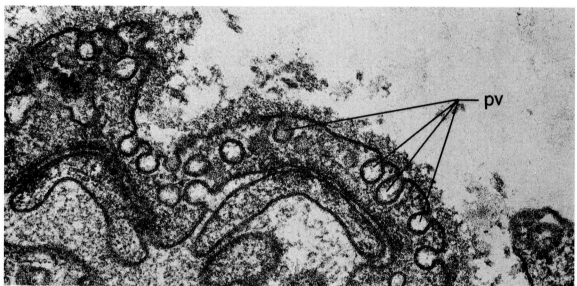

FIGURE 2.10.   Pinocytosis. *(A)* The particle enters a cleft and becomes enclosed in a vesicle. *(B)* The particle is adsorbed on the surface of the membrane and is enclosed in a vesicle. *(C)* Electron micrograph of formation of vesicles in skeletal muscle capillary. pv, pinocytic vesicle. × 22,000.

FIGURE 2.11.   Phagocytosis.   Pseudopods   are formed that engulf the particle.

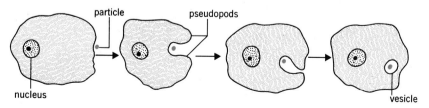

TABLE 2.3 A Summary of Methods of Acquiring Materials

Process	Type of material moved	Comments
PASSIVE PROCESSES		
Diffusion	Any solute	Most efficient for small particles. Rate decreases as size increases. No membrane necessary.
Osmosis (diffusion of water)	Water	Rate and direction determined by solute concentration. A selective membrane is required. One nonpermeable particle is required.
Filtration	All substances small enough to pass through the membrane	Force causing it is pressure. Membrane acts as a sieve.
ACTIVE PROCESSES		
Active transport	Ions, larger molecules, sugars, amino acids	Require carrier, energy, enzymes, all from cell. A membrane is required. A threshold or gradient is overcome.
Pinocytosis	Water and large molecules such as proteins, and lipids	"Sinking in" of membrane. Only method for intake of very large molecules.
Phagocytosis	Particulate materials are engulfed	Protects against bacterial invasion and rids the body of cellular debris.

is responsible for cellular transport of amino acids, glucose, vitamins, and inorganic salts. PINOCYTOSIS (Fig. 2.10) involves the formation of vacuoles by a "sinking in" of the cell membrane. The process "takes in a chunk" of the fluid environment of the cell, including any molecules too large to diffuse or be actively transported. Proteins, large lipid molecules, and nucleic acids may be taken into the cell by this process. A "reverse pinocytosis" or emeiocytosis may occur when a vacuole moves to the cell membrane, fuses with it, and discharges its contents to the outside of the cell. This process may be used to excrete materials from the cell. PHAGOCYTOSIS (Fig. 2.11) involves the engulfing or surrounding

of a particle or large molecule by the cell. The cell then usually digests the engulfed material. Protection may be afforded by this process, as in the engulfing of bacteria by white blood cells; clearing of a site of tissue damage may also occur by phagocytosis. A summary of methods of acquisition of materials is given in Table 2.3.

We note again that the passage of substances through cell membranes and through the fluids around the cells has its counterpart in the ingestion and processing of foods by the body as a whole. A bit of time spent in learning what the basic activities of a cell are, and how it carries out these activities will pay dividends later.

Summary

1. Most cells have three basic parts: a membrane, cytoplasm, and nucleus.

    a. The membrane is a selective structure surrounding the cell. It is composed of lipid and protein components.

    b. The cytoplasm contains a number of organelles that carry out the cell's functions of metabolism.

        (1) Endoplasmic reticulum distributes substances through the cell and synthesizes proteins and lipids.

        (2) Ribosomes assemble proteins.

        (3) Mitochondria produce energy for cell activity.

        (4) Golgi bodies produce large molecules and package secretions.

        (5) Lysosomes break large molecules into smaller ones.

        (6) Central bodies are essential for cell division.

    c. The nucleus, by its content of nucleic acids, controls cellular activity, mainly through protein synthesis.

2. The chemicals found in the cell consist of elements, molecules, and compounds.

    a. C, H, O, and N form over 95 percent of the body. The remainder is formed of some six other elements.

    b. Water is a nontoxic solvent making up more than one half of the body substance. It also transfers heat and causes many substances to ionize and be more chemically reactive.

    c. Inorganic substances (Na, K, Ca, Cl, HCO_3) create osmotic gradients, aid in buffering, and aid in the creation of excitability.

    d. Carbohydrates are composed of C, H, and O, with H and O in a 2:1 ratio. They are the preferred sources of energy for body activity.

    e. Lipids are water insoluble substances composed of C, H, and O. They serve as energy sources, are found in cell membranes, and serve a variety of mechanical functions in the body.

    f. Proteins are structural materials composed of C, H, O, and N.

    g. Enzmes are proteins that control the rate of chemical reactions in the body.

    h. Nucleic acids, DNA and RNA, control cellular activity.

    i. Trace substances (metals, vitamins, hormones) are required by the body in small amounts to function normally.

3. Cells acquire materials by passive and active processes.

    a. Passive processes occur by physical principles, cause substances to move from areas of high pressure or concentration to areas of lower pressure or concentration, and do not require activity by the cell.

        (1) Diffusion, osmosis, and filtration are passive processes.

(a) Diffusion results from molecular movement, and distributes a solute evenly in a solvent.

(b) Osmosis involves only water passage through a membrane. Net movement of water occurs only if a solute concentration difference exists on the two sides of a membrane. A hypotonic solution causes net movement of water *into* a cell; a hypertonic solution causes a net movement of water *from* a cell.

(c) Filtration is movement of substances through a membrane by pressure.

b. Active processes require cell participation and work.

(1) Active transport, pinocytosis, and phagocytosis may cause substances to pass membranes at any time and in any direction.

(a) Active transport uses carriers, ATP, and enzymes.

(b) Pinocytosis involves membrane sinking and vacuole formation.

(c) Phagocytosis is engulfing of something by a cell.

Questions

1. What cellular structures and organelles would one expect to find in all body cells? Give a function for each structure listed.

2. What do you consider to be the four most important classes of chemicals in the body? Defend your choices with evidence of function and method of function.

3. Compare and contrast passive and active processes of acquisition of materials by cells. Concentrate on what determines rate and extent of acquisition by these two broad categories.

4. Explain what would happen and why, to a human cell (solute concentration about 1.0 percent) placed in a 2.0 percent salt solution.

Readings

Allison, A. "Lysosomes and Disease." *Sci. Amer. 217*:62 (Nov) 1967.

Bittar, E. Edward (ed). *Cell Biology in Medicine.* Wiley. New York, 1973.

Bretscher, Mark S. "Membrane Structure, Some General Principles." *Science 181*:622, 1973.

Fawcett, Don W. *An Atlas of Fine Structure. The Cell.* W. B. Saunders. Philadelphia, 1966.

Fox, C. Fred. "The Structure of Cell Membranes." *Sci. Amer. 226*:30 (Feb) 1972.

Frieden, Earl. "The Chemical Elements of Life." *Sci. Amer. 227*:52 (July) 1972.

Loewy, A. G., and Philip Siekevitz. *Cell Structure and Function.* 2nd. ed. Holt, Rinehart and Winston. 1970.

Neutra, M., and C. P. LeBland. "The Golgi Apparatus." *Sci. Amer. 220*:100 (Feb) 1969.

Nomura, M. "Ribosomes." *Sci. Amer. 221*:28 (Oct) 1969.

Oldfield, E., and S. J. Singer. "Are Cell Membranes Fluid?" *Science 180*:982, 1973.

Pryor, William A. "Free Radicals in Biological Systems." *Sci. Amer. 223*:70 (Aug) 1970.

Racker, E. "The Membrane of the Mitochondrion." *Sci. Amer. 218*:32 (Feb) 1968.

"Readings from Scientific American." *The Living Cell.* W. H. Freeman. San Francisco, 1965.

Stent, Gunther S. "Cellular Communication." *Sci. Amer. 227*:42 (Sept) 1972.

Wessells, Norman K. "How Living Cells Change Shape." *Sci. Amer. 225*:76 (Oct) 1971.

chapter 3
Cell Reproduction and Body Development

chapter 3

Cell Reproduction

The body does not spend its life with the same cells it was born with. Cells constantly die; the body is injured and must be repaired. The replacement of worn out cells or production of new cells, for purposes of repairing injury, and continuance of the species, is common in most organisms. Two processes occur in the body that result in the production of new cells by division of preexisting cells.

Mitosis (Fig. 3.1)

This is a type of cell division producing two daughter cells that are GENETICALLY IDENTICAL to the parent cell. Mitosis occurs in all body areas at some time during life. Mitosis takes place in several stages, each arbitrarily separated according to what is occurring in that stage.

In INTERPHASE, the cell appears typical for its type and is not undergoing division. It is engaged in accumulating chemicals for its own use, synthesis of proteins, metabolism of fuels, and *replication of DNA and RNA*. The chromatin material, and therefore each chromosome, is duplicated in this stage, as are the centromeres to which the DNA helices are attached. In effect, the cell contains a double set of chromosomes. (Geneticists "count" centromeres to determine numbers of chromosomes. If there is one centro-

FIGURE 3.1. The major events occurring in mitosis. A cell having five chromosomes is shown.

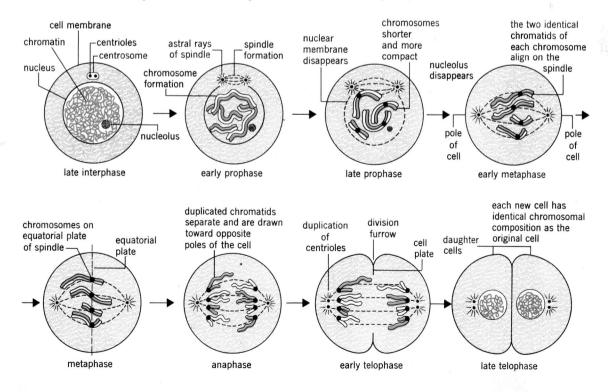

mere, it is equivalent to one chromosome regardless of how many strands of DNA are attached to it.)

In PROPHASE, the paired centrioles separate and migrate to opposite poles (ends) of the cell. The nuclear membrane disappears, and a spindle of fibrils is formed between the centrioles. Visible chromosomes are formed from the chromatin material.

In METAPHASE, the duplicated chromosomes *align* themselves on the equatorial plate of the cell between the centrioles.

In ANAPHASE, *separation* of duplicated chromosomes occurs (possibly by the centromeres being pulled by the spindle fibers).

In TELOPHASE, the chromosomes return to the

granular chromatin state, the nuclear membranes reorganize, the centrioles divide, and the cytoplasm and its organelles undergo a nearly equal separation into two cells. Since the nuclear material of the two resulting cells was duplicated in the original cell, and then separated, the chromosome number and identity of a given cell line is maintained.

Meiosis (Fig. 3.2)

This results in the production of four daughter cells from each parent cell, each of which contains one half the original number of chromosomes found in the parent cell. It occurs in the ovaries

FIGURE 3.2.   The events occurring in meiosis. Solid "chromosomes" indicate those from one parent, open ones from the other parent. Note the different chromosome constitution that may result according to how chromatids separate.

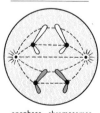

prophase—formation of chromosomes

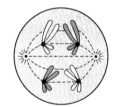

prophase—pairing of homologous chromosomes (crossing over may occur at this stage)

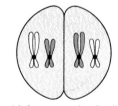

prophase—duplication of DNA helices without centromere duplication

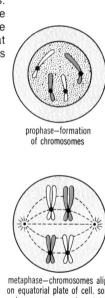

metaphase—chromosomes align on equatorial plate of cell. some chromosomes may be rotated relative to others on plate

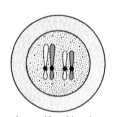

anaphase—chromosomes are separated

1st meiotic division

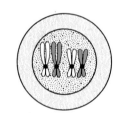

telophase—one member of each chromosome pair is found in each daughter cell

metaphase—centromeres divide

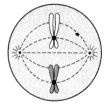

anaphase—chromosomes are separated

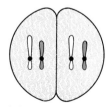

telophase—only one member of each original pair is present— haploid cells

2nd meiotic division—only one daughter cell is depicted

and testes as these organs form ova (eggs) and sperm.

The basic differences between this process and mitosis are: in meiosis a series of two cell divisions occurs, one without previous duplication of chromosomes; and, exchange of portions of chromosomes may occur during division. Thus, the daughter cells contain a DIFFERENT GENETIC MAKEUP than the parent cell. The stages of meiosis are similar to those of mitosis.

In MEIOTIC INTERPHASE, duplication of chromatin material occurs, without duplication of centromeres. In effect, the number of chromosomes remains the same but, "inside," each contains a duplicate set of chromatin, or DNA helices.

In FIRST MEIOTIC PROPHASE, visible chromosomes appear, and the two members of a paired set of chromosomes come to lie parallel to one another (synapsis) to form "bivalents." *This step does not occur in mitosis.* During synapsis, sections of paired chromosomes may be exchanged between the two chromosomes.

In FIRST MEIOTIC METAPHASE, ANAPHASE, and TELOPHASE, separation of paired chromosomes and cytoplasmic division occurs, and the number of centromeres (chromosomes) is reduced to one half the original number. The first cell division in meiosis is, therefore, a *reductional division*, and two cells are formed.

A SECOND MEIOTIC DIVISION next occurs, which is really a mitotic division, that is to say, the centromeres divide to form two separate chromosomes. These then separate as in mitosis, without further change of chromosome number. Each of the two cells resulting from the first division gives rise to two cells; a total of four cells is thus formed.

Because of pairing of chromosomes and exchange of parts, the process of MEIOSIS INTRODUCES VARIATION in the genetic makeup of the daughter cells as well as reducing the normal chromosome number to one half the number characteristic of the species.

Neoplasms (*cancers*)

Cell division by mitosis is usually balanced to the needs of the body for cell replacement and repair. Sometimes, uncontrolled cell division may result in the development of neoplasms or cancer.

Two major types of neoplasms are recognized. BENIGN neoplasms usually grow slowly, are limited from surrounding normal tissue by connective tissue capsules and are not considered grave threats to the body unless they are in an area that permits their undetected growth to large sizes. In such areas (e.g., brain), mechanical pressure forms the major threat the neoplasm poses as it grows. MALIGNANT neoplasms grow rapidly, are not usually limited by capsules, and *metastasize* easily, that is, shed cells into lymphatic or blood vessels. Such cells may be carried throughout the body to create neoplasms in many other body areas. While the causes of most neoplasms are not yet definitely known, many agents, including viruses, chemicals, and radiation have been shown to be carcinogenic, or capable of causing neoplasms. The American Cancer Society suggests that there are seven "danger signals" that may indicate the presence of a neoplastic process in the body. These are:

Any unusual discharge or bleeding from a body opening (e.g.: anus, vagina, mouth, nose).

A sore which does not heal.

A change in bowel habits (including frequency and consistency) which persists more than 3 weeks.

Chronic indigestion or difficulty in swallowing.

Any change in size of a wart or mole.

A lump or thickening in the breast or elsewhere.

Persistent hoarseness or coughing.

Cancer is the second leading cause of death in the United States; early detection utilizing awareness of the "danger signals" could reduce cancer deaths by one third. Although cancers as a group cannot yet be cured, the treatments available are much more effective when begun early, thus prognosis is improved.

The basic development of the individual

Human sperm and ova, produced by meiosis, each contain one half the chromosome number characteristic of the species. This halved number of chromosomes is termed the HAPLOID number and is 23 for the human. The haploid number of chromosomes consists of 22 autosomes, and an X or Y (sex) chromosome. The meioses producing sperm create cells with either a 22 + X or 22 + Y content; those producing ova result in a 22 + X content. Restoring normal chromosome number (46) depends on the joining of an egg and sperm to form a ZYGOTE in the process of FERTILIZATION. Fertilization results in a 44 + XY (male) or 44 + XX (female) chromosome constitution in the zygote. Fertilization is the time of conception. Development from conception onward may be described "by the week" in terms of the major processes concerned.

The first week

After fertilization, the zygote undergoes a series of mitotic divisions known as CLEAVAGE. A solid mass of cells, known as a MORULA results. (The morula is about the size of a period in this text.) Little time for growth of cells occurs between cleavages, and the morula is only a little larger than the zygote. Cleavage results in a doubling of cell number for the first 8–10 divisions, and thereafter becomes irregular. As cleavage is occurring, the zygote is passing down the uterine (Fallopian) tube from ovary to uterus, a journey that takes about 3 days. Arriving in the uterine cavity, the morula adheres to the lining of the uterus (endometrium) and undergoes a reorganization into a BLASTOCYST. The blastocyst is a hollow structure containing a *cavity*, and an *inner mass* of cells. This reorganization takes 3–4 days. The events of the first week are shown in Figure 3.3.

The second week

The blastocyst undergoes further reorganization with the appearance of two cavities in the inner cell mass. The upper one is the AMNIOTIC CAVITY,

the lower one, the cavity of the YOLK SAC. A two-layered plate of cells separates the two cavities, and is known as the EMBRYONIC PLATE or DISC. Development to this stage is said to result in the formation of a TWO-LAYERED EMBRYO. The upper layer of cells in this embryo is termed ECTODERM, the lower layer, ENDODERM. These layers form two of the three "germ layers" from which all body structures will develop. Finally, the embryo undergoes IMPLANTATION in the wall of the uterus. In this process, the blastocyst "digests" its way into the vascular and glandular endometrium of the uterus to establish a relationship with the mother's blood supply. This relationship assures the nutrients necessary for further growth and development. These processes are shown in Figure 3.4.

The third week

The major event occurring at this time is the formation of the third basic germ layer. By mitotic cell division, a layer of cells termed the MESODERM is formed between the ectoderm and endoderm of the two-layered embryo. Table 3.1 shows the ultimate derivatives of these three germ layers.

The fourth week

The fourth week is associated with the formation of the NEURAL TUBE, the beginning of the nervous system, and of SOMITES or blocks of mesoderm along the backbone of the embryo. These somites are the forerunners of the bones and muscles of the back. The embryo appears as is shown in Figure 3.5. Notice that development occurs in a head-to-tail (cephalo-caudal) direction, as does maturation of function, once an organ or system is formed.

The fifth through eighth weeks

COMPLETION OF THE EMBRYO occurs during this

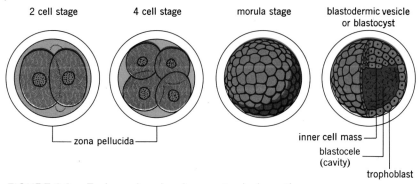

2 cell stage   4 cell stage   morula stage   blastodermic vesicle
                                                 or blastocyst

zona pellucida

inner cell mass
blastocele
(cavity)
trophoblast

FIGURE 3.3. Embryonic development during the
first week; cleavage and blastocyst formation.

FIGURE 3.4. Embryonic development during the
second week; implantation, and the formation of a
two-layered embryo.

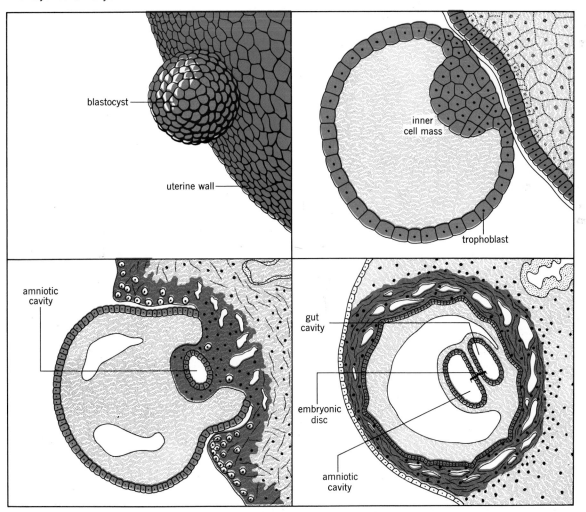

blastocyst

uterine wall

inner
cell mass

trophoblast

amniotic
cavity

gut
cavity

embryonic
disc

amniotic
cavity

TABLE 3.1   Derivatives of Germ Layers

Ectoderm	Mesoderm	Endoderm
1. Outer layer (epidermis) of skin: 　Skin glands 　Hair and nails 　Lens of eye	1. Muscle: 　Skeletal, 　Cardiac, 　Smooth	1. Epithelium of: 　Pharynx 　Auditory (ear) tube 　Tonsils 　Thyroid 　Parathyroid 　Thymus
2. Lining tissue (epithelium) of: 　Nasal cavities 　Sinuses 　Mouth: 　　Oral glands 　　Tooth enamel 　Sense organs 　Anal canal	2. Supporting (connective) tissue: 　Cartilage 　Bone 　Blood 3. Bone marrow 4. Lymphoid tissue	Larynx 　Trachea 　Lungs 　Digestive tube and its glands 　Bladder 　Vagina and vestibule 　Urethra and glands
3. Nervous tissues 4. Pituitary (posterior lobe) 5. Adrenal medulla	5. Epithelium of: 　Blood vessels 　Lymphatics 　Celomic cavity 　Kidney and ureters 　Gonads and ducts 　Adrenal cortex 　Joint cavities	2. Pituitary (anterior and middle lobes)

FIGURE 3.5.   The human embryo at four weeks of development.

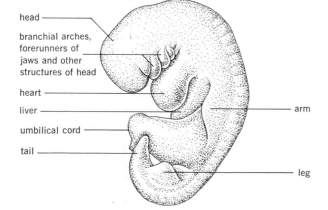

head

branchial arches, forerunners of jaws and other structures of head

heart

liver

umbilical cord

tail

arm

leg

period, and it assumes a clearly human form. The brain is enclosed, the digestive system forms, a heart is formed and circulation of blood in vessels is established. The limbs, eyes, ears, and other features are evident (Fig. 3.6).

The ninth week to birth

This period constitutes the period of the FETUS, a creature that has definite human form and all basic body systems. The fetus is, in some cases, capable of independent survival after about 26 weeks of development, a time that normally permits lung development to be far enough advanced to support life. Table 3.2, Correlated Human Development, indicates that the fetal period is one mainly of growth of the body; by 8–12 weeks, all body systems are present and need only to grow

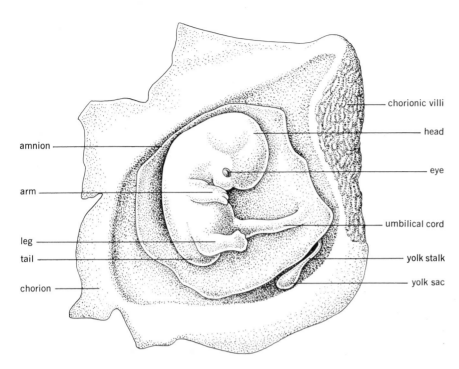

FIGURE 3.6.   The human embryo at about eight weeks of development (twice natural size).

FIGURE 3.7.   The changes in the body proportions from before birth to adulthood.

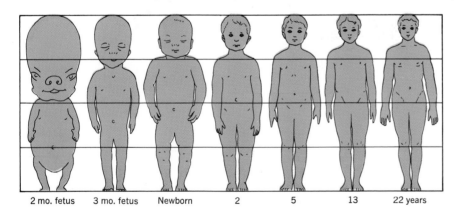

2 mo. fetus   3 mo. fetus   Newborn   2   5   13   22 years

and undergo final refinement. The first 12 weeks of development is thus one of critical importance for establishing normal organs. Drugs used or diseases (e.g., measles) contracted during this time period may lead to malformations in the development of body organs and systems. Congenital (born with) anomalies or malformations are the result of developmental errors. They may be evident at birth, or may cause disorders later in life. Such malformations are the third leading cause of death in the United States between birth and 4 years of age, and the fourth leading cause of death between the ages of 5 to 14 years. Since several organs and systems develop at the same time, a process affecting one system may result in changes in a simultaneously developing system, so that malformations are usually seen in more than one system.

During fetal growth, and after birth, body proportions change (Fig. 3.7). Thus a newborn has a relatively larger head and shorter legs than does the child and adult. Growth in size is the most

TABLE 3.2. A Reference Table of Correlated Human Development (from Arey).

Age in Weeks	Size (C R) in Mm	Body form	Mouth	Pharynx and derivatives	Digestive tube and glands	Respiratory system	Coelom and mesenteries
2.5	1.5	Embryonic disc flat. Primitive streak prominent. Neural groove indicated.	–	–	Gut not distinct from yolk sac.	–	Extra-embryonic coelom present. Embryonic coelom about to appear.
3.5	2.5	Neural groove deepens and closes (except ends). Somites 1 –16± present. Cylindrical body constricting from yolk sac. Branchial arches 1 and 2 indicated.	Mandibular arch prominent. Stomodeum a definite pit. Oral membrane ruptures.	Pharynx broad and flat. Pharyngeal pouches forming. Thyroid indicated.	Fore- and hind-gut present. Yolk sac broadly attached at mid-gut. Liver bud present. Cloaca and cloacal membrane present.	Respiratory primordium appearing as a groove on floor of pharynx.	Embryonic coelom a U-shaped canal, with a large pericardial cavity. Septum transversum indicated. Mesenteries forming. Mesocardium atrophying.
4	5.0	Branchial arches completed. Flexed heart prominent. Yolk stalk slender. All somites present (40). Limb buds indicated. Eye and otocyst present. Body flexed; C-shape.	Maxillary and mandibular processes prominent. Tongue primordia present. Rathke's pouch indicated.	Five pharyngeal pouches present. Pouches 1-4 have closing plates. Primary tympanic cavity indicated. Thyroid a stalked sac.	Esophagus short. Stomach spindle-shaped. Intestine a simple tube. Liver cords, ducts and gall bladder forming. Both pancreatic buds appear. Cloaca at height.	Trachea and paired lung buds become prominent. Laryngeal opening a simple slit.	Coelom still a continuous system of cavities. Dorsal mesentery a complete median curtain. Omental bursa indicated.
5	8.0	Nasal pits present. Tail prominent. Heart, liver and mesonephros protuberant. Umbilical cord organizes.	Jaws outlined. Rathke's pouch a stalked sac.	Phar. pouches gain dors. and vent. diverticula. Thyroid bilobed. Thyro-glossal duct atrophies	Tail-gut atrophies. Yolk stalk detaches. Intestine elongates into a loop. Caecum indicated.	Bronchial buds presage future lung lobes Arytenoid swellings and epiglottis indicated.	Pleuro-pericardial and pleuro-peritoneal membranes forming. Ventral mesogastrium draws away from septum.
6	12.0	Upper jaw components prominent but separate. Lower jaw-halves fused. Head becomes dominant in size. Cervical flexure marked. External ear appearing. Limbs recognizable as such.	Lingual primordia fusing. Foramen caecum established. Labio-dental laminae appearing. Parotid and submaxillary buds indicated.	Thymic sacs, ultimo-branchial sacs and solid parathyroids are conspicuous and ready to detach. Thyroid becomes solid and converts into plates.	Stomach rotating. Intestinal loop undergoes torsion. Hepatic lobes identifiable. Cloaca subdividing.	Definitive pulmonary lobes indicated. Bronchi sub-branching. Laryngeal cavity temporarily obliterated.	Pleuro-pericardial communications close. Mesentery expands as intestine forms loop.
7	17.0	Branchial arches lost. Cervical sinus obliterates. Face and neck forming. Digits indicated. Back straightens. Heart and liver determine shape of body ventrally. Tail regressing.	Lingual primordia merge into single tongue. Separate labial and dental laminae distinguishable. Jaws formed and begin to ossify. Palate folds present and separated by tongue.	Thymi elongating and losing lumina. Parathyroids become trabeculate and associate with thyroid. Ultimobranchial bodies fuse with thyroid. Thyroid becoming crescentic.	Stomach attaining final shape and position. Duodenum temporarily occluded. Intestinal loops herniate into cord. Rectum separates from bladder-urethra. Anal membrane ruptures. Dorsal and ventral pancreatic primordia fuse.	Larynx and epiglottis well outlined; orifice T-shaped. Laryngeal and tracheal cartilages foreshadowed. Conchae appearing. Primary choanae rupturing.	Pericardium extended by splitting from body wall. Mesentery expanding rapidly as intestine coils. Ligaments of liver prominent.
8	23.0	Nose flat; eyes far apart. Digits well formed. Growth of gut makes body evenly rotund. Head elevating. Fetal state attained.	Tongue muscles well differentiated. Earliest taste buds indicated. Rathke's pouch detaches from mouth. Sublingual gland appearing.	Auditory tube and tympanic cavity distinguishable. Sites of tonsil and its fossae indicated. Thymic halves unite and become solid. Thyroid follicles forming.	Small intestine coiling within cord. Intestinal villi developing. Liver very large in relative size.	Lung becoming gland-like by branching of bronchioles. Nostrils closed by epithelial plugs.	Pleuro-peritoneal communications close. Pericardium a voluminous sac. Diaphragm completed, including musculature. Diaphragm finishes its 'descent.'

Urogenital system	Vascular system	Skeletal system	Muscular system	Integumentary system	Nervous system	Sense organs	Age in weeks
Allantois present.	Blood islands appear on chorion and yolk sac. Cardiogenic plate reversing.	Head process (or notochordal plate) present.	–	Ectoderm a single layer.	Neural groove indicated.	–	2.5
All pronephric tubules formed. Pronephric duct growing caudad as a blind tube. Cloaca and cloacal membrane present.	Primitive blood cells and vessels present. Embryonic blood vessels a paired symmetrical system. Heart tubes fuse, bend S-shape and beat begins.	Mesodermal segments appearing (1 – 16 ±). Older somites begin to show sclerotomes. Notochord a cellular rod.	Mesodermal segments appearing (1 – 16 ±). Older somites show myotome plates.	–	Neural groove prominent; rapidly closing. Neural crest a continuous band.	Optic vesicle and auditory placode present. Acoustic ganglia appearing.	3.5
Pronephros degenerated. Pronephric (mesonephric) duct reaches cloaca. Mesonephric tubules differentiating rapidly. Metanephric bud pushes into secretory primordium.	Hemopoiesis on yolk sac. Paired aortae fuse. Aortic arches and cardinal veins completed. Dilated heart shows sinus, atrium, ventricle, and bulbus.	All somites present (40). Sclerotomes massed as primitive vertebrae about notochord.	All somites present (40).	–	Neural tube closed. Three primary vesicles of brain represented. Nerves and ganglia forming. Ependymal, mantle and marginal layers present.	Optic cup and lens pit forming. Auditory pit becomes closed, detached otocyst. Olfactory placodes arise and differentiate nerve cells.	4
Mesonephros reaches its caudal limit. Ureteric and pelvic primordia distinct. Genital ridge bulges.	Primitive vessels extend into head and limbs. Vitelline and umbilical veins transforming. Myocardium condensing. Cardiac septa appearing. Spleen indicated.	Condensations of mesenchyme presage many future bones.	Premuscle masses in head, trunk and limbs.	Epidermis gaining a second layer (periderm).	Five brain vesicles. Cerebral hemispheres bulging. Nerves and ganglia better represented. [Suprarenal cortex accumulating.]	Chorioid fissure prominent. Lens vesicle free. Vitreous anlage appearing. Octocyst elongates and buds endolymph duct. Olfactory pits deepen.	5
Cloaca subdividing. Pelvic anlage sprouts pole tubules. Sexless gonad and genital tubercle prominent. Müllerian duct appearing.	Hemopoiesis in liver. Aortic arches transforming. L. umbil. vein and d. venosus become important. Bulbus absorbed into right ventricle. Heart acquires its general definitive form.	First appearance of chondrification centers. Desmocranium.	Myotomes, fused into a continuous column, spread ventrad. Muscle segmentation largely lost.	Milk line present.	Three primary flexures of brain represented. Diencephalon large. Nerve plexuses present. Epiphysis recognizable. Sympathetic ganglia forming segmental masses. Meninges indicated.	Optic cup shows nervous and pigment layers. Lens vesicle thickens. Eyes set at 160°. Naso-lacrimal duct. Modeling of ext., mid. and int. ear under way. Vomero-nasal organ.	6
Mesonephros at height of its differentiation. Metanephric collecting tubules begin branching. Earliest metanephric secretory tubules differentiating. Bladder-urethra separates from rectum. Urethral membrane rupturing.	Cardinal veins transforming. Inf. vena cava outlined. Atrium, ventricle and bulbus partitioned. Cardiac valves present. Stem of pulm. vein absorbed into l. atrium. Spleen anlage prominent.	Chondrification more general. Chondrocranium.	Muscles differentiating rapidly throughout body and assuming final shapes and relations.	Mammary thickening lens-shaped.	Cerebral hemispheres becoming large. Corpus striatum and thalamus prominent. Infundibulum and Rathke's pouch in contact. Chorioid plexuses appearing. Suprarenal medulla begins invading cortex.	Chorioid fissure closes, enclosing central artery. Nerve fibers invade optic stalk. Lens loses cavity by elongating lens fibers. Eyelids forming. Fibrous and vascular coats of eye indicated. Olfactory sacs open into mouth cavity.	7
Testis and ovary distinguishable as such. Müllerian ducts, nearing urogenital sinus, are ready to unite as uterovaginal primordium. Genital ligaments indicated.	Main blood vessels assume final plan. Primitive lymph sacs present. Sinus venosus absorbed into right atrium. Atrio-ventricular bundle represented.	First indications of ossification.	Definitive muscles of trunk, limbs and head well represented and fetus capable of some movement.	Mammary primordium a globular thickening.	Cerebral cortex begins to acquire typical cells. Olfactory lobes visible. Dura and pia-arachnoid distinct. Chromaffin bodies appearing.	Eyes converging rapidly. Ext., mid. and int. ear assuming final form. Taste buds indicated. External nares plugged.	8

TABLE 3.2 (continued)

Age in Weeks	Size (C R) in Mm	Body form	Mouth	Pharynx and derivatives	Digestive tube and glands	Respiratory system	Coelom and mesenteries
10	40.0	Head erect. Limbs nicely modeled. Nail folds indicated. Umbilical hernia reduced.	Fungiform and vallate papillae differentiating. Lips separate from jaws. Enamel organs and dental papillae forming. Palate folds fusing.	Thymic epithelium transforming into reticulum and thymic corpuscles. Ultimobranchial bodies disappear as such.	Intestines withdraw from cord and assume characteristic positions. Anal canal formed. Pancreatic alveoli present.	Nasal passages partitioned by fusion of septum and palate. Nose cartilaginous. Laryngeal cavity reopened; vocal folds appear.	Processus (saccus) vaginales forming. Intestine and its mesentery withdrawn from cord.
12	56.0	Head still dominant. Nose gains bridge. Sex readily determined by external inspection.	Filiform and foliate papillae elevating. Tooth primordia form prominent cups. Cheeks represented. Palate fusion complete.	Tonsillar crypts begin to invaginate. Thymus forming medulla and becoming increasingly lymphoid. Thyroid attains typical structure.	Muscle layers of gut represented. Pancreatic islands appearing. Bile secreted.	Conchae prominent. Nasal glands forming. Lungs acquire definitive shape.	Omentum an expansive apron partly fused with dorsal body wall. Mesenteries free but exhibit typical relations. Coelomic extension into umbilical cord obliterated.
16	112.0	Face looks 'human.' Hair of head appearing. Muscles become spontaneously active. Body outgrowing head.	Hard and soft palates differentiating. Hypophysis acquiring definitive structure.	Lymphocytes accumulate in tonsils. Pharyngeal tonsil begins development.	Gastric and intestinal glands developing. Duodenum and colon affixing to body wall. Meconium collecting.	Accessory nasal sinuses developing. Tracheal glands appear. Mesoderm still abundant between pulmonary alveoli. Elastic fibers appearing in lungs.	Greater omentum fusing with transverse mesocolon and colon. Mesoduodenum and ascending and descending mesocolon attaching to body wall.
20-40 (5-10 mo.)	160.0-350.0	Lanugo hair appears (5). Vernix caseosa collects (5). Body lean but better proportioned (6). Fetus lean, wrinkled and red; eyelids reopen (7). Testes invading scrotum (8). Fat collecting, wrinkles smoothing, body rounding (8-10).	Enamel and dentine depositing (5). Lingual tonsil forming (5). Permanent tooth primordia indicated (6-8). Milk teeth unerupted at birth.	Tonsil structurally typical (5).	Lymph nodules and muscularis mucosae of gut present (5). Ascending colon becomes recognizable (6). Appendix lags behind caecum in growth (6). Deep esophageal glands indicated (7). Plicae circulares represented (8).	Nose begins ossifying (5). Nostrils reopen (6). Cuboidal pulmonary epithelium disappearing from alveoli (6). Pulmonary branching only two-thirds completed (10). Frontal and sphenoidal sinuses still very incomplete (10).	Mesenterial attachments completed (5). Vaginal sacs passing into scrotum (7-9).

Urogenital system	Vascular system	Skeletal system	Muscular system	Integumentary system	Nervous system	Sense organs	Age in weeks
Kidney able to secrete. Bladder expands as sac. Genital duct opposite sex degenerating. Bulbo-urethral and vestibular glands appearing. Vagina sacs forming.	Thoracic duct and peripheral lymphatics developed. Early lymph glands appearing. Enucleated red cells predominate in blood.	Ossification centers more common. Chondrocranium at its height.	Perineal muscles developing tardily.	Epidermis adds intermediate cells. Periderm cells prominent. Nail field indicated. Earliest hair follicles begin developing on face.	Spinal cord attains definitive internal structure.	Iris and ciliary body organizing. Eyelids fused. Lacrimal glands budding. Spiral organ begins differentiating.	10
Uterine horns absorbed. External genitalia attain distinctive features. Meson. and rete tubules complete male duct. Prostate and seminal vesicle appearing. Hollow viscera gaining muscular walls.	Blood formation beginning in bone marrow. Blood vessels acquire accessory coats.	Notochord degenerating rapidly. Ossification spreading. Some bones well outlined.	Smooth muscle layers indicated in hollow viscera.	Epidermis three-layered. Corium and subcutaneous now distinct.	Brain attains its general structural features. Cord shows cervical and lumbar enlargements. Cauda equina and filum terminale appearing. Neuroglial types begin to differentiate.	Characteristic organization of eye attained. Retina becoming layered. Nasal septum and plate fusions completed.	12
Kidney attains typical shape and plan. Testis in position for later descent into scrotum. Uterus and vagina recognizable as such. Mesonephros involuted.	Blood formation active in spleen. Heart musculature much condensed.	Most bones distinctly indicated throughout body. Joint cavities appear.	Cardiac muscle appearing in earlier weeks, now much condensed. Muscular movements in utero can be detected.	Epidermis begins adding other layers. Body hair starts developing. Sweat glands appear. First sebaceous glands differentiating.	Hemispheres conceal much of brain. Cerebral lobes delimited. Corpora quadrigemina appear. Cerebellum assumes some prominence.	Eye, ear and nose grossly approach typical appearance. General sense organs differentiating.	16
Female urogenital sinus becoming a shallow vestibule (5). Vagina regains lumen (5). Uterine glands appear (7). Scrotum solid until sacs and testes descend (7-9). Kidney tubules cease forming at birth.	Blood formation increasing in bone marrow and decreasing in liver. (5-10). Spleen acquires typical structure (7). Some fetal blood passages discontinue (10).	Carpal, tarsal and sternal bones ossify late; some after birth. Most epiphyseal centers appear after birth; many during adolescence.	Perineal muscles finish development (6).	Vernix caseosa seen (5). Epidermis cornifies (5). Nail plate begins (5). Hairs emerge (6). Mammary primordia budding (5); buds branch and hollow (8). Nail reaches finger tip (9). Lanugo hair prominent (7); sheds (10).	Commissures completed (5). Myelinization of cord begins (5). Cerebral cortex layered typically (6). Cerebral fissures and convolutions appearing rapidly (7). Myelinization of brain begins (10).	Nose and ear ossify (5). Vascular tunic of lens at height (7). Retinal layers completed and light perceptive (7). Taste sense present (8). Eyelids reopen (7-8). Mastoid cells unformed (10). Ear deaf at birth.	20-40 (5-10 mo.)

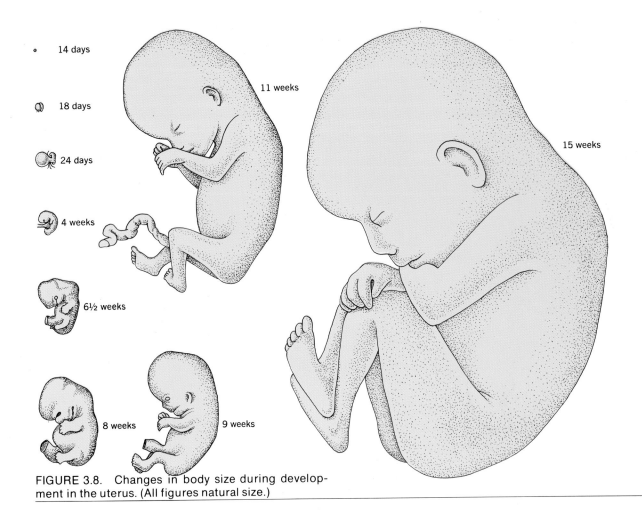

14 days

18 days

24 days

4 weeks

6½ weeks

8 weeks   9 weeks

11 weeks

15 weeks

FIGURE 3.8.   Changes in body size during development in the uterus. (All figures natural size.)

obvious change (Fig. 3.8) occurring after about 2
months *in utero*.

These sections have given a general view of
body formation. The basic development of each
system is discussed at the beginning of appropriate chapters that follow.

Summary

1. Cell reproduction produces new cells for maintenance of cell lines, repair,
and reproduction of the species.

   a. Mitosis perpetuates a cell line, causes growth in size, and is used in repair of injury. It results in two cells that are genetically the same as the
parent cell. Mitosis occurs in five stages: interphase, prophase, metaphase, anaphase, and telophase.

b. Meiosis occurs in ovaries and testes, and produces four sex cells that are haploid (one half normal chromosome number), and different in genetic makeup from the parent cell.

2. Neoplasms (cancers) are the result of uncontrolled mitotic cell division. They may be benign or malignant. Seven "danger signals" may suggest the presence of a neoplastic process in the body.

3. Basic human development may be regarded as occurring in "weekly subdivisions."

   a. The first week of life is concerned with fertilization (union of egg and sperm), cleavage (mitotic division), and formation of a blastocyst.

   b. The second week is concerned with implantation (burrowing of blastocyst into uterus wall), and formation of a two-layered embryo composed of ectoderm and endoderm.

   c. Mesoderm formation occurs during the third week, and differentiation of cells into lines that will form specific tissues and organs takes place.

   d. The nervous system and somites begin formation during the fourth week.

   e. The laying down of the body systems and assumption of human form (fetus) occurs during the fifth to eighth weeks.

   f. Growth, and refinement of the body systems, occurs from the ninth week to birth. Body proportions and size change.

Questions

1. Outline the major characteristics of the five stages of mitosis.

2. Account for the differences meiosis produces in its daughter cells.

3. What is a neoplasm? How does such a growth differ from normal cellular reproduction?

4. What are the main processes that occur in the first four weeks of life? Comment briefly about the significance of each process to the later development of the body.

5. How may development of organs or systems be affected? How can we account for the effect of an agent on more than one system?

Readings

Arey, L. B. *Developmental Anatomy*. 7th ed., Saunders. Philadelphia, 1965.

Moore, Keith L. *The Developing Human. Clinically Oriented Embryology*. Saunders. Philadelphia, 1973.

Rafferty, Keen A., Jr. "Herpes Virus and Cancer." *Sci. Amer. 229:26* (Oct) 1973.

Rugh, Roberts, and L. B. Shettles. *From Conception to Birth.* Harper and Row. New York, 1971.

Science News. "Chemical for Immunity." p 86, Feb. 9, 1974. Vol. 105, No. 6.

Science News. "Dietary Attack on Cancer." p. 177, March 16, 1974. Vol. 105, No. 11.

Science News. "Cigarette Gases and Tars Harmful." p. 177, March 16, 1974. Vol. 105, No. 11.

Stern, H. and D. L. Nanney. *The Biology of Cells.* Wiley. New York, 1965.

chapter 4
Control of Cellular Activity; Energy Sources for Cellular Activity

chapter 4

Growth and development, differentiation of tissues, and metabolism in general, depend on the production of specific chemical compounds to control these processes. The basis of this control appears to be genetic, in that the DNA molecules of the nucleus produce RNA molecules that pass to the cytoplasm and govern the production of proteins. These proteins, as enzymes and hormones, determine the rates and directions of chemical reactions in the body.

Energy for producing such chemicals, and for sustaining cellular activity in general, comes from the metabolism of the three basic foodstuffs (carbohydrates, lipids, proteins), and the synthesis and use of adenosine triphosphate (ATP).

Genetic control by nucleic acids

Both of the nucleic acids (DNA and RNA) contain units known as NUCLEOTIDES (Fig. 4.1). Each nucleotide is composed of a nitrogenous base, a sugar, and phosphoric acid. In DNA, four bases are commonly found: adenine (A); guanine (G); cytosine (C); and thymine (T). The sugar is a five-carbon unit called deoxyribose ($C_5H_{10}O_4$). In RNA, uracil (U) substitutes for thymine, and the sugar ribose ($C_5H_{10}O_5$) is present. The nucleotides are bound together by sugar-phosphate bonds into nucleic acid molecules.

In 1953, Watson and Crick proposed that DNA consisted of two nucleotide chains, coiled as a double helix, and bound to one another by base to base linkages (Fig. 4.2). Only certain bases are capable of linking to one another (A-T and C-G), in what is called *base pairing*. Thus specific configurations or sequences of nucleotides are created. A particular sequence of nucleotides may be a gene, the unit controlling the expression of a body characteristic. Watson and Crick further proposed that the four bases of DNA carried information that was coded according to the sequence of bases in the molecule. A sequence of three bases, or a TRIPLET, was suggested to form a code word corresponding to a particular amino acid. The position of that amino acid in a protein molecule is then determined by the position of the triplet in the nucleic acid.

DNA remains within the nucleus, and directs the production of proteins through the synthesis of RNA molecules. In the process known as TRANSCRIPTION, the two joined bases of the double DNA helix are believed to attract a particular nucleotide, free in the nucleus, to a given point along the DNA molecule. Many nucleotides are thus "lined up" in a specific order determined by base pairing between DNA and the nucleotide. These nucleotides are then joined to form a molecule of what is termed messenger-RNA (m-RNA). There must be as many m-RNA molecules as there are proteins in the body, since these molecules govern the production of proteins. With three bases available, 64 code words may be created. These 64 code words can form a nearly infinite number of combinations (enough to code

FIGURE 4.1. A diagram of the structure of a nucleotide.

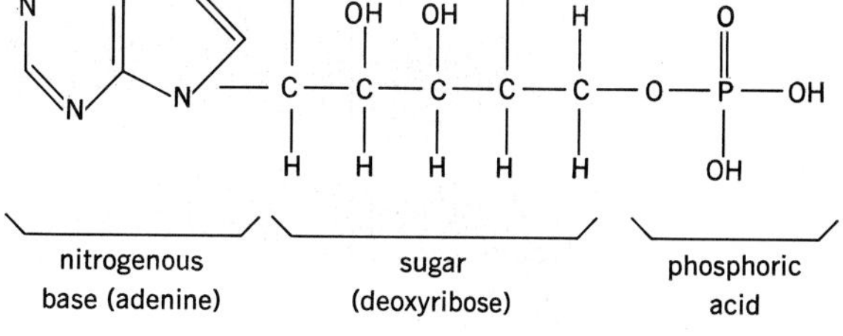

nitrogenous base (adenine)   sugar (deoxyribose)   phosphoric acid

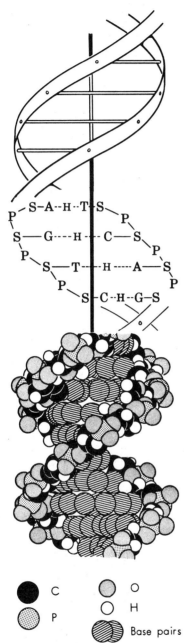

C ●   O ◉

P ▦   H ○

Base pairs

FIGURE 4.2. A three-way representation of DNA, showing the double helix, base pairing held by hydrogen bonds, and the arrangement of the atoms in the helices.

for the 100,000± proteins of the body). Messenger-RNA, formed in the nucleus, moves to the cytoplasm through nuclear pores and perhaps the endoplasmic reticulum, and becomes associated with the ribosomes. Another type of RNA, named transfer-RNA (t-RNA) is produced in similar fashion by DNA. T-RNA is a cloverleaf-shaped single helix. It is collected on the nucleolus and passes into the cytoplasm through the nuclear pores. T-RNA attaches to amino acids (the particular one determined by the three bases in the end of the t-RNA molecule) and carries them to the ribosomes for synthesis into proteins. There are as many t-RNA molecules as there are amino acids to be carried, about 20.

Arriving at an m-RNA molecule, the t-RNA molecule, with its attached amino acid, is attracted to a particular point on the m-RNA according to base-pairing properties (only A-U and C-G pairings can occur in RNA). Thus the amino acid is "inserted" in proper order for that particular protein. This process is called TRANSLATION (decoding of instructions). Joining of the amino acids forms the protein, which leaves the ribosome, allowing more molecules of protein to be synthesized. T-RNA molecules are also released and pick up more amino acids. These events are presented in Figure 4.3.

MUTATIONS, or changes in the genes and, therefore, in the proteins they produce, may be interpreted, in the light of this "genetic code," as misspelled words. An abnormal or misplaced triplet will obviously code for an amino acid that does not belong in that region, or for a different amino acid, and an abnormal protein will be produced. Thus the reaction that the protein is concerned with will not occur, or will occur abnormally. J. C. Kendrew illustrates the manner of production of mutations by the following passages:

A mutation, interpreted in the light of the genetic code, may be likened to a misprint in a line of newspaper type (with a letter representing an amino acid):

"Say it with *glowers*." In this example, there has been a substitution of an incorrect for a correct letter.

"The prime minister spent the weekend in the coun-

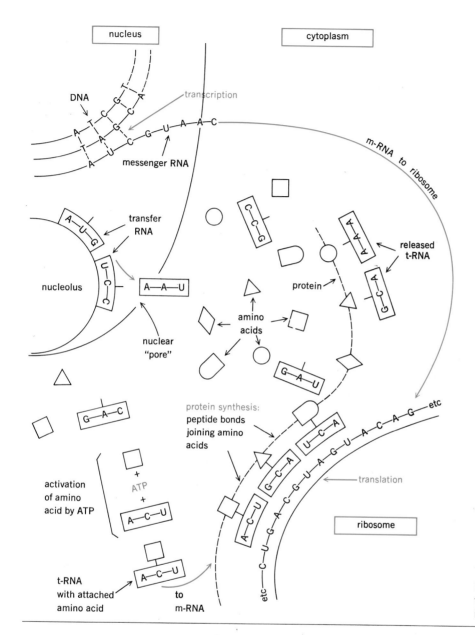

nucleus

cytoplasm

DNA

transcription

messenger RNA

m-RNA to ribosome

transfer RNA

released t-RNA

nucleolus

A—A—U

protein

amino acids

nuclear "pore"

G—A—C

protein synthesis: peptide bonds joining amino acids

translation

activation of amino acid by ATP

ATP

ribosome

t-RNA with attached amino acid

to m-RNA

etc

FIGURE 4.3. Transcription, translation, and protein synthesis in a cell.

try, shooting p_easants." In this case a letter is missing; a deletion has occurred.

"The treasury controls the public monkeys." Here, an additional letter is present in what is termed an insertion.

"We put our trust in the Untied Nations." Two letters have been reversed in what is termed an inversion.

"Each gene consists of *xgmdonrsd ad zbqpt ytrs.*" Suddenly, a sequence becomes gibberish, forming nonsense words.

It is easy to suspect the havoc that could be wrought in the code by these types of errors which result in abnormal proteins being produced. Most mutations usually affect only one representative of the pair of genes controlling expression of a given characteristic. The normal gene exerts its effect more strongly (dominance), and the mutation is not expressed. The defect is carried, however, and may express itself at some future time.

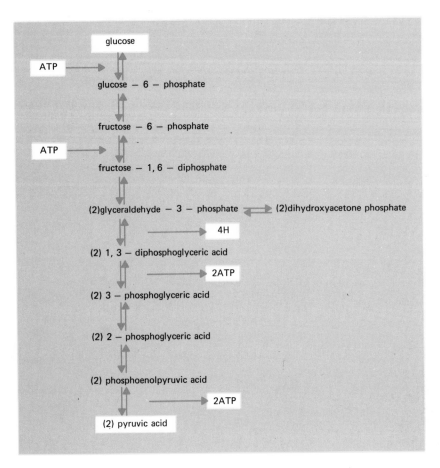

net reaction:

glucose ⟶ 2 pyruvic acid + 2ATP + 4H + 56 kcal/mole

FIGURE 4.4.   Glycolysis.

Energy sources for cellular activity

Energy is necessary to support the many activities the body cells carry out. Energy is supplied by the combustion of basic foodstuffs such as glucose and fats. Amino acids are usually not combusted to any great extent for energy, since they form the proteins that are the structural and controlling substances of the body. Carbohydrates and fats are degraded in stepwise fashion by metabolic cycles, and the energy released is used to synthesize ATP molecules. ATP (adenosine triphosphate) molecules contain what are called "high energy phosphate bonds." Actually, the last phosphate group in the molecule is transferred to another molecule to raise its energy level and to enable it to continue or start its metabolism. ATP thus acts as a means of transferring energy to a particular physiological process such as muscle contraction, or production of secretions by a cell.

Among the more important metabolic cycles that provide energy for the synthesis of ATP are glycolysis, the Krebs cycle, beta-oxidation, and oxidative phosphorylation.

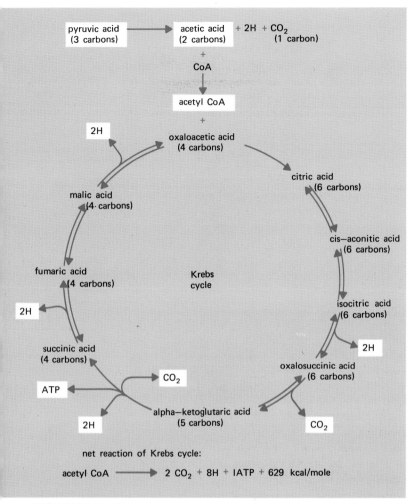

FIGURE 4.5. The conversion of pyruvic acid to acetyl CoA and the combustion of acetyl CoA in the Krebs cycle.

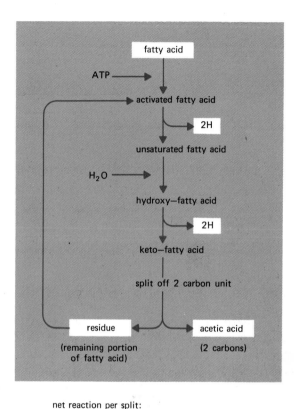

net reaction per split:

fatty acid ⟶ acetic acid + 4H + residue to other cycles (e.g. Krebs)

FIGURE 4.6. Beta-oxidation, the oxidation of fatty acids.

Glycolysis (Fig. 4.4)

Glycolysis is a scheme by which glucose is broken down and combusted in stepwise fashion to two molecules of pyruvic acid, without the presence of oxygen. In the course of the reactions, a total of about 56,000 calories of energy are produced. About 20,000 calories are used to synthesize two new ATP molecules, the remainder being released as heat to maintain body temperature. The cycle also releases four hydrogen atoms. Glycolysis normally catabolizes about three fourths of the glucose used by the body cells.

The Krebs cycle (Fig. 4.5)

This cycle, after the removal of carbon dioxide from pyruvic acid, combusts the resulting acetic acid to carbon dioxide in the presence of oxygen. One new ATP, 2 CO_2, and 8 hydrogens are produced for each molecule of acetic acid that goes through the cycle. A large mass of heat (600,000 calories/mole of pyruvic acid) is also produced.

Beta oxidation (Fig. 4.6)

Beta oxidation reduces fatty acids to two carbon units of acetic acid that may then enter the Krebs cycle. Hydrogens are also released by this cycle.

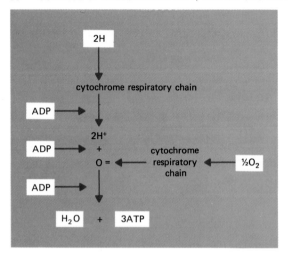

net reaction:

$$2H + \tfrac{1}{2}O_2 + 3ADP \longrightarrow H_2O + 3ATP$$

FIGURE 4.7. Oxidative phosphorylation, the biological oxidation of hydrogen.

Oxidative phosphorylation (Fig. 4.7)

Hydrogens released from other cycles are carried to this cycle on large molecules known as hydrogen acceptors. Utilizing compounds known as cytochromes, this scheme converts the hydrogens, in the presence of oxygen, to water. It liberates sufficient energy to synthesize either two or three ATP molecules per pair of hydrogens passing through the scheme.

Found within cells are other metabolic schemes for synthesis of glycogen and recovery of glucose from glycogen, and for production of urea, amino acids, and other compounds. Their roles in the body economy are presented in later chapters.

FIGURE 4.8.   The "metabolic mill," the interrelationships in the metabolism of protein, fatty acids, and glucose.

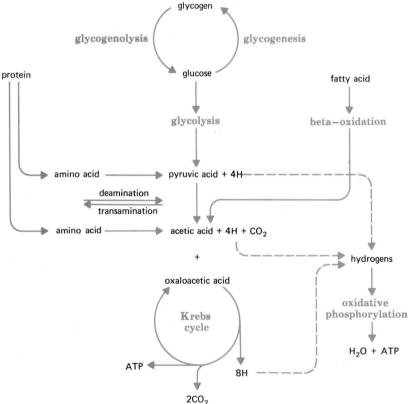

The metabolic mill

The schemes mentioned above interlock in what is known as the "metabolic mill" (Fig. 4.8). The mill emphasizes that the metabolism of the three basic foodstuffs is interrelated, and that it is possible to make one substance from another, that is,

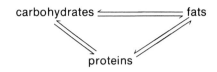

Since ATP may be produced by utilizing the energy from the breakdown of any of the three basic foodstuffs, the cells need never suffer a shortage of ATP.

Summary

1. Nucleic acids govern protein synthesis and thus the proteins, as enzymes and hormones, control the growth, development, and chemical reactions that occur within the body.

    a. DNA composes the chromosomes and genes and causes the synthesis of RNA molecules.

    b. Messenger RNA (m-RNA) is produced by the DNA and goes to the ribosomes to direct protein synthesis.

    c. Transfer-RNA (t-RNA) is produced by the DNA and enters the cytoplasm to attach to specific amino acids.

    d. t-RNA carries amino acids to m-RNA on the ribosomes, where they are joined to form proteins to govern cell activity.

2. Mutations represent changes in DNA or RNA molecules, which result in synthesis of faulty or abnormal proteins.

3. Energy sources for cellular activity include ATP, and the carbohydrates, fats, and amino acids that are combusted to give energy to synthesize ATP. Metabolic cycles are utilized to break down the foodstuffs.

    a. Glycolysis takes glucose to pyruvic acid, produces two new ATPs and releases four hydrogens and heat.

    b. The Krebs cycle combusts pyruvic acid to CO_2, and produces eight hydrogens and one ATP for each molecule of pyruvic acid.

    c. Beta-oxidation combusts fatty acids to four hydrogens and a two-carbon unit that may enter the Krebs cycle.

    d. Oxidative phosphorylation converts hydrogens to ATP and water, in the presence of oxygen.

    e. The metabolic mill shows the interrelationships of fat, protein, and carbohydrate metabolism, and the fact that ATP may be synthesized from any of the three basic foodstuffs.

Questions

1. What is the role of DNA in control of cellular activity?

2. What does m-RNA do? What does t-RNA do?

3. What is a mutation? How does it affect cellular activity?

4. What is transcription? Translation?

5. Explain the contribution of the Krebs cycle and oxidative phosphorylation to ATP synthesis.

6. What is the importance of ATP to the cell?

Readings

Brown, Donald D. "The Isolation of Genes." *Sci. Amer. 229*:20 (Aug) 1973.

Clark, Brian F. C., and Kjeld A. Marcker. "How Proteins Start." *Sci. Amer. 218*:36 (Jan) 1968.

Darnell, J. E. et al. "Biogenesis of m-RNA: Genetic Regulation in Mammalian Cells." *Science 181*:1215, 1973.

Harper, Harold A. *Review of Physiological Chemistry.* 13th ed. Lange Medical Pubs. Los Altos, Cal., 1971.

Hurwitz, J., and J. J. Furth. "Messenger RNA." *Sci. Amer. 206*:41 (Feb) 1962.

Kendrew, John C. *The Thread of Life.* Harvard Univ. Press. 1968.

Kornberg, A. "The Synthesis of DNA." *Sci. Amer. 219*:64 (Oct) 1968.

McKusick, Victor A. "The Mapping of Human Chromosomes." *Sci. Amer. 224*:104 (Apr) 1971.

Merrifield, R. B. "The Automatic Synthesis of Proteins." *Sci. Amer. 219*:56 (Mar) 1968.

Miller, O. L. "The Visualization of Genes in Action." *Sci. Amer. 228*:34 (Mar) 1973.

Mills, D. R. et al. "Complete Nucleotide Sequences of a Replicating RNA Molecule." *Science 180*:916, 1973.

Mirsky, Alfred E. "The Discovery of DNA." *Sci. Amer. 218*:78 (June) 1968.

Ptashne, Mark, and Walter Gilbert. "Genetic Repressors." *Sci. Amer. 222*:36 (June) 1970.

Tenin, Howard M. "RNA Directed DNA Synthesis." *Sci. Amer. 226*:24 (Jan) 1972.

Thompson, J. S., and M. W. Thompson. *Genetics in Medicine.* Saunders. Philadelphia, 1973.

Zubay, G. A Theory on the Mechanism of Messenger-RNA Synthesis. Proceedings, Natl. Acad. Sci. *48*:456, 1962.

chapter 5
Tissues of the Body; The Skin

chapter 5

Chapter 1 has indicated that similar cells form tissues, and that four basic groups of tissues compose the body. The groups are: epithelial, connective (including blood and blood-forming tissues), muscular, and nervous tissues.

Epithelial tissues

Epithelial tissues are the cellular layers that cover and line the outer body surfaces, the interior of hollow organs, and the body cavities. They also act as absorptive and secretory tissues, and generally contain nervous structures, called receptors, for receiving stimuli. Many glands, which secrete a variety of products, are derivatives of epithelial tissues.

Characteristics of epithelia

To form an effective covering, epithelial cells must FIT VERY CLOSELY TOGETHER, or our bodies would be easily invaded from the outside or suffer loss of vital substances from the inside. Epithelia usually attach to underlying connective tissues by a BASEMENT MEMBRANE. Epithelia DO NOT CONTAIN BLOOD VESSELS, but are nourished by vessels located in the underlying connective tissue. The farther the epithelial cells are from the blood vessels, the less will be their chances of survival. Epithelia may HAVE NERVES that pass between their cells from the underlying connective tissue, and typically show GREAT POWERS OF REGENERATION. For example, it has been estimated that the skin epidermis (outer epithelial layer) is replaced every three weeks, and that intestinal epithelial cells are replaced every two to three days. Epithelia may arise from all three germ layers.

Types of epithelia

If an epithelium contains only one layer of cells, it is called a SIMPLE epithelium. Simple epithelia are well nourished because all cells lie as close as possible to the vessels in the underlying connective tissue. Because of their thinness simple epithelia are generally the ones most involved in active and passive transfer of substances. If the epithelium contains two or more layers of cells, it is called a STRATIFIED epithelium. The upper layers of cells in a stratified epithelium are far removed from the blood vessels in the underlying connective tissue, and usually show changes characteristic of cell degeneration and death. In many cases, the upper layers of cells in a stratified epithelium are toughened or modified in some way to make the epithelium a good protective layer.

The shape of the cells in an epithelium may be SQUAMOUS or flat, with the cell resembling a piece of tile. Shape may be CUBOIDAL, like a cube of sugar, or it may be COLUMNAR, with the cell shaped like a short pencil. If the epithelium is stratified, it is named according to the shape of the top layer of cells.

There are several types of epithelia that result if we combine the terms describing the layers and shapes given above. They are shown in Figure 5.1, and are described below.

SIMPLE SQUAMOUS EPITHELIUM is one layer of flattened cells. It lines the thoracic and abdomino-pelvic cavities (*mesothelium*), blood and lymph vessels (*endothelium*), the filtering membrane of the kidney (*glomerulus*), and the air sacs (*alveoli*) of the lungs. It is a tissue adapted, by its thinness, to allow passage of solutes and water across it by physical processes. It is not adapted to resist wear and tear.

SIMPLE CUBOIDAL EPITHELIUM is one layer of

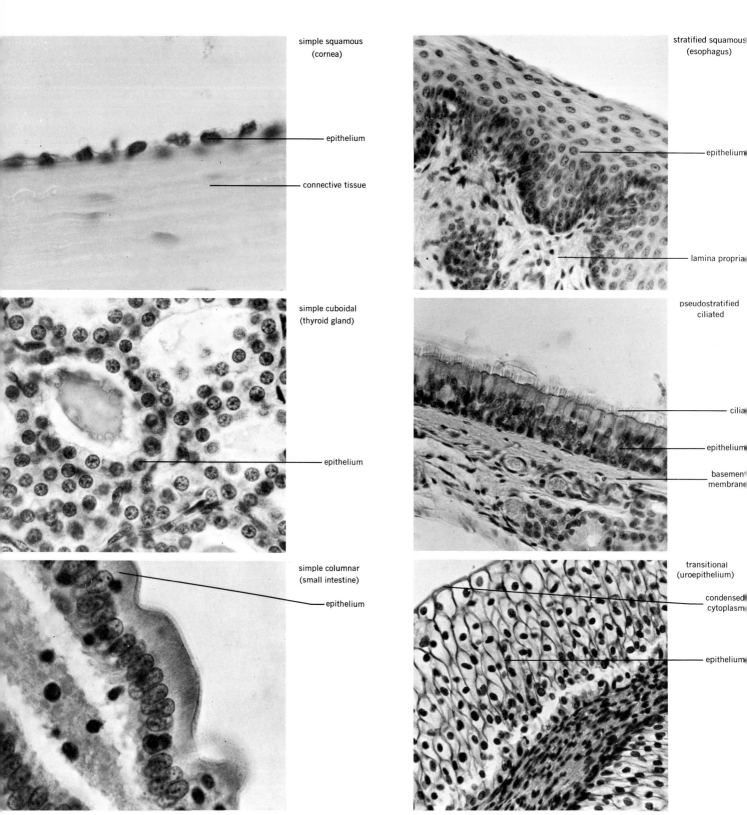

simple squamous
(cornea)

epithelium

connective tissue

simple cuboidal
(thyroid gland)

epithelium

simple columnar
(small intestine)

epithelium

stratified squamous
(esophagus)

epithelium

lamina propria

pseudostratified
ciliated

cilia

epithelium

basement
membrane

transitional
(uroepithelium)

condensed
cytoplasm

epithelium

FIGURE 5.1.   Common types of epithelia.

cube-shaped cells. It lines many of the kidney tubules, where it actively absorbs and secretes many materials. It covers the surface of the ovary, and lines the follicles of the thyroid gland (an endocrine gland producing hormones).

SIMPLE COLUMNAR EPITHELIUM is one layer of elongated cells. It lines the digestive tube from stomach to anal canal, and forms a secretory and absorptive membrane. It also lines the small tubes of the respiratory tree.

STRATIFIED SQUAMOUS EPITHELIUM consists of several layers of cells, with the top layer flattened. It covers those body surfaces or organs subjected to mechanical wear and tear. Thus the mouth, throat, esophagus, skin, and anal canal have this type of epithelium. The surface cells are easily removed (desquamated) and are rapidly replaced from below, affording protection against wear and tear. The cells may contain a protein, *keratin*; if present, keratin toughens and waterproofs the epithelium, as in the skin. If keratin is present, the epithelium is said to be keratinized or cornified; if absent, the epithelium is said to be non-cornified or "soft." This epithelium may become very thick (e.g., calluses) if subjected to great mechanical wear and tear. However, even a newborn has "thick skin" on the palms of his hands and soles of his feet. Such skin is thick because of genetic influences and not "wear and tear."

Stratified cuboidal and columnar epithelia are of rare occurrence (see Table 5.1).

It is easy to name the epithelia described above on the basis of number of cell layers and shape of cells. Five other varieties of epithelia are found

FIGURE 5.2. Microvilli, a surface modification to increase area for absorption and secretion.

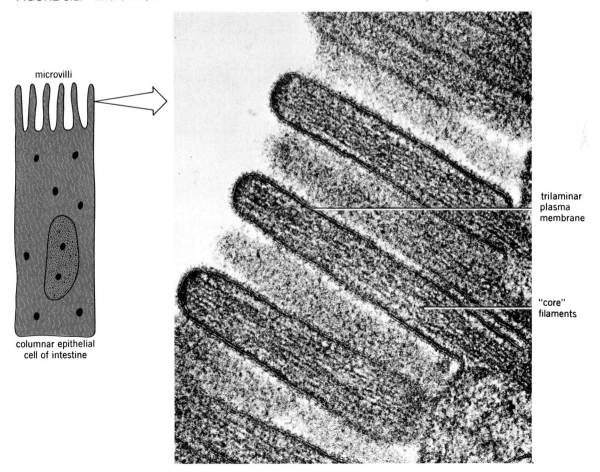

microvilli

columnar epithelial
cell of intestine

trilaminar
plasma
membrane

"core"
filaments

in the body, and they are named by characteristics other than those of cell layers and shape.

PSEUDOSTRATIFIED EPITHELIUM appears stratified, as judged by the many nuclei at different levels. However, all cells touch the basement membrane, but not all reach the surface. This epithelium lines most of the respiratory system. It is a type that can change into stratified squamous; this change often occurs in heavy smokers, and may give rise to cancers.

TRANSITIONAL EPITHELIUM (*uroepithelium*) is a stratified tissue lining the ureters and urinary bladder. The surface cells are not uniform in shape, and are capable of stretching as the organs expand to pass or contain fluids. The name transitional means "to pass from one state to another"; the cells change from rounded to flattened shape as the epithelium stretches.

SYNCYTIAL EPITHELIUM is a continuous, multi-nucleated mass with no cell membranes between

TABLE 5.1 Summary of Epithelium

Epithelial type	Characteristics	Examples of locations	Surface modification commonly present	Functions and comments
Simple squamous	One row flat cells, nuclei flattened and parallel to surface	Glomerular capsule, endothelium, mesothelium, Henle's loop of kidney	Microvilli (on meso-thelium)	Exchange, because of thinness
Simple cuboidal	One row isodia-metric cells, nuclei rounded	Kidney tubules, thyroid, sur-face of ovary	Microvilli (in kidney)	Secretion and absorption
Simple columnar	One row tall cells nuclei elongated perpendicular to surface	Stomach, small and large in-testines, gall bladder, ducts, bronchioles, uterus	Microvilli (in gut) Cilia (in respiratory system)	Secretion and absorption (gut) movement of substances across sur-face if ciliated
Stratified squamous	Several layers of cells, top layer flattened	Skin (fully corni-fied), vagina (partially cornified), mouth, esoph-agus (slight or no cornifica-tion)	None	Protection, since cells are easily removed from the sur-face and replaced rapidly from below
Stratified cuboidal	Several layers of cells, top layer cuboidal	Sweat glands	None	Secretory

cells. It covers the villi of the placenta where it is very active in absorption and secretion of materials.

GERMINAL EPITHELIUM lines the tubules of the testis and covers the ovary. In the testes, it is responsible for the production of sperm. In the ovary, it merely forms a covering for the organ.

NEUROEPITHELIUM consists of specialized epithelium in the nose, eye, ear, and tongue. It contains nerve cells specialized for reception of

stimuli (environmental changes inside or outside the body).

Surface modifications of epithelia

The term surface modification refers to the presence in the epithelium of something other than the regular epithelial cells, or to a special structure on the surface of the cells. Epithelia may

Epithelial type	Characteristics	Examples of locations	Surface modification commonly present	Functions and comments
Stratified columnar	Several layers of cells, top layer columnar	Larynx, upper pharynx	Cilia	Transition form between stratified squamous and pseudo-stratified
Pseudo-stratified	All cells reach basement membrane, not all reach surface. Nuclei at many levels	Nasal cavity, trachea, bronchi, male and female reproductive systems	Cilia	Movement of materials across surface
Transitional	Stratified, top layer not uniform in shape	Kidney pelvis, ureter, bladder	Condensed cytoplasm	Stretches and protects cells from acidic urine
Syncytial	Simple, no membranes between cells	Placental villus	Microvilli	Secretion and protects
Germinal	Several layers showing stages of sperm formation	Tubules of testis	None	Produces sperm
Neuro-epithelium	Nerve cells form part of epithelium	Taste buds, olfactory area, retina, cochlea	None	Sensory, usually as receptors of stimuli

contain one-celled mucus secreting glands known as GOBLET CELLS. Mucus is a thick, moist, and sticky fluid that helps to moisten the epithelium and to protect it from enzyme action (as in the digestive tube), or to aid in cleansing the organ (as in the respiratory system where it "traps" dust and pollen). CILIA, motile (movable) hairlike structures, are found on epithelia where something is to be moved across its surface (e.g., respiratory system, where mucus is moved; uterine tubes, where ova are moved). MICROVILLI (Fig. 5.2), are tiny fingerlike extensions of the cell surface, and are common on simple cuboidal and simple columnar epithelia (kidney, gut). The extra surface area created insures rapid absorption and secretion of substances. The epithelium of the urinary system has a thickened layer of cytoplasm (condensed cytoplasm) that protects the cells against the acid urine.

The epithelia are summarized in Table 5.1.

Epithelial membranes

The combination of certain epithelia and their underlying connective tissues, form an epithelial

Type of gland*	Diagram*	Characteristics	Examples
UNICELLULAR			
		One celled, mucus secreting	Goblet cells of resp. and diges. system
MULTICELLULAR			
Simple tubular		One duct—secretory portion straight tube	Crypts of Lieberkuhn of intestines
Simple branched tubular		One duct—secretory portion branched tube	Gastric glands, uterine glands
Simple coiled tubular		One duct—secretory portion coiled	Sweat glands

TABLE 5.2 A Summary of Exocrine Glands

*Simple implies one duct in the gland; compound implies a series of branching ducts in the gland. The heavy shaded portion of the gland indicates the extent of its duct or duct system.

membrane. Two types of these membranes are found in the body.

MUCOUS MEMBRANES form linings of body cavities opening on the body surface (mouth, nose, anus), or of organs that form part of the tube opening on a surface (e.g., stomach, intestines, respiratory system). Epithelial cell type is usually pseudostratified, simple columnar, or stratified squamous, and the membrane is always moistened by mucus. The mucus is secreted by goblet cells, or by glands in the connective tissue.

SEROUS MEMBRANES cover the visceral organs and line the true body cavities (those that do not open onto a body surface, e.g., thoracic and abdominopelvic cavities). The epithelium is always simple squamous, and is moistened by a watery (serous) secretion formed from the blood vessels in the membrane.

Other hollow organs (e.g., heart, blood vessels) are lined with an epithelium, which may be underlain with a few connective tissue fibers.

Glands

As mentioned above, secretion is a primary func-

Type of gland*	Diagram*	Characteristics	Examples
Simple alveolar		One duct—secretory portion saclike	Sebaceous glands
Simple branched alveolar		One duct—secretory portion branched and saclike	Sebaceous glands
Compound tubular		System of ducts— secretory portion tubular	Testes, liver
Compound alveolar		System of ducts— secretory portion alveolar	Pancreas, salivary glands, mammary
Compound tubulo— alveolar		System of ducts— secretory portion both tubular and alveolar	Salivary glands

TABLE 5.3   Manner of Production of Secretion by Exocrine Glands

Type of gland	Examples
Merocrine—synthesized product independent of basic cell structure	Pancreas, salivary glands
Apocrine—some portion of the cell issued as part of the secretion	Mammary gland
Holocrine—product of gland is a cell, or cell as a whole is shed in the secretion	Testis, sebaceous glands

FIGURE 5.3.   Common types of connective tissue.

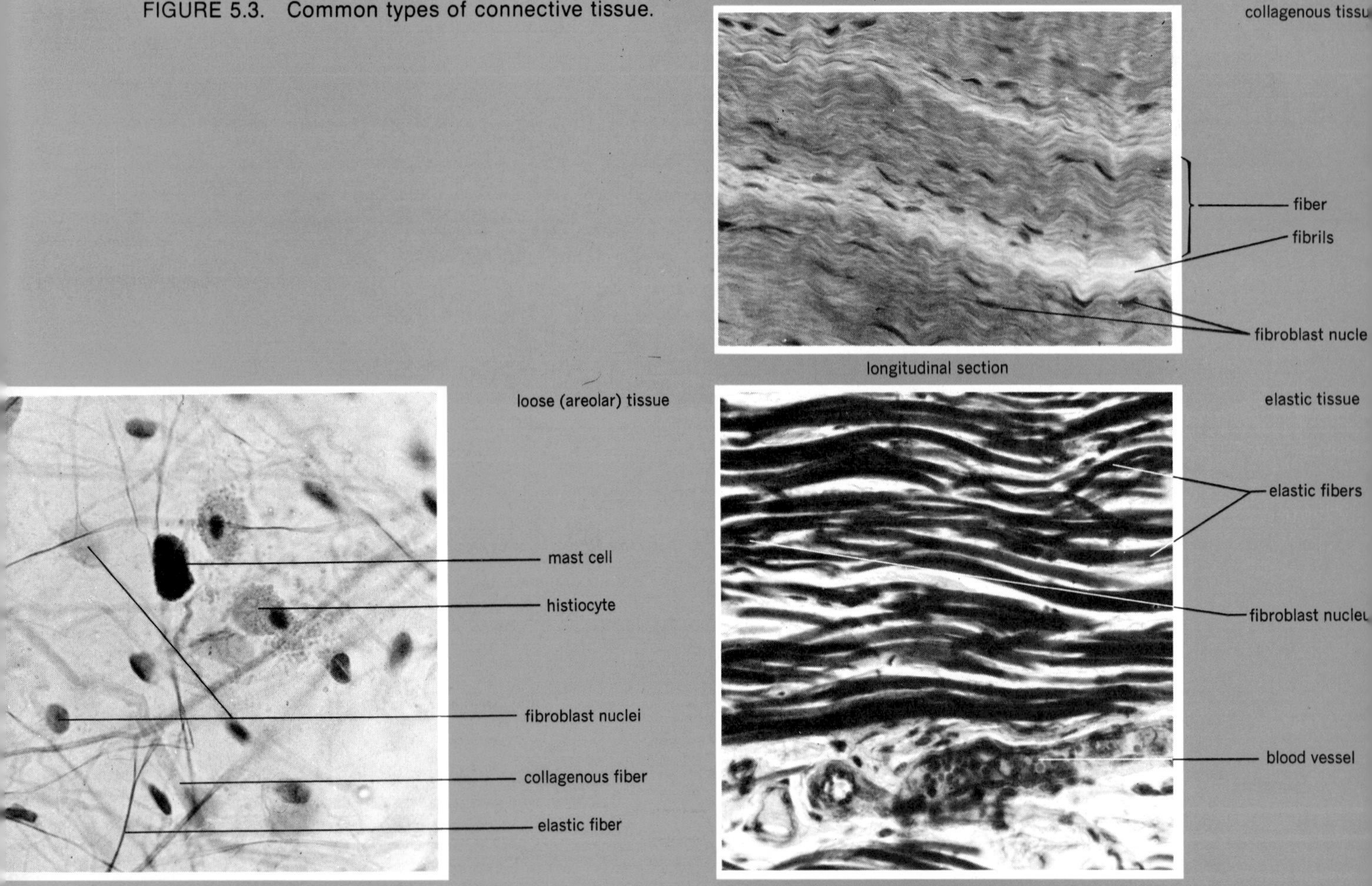

tion of epithelia. Glands are specialized cells in the epithelium, groups of epithelial cells that are buried in the connective tissue underlying the epithelium, or groups of cells that originated from an epithelium, but that have lost any obvious or visible relationship to the epithelium. Glands produce substances different from the fluids available to them, and thus utilize active processes to synthesize and release their products. If the gland retains a connection, by a duct or ducts, with the epithelial surface, it is termed an EXOCRINE GLAND. If it has lost its epithelial connection and secretes into the bloodstream, it is termed an ENDOCRINE GLAND. Tables 5.2 and 5.3 summarize facts about exocrine glands. The endocrine glands are described in Chapter 26.

Connective tissues

Connective tissues are the most abundant and widespread tissues of the body. They connect, support, and offer protection to body organs. They appear to be continuous throughout the body, and if they could be caused to disappear from the body in a wink of the eye, the body

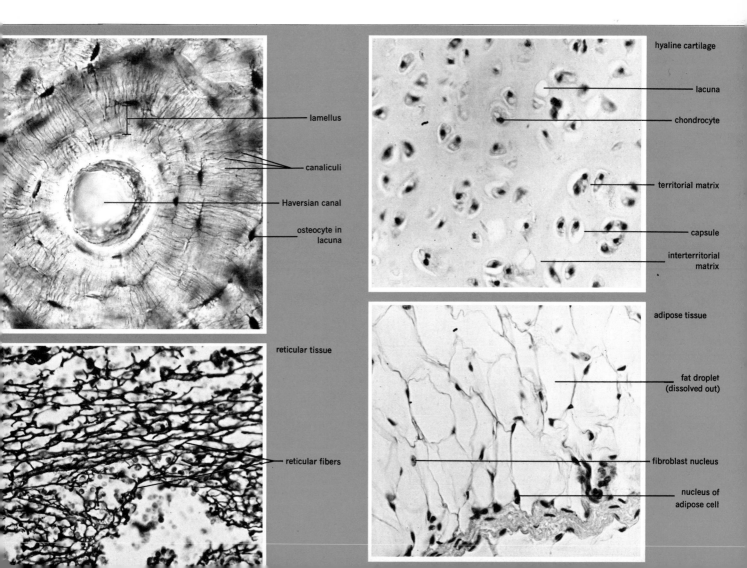

would collapse into a pile of many types of cells; all organization would be lost.

Characteristics of connective tissues

Connective tissues consist of LARGE AMOUNTS OF INTERCELLULAR MATERIAL with rather widely scattered cells. The intercellular material is often FIBROUS, and contains collagenous (tough) and elastic (stretchy) fibers. Connective tissues are not generally found on surfaces (except in joints), are usually vascular, and have a cell that produces and maintains the intercellular material.

Types of connective tissue (Fig. 5.3)

Of the many types of connective tissue present in the body, those of most widespread occurrence, or of greatest importance, will be described.

LOOSE (*areolar*) CONNECTIVE TISSUE consists of a three-dimensional network of collagenous (white) fibers, and elastic fibers. Collagenous fibers are composed of a protein called collagen, are very strong, and resist stretching; elastic fibers are composed of a protein called elastin, are stretchable, but not very strong. The tissue thus possesses both strength and elasticity "in all directions." FIBROBLASTS produce the fibers of this tissue. Loose connective tissue is found beneath the skin as the superficial fascia or subcutaneous tissue, and occurs around many body organs as a "packing material."

ADIPOSE TISSUE is a connective tissue consisting of large numbers of adipose (fat) cells, which are fibroblasts specialized to store fat. It is found beneath the skin and around many body organs. It serves as an insulating material to help prevent loss of body heat, gives rounded form and shape to the body, stores food energy, and aids in protecting those organs it surrounds (e.g., kidney).

RETICULAR CONNECTIVE TISSUE forms the internal supporting framework of many body organs (e.g., liver, spleen). Its delicate fibers form a three-dimensional support for the cells of these organs.

"FIBROUS CONNECTIVE TISSUE" includes the dense *collagenous tissue* of tendons and ligaments, and the *elastic connective tissue* of the large ar-

teries. These tissues are composed almost entirely of one type of fiber in a "one-way" pattern of arrangement.

CARTILAGE (*gristle*) is a tough, semisolid tissue, which contains CHONDROCYTES that produce the intercellular material. The intercellular material is called the MATRIX, and contains collagenous or elastic fibers. Cartilages are surrounded by a membrane known as the PERICHONDRIUM. The membrane contains cells that can turn into chondrocytes and form new cartilage if the cartilage is broken or damaged. Replacement or healing of cartilage is very slow, probably because of the small number of blood vessels in the tissue. There are three varieties of cartilage.

Hyaline cartilage is a translucent mass forming most of the embryonic skeleton, the cartilages connecting ribs to sternum (costal cartilages), and the covering of bones of freely movable joints.

Fibrous cartilage is very tough, because of the presence of many collagenous fibers in the matrix. It is found as the intervertebral discs between the bones of the spine.

Elastic cartilage is a rubbery mass making up the pinna of the ear (the "flap" on the side of the head), and the epiglottis of the larynx. It contains many elastic fibers in the matrix.

BONE is a hard substance forming the major part of the skeleton. Its development, form, structure, and properties are described in Chapter 6. A summary of the main connective tissues appears in Table 5.4.

Formation and physiology of connective tissues

Since the primary component of a connective tissue is the intercellular material, production of that material becomes the primary concern in the development and maintenance of the tissue. In loose connective tissue, and other fibrous tissues, the fibroblast is the cell that produces both the fibers and semiliquid components (ground substance) of the intercellular material. The ribosomes and Golgi apparatus of the fibroblasts synthesize the basic molecules (collagen, elastin, protein polysaccharides) necessary for intercellular materials, and these are then secreted from the cell and formed into fibers or ground substance. The fibroblasts of these tissues are also stimulated to greater activity by foreign substances and by the products of cell injury, so that they become very important in healing of injuries. They also are stimulated by adrenal cortical hormones, and play a role in the inflammatory process.

Cartilage relies on its chondrocytes to produce matrix and fibers. Again, the ribosomes and Golgi apparatus appear to be the sites of production of the necessary substances. When damaged, cartilage heals by forming new cartilage at the surface (*appositional growth*) through activity of cells in the perichondrium, or by formation within the mass (*interstitial growth*). Formation of new cartilage is greatly influenced by several vitamins and hormones.

Vitamins A and C are necessary for formation of fibers and ground substance.

Vitamin D is necessary for the absorption of calcium for formation of bone from cartilage.

Growth hormone (from the pituitary gland) and thyroxin (from the thyroid gland) cause stimulation of production of cartilage during body growth.

Clinical considerations

The fibers or matrix or both of connective tissue may be formed abnormally, leading to deformities or abnormalities of the organs in which the connective tissue is found. SCLERODERMA is a condition affecting the skin and blood vessels, in which large amounts of collagenous fibers are formed in the organs. The tissues become hardened and lose elasticity. ACHONDROPLASIA (*chondrodystrophy*) results when cartilage matrix is not formed properly, particularly during skeletal growth. Dwarfing, and misshapen bones result. Such disorders appear to have a hereditary basis.

TABLE 5.4   A Summary of the Main Connective Tissues (c.t.)

Tissue type	Characteristics	Examples of locations
Loose c.t.	Feltwork of collagenous and elastic fibers	Around organs, subcutaneous tissue
Fibrous c.t. (collagenous)	Wavy bundle of fibers, fibrils present	Ligaments, tendons, sclera of eye
Elastic c.t.	Branching, homogeneous fibers	Aorta, ligamentum nuchae, ligamenta flava
Reticular c.t.	Fine, irregular, branching fibers	Interior of liver, spleen, lymph nodes
Adipose tissue	"signet ring" cells with fat drops inside	Anywhere (almost): subcutaneous tissue of skin
Hyaline cartilage	Clear intercellular material (ICM), cells in lacunae; capsule around lacunae	Embryonic skeleton, costal cartilages, nasal cartilages, articular surfaces
Fibrocartilage	Visible collagenous fibers in ICM; other features same as in hyaline	Intervertebral discs, symphyses
Elastic cartilage	Visible elastic fibers in ICM; other features same as in hyaline	Epiglottis, external ear

FIGURE 5.4.   The three types of muscle.
(A) Smooth muscle. (B) Cardiac muscle.
(C) Skeletal muscle.

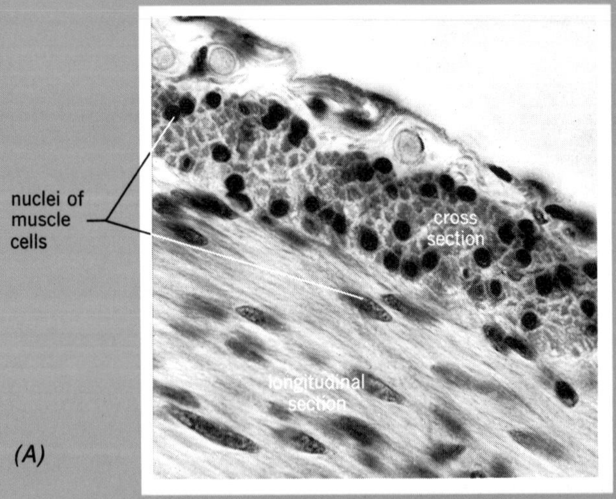

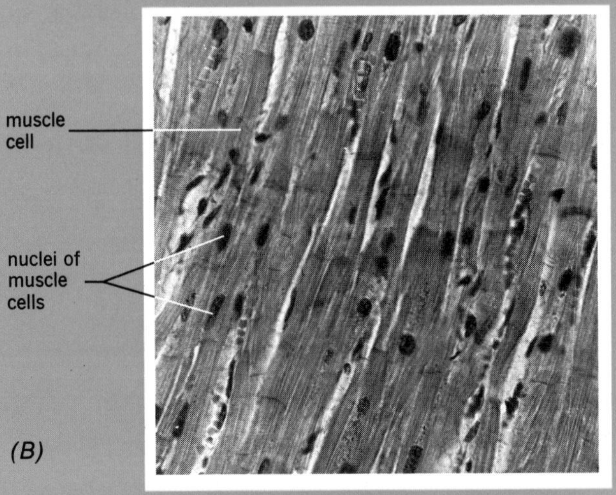

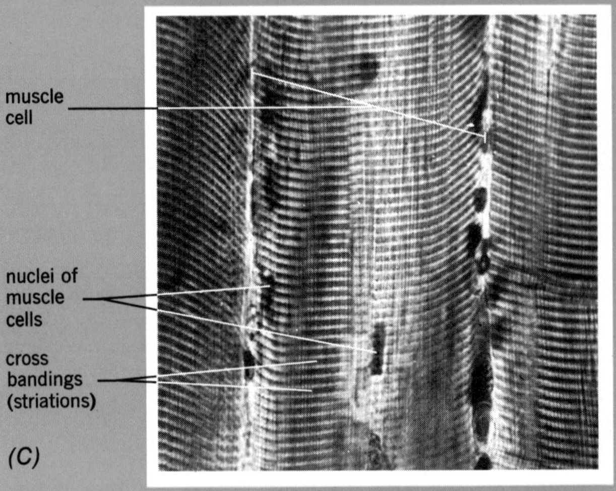

Muscular tissues (Fig. 5.4)

While the structure and functions of muscular tissues are considered in greater detail in subsequent chapters, it may be stated here that the muscular tissues are the CONTRACTILE tissues of the body. By shortening, they cause movement of the body as a whole, or of substances through the body. SKELETAL MUSCLE attaches to, and moves, the bones of the skeleton. CARDIAC MUS-CLE is found only in the heart, and causes the circulation of the blood. SMOOTH or VISCERAL MUSCLE is found in internal body organs such as blood vessels, and the stomach and intestines. It is important in control of blood pressure, blood flow, and movement of materials through the alimentary tract.

Nervous tissue (Fig. 5.5)

Nervous tissue is the highly EXCITABLE and CON-DUCTILE tissue of the body. It responds to changes (stimuli) in the body's internal and external en-vironments, and forms the organs for interpreta-tion and integration of body response to such stimuli.

The skin

The skin and its derivatives, such as hair, nails, and several types of glands, form the integumentary system. The skin is a tough but pliable, durable, and resistant covering for the body. It has, in the adult, a surface area averaging 1.75 square meters (about 3000 square inches), composes about 7 percent of the body weight, and receives about one third of the blood pumped by the left ventricle of the heart. It is the largest organ of the body!

Structure of the skin (Fig. 5.6)

Skin has three major layers: the EPIDERMIS is the outer epithelial layer; the DERMIS (*corium*) is a layer of connective tissue beneath the epidermis, and contains blood vessels, nerves, and lymphatics; the SUBCUTANEOUS LAYER (*hypodermis*) lies beneath the dermis and contains much adipose tissue.

The epidermis of the skin contains 30–50 layers of cells arranged in several sublayers. From the deepest layer outward, these layers are:

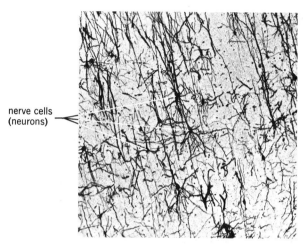

nerve cells (neurons)

FIGURE 5.5.   Nervous tissue, cerebrum.

FIGURE 5.6.   Diagram showing the structure of hairy skin.

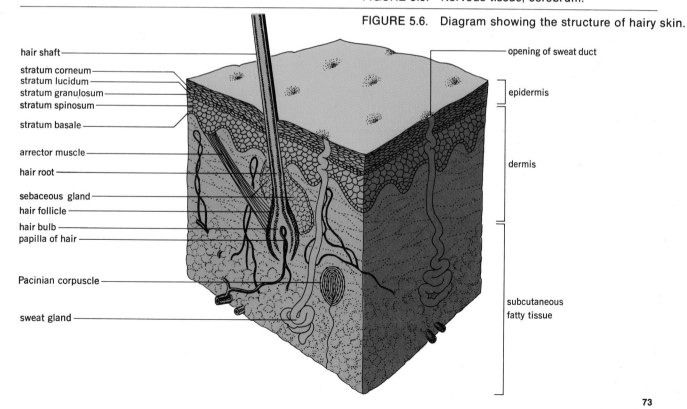

hair shaft

stratum corneum
stratum lucidum
stratum granulosum
stratum spinosum

stratum basale

arrector muscle

hair root

sebaceous gland
hair follicle
hair bulb
papilla of hair

Pacinian corpuscle

sweat gland

opening of sweat duct

epidermis

dermis

subcutaneous fatty tissue

1. *Statum basale, also known as the stratum cylindricum or stratum germinativum ("basal, cylindrical, or germinal layer").* The cells of this layer form a single layer of columnar-shaped cells that rest on the basement membrane. They undergo continual mitotic division to renew the epidermis. Cells push toward the surface from this layer and are shed (40 to 50 pounds in a lifetime) from the surface.

2. *Stratum spinosum ("prickle cell layer").* The cells of this layer show spinelike projections that attach the cells to one another.

3. *Stratum granulosum ("granular cell layer").* The cells in this layer show changes characteristic of death (lack of nuclei, loss of organelles), and contain dark staining granules. The process of cornification (keratinization) by which the cells become filled with keratin, is begun in this layer. Death of cells occurs because of the inability of nutrients to diffuse to the cells

4. *Stratum lucidum ("clear layer").* The cells of this layer show no nuclei or membranes, and the layer appears as a clear shiny line. The cells contain a chemical known as eleidin, which represents an intermediate stage in formation of keratin.

5. *Stratum corneum ("horny layer").* This is the surface layer of the epidermis, and consists of many layers of flat, dead, scalelike cells filled with a protein called *keratin.* The keratin waterproofs and toughens the skin, and renders it an effective protective tissue.

The lower surface of the epidermis is folded and interlocks with folds of the underlying dermis. The folds are reflected in the surface as the fingerprints (friction ridges) that are specific for each individual and that aid in picking up or grasping objects.

The DERMIS averages about 3 millimeters in thickness and is composed of a dense connective tissue, containing both collagenous and elastic fibers, many blood vessels, glands, lymphatics, and sensory corpuscles. It is divided into an upper PAPILLARY (folded) LAYER and a lower RETICULAR LAYER. The folded layer contains many capillary networks that nourish the germinal layer, and that allow radiation of heat through the surface. This layer also contains touch receptors. The reticular layer contains many arteries and veins that may connect directly with one another

to form an arteriovenous anastomosis. These anastomoses cause blood to bypass the capillary beds, and conserve body heat. Receptors for pressure are found in this layer, as are the sweat and sebaceous glands. Loss of elastic fibers of the dermis occurs with aging, and accounts for the folding (wrinkling) of the skin in older individuals.

The SUBCUTANEOUS LAYER (hypodermis) is composed of loose connective tissue heavily loaded with FAT cells. The layer aids in insulating the body against loss of deep heat, and is loose enough to permit the injection of several milliliters of fluid without development of significant pain (subcutaneous injection). The fat in this layer is what gives the general form and shape to the body. It cushions and protects muscles and nerves, and is an area of food storage.

Appendages of the skin

HAIRS (see Fig. 5.6) develop as a downgrowth of epidermal cells into the dermis to form a HAIR FOLLICLE that is set at an angle into the skin. Mitosis of the cells in the base of the follicle produces the hair. A HAIR has an expanded *bulb* at its lower end, and the bulb is indented by a *dermal papillum* that supplies blood vessels and nutrients to insure growth of the hair. The *root* is imbedded in the skin, and the *shaft* is the visible portion of the hair above the skin. *Muscles* (arrector pili) attach to the follicles and pull them into a more vertical position. In furry animals, these muscles erect the hair, thickening the layer, and increasing its capacity to act as an insulating layer. In man, "goose flesh" occurs as the hair pushes a mound of skin to one side as it is pulled into a straight position. Hair protects the head from sunlight and the eyelashes and eyebrows aid in protecting the eyes from light. When the eyelids are partly closed the eyelashes filter and scatter the light rays entering the eye, reducing their intensity.

NAILS (Fig. 5.7) are appendages useful in grasping and picking up small objects. The nail represents a hardened corneal layer and has a *free edge, body,* and *root.* The nail rests on the *nailbed* (derived from the basal and spiny layers of the epidermis), and has a *cuticle* (eponychium) above the root of the nail, and a *quick* (hyponychium) beneath the free edge. A white half-moon or *lunula,* lies under the root, and represents the active, growing region of the nail.

The GLANDS OF THE SKIN (see Fig. 5.6) are of three types: SWEAT (sudoriferous) glands; SEBACEOUS (oil) glands secrete a substance (sebum) that aids in keeping skin and hairs lubricated, pliable, and waterproofed; in the ear canal, CERUMINOUS glands secrete a bitter, brownish, waxlike material that repels insects and lubricates the skin of that area.

SWEAT GLANDS are of two types. *Eccrine glands* are simple coiled tubular glands of widespread occurrence over the body. Their secretion is watery and aids in cooling the body as it evaporates. Eccrine sweat contains inorganic salts. *Apocrine* sweat *glands* are large glands found in the armpits and around the anus. They produce a fluid rich in organic substance, which may be acted on by bacteria to produce odors. Apocrine sweat glands are associated with hairs and begin to function at puberty.

Functions of the skin

The skin is a versatile and active organ. As a cover-ing, it protects the body it encloses, and this forms its primary function. It also acts to aid in regulation of the body temperature, as an absorptive and excretory organ, and in the reception of stimuli that impinge on the body.

PROTECTION. The skin presents several lines of defense against the environment.

A SURFACE FILM composed of water, lipids, amino acids, and polypeptides is derived from the secretions of the sweat and sebaceous glands, and from the breakdown of the cornified surface cells. It has a pH of 4–6.8 and forms an effective antiseptic layer, retarding the growth of bacteria and fungi on the skin surface. It also reacts with many potentially toxic materials, preventing their entry through the skin surface. Its water content moistens the skin, and the lipids lubricate and aid in waterproofing the surface. The water-proofing function is not perfect as evidenced by the wrinkles that our skin develops (fingers) on long exposure to water. This is caused by water entering the upper layers of the skin and caus-ing it to swell and fold.

The presence of the horny layer of the skin, with its keratinized cells, acts as a PHYSICAL BAR-RIER to the penetration of most chemicals, bac-teria, parasites, and a good part of the environ-mental radiation (e.g., sunlight). At the junction

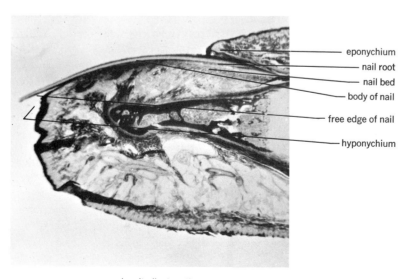

FIGURE 5.7. A longitudinal section of the fingertip showing the nail and its associated structures.

eponychium
nail root
nail bed
body of nail
free edge of nail
hyponychium

longitudinal section

between the cornified and noncornified layers of the skin, is a region of positive and negatively charged ions that will repel charged substances trying to enter the body.

These devices, which prevent entry of substances into the body from outside, also prevent the loss of essential body constituents from the inside. The body does not "dry out," or lose essential proteins, salts, and other chemicals.

TEMPERATURE REGULATION. Heat production by metabolic processes must constantly be balanced by heat loss to insure maintenance of temperature homeostasis. Skin contains capillary networks through which blood may be circulated close to the body surface. If the environmental temperature is less than body temperature, RADIATION OF HEAT will occur from the warmer to cooler area. The amount of blood reaching the capillary beds is controlled by the diameter of the arteries leading into the capillaries. The nervous system is responsible for controlling the size of these arteries. Skin also contains the arteriovenous junctions described previously that can bypass many capillary beds. Thus if the arteries narrow, less blood passes to the upper skin layers, and heat is conserved; opening of these vessels allows greater heat loss.

A second means of temperature regulation is afforded by the SECRETION OF THE ECCRINE SWEAT GLANDS. The watery solution poured on the body surface requires heat to evaporate it. This causes heat loss from the body, and becomes an important method of heat loss when environmental temperature is close to, or above, body temperature.

ABSORPTION. It has been claimed for years that the skin "breathes." Indeed, oxygen, carbon dioxide, and nitrogen pass relatively easily through the skin. Any substance that can dissolve in the surface film of the skin will pass more easily through the skin. A substance that can dissolve the keratin (keratolytic) of the horny layer will also enter the body through the skin. Sex hormones (e.g., estrogens) are fat soluble, and are found in some creams that are applied to the body. Aspirin is a keratolytic, and in salves, may be applied to the skin surface for relief of "aches and pains."

EXCRETION. The excretory function of the skin depends on the skin glands. The eccrine sweat glands continually secrete SWEAT that does not come to our conscious attention (insensible perspiration). This excretion aids in maintaining the body's water balance. The secretion also contains salt (sodium chloride), which if exerted in large amounts (as during heavy work in a hot environment) can lead to muscle cramps. Measurable and significant amounts of METABOLIC WASTES (urea, ammonia) are also eliminated in the sweat. Apocrine sweat glands and sebaceous glands secrete a product rich in organic compounds that may be odoriferous. These compounds are readily attacked and decomposed by the skin bacteria to produce additional compounds that may contribute to "body odor." Suppression or removal of the odors requires removal of the compounds, usually by the use of soap and water.

RECEPTION OF STIMULI. Sensory receptors for heat, cold, touch, pressure, and pain are found in the skin. Their structure and function are described in Chapter 14. These receptors are essential to the recognition of changes in the environment so that adjustments can be made to maintain homeostasis.

Skin and hair color

Most of us have some color to our skin and hair. The color itself, and the intensity of the color is determined by several factors.

PIGMENTS in the skin and hair are produced in skin or hair cells by a variety of biochemical reactions that are genetically determined. *Melanin* is a yellow to black pigment found only in the basal layer of the skin of white people, but in all epidermal layers in the skin of black people. Amount of pigment and, therefore, darkness of skin color is not determined by a single gene, but by as many as eight genes. Thus the range of skin color is very great. In albinos (L. *albus*, white) a gene has mutated, and melanin cannot be synthesized because of a missing enzyme. Such individuals have white skin (*no* pigment), white hair, and no pigment in the irises of their eyes. Their eyes are very sensitive to light because the lack of iris pigmentation allows much light to enter. *Carotene* is a yellow pigment found in the horny layer of the skin. It is more abundant in the skin of certain Asian races. Mixtures of the two pigments may give yellows and reds of various intensities.

Pigments in the skin afford protection against solar radiation. Humans are believed to have originated in Africa, where sun and heat is great. Dark skin absorbs much heat, but radiates that heat more efficiently than lighter colored skin, and this gives protection against the effects of heat.

Hair color is determined by the same pigments as are found in the skin, and intensity of color in both skin and hair are correlated. As we age, pigment synthesis in the hair follicles diminishes, and many of the hair cells contain air; this causes graying or whitening of the hair.

The AMOUNT OF BLOOD circulating through the skin and the thickness of the skin determines its "pinkness." Thinner skin allows more color of the blood to show through. The amount of hemoglobin with oxygen on it also determines skin color. Richly oxygenated hemoglobin is bright red; hemoglobin with little oxygen on it is bluish in color. If sufficient "blue hemoglobin" is present in the capillary networks of the skin, a blue color (cyanosis) is imparted to the whole skin. Jaundice is a yellow color imparted to the skin by excessive red blood cell destruction, or liver malfunction. It results from excessive bilirubin (a pigment) in the bloodstream.

Exposure to ENVIRONMENTAL FACTORS, particularly radiation, can also influence skin color. The ultraviolet radiation of sunlight is damaging to the blood vessels and other tissues of the skin. One response to the radiation is *tanning*, in which the amount of melanin is temporarily increased in the epidermis. The pigment then absorbs more of the radiation. Sunlight also dehydrates the skin and makes it more "leathery," so that a tan may look nice but can be damaging to the health of the skin.

Clinical considerations

The skin is a very sensitive indicator of the presence of abnormal processes at the skin surface, and within the body itself. Rashes or eruptions of the skin may occur as a result of infection or other processes.

LESIONS. Skin eruptions ("breaking out" with a visible change in the skin surface) accompany many diseases. The changes that occur follow a similar series of events regardless of the agent causing them. A given condition may stop in any one of the stages, or progress through them all. The terms below are ones that health personnel use most generally, and are given in order of their appearance. A good example on which to observe the follow-through of each stage is the smallpox vaccination. The various stages through which skin lesions progress are:

Macule A discolored area on the skin, neither raised nor depressed.

Papule A red, elevated area on the skin.

Vesicular Stage
{
Vesicle A pinhead to split-pea sized elevation filled with fluid.

Pustule A vesicle containing pus.

Bulla A large vesicle or blister.
}

Crust Dry exudate (weeping of fluid and/or pus) adhering to the skin.

Lichinification Thickening and hardening of the skin.

Scar Replacement of cells by fibrous tissue as healing progresses.

Keloid A large, elevated scar.

COMMON DISORDERS OF THE SKIN. INFECTIONS are perhaps the most common disorders affecting the skin. *Bacteria* (mostly staphylococcus and streptococcus) are the most common infectious agents. Boils are a result of coccal infection. *Viruses* may produce warts or herpes (itching vesicles). *Fungi* produce cracks or fissuring as in athletes foot, and scalp infections (ringworm). The fungi thrive in warm, moist environments that makes for an easy transfer from the tennis shoe to the locker room and showers or swim areas. *Insects* such as mites, lice, bees, and spiders may bite or sting, causing eruptions on the skin.

Acne or "pimples" is a physiological disturbance of puberty and early adulthood and affects more than 80 percent of teenagers. It is caused by the effect of androgenic hormones (male sex hormones) on the sebaceous glands and hair follicles. It occurs in both sexes, because androgens are produced by the adrenal glands, testes, and ovaries. Acne occurs in four grades, depending on severity.

Grade I acne consists of the formation of a fatty-keratin plug, a blackhead, in the opening of a follicle on the skin.

Grade II acne results when the plugged duct ruptures and spills sebum into the surrounding tissues. The sebum irritates the tissue and a pustule forms. Bacteria may also be trapped in the tissue.

Grade III acne occurs when there is tissue destruction by the inflammation, and scarring results.

Grade IV acne results in lesions extending to the shoulders, arms and trunk, with severe scarring occurring.

FIGURE 5.8. The "rule of 9's" used to calculate extent of burns.

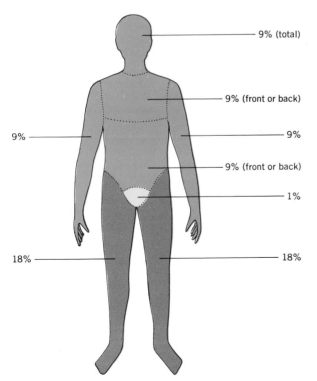

9% (total)

9% (front or back)

9%

9%

9% (front or back)

1%

18%

18%

Treatment involves washing with a mild soap and warm water as needed, and avoiding foods that may aggravate the condition (e.g., chocolate, nuts, cola drinks). Sun may be helpful, and some of the newer medications are promising. Squeezing or "popping" the pustules should not be attempted since this may increase chances of scarring. Too frequent washing also may traumatize the skin. If the face is disfigured by acne, psychologically sensitive individuals may withdraw from social contact with their fellows, with resultant need for counseling. *Dermabrasion*, in which the skin is "ground down" to the level of the dermal upfolds, may be utilized to remove scars. This method of treatment is done only with the advice of a skin specialist, because not all types of acne scarring will be improved by the procedure.

Eczema, seborrhea, and psoriasis are other disorders of the skin whose names are commonly read or heard in the mass media.

Eczema refers to a chronic irritation of the skin in which there is redness, eruptions, watery discharge, and formation of crusts and scabs. Its cause is not known, and it probably represents a symptom rather than being a disorder in itself.

Seborrhea is a disorder of the sebaceous glands in which there is excessive production of sebum that may accumulate as crusts and scales on the head, face, or trunk.

Psoriasis results in the formation of silvery scales on scalp, elbows, knees, or the body generally. The cause is unknown, and the disease can occur at any age in both sexes. Treatment is directed toward relief of symptoms, as there appears to be no cure.

BURNS. One of the first hazards man recognized in his efforts to change the quality of his survival was burns. Today nearly one million people are burned severely enough to be hospitalized annually, and about 7000 of these people expire. Burns are one of the major categories included under accidents, the first leading cause of death between 1--44 years of age.

Exposure to flame, scalding, hot objects, or some chemicals destroys the skin, and its protective functions are lost. Infection and loss of fluids, electrolytes, and blood proteins follow as a result of skin destruction. A burn is described in two ways: first, by the extent of destruction of surface area (the "rule of nines," Fig. 5.8, is a convenient method of estimating the extent of destruction); second, the depth of the burn is described by "degrees."

A FIRST DEGREE BURN involves only the surface layers of the epidermis. The skin is reddened and tender to the touch. A sunburn is a good example of a first degree burn.

A SECOND DEGREE BURN is one in which much of the epidermis is destroyed, but some epidermal remnants are present. It extends into the dermis. There is redness, and blisters are usually present. A burn resulting from briefly touching a hot object is usually a second degree burn. This type of burn is usually very painful because of irritation of the nerves of the dermis by products of cell destruction.

A THIRD DEGREE BURN involves all skin layers, with no epidermal remnants present in the burned area. Charring of the skin is common, and

destruction may involve muscles, tendons, and bones. The skin is insensitive to stimuli because of destruction of nerves in the dermis and hypodermis. On healing, fibrous masses of dead tissue (eschars) may form, which limit movement at joints. Third degree burns may result from prolonged contact with hot objects, from open flame, steam (car radiators), chemicals (strong acids or alkalies), and hot liquids (paraffin, oil).

There are three considerations of primary importance in the treatment of burns.

Prevent, as best as possible, secondary infection in the burned area.

Maintain water and electrolyte homeostasis by administration of appropriate solutions. Urine output is a convenient measure of adequacy of fluid intake.

Encourage adequate nutrition. A burned individual is one who is not interested in eating. Intravenous (IV) feeding may be required. Mobilizing the patient as early as his burns permit encourages recovery and feeding.

Summary

1. Cells that are similar form tissues. Four tissue groups compose the body: epithelial, connective, muscular, and nervous tissues.

2. Epithelia have special characteristics.
   a. They cover and line free surfaces of the body.
   b. They are composed of closely packed cells.
   c. They are active as secretory and absorptive tissue.
   d. They rest on a basement membrane and are nourished from blood vessels beneath the epithelium.
   e. They show great powers of regeneration.

3. Epithelia are either one layered (simple) or many layered (stratified), and contain cells that are flat (squamous), cuboidal (cube shaped), or columnar (tall).
   a. Simple squamous epithelium is one layer of flat cells. It is adapted for exchange of materials.
   b. Simple cuboidal epithelium is one layer of cube-shaped cells. It is adapted for absorption and secretion.
   c. Simple columnar epithelium is one layered of tall cells. It is an absorptive and secretory tissue.

d. Stratified squamous epithelium is several layers of cells, with the top layer flat. It is adapted to resist mechanical wear and tear.

e. Pseudostratified epithelium lines the respiratory and parts of the reproductive systems. It generally has cilia.

f. Transitional epithelium lines the urinary system and stretches as the organs expand.

g. Syncytial epithelium has no membranes between cells; it covers the placenta.

h. Germinal epithelium lines the testis tubules, and covers the ovaries. It produces sperm in the testes.

i. Neuroepithelium is specialized for reception of stimuli. It is found in the organs of special sense (eye, ear, nose, tongue).

4. Epithelia and their underlying connective tissue form epithelial membranes.

a. Mucous membranes line cavities that open on a body surface. They are moistened by mucus.

b. Serous membranes line internal body cavities. They are moistened by a watery secretion.

5. Glands are derivatives of epithelia.

a. Exocrine glands secrete onto a body surface.

b. Endocrine glands secrete into the bloodstream.

6. Connective tissues have special characteristics.

a. They connect, support, and protect body structures.

b. They have large amounts of intercellular material that is usually fibrous.

c. They are vascular.

d. They contain scattered cells.

7. There are several important types of connective tissue.

a. Loose connective tissue contains strong white fibers and stretchy elastic fibers. The cells called fibroblasts produce fibers. Loose connective tissue occurs under the skin.

b. Adipose tissue contains many fat cells. It stores energy, insulates and gives form to the body.

c. Reticular tissue forms the internal framework of body organs.

d. Fibrous connective tissue is dense and forms tendons, ligaments, and is found in blood vessel walls.

e. Cartilage is semisolid and forms part of the skeleton (hyaline), discs between vertebrae (fibrous), and the outer ear (elastic).

f. Bone is hard, forms the skeleton, and protects body organs.

8. The components of the intercellular substance of a connective tissue are produced by the characteristic cells of that tissue. Connective tissues reflect nutritional or hormonal deficiencies very quickly.

9. Connective tissues may form abnormally and result in malformed skeletal structures.

10. Muscular tissue is contractile and occurs in three varieties.

  a. Skeletal muscle attaches to and moves the skeleton.

  b. Cardiac muscle is found in the heart and circulates the blood.

  c. Smooth muscle is found in internal organs and controls blood pressure and movement through hollow organs.

11. Nervous tissue is excitable and conductile and forms sensory, interpretive, and motor pathways.

12. The skin is the major organ of the integumentary system and has three main layers.

  a. Five layers occur in the epidermis.

  b. The dermis contains fibrous connective tissue, sensory corpuscles, blood vessels, and glands.

  c. The subcutaneous tissue contains much fat.

13. Hairs, nails, and glands are appendages (derivatives) of the skin.

  a. Hairs lie in a follicle and grow from the follicle.

  b. Nails are hardened structures useful in grasping small objects.

  c. Sweat glands help cool the body; sebaceous glands lubricate the hair and skin.

14. The functions of the skin include:

  a. Protection. A surface film of chemicals acts as an antiseptic, and reacts with toxic materials. The horny layer forms a physical barrier to entry of microorganisms and radiation. Loss of body components is also prevented.

  b. Temperature regulation. Heat radiates from the blood passing through the skin. Sweat glands produce a fluid that evaporates and cools the skin.

  c. Absorption. Gases and lipid soluble materials pass through the skin, as do those substances that dissolve the keratin in the epidermis.

  d. Excretion. Water, salts, urea, and ammonia are excreted by the skin glands.

  e. Reception of stimuli. Sensory receptors sensitive to heat, cold, touch, pressure, and pain are found in the skin.

15. Skin color depends on:

  a. Pigment cells

  b. Blood flow

  c. Oxygen levels of the blood

  d. Bilirubin levels

  e. Exposure to sunlight

16. Hair color is determined by pigments or air in the hair cells.

17. The skin is a sensitive indicator of whole body physiology.

  a. It may develop lesions; a common series of phases occurs.

  b. It may develop disorders including infections, acne, eczema, seborrhea, and psoriasis.

  c. It may be burned.

18. Burns are described as to the extent of burning and depth.

  a. Extent is referred to by a percent.

  b. Depth is referred to as first, second, and third degree. Reddening and tenderness characterizes a first degree burn; destruction, but some epithelial remnants characterizes a second degree burn; no epithelial remnants characterizes a third degree burn.

19. Treatment of burns is centered around prevention of secondary infection, maintenance of fluid and electrolyte balance, and maintenance of adequate nutrition.

Questions

1. Compare and contrast epithelia as to general characteristics, locations, and functions.

2. Describe the relationships between the type of epithelium lining an organ and the function(s) the organ serves. Give examples.

3. What is an epithelial membrane? What function(s) do they serve?

4. How are glands related to epithelia? How are they classified?

5. How does heredity influence connective tissue formation?

6. Compare muscular and nervous tissues as to basic properties.

7. Describe the structure of hairy skin.

8. How is the skin involved in temperature regulation?

9. What changes occur in the skin as it ages?

10. Define: macule, vesicle, scar.

11. What is meant by a "third degree burn over 80 percent of the body"?

Readings

Bloom, William, and Don W. Fawcett. *A Textbook of Histology.* 9th ed. Saunders. Philadelphia, 1968.

Elden, H. R. (ed). *Biophysical Properties of the Skin.* Wiley. New York, 1971.

Ferriman, D. *Human Hair Growth in Health and Disease.* Thomas Pubs. Springfield, Ill., 1965.

Hardy, J. D., A. P. Gagge, and A. J. Stolwijk. *Physiological and Behavioral Temperature Regulation.* Thomas Pubs. Springfield, Ill., 1970.

Jeghers, H., and L. M. Edelstein. "Pigmentation of the Skin," *Signs and Symptoms.* Edited by C. M. MacBryde, and R. S. Blacklow. Lippincott. Philadelphia. 1970. 12th ed. pp 916–959.

Marback, H. I., and H. S. Gavin. *Skin Bacteria and Their Role in Infection.* McGraw-Hill. New York, 1965.

Marples, M. S. *The Ecology of the Human Skin.* Thomas Pubs. Springfield, Ill., 1965.

Montagna, W. *The Epidermis.* Academic. New York, 1964.

Nicoll, P. A., and T. A. Cortese, Jr. "The Physiology of Skin" in: *Annual Review of Physiology,* Vol. 34. Annual Reviews. Palo Alto, Cal., 1972.

Polk, H. C., Jr., and Stone, H. H. (eds). *Contemporary Burn Management.* Little, Brown. Boston, 1971.

Rook, A. S., and G. S. Walton (eds). *Comparative Physiology and Pathology of Skin.* Blackwell. Oxford, 1965.

Tregear, R. T. *Physical Functions of skin.* Oxford Univ. Press. New York, 1966.

chapter 6
The Skeleton

chapter 6

The human skeleton is a living, dynamic structure, which serves as a supporting and protective framework for the body. Muscles attach to the skeleton and cause skeletal movement. Many body organs are suspended from the skeleton. The skeleton provides a storehouse for calcium and phosphate, essential substances for body function. It contains a tissue known as bone marrow, which serves as a reservoir of nutrients and acts as the area of production of several types of blood cells. The bones are thus sites of intense activity, and are not the dry lifeless structures commonly studied in the anatomy laboratory.

The skeleton is composed of osseous tissue (bone), a tissue made hard by the deposition of inorganic substances in it, in the process known as calcification.

The structure of bone

Bone contains about 65 percent inorganic substance, chiefly calcium phosphate $[Ca_3(PO_4)_2]$, and about 35 percent organic substance, consisting of cells and a fibrous protein known as bone collagen. According to the arrangement of these components, three types of bony tissue are described.

CANCELLOUS or SPONGY BONE (Fig. 6.1) consists of interlacing bars and plates of bony tissue with many spaces between. It is found inside many bones, such as, the interior of ribs, skull bones, and vertebrae, and the ends of the long bones. Bone cells, or OSTEOCYTES, lie within cavities or LACUNAE in the bony tissue.

COMPACT BONE is a very dense material with microscopic subunits of structure (Fig. 6.2), known as OSTEONS or Haversian systems. These units are oriented lengthwise in the compact bone. The center of the osteon is formed by the HAVERSIAN CANAL, which carries the blood vessels into the bones. Rings of bony tissue known as LAMELLAE lie around the canal, more or less alternating with rows of LACUNAE containing OSTEOCYTES. Tiny canals or CANALICULI connect lacunae with one another and with the canal. Compact bone forms the shafts of the long bones.

DENSE BONE, without osteons, but containing lacunae and osteocytes, forms the outer covering of the bones.

Formation of bone

Types of formation

Bone is formed two ways in the body.

In INTRAMEMBRANOUS FORMATION, bone is formed "directly" from connective tissue within a fibrous membrane. Mesoderm cells increase in number and size, and the number of blood vessels increases. Some mesodermal cells change into bone-forming cells or OSTEOBLASTS. These osteo-blasts form a semisolid mass containing collagenous fibrils known as OSTEOID. The osteoid is then CALCIFIED by deposition of inorganic material to form bone.

In "indirect" bone formation or INTRACARTILAGENOUS (*endochondral*) FORMATION (Fig. 6.3), a MODEL of the bone is first formed in hyaline cartilage. The bones are initially very small in the embryo (see Fig. 3.8). Cartilage cells in the

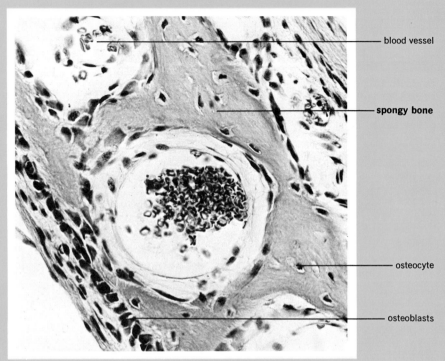

FIGURE 6.1.   Intramembranous bone formation, illustrating spongy bone.

FIGURE 6.2.   A cross section of ground compact bone, showing the parts of the osteon.

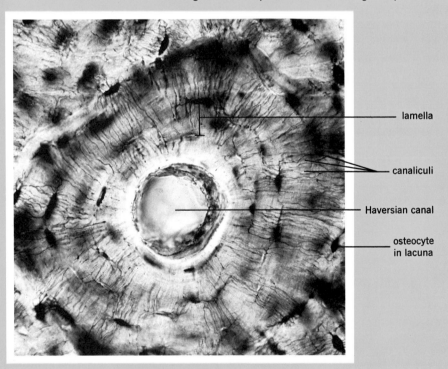

center of the model enlarge and become more numerous, squeezing the matrix into thin bars and plates. Outside the model, perichondrial cells are changing into osteoblasts and laying down (intramembranously) a BONE COLLAR around the center of the model to support it. The cartilage matrix is then CALCIFIED. Next, blood vessels penetrate the collar, bringing OSTEO-BLASTS into the model. The vessels destroy the cartilage cells; the osteoblasts brought in lay down OSTEOID (see above) on the calcified cartilage plates, and this is CALCIFIED to form bone. Formation of bone from cartilage thus involves two calcifications, with tissue destruction occurring.

In the long bones, formation from cartilage usually begins in the center of the bone, and later in the two ends, so that plates of cartilage (epiphyseal cartilages) remain between the shaft and ends of the bone. As long as these plates are cartilagenous, the bone may grow in length (see below).

Calcification

Deposition of inorganic substance in osteoid to form bone is called calcification. Although it is not definitely known how calcification occurs, it is clear that the calcium and phosphate necessary for the process must first be absorbed from the gut or placenta, until some critical blood value

FIGURE 6.3. The formation of bone from cartilage (intracartilagenous bone formation).

Diagram of the development of a typical long bone as shown in longitudinal sections. *Green*, bone, *blue*, calcified cartilage, *red*, arteries. *a'*, *b'*, *c'*, *d'*, *e'*, cross sections through the centers of *a*, *b*, *c*, *d*, *e*, respectively. *a*, cartilage model, appearance of the periosteal bone collar; *b*, before the development of calcified cartilage; *c*, or after it, *d*; *e*, vascular mesenchyme has entered the calcified cartilage matrix and divided it into two zones of ossification, *f*; *g*, blood vessels and mesenchyme enter upper epiphyseal cartilage; *h*, epiphyseal ossification center develops and grows larger; *i*, ossification center develops in lower epiphyseal cartilage; *j*, the lower and, *k*, the upper epiphyseal cartilages disappear as the bone ceases to grow in length, and the bone marrow cavity is continuous throughout the length of the bone. After the disappearance of the cartilage plates at the zones of ossification, the blood vessels of the diaphysis, metaphysis, and epiphysis intercommunicate. (Bloom and Fawcett. *A Textbook of Histology*, courtesy of W. B. Saunders Co.)

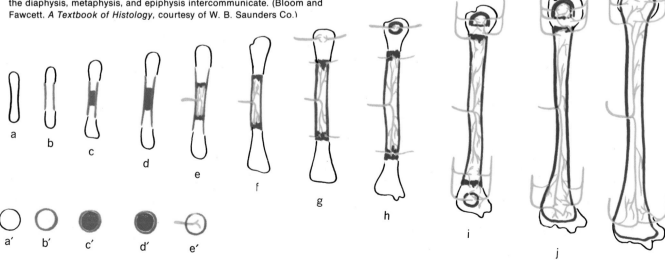

is reached. Calcification requires vitamin D, which accelerates absorption of calcium. Second, the tissue to be calcified must be made calcifyable, that is, capable of having the salts deposited in it. This latter task appears to be carried out by the osteoblasts, which secrete a substance causing binding of phosphate to collagen fibrils. This combination then acts as a center for crystal formation, and deposition of salts may proceed.

Growth of bones

Long bones grow in length by retaining a plate of cartilage between the shaft (diaphysis) and the end (epiphysis) of the bone. Cartilage production must occur more rapidly than bone formation to cause length increase, as new bone is laid down on either side of the epiphyseal cartilage. When the plate itself becomes bony, growth in length ceases. Growth in diameter of bones occurs by intramembranous formation of new bone tissue on the surface of the bone, through the activity of cells located in the PERIOSTEUM, a membrane surrounding all bones. The periosteum forms bone in the intramembranous manner. Growth of bones involves the activity of bone-forming cells, or osteoblasts, and bone remodeling or destroying cells, OSTEOCLASTS. Typically, as bone is added in one area, it is removed in another, so that the thickness of the wall of the bone, or its weight, tends to remain within normal limits, and it usually does not become extremely heavy or thick.

Clinical considerations

Without vitamin D the inorganic constituents are not available for deposition, and bones remain soft or only partially calcified (rickets). Vitamin A deficiency retards maturation and growth of cartilage cells in intracartilagenous formation; vitamin A excess causes accelerated destruction of bone. Vitamin C deficiency results in production of a noncalcifyable matrix.

As bone ages, formation of the organic portion is slowed, and the bones become more brittle. Density decreases, the bones may appear "moth-eaten," and are easily fractured.

Growth hormone from the pituitary gland is important in controlling production of new cartilage, and in controlling the activity of the periosteum. Parathyroid hormone (PTH) controls the activity of bone-destroying cells (osteoclasts); increased PTH levels are associated with increased bone destruction. Calcitonin, a hormone of the thyroid gland, increases deposition of inorganic salts in bone.

Classification of bones

Bones fall into four general classes according to their shape:

LONG BONES have a greater length than width, have a central SHAFT (diaphysis) composed of compact bone, and proximal and distal ENDS (epiphyses) composed of spongy bone. A central MEDULLARY CAVITY contains bone marrow, and is lined with ENDOSTEUM, a thin membrane. The bones of the fingers and toes (phalanges), upper arm (humerus), lower arm (radius and ulna), thigh (femur), leg (tibia and fibula) are examples of long bones. Figure 6.4 shows a section of a long bone to illustrate its parts.

SHORT BONES are ones in which length and width are not greatly different. The bones of the wrist (carpals) and ankle (tarsals) are short bones.

FLAT BONES consist of inner and outer layers of dense bone (tables) and a central layer of spongy bone (diploë). The cranial bones and scapulae (shoulder blades) are flat bones.

All other bones are classed as IRREGULAR BONES. They are of complicated shape. The vertebrae and certain skull bones are of this type.

Two other "types of bones" may be mentioned. WORMIAN BONES (*sutural bones*) are small, irregular pieces of bone found in the course of the major sutures of the cranium. They are the result of isolated areas of bone formation separate

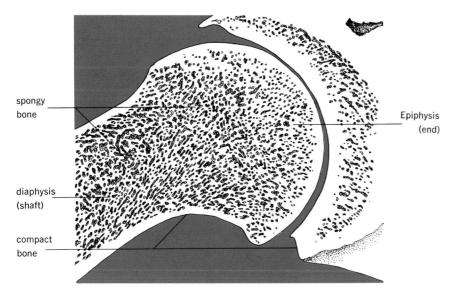

FIGURE 6.4. A section of the proximal end of the humerus and scapula to illustrate spongy and compact bone.

from those that create the cranial bones, and are of no consequence.

SESAMOID BONES are bones that develop in tendons. The kneecap is a sesamoid bone formed in the patellar tendon. Sesamoid bones often develop where pressure is put on a tendon, as under the "ball" of the foot.

Naming bones and features of bones

Our understanding of bones and their parts can be aided by remembering that most bone names are descriptive of their shape or resemblance to objects around us. For example, the word parietal means, "a wall," and this bone is a major component of the wall of the cranial valut; a bone of the wrist is called the lunate, and it bears a striking resemblance to a (half) moon. Some terms describing surface features are applicable to many bones, and are presented in alphabetical order:

CONDYLE (G. *kondylos*, knuckle). A rounded protuberance at the end of a bone forming an articulation or joint.

CREST (L. *crista*, tuft). A ridge or elongated prominence on a bone.

EPICONDYLE (G. *epi-*, above, + condyle). A projection from a bone, above a condyle.

FACET (Fr. *facette*, small face). A small, smooth, flat or shallow surface on a bone, particularly a vertebra.

FISSURE (L. *fissus*, a cleft). A groove or narrow cleft-like opening.

FORAMEN (L. *foramen*, an opening). A passageway through a bone; a hole.

FOSSA (L. *fossa*, a ditch). A furrow or shallow depression.

FOVEA (L. *fovea*, a pit). A small cuplike depression.

HEAD (A. S. *heafod*, head). The proximal end, or larger extremity of a bone.

LINE (L. *linea*, line). A long, narrow ridgelike structure.

MEATUS (L. *meatus*, passage). A passageway or short canal running within a bone.

NOTCH (A. S. *nocke*, notch). A deep indentation or narrow gap in the edge (or margin) of a bone.

PROCESS (L. *processus*, a going before). A projection or outgrowth from a bone.

SINUS (L. *sinus*, a curve). A cavity within a bone.

SPINE (L. *spina*, spine). A somewhat sharp process from a bone.

SULCUS (L. *sulcus*, a groove). A shallow furrow or groove.

TROCHANTER (G. *trochanter*, a runner). Large processes found only on the proximal end of the femur.

TUBERCLE (L. *tuberculum*, a little swelling). A small rounded eminence on a bone.

TUBEROSITY (L. *tuberositas*, a swelling). An elevated rounded process from a bone. Usually larger than a tubercle (word often used synonomously with tubercle).

One should be aware that it is possible to combine some of these basic terms to give an actual bony part. For example, the condyloid process (of mandible) is a knucklelike process forming a joint.

The organization of the skeleton*

The skeleton (Fig. 6.5) is composed of 206 bones, divided into an AXIAL SKELETON, and an APPENDICULAR SKELETON. The axial skeleton contains 80 bones and consists of the skull, vertebral column, and thorax (ribs and sternum). The appendicular skeleton contains 126 bones, and consists of the limbs and the bones supporting the limbs.

The skull

The skull bones may be divided into two groups: those forming the CRANIUM (the "brain box"), and those forming the FACE. The cranial bones enclose the cranial cavity which is divided into anterior, middle, and posterior portions; those of the face form mainly the anterior part of the skull.

The CRANIAL BONES are 8 in number:

1 frontal	2 temporals
2 parietals	1 sphenoid
1 occipital	1 ethmoid

The features of each bone appear in Table 6.1.

The FRONTAL BONE (Fig. 6.6) forms the forehead, the floor of the anterior cranial cavity (or roof of the orbit which houses the eye), and part of the

roof of the nasal cavities. The bone contains the FRONTAL SINUSES (Fig. 6.7). In the fetus, its two halves are separated by the FRONTAL (*metopic*) SUTURE.

The PARIETAL BONES (Fig. 6.8) form the larger part of the sidewalls and roof of the cranium. They join the frontal bone at the CORONAL SUTURE, and are separated from one another by the SAGITTAL SUTURE.

The OCCIPITAL BONE (Fig. 6.9) forms the posterior and basal portion of the posterior cranial cavity. The LAMBDOIDAL SUTURE separates the occipital bone from the parietal bones, and commonly contains one or more Wormian bones.

The TEMPORAL BONES (Fig. 6.10) form part of the lateral walls and floor of the cranium. The SQUAMOUS PORTION of the temporal is separated from the parietal bones by the SQUAMOUS SUTURE. The TYMPANIC PORTION lies inferior to the squama, and contains the opening of the ear. The MASTOID PORTION lies posterior to the tympanic portion. The PETROUS PORTION lies mainly within the cranial cavity (Fig. 6.11), where it appears as a ridge of bone housing the delicate structures of the inner ear (cochlea, semicircular canals).

The SPHENOID BONE (Fig. 6.12) is shaped like a butterfly, and articulates with all other cranial bones.

The ETHMOID BONE (Fig. 6.13) is an irregularly shaped bone that lies between the orbits, and that forms a small part of the floor of the anterior cranial cavity.

*The figure numbers that appear in the following discussion indicate that picture in which the feature described is most clearly shown. It may appear in other figures as well.

TABLE 6.1   A Summary of the Skull Bones (Italics Indicate the Most Important Features).

Name of bone and number	Features on the bone	Description/location of bone/feature
Frontal (1)		Forms forehead, roof of orbit and nasal cavities
	Supraorbital margins	Upper edge of orbits
	Superciliary ridges	Ridge above orbits
	Supraorbital foramen/notch	Hole/gap in upper margin
	Frontal sinuses	Cavity inside bone behind ridges
	Glabella	Central flat part of forehead
Parietal (2)		Roof and side walls of cranium
	Temporal lines	Arched lines on side of bone
Occipital (1)		Back and base of skull
	Foramen magnum	Large hole in base of bone
	Occipital condyles	Smooth joint surfaces each side of foramen magnum
	Nuchal lines	Ridges horizontally and vertically on back of bone
	External occipital pro-tuberance	"Bump" on outside of back of bone
	Internal occipital pro-tuberance	"Bump" on inside of bone
Temporals (2)		Side walls and floor of cranium
	Squamous portion (squama)	"Temple" (side) of cranium
	Zygomatic process	Projection from squama, anterior
	Mandibular fossa	At base of zygomatic process; mandible attaches here
	Tympanic portion	Below squama
	External auditory meatus	Opening of ear
	Styloid process	"Spike" extending down from skull
	Mastoid portion	Posterior to ear opening
	Mastoid process	Extends downward behind ear
	Mastoid air cells	Air cavities in mastoid process
	Stylomastoid foramen	Hole between styloid and mastoid processes
	Petrous portion	Ridge inside cranium
	Internal auditory meatus	On anterior medial border of petrous portion
	Jugular foramen	Halfway down joint between petrous temporal and occipital bone
	Carotid canal	Center of base of petrous portion
Sphenoid (1)		"Butterfly shaped" bone in center of skull
	Body	Central part of bone
	Sphenoid sinus	Cavity in body
	Sella turcica	Depression on upper surface of body

(TABLE 6.1, Continued)

Name of bone and number	Features on the bone	Description/location of bone/feature
Sphenoid (1) (Continued)	*Greater wings*	Extend laterally from body
	Lesser wings	Superior to body and greater wings
	Pterygoid processes	Inferiorly from body
	Optic foramen	Hole at base of lesser wing
	Superior orbital fissure	Cleft between greater and lesser wings
	Foramen ovale	Hole at base of greater wing
	Foramen spinosum	Lateral to foramen ovale
	Foramen rotundum	In posterior side of base of greater wing
Ethmoid		Lies between orbits
	Horizontal (cribriform) plate	Roof of nasal cavity; full of holes (passes olfactory nerves)
	Crista galli	Crest above horizontal plate
	Perpendicular plate	Separates nasal cavities (upper 2/3)
	Lateral masses (conchae)	Carry two of three turbinates for increasing area of nasal cavities
Nasals (2)		Bridge of the nose
Zygomatic (malar) (2)		Cheekbones
Maxillae (2)		Floor of orbits, upper jaws, part of hard palate
	Orbital plates	Floor of orbits
	Palatine processes	Anterior three fourths of hard palate
	Infraorbital foramen	Hole below orbit
	Maxillary sinus (antrum of Highmore)	Cavity in bone
	Alveolar process	Bears teeth of upper jaw
Mandible (1)		Lower jaw
	Body	Horseshoe shaped part
	Rami (sing.: ramus)	Upwards directed processes
	Angle	Where body and rami meet
	Coronoid processes	Anterior processes of rami
	Condyloid processes	Posterior processes of rami; form joint with temporal
	Mandibular foramen	Hole in medial side of ramus
	Mental foramen	Hole in anterior body
	Mylohyoid line	Ridge from mandibular foramen
	Alveolar margin	Bears teeth of lower jaw
Lacrimal (2)		Fingernail-sized bones in front of medial orbit wall
	Lacrimal canal	Passage through bone

Name of bone and number	Features on the bone	Description/location of bone/feature
Vomer (1)		Blade-like bone forming lower ⅓ of nasal septum
Palatine (2)		L-shaped; form posterior ¼ of hard palate and sides of nasal cavities
	Horizontal plates	Posterior ¼ of hard palate
	Perpendicular plate	Forms part of lateral nasal wall
Inferior conchae (2)		Lower part of lateral walls of nasal cavities
Hyoid (1)		U-shaped bone in upper neck
	Body	Front part of bone
	Greater cornua (horns)	Large, posteriorly projecting processes
	Lesser cornua (horns)	Small, superiorly projecting processes
Ear ossicles (6) [malleus (2), incus (2), stapes (2)]		Found in middle ear cavities
Sutures	Frontal (metopic)	Separates two halves of frontal bone in newborn
	Coronal	Separates frontal and parietal bones
	Sagittal	Separates parietal bones
	Lambdoidal	Separates parietals from occipital bone
	Squamous	Separates parietal and temporal bones
Sinuses or air cells	Frontal sinus	Behind medial parts of superciliary ridges
	Maxillary sinus	Inside maxillary bone
	Sphenoid sinus	In body of sphenoid
	Mastoid air cells	In mastoid process
	Ethmoid air cells	In ethmoid bone
Fontanels (6)		
Frontal (1)		At junction of coronal and sagittal sutures
Occipital (1)		At junction of sagittal and lambdoidal sutures
Sphenoid (2)		At junction of coronal and squamous sutures
Mastoid (2)		At junction of squamous and lambdoidal sutures

FIGURE 6.5. Anterior and posterior views of the skeletal system.

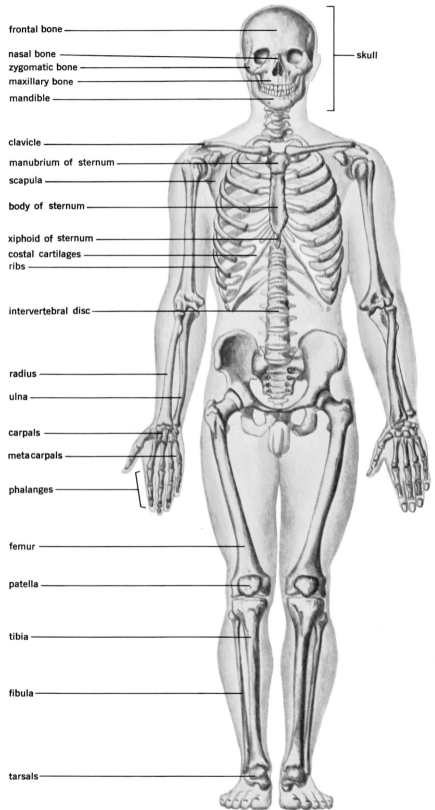

frontal bone

nasal bone
zygomatic bone
maxillary bone
mandible

skull

clavicle
manubrium of sternum
scapula

body of sternum

xiphoid of sternum
costal cartilages
ribs

intervertebral disc

radius

ulna

carpals

metacarpals

phalanges

femur

patella

tibia

fibula

tarsals

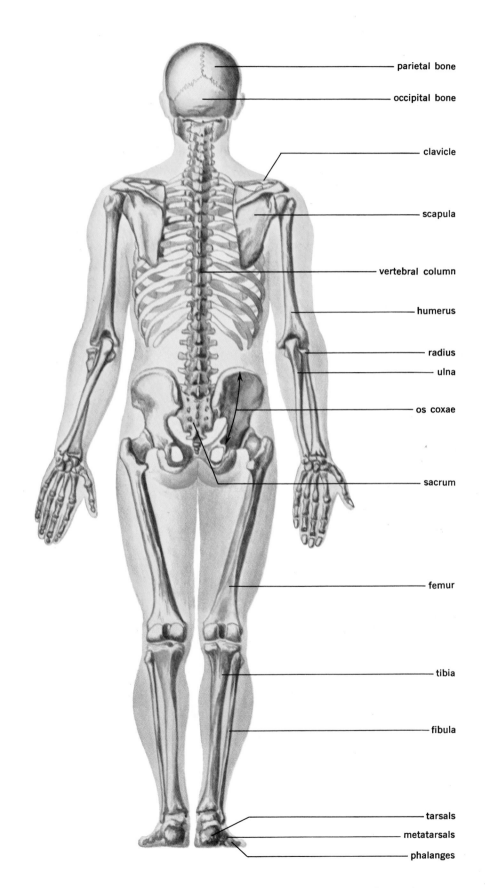

parietal bone

occipital bone

clavicle

scapula

vertebral column

humerus

radius

ulna

os coxae

sacrum

femur

tibia

fibula

tarsals

metatarsals

phalanges

97

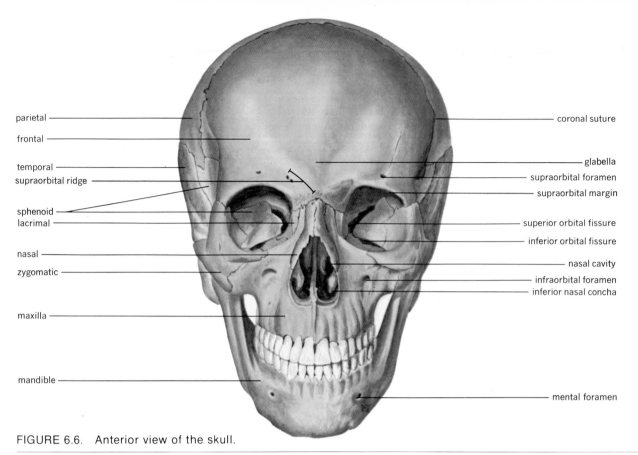

parietal

frontal

temporal

supraorbital ridge

sphenoid

lacrimal

nasal

zygomatic

maxilla

mandible

coronal suture

glabella

supraorbital foramen

supraorbital margin

superior orbital fissure

inferior orbital fissure

nasal cavity

infraorbital foramen

inferior nasal concha

mental foramen

FIGURE 6.6.   Anterior view of the skull.

FIGURE 6.7.   View of a median section of the skull.

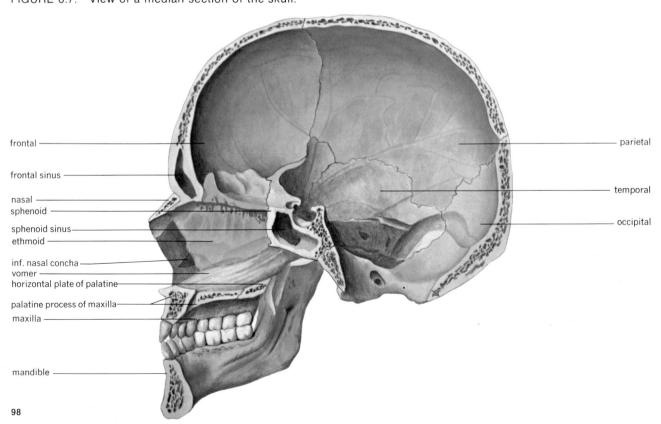

frontal

frontal sinus

nasal

sphenoid

sphenoid sinus

ethmoid

inf. nasal concha

vomer

horizontal plate of palatine

palatine process of maxilla

maxilla

mandible

parietal

temporal

occipital

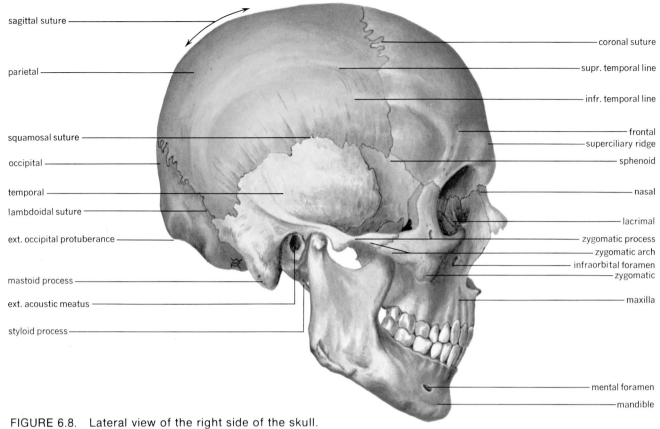

sagittal suture

parietal

squamosal suture

occipital

temporal

lambdoidal suture

ext. occipital protuberance

mastoid process

ext. acoustic meatus

styloid process

coronal suture

supr. temporal line

infr. temporal line

frontal

superciliary ridge

sphenoid

nasal

lacrimal

zygomatic process

zygomatic arch

infraorbital foramen

zygomatic

maxilla

mental foramen

mandible

FIGURE 6.8. Lateral view of the right side of the skull.

FIGURE 6.9. The skull viewed from below.

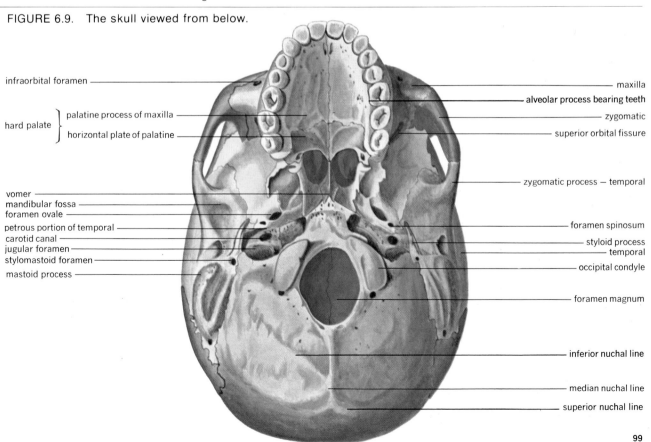

infraorbital foramen

hard palate {
palatine process of maxilla

horizontal plate of palatine
}

vomer

mandibular fossa

foramen ovale

petrous portion of temporal

carotid canal

jugular foramen

stylomastoid foramen

mastoid process

maxilla

alveolar process bearing teeth

zygomatic

superior orbital fissure

zygomatic process — temporal

foramen spinosum

styloid process

temporal

occipital condyle

foramen magnum

inferior nuchal line

median nuchal line

superior nuchal line

99

FIGURE 6.10. The right temporal bone. Lateral view (above), medial view (below).

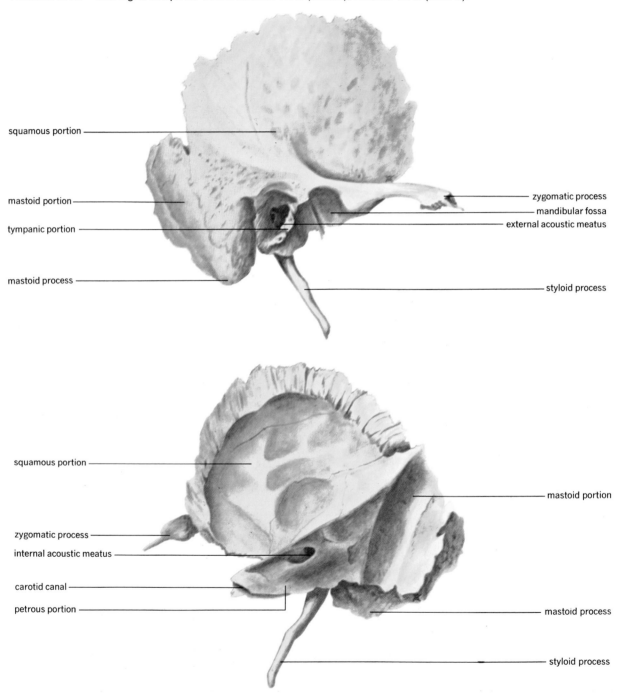

squamous portion

mastoid portion

tympanic portion

mastoid process

zygomatic process

mandibular fossa

external acoustic meatus

styloid process

squamous portion

zygomatic process

internal acoustic meatus

carotid canal

petrous portion

mastoid portion

mastoid process

styloid process

The FACIAL BONES are 8 in number:

2 nasals	2 lacrimals
2 zygomatics (malars)	1 vomer
2 maxillae	2 palatines
1 mandible	2 inferior conchae (turbinates)

Table 6.1 presents the features of each bone.

The NASAL BONES (see Figs. 6.6 and 6.8) are small bones forming the bridge of the nose. What is commonly referred to as the "nose" is mostly cartilage and flesh.

The ZYGOMATIC (*malar*) BONES form the cheek, and part of the ZYGOMATIC ARCH which reaches from cheek to ear. The bone also forms part of the lateral wall and floor of the orbit. Because of its prominence, the zygomatic is very likely to be fractured when blows are delivered to the face (e.g., automobile crashes).

The MAXILLAE (Fig. 6.14) help form the floor of the orbit, form the greater part of the hard palate that roofs the oral cavity (floor of nasal cavity), and carry the teeth of the upper jaw.

The MANDIBLE (Fig. 6.15) is the largest of the facial bones, and forms the lower jaw. It also carries teeth.

The LACRIMAL BONES (Fig. 6.16) are the smallest bones of the face, about the size of a fingernail. They lie in the anteromedial wall of the orbit, and contain a lacrimal canal through which the nasolacrimal duct passes. The duct is part of the system for draining tears from eye to nasal cavity.

The VOMER (Fig. 6.17) is shaped like the blade of a plow. It forms about the lower one third of the bony septum of the nasal cavities.

The PALATINE BONES (Fig. 6.18) are L-shaped bones that form the posterior one fourth of the

FIGURE 6.11.   The interior of the skull as viewed from above.

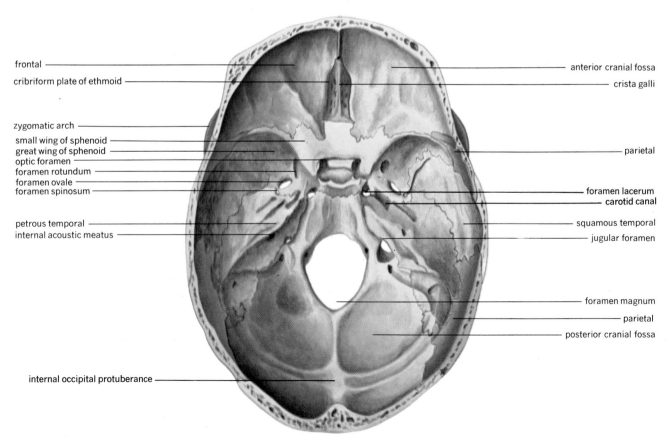

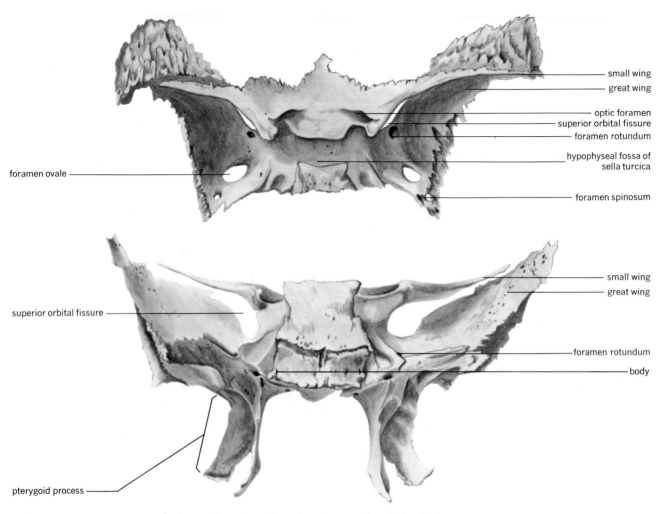

small wing
great wing
optic foramen
superior orbital fissure
foramen rotundum
hypophyseal fossa of sella turcica
foramen ovale
foramen spinosum

superior orbital fissure
small wing
great wing
foramen rotundum
body
pterygoid process

FIGURE 6.12.   The sphenoid bone. Superior (above) and posterior (below) views.

FIGURE 6.13.   The ethmoid bone. Anterior and posterior views.

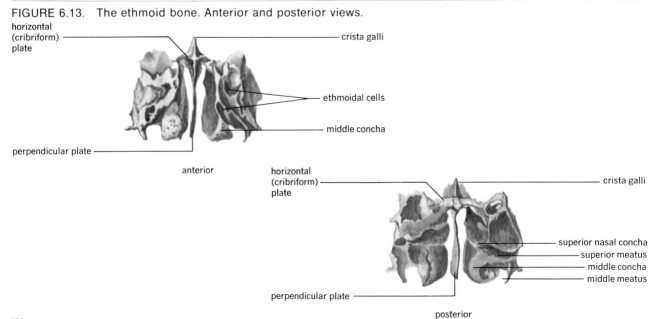

horizontal (cribriform) plate
crista galli
ethmoidal cells
middle concha
perpendicular plate

anterior

horizontal (cribriform) plate
crista galli
superior nasal concha
superior meatus
middle concha
middle meatus
perpendicular plate

posterior

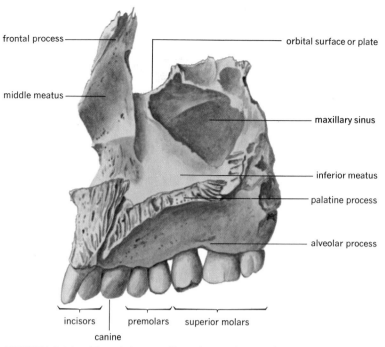

FIGURE 6.14. The right maxillary bone viewed from the medial side.

FIGURE 6.15. The right half of the mandible. Lateral view (above), medial view (below).

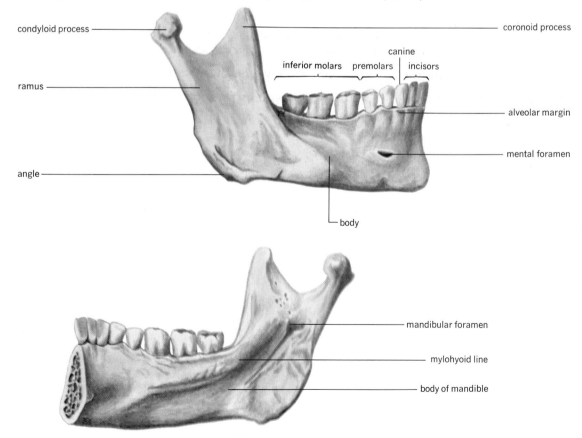

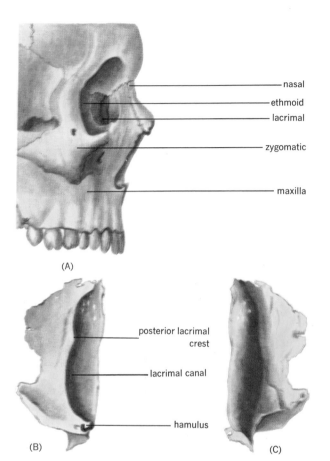

FIGURE 6.16.   The right lacrimal bone. *(A)* The bone *in situ*, *(B)* Lateral view, *(C)* Medial view.

hard palate, part of the lateral walls of the nasal cavities, and a small part of the orbit.

The INFERIOR CONCHAE (Fig. 6.19), or turbinates, extend from the lateral walls of the nasal cavities into the lower portion of the nasal cavities. They aid in creating a greater surface area for cleansing inhaled air.

Other bones associated with the skull include the hyoid bone and the ear ossicles. The HYOID BONE (Fig. 6.20) is a U-shaped bone lying in the neck at the level of the mandibular angle. The bone affords attachment for several swallowing muscles. The 6 EAR OSSICLES (3 in each middle ear cavity) (Fig. 6.21) aid in transmission of sound waves from ear drum to cochlea. Their names are MALLEUS (hammer), INCUS (anvil), and STAPES (stirrup). Their functions are described in Chapter 15.

A grand total of 29 bones thus compose or are associated with the skull.

FONTANELS. Bone formation is not complete at birth. There are, where the major sutures join one another, large membranous areas known as fontanels. They permit brain growth, since the

FIGURE 6.17.   The vomer. The bone *in situ* (above), and in lateral view (below).

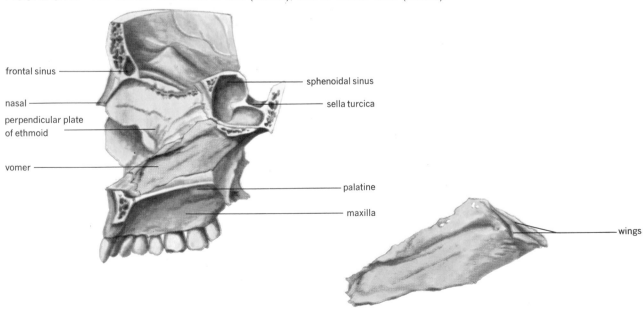

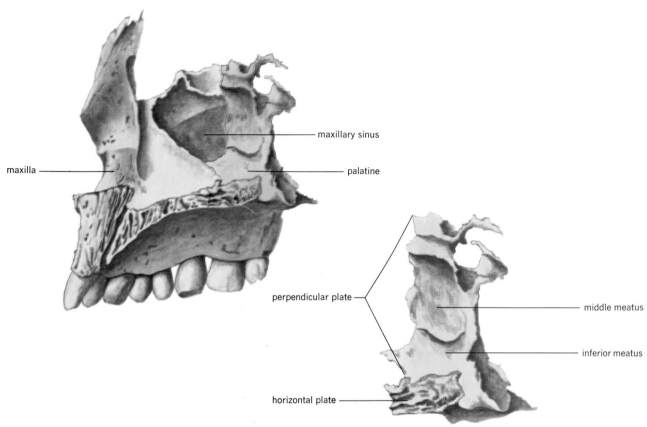

maxillary sinus

maxilla

palatine

perpendicular plate

middle meatus

inferior meatus

horizontal plate

FIGURE 6.18.   The right palatine bone. The bone *in situ* (above), and in medial view (below).

FIGURE 6.19.   The right inferior concha. *(A)* The bone *in situ*, *(B)* Medial view, *(C)* Lateral view.

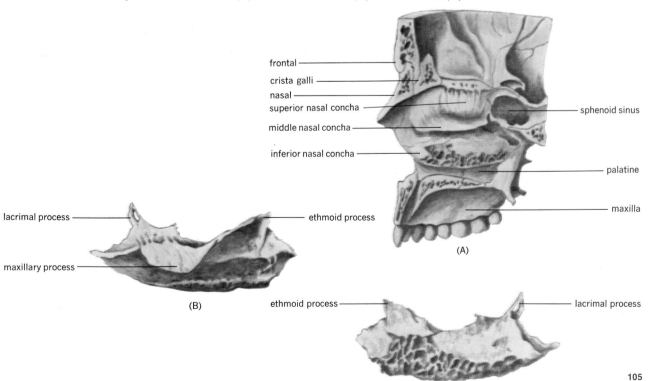

frontal

crista galli

nasal

superior nasal concha

middle nasal concha

inferior nasal concha

sphenoid sinus

palatine

maxilla

(A)

lacrimal process

ethmoid process

maxillary process

(B)

ethmoid process

lacrimal process

(C)

105

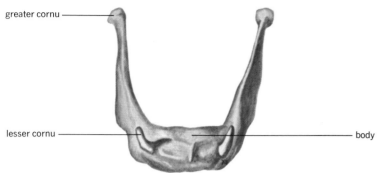

greater cornu

lesser cornu

body

FIGURE 6.20. The hyoid bone, anterior view.

FIGURE 6.21. The ear ossicles *in situ* in the temporal bone.

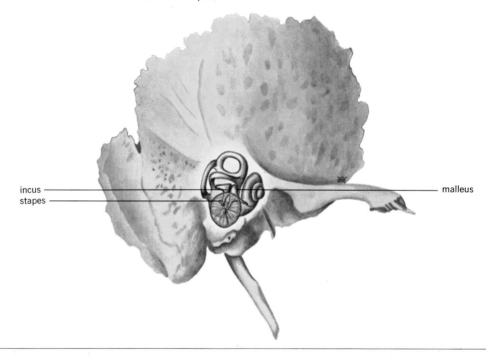

incus
stapes

malleus

bones have not yet knit together. The locations and names of the fontanels are shown in Figure 6.22. Some synonomy of terminology exists as is shown below:

Frontal or anterior fontanel or bregma
Occipital or posterior fontanel or lambda
Sphenoid or anterolateral fontanel
Mastoid or posterolateral fontanel

The anterior fontanel measures 4–6 centimeters in greatest diameter at birth, and closes between 4 and 26 months of age. Ninety percent close between 7 and 19 months of age. The posterior fontanel measures 1–2 centimeters in diameter at birth and usually closes by 2 months of age. The lateral fontanels are usually closed at birth.

The anterior fontanel is commonly observed and palpated (felt) in the infant to give clues as to the state of hydration and intracranial pressure. In a quiet, sitting infant, the fontanel should lie nearly even with the skull bones. An obviously depressed fontanel suggests dehydration, while a bulging fontanel suggests increased intracranial pressure. It should be noted that the fontanel bulges (normally) when the infant cries, coughs, or vomits.

FIGURE 6.22. The infant skull. Lateral view (above) and basal view (below).

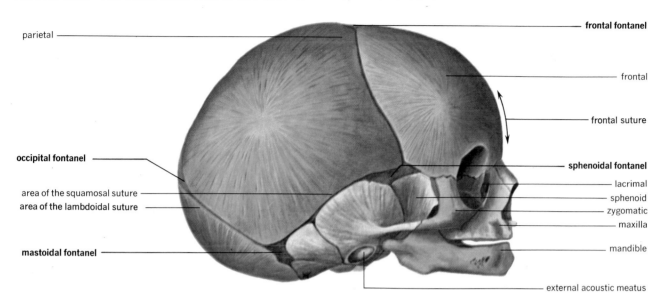

parietal

frontal fontanel

frontal

frontal suture

occipital fontanel

sphenoidal fontanel

lacrimal

area of the squamosal suture

sphenoid

area of the lambdoidal suture

zygomatic

maxilla

mastoidal fontanel

mandible

external acoustic meatus

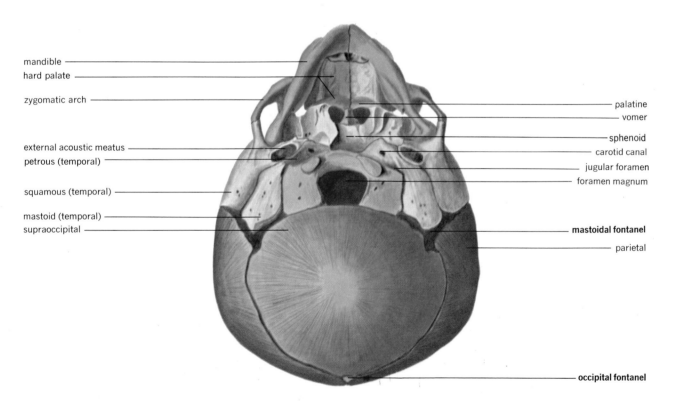

mandible

hard palate

zygomatic arch

palatine

vomer

sphenoid

external acoustic meatus

carotid canal

petrous (temporal)

jugular foramen

foramen magnum

squamous (temporal)

mastoid (temporal)

mastoidal fontanel

supraoccipital

parietal

occipital fontanel

FIGURE 6.23. Lateral view of the vertebral column.

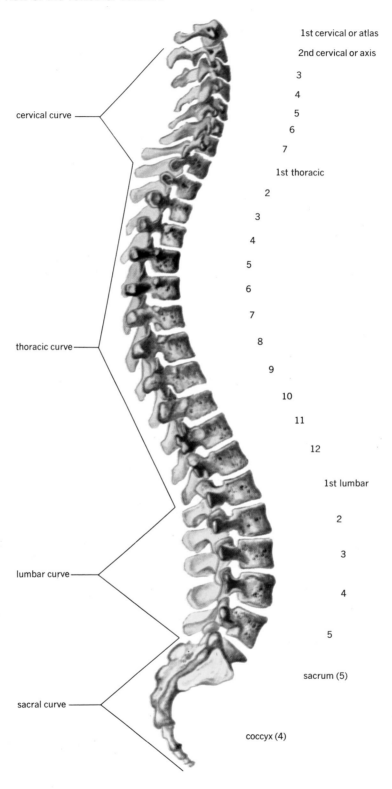

1st cervical or atlas

2nd cervical or axis

3

4

5

6

7

1st thoracic

2

3

4

5

6

7

8

9

10

11

12

1st lumbar

2

3

4

5

sacrum (5)

coccyx (4)

cervical curve

thoracic curve

lumbar curve

sacral curve

The vertebral column

The vertebral column (backbone or spine) of the adult consists of 24 separate bones known as VERTEBRAE, plus the SACRUM and COCCYX. The vertebrae are divided into three groups: 7 CERVICAL (neck); 12 THORACIC (chest); and 5 LUMBAR (small of the back). The sacrum is a single bone composed of 5 fused, modified vertebrae, while the coccyx ("tail bone") consists of 3–5 fused, small bones. Viewed from the side (Fig. 6.23), the adult column shows a number of CURVATURES. The cervical and lumbar curves are convex anteriorly, while the thoracic and sacral curves are concave anteriorly. The latter two are termed primary curves, the first two secondary curves. At birth there is only one curvature, an anteriorly concave curve extending from head to buttocks. As the child raises his head, and then walks, the secondary curves develop. The curves allow the column to absorb the shocks of locomotion, acting like a "spring." Exaggeration of the thoracic curve is termed *kyphosis* ("hunchback"); of the lumbar curve, *lordosis* ("swayback"); a deviation laterally from the normal straight-line of the spine is *scoliosis* (Fig. 6.24).

Most vertebrae resemble each other in having a BODY, PEDICLES, LAMINAE, TRANSVERSE and SPINOUS PROCESSES, and ARTICULATING PROCESSES (Fig. 6.25). Regionally, there are certain variations (Fig. 6.26).

All cervical vertebrae have, in their transverse processes, a TRANSVERSE FORAMEN that passes an artery to the brain. The first cervical vertebra or

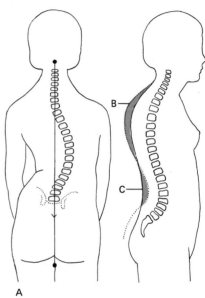

A

FIGURE 6.24. Abnormal curvatures of the spine. *(A)* Scoliosis, *(B)* Kyphosis, *(C)* Lordosis.

FIGURE 6.25. The components of a vertebra, superior view.

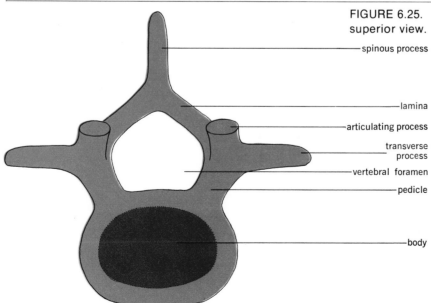

spinous process

lamina

articulating process

transverse process

vertebral foramen

pedicle

body

FIGURE 6.26. Regional variations of vertebrae, to illustrate major differences. (Drawings to the same scale.)

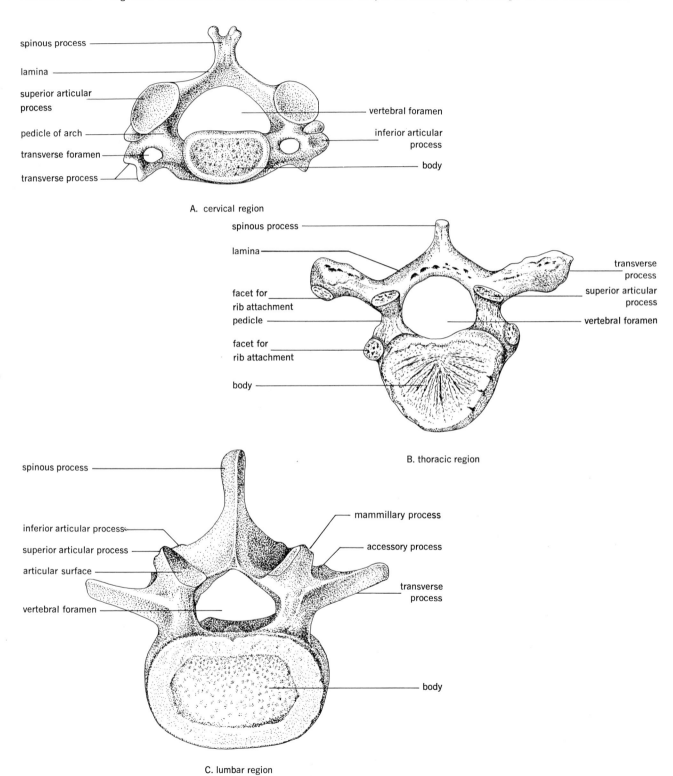

spinous process

lamina

superior articular
process

pedicle of arch

transverse foramen

transverse process

vertebral foramen

inferior articular
process

body

A. cervical region

spinous process

lamina

facet for
rib attachment

pedicle

facet for
rib attachment

body

transverse
process

superior articular
process

vertebral foramen

B. thoracic region

spinous process

inferior articular process

superior articular process

articular surface

vertebral foramen

mammillary process

accessory process

transverse
process

body

C. lumbar region

ATLAS, lacks a body and is ringlike in shape. It articulates with the skull and permits "yes" movements of the head. The second cervical vertebra, or AXIS (*epistropheus*), has a DENS (*odontoid process*) attached to its body. With the atlas, the dens forms a pivot joint permitting the "no" movements of the head.

All thoracic vertebrae have, on body and transverse processes, FACETS for rib attachments.

Lumbar vertebrae are very HEAVY, lack transverse foramina and facets, and have articulating processes that are vertically oriented for strength and weight bearing.

The SACRUM (Fig. 6.27) is designed for strength and support of the body weight above and the lower limbs below. The COCCYX is regarded as the remnant of a tail.

The vertebrae are separated from one another by pads of fibrous cartilage called INTERVERTEBRAL DISCS (see Fig. 6.5). Each disc has a soft center (nucleus pulposus) and an outer fibrous coat. Rupture of the fibrous coat may allow protrusion of the soft center into the vertebral canal to press on the spinal cord; also, some compression may occur as the vertebra above the ruptured disc "settles" on the one below. Compression of spinal nerves, which exit from between the vertebrae, may occur.

Failure of the laminae of the vertebrae to fuse as the bones develop creates SPINA BIFIDA. The membranes surrounding the cord may protrude through the opening (meningocele), or the membranes and the cord itself may protrude (meningomyelocele).

FIGURE 6.27. Sacrum and coccyx, posterior view.

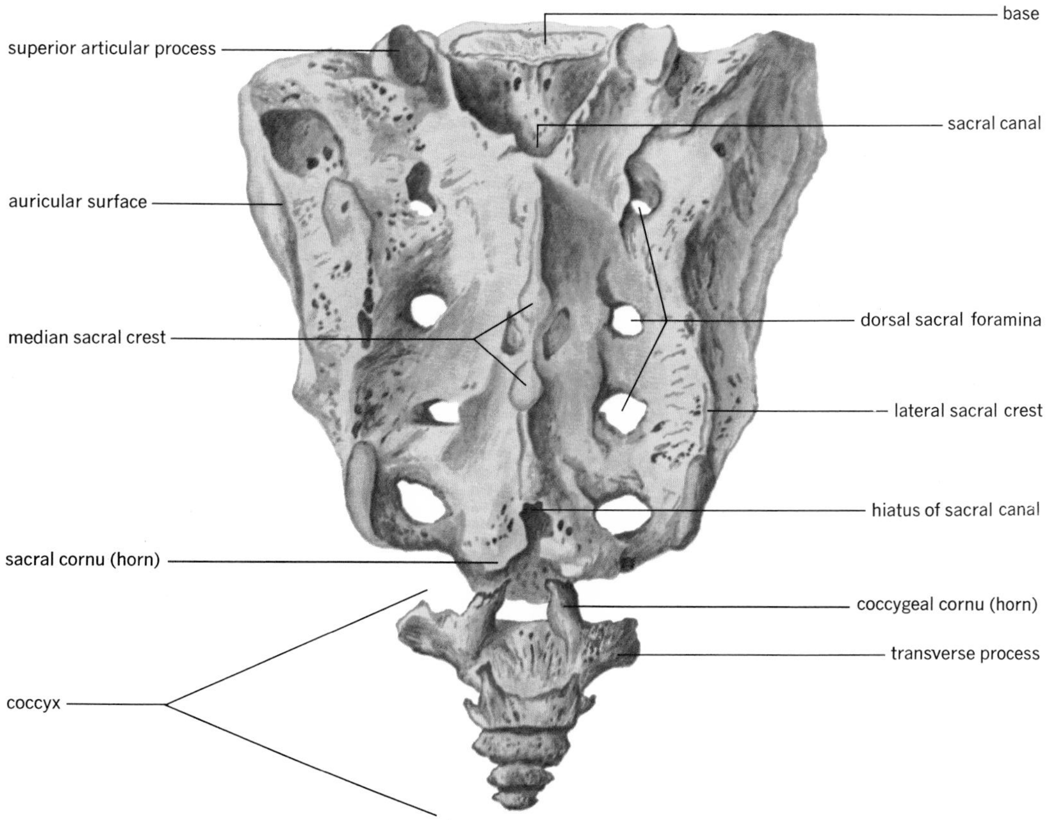

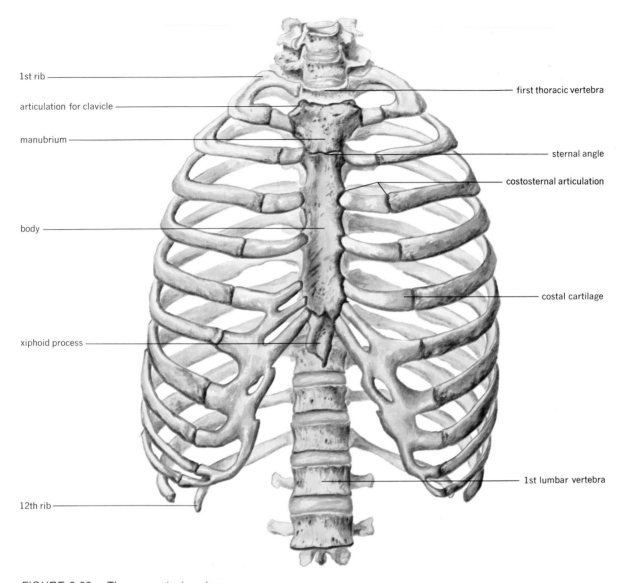

1st rib

articulation for clavicle

manubrium

body

xiphoid process

12th rib

first thoracic vertebra

sternal angle

costosternal articulation

costal cartilage

1st lumbar vertebra

FIGURE 6.28. Thorax, anterior view.

FIGURE 6.29. A right central rib, posterior view.

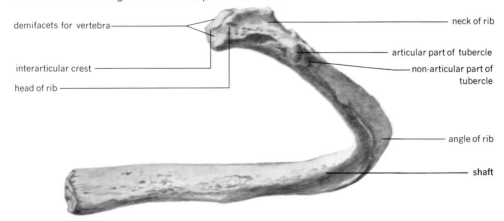

demifacets for vertebra

interarticular crest

head of rib

neck of rib

articular part of tubercle

non-articular part of tubercle

angle of rib

shaft

The thorax

The thorax (Fig. 6.28) is the roughly cone-shaped bony cage formed by the sternum, 12 pairs of ribs, and the costal cartilages. The STERNUM lies anteriorly, and consists of an upper MANUBRIUM, joined to a central BODY at the sternal angle, and a lower XIPHOID (*ensiform*) PROCESS. The cartilage of the second rib attaches at the angle and enables easy numbering of ribs. The manubrium is subcutaneous (just beneath the skin) on the chest, and affords an easy site for removal of bone mar-

row samples to aid in diagnosis of some blood disorders.

The 24 RIBS attach posteriorly to the thoracic vertebrae, and curve forward and down. The cartilages of the first 7 pairs attach directly to the sternum and these are known as *true ribs*. Those of the next 5 pairs attach to the preceding cartilage or not at all, and are called *false ribs*. The last 2 pairs of false ribs, which have no anterior attachment to the sternum, are *floating ribs*. A typical rib (Fig. 6.29) has a HEAD, NECK, TUBERCLE, SHAFT, ANGLE, and FACET for a costal cartilage.

FIGURE 6.30.   Right clavicle, anterior view.

FIGURE 6.31.   Right scapula. Anterior view (left), posterior view (right).

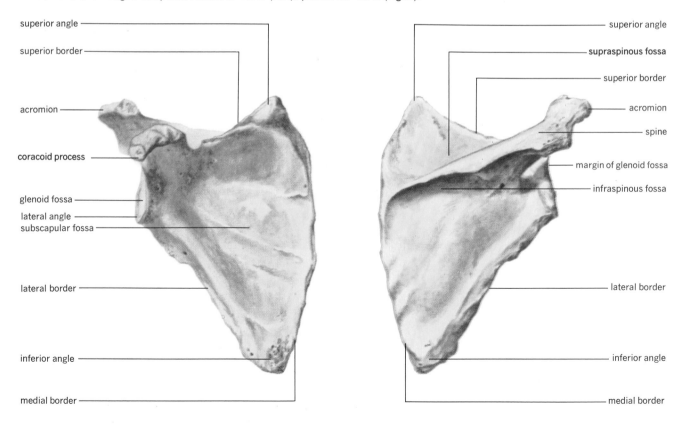

FIGURE 6.32. The right humerus. Anterior view (left), posterior view (right).

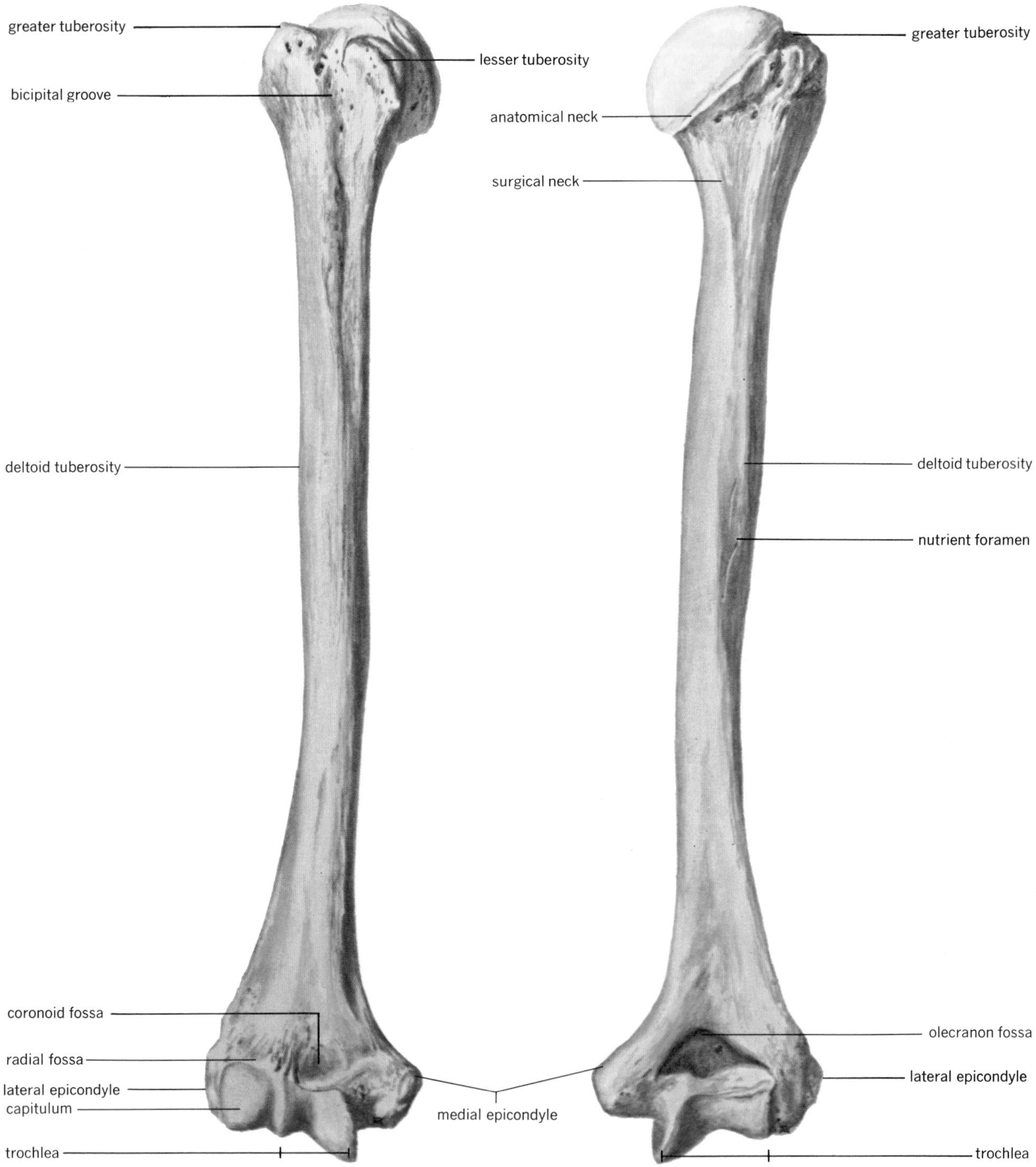

greater tuberosity

lesser tuberosity

bicipital groove

greater tuberosity

anatomical neck

surgical neck

deltoid tuberosity

deltoid tuberosity

nutrient foramen

coronoid fossa

radial fossa

lateral epicondyle

capitulum

olecranon fossa

lateral epicondyle

trochlea

medial epicondyle

trochlea

FIGURE 6.33. The right radius and ulna viewed anteriorly. The connective tissue membranes form a syndesmosis between the bones.

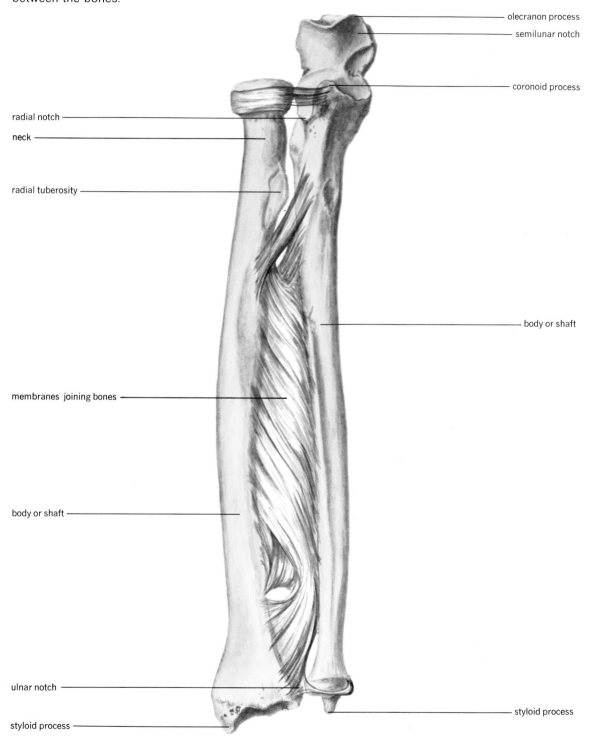

olecranon process

semilunar notch

coronoid process

radial notch

neck

radial tuberosity

body or shaft

membranes joining bones

body or shaft

ulnar notch

styloid process

styloid process

The appendicular skeleton

THE UPPER APPENDAGE. Each upper appendage contains 32 bones and consists of: the shoulder (pectoral) girdle, composed of scapula and clavicle; the humerus (upper arm); radius and ulna (lower or forearm); carpals (wrist); metacarpals (hand); and phalanges (fingers and thumb). The features on the bones of the upper appendages are listed in Table 6.2.

The CLAVICLE (Fig. 6.30) is a doubly curved bone that affords the only direct attachment of the upper limb to the axial skeleton. This attachment occurs at the sternoclavicular joint, and allows great mobility of the upper limb. Bracing the shoulder, the clavicle is often fractured when falling upon an extended arm.

The SCAPULA (Fig. 6.31) floats in muscle, attaching to the clavicle at the acromioclavicular joint, and to the humerus.

The HUMERUS (Fig. 6.32) is the largest bone of the upper limb, and forms the upper arm. It articulates proximally with the scapula, distally with the radius and ulna.

The ULNA (Fig. 6.33) is the *medial* bone of the forearm, and articulates proximally with the

FIGURE 6.34.   The right hand and wrist, posterior view.

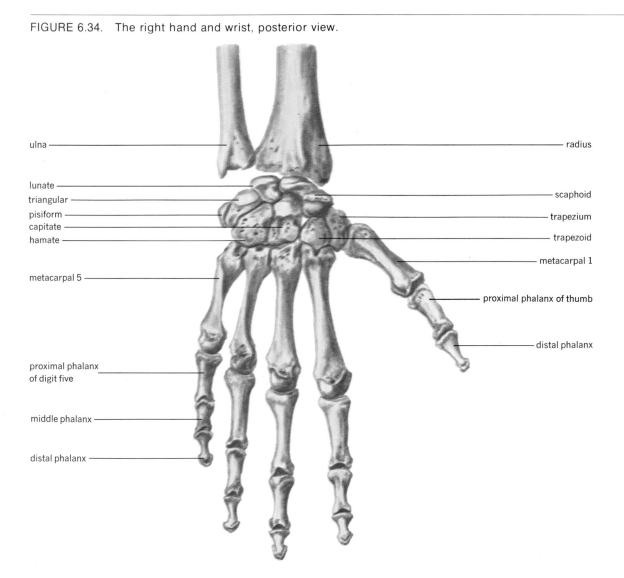

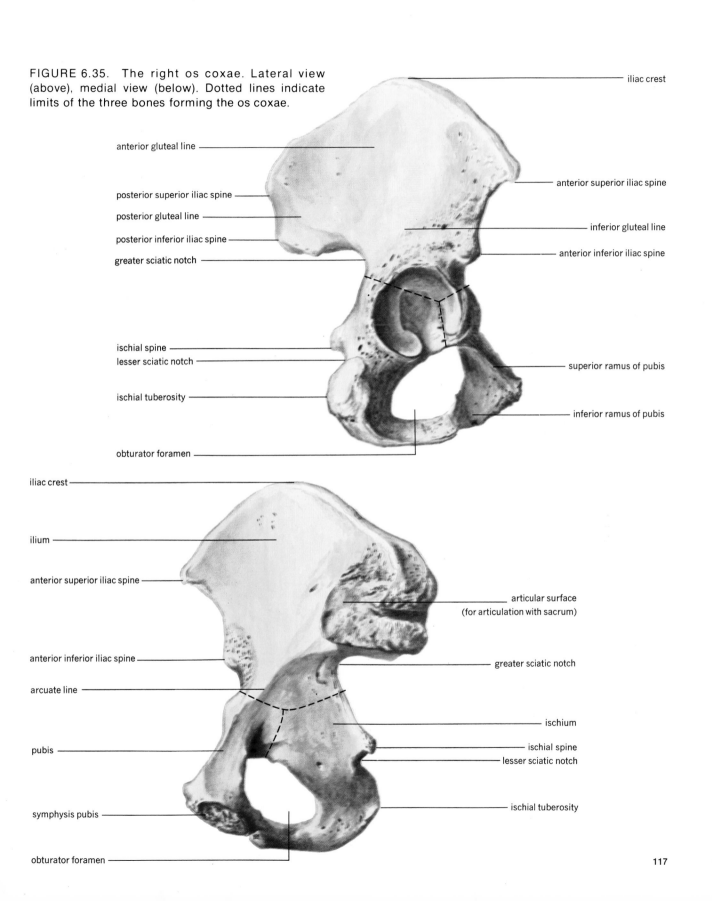

FIGURE 6.35. The right os coxae. Lateral view (above), medial view (below). Dotted lines indicate limits of the three bones forming the os coxae.

iliac crest

anterior gluteal line

posterior superior iliac spine

posterior gluteal line

posterior inferior iliac spine

greater sciatic notch

anterior superior iliac spine

inferior gluteal line

anterior inferior iliac spine

ischial spine

lesser sciatic notch

ischial tuberosity

superior ramus of pubis

inferior ramus of pubis

obturator foramen

iliac crest

ilium

anterior superior iliac spine

articular surface
(for articulation with sacrum)

anterior inferior iliac spine

arcuate line

greater sciatic notch

ischium

pubis

ischial spine

lesser sciatic notch

symphysis pubis

ischial tuberosity

obturator foramen

117

trochlea of the humerus, distally with the carpals.

The RADIUS (see Fig. 6.33) is a bit shorter than the ulna, and forms the *lateral* bone of the forearm. It articulates proximally with the capitulum of the humerus, distally with the carpals. The most common fracture of the radius is a break in the distal end (Colles' fracture) caused by falling on the hand.

The wrist and hand (Fig. 6.34). Eight (8) CARPAL BONES form each wrist. They are arranged in two rows of four bones each. From the thumb side, the proximal row contains the SCAPHOID or navicular ("boat"), LUNATE (moon), TRIQUETRUM or triquetral (triangular), PISIFORM ("pealike"). The distal row, also from the thumb side contains the TRAPEZIUM (greater multangular, "having many angles"), the TRAPEZOID (lesser multangular), CAPITATE ["having a (rounded) head"], and HAMATE ("hooked" or "hooklike").

Five METACARPALS, numbered 1–5 beginning with the thumb, make up each hand, while 14 PHALANGES compose the 5 digits in each hand. There are 2 phalanges in the thumb, while the remaining digits contain 3 each. Strictly speak-

TABLE 6.2   A Summary of the Features of the Bones of the Upper Appendages and Girdles (Italics Indicate the Most Important Features)

Name and number of bones in *each* appendage	Features on the bone	Description/location of bone/feature
Clavicle (1)		Collarbone
Scapula (1)	*Body*	Triangular, flat part of the bone carries most other features
	Superior border	Upper border
	Axillary (lateral) border	Faces the armpit
	Vertebral (medial) border	Faces the backbone
	Spine	Horizontal ridge on posterior surface; carries acromion process
	Supraspinous fossa	Depression above spine
	Infraspinous fossa	Depression below spine
	Subscapular fossa	Depression on anterior surface
	Glenoid fossa	Lateral; joins to humerus
	Acromion process	Projects above glenoid fossa
	Coracoid process	Beaklike; anterior
	Inferior angle	Lower "point"
	Superior angle	Upper and medial "point"
	Lateral angle	"Point" beneath glenoid fossa
Humerus (1)	*Head*	Rounded upper end
	Anatomical neck	Just below head
	Surgical neck	Upper part of shaft
	Greater tuberosity	Lateral to anatomical neck
	Lesser tuberosity	Anterior, below anatomical neck
	Bicipital groove	Lies between tuberosities
	Deltoid tuberosity	Lateral, about ⅓ the way down bone
	Medial epicondyle	Distal and medial
	Lateral epicondyle	Distal and lateral
	Capitulum } are	Round surface on distal end
	Trochlea } condyles	Spool-shaped surface on distal end

ing, each hand thus contains 4 fingers and a thumb, not 5 fingers.

THE LOWER APPENDAGE. Each lower appendage contains 30 bones. The os coxae (hip bone) joins the sacrum at the sacroiliac joint, and affords attachment for the lower limb. Each lower limb, in turn consists of a femur (thigh bone), patella (kneecap), tibia and fibula (leg bones), tarsals (ankle bones), metatarsals (foot bones), and phalanges (toe bones). The features on the bones of the lower appendage are shown in Table 6.3.

The os COXAE or innominate bone (Fig. 6.35) is the bone of the pelvic girdle, and is formed by the fusion of three fetal bones. A superior ILIUM, a posterior ISCHIUM, and an anterior PUBIS form the bone. The suture lines meet in the cuplike ACETABULUM, which articulates with the femur.

The sacrum plus the two os coxae comprise the PELVIS. Males and females show differences in shape and size of the pelvis. These differences (Fig. 6.36) appear correlated with the strength usually associated with the male, and the child-bearing function usually associated with the fe-

Name and number of bones in *each* appendage	Features on the bone		Description/location of bone/feature
Humerus (1) (Continued)	*Coronoid fossa*		Depression on posterior distal end
	Olecranon fossa		Depression on anterior distal end
	Nutrient foramina		Holes in shaft of bone; allow blood vessels inside
Ulna (1)	*Semilunar notch*		Half-moon-shaped surface, proximal end; joins trochlea
	Olecranon process		"Point" of elbow (posterior)
	Coronoid process		Anterior, above semilunar notch
	Radial notch		Lateral on proximal end; joins radius head
	Styloid process		Distal pointed end
Radius (1)	*Head*		Rounded surface at proximal end; joins capitulum
	Neck		Below head
	Tuberosity		Below neck on anterior surface
	Styloid process		Distal pointed end
Carpals (8)	Navicular		"Boat"
	Lunate	Proximal row,	"Moon" (halfmoon)
	Triangular	lateral to medial	Shaped like a right triangle
	Pisiform		"Pea"
	Trapezium		"Little table"
	Trapezoid	Distal row,	"Table shaped"
	Capitate	lateral to medial	"Having a head" (rounded)
	Hamate		"Hooked"
Metacarpals (5)	Form the palm of the hand; numbered 1–5 starting with thumb.		
Phalanges (14)	Form the five digits; 2 in the thumb, 3 in each of the 4 other digits.		

TABLE 6.3 A Summary of the Features of the Bones of the Lower Appendage and Girdles
(Italics Indicate the Most Important Features)

Name and number of bones in each appendage	Features on the bone	Description/location of bone/feature
Os coxae (1)	Formed from 3 bones	Bone of the pelvic girdle; attaches to sacrum
Ilium		Superior bone of os coxae
	Crest	Upper border of bone
	Anterior superior spine	"Point" of the hip; anterior
	Anterior inferior spine	Below anterior superior spine
	Posterior superior spine	At posterior end of crest
	Posterior inferior spine	Below posterior superior spine
	Gluteal lines	Three curved lines on posterior surface
	Greater sciatic notch	Notch below posterior inferior spine
Ischium		Posterior bone of os coxae
	Tuberosity	"We sit on these"
	Spine	Pointed process above tuberosity
	Lesser sciatic notch	Notch below spine
Pubis		Anterior bone of os coxae
	Symphysis	Joint between the 2 pubic bones
	Rami	Arches above and below symphysis
	Obturator foramen	Large hole enclosed by ischium and pubis
	Acetabulum	The cuplike joint surface on the lateral side of the os coxae (all 3 bones contribute to its formation)
Femur (1)		Bone of the thigh
	Head	Rounded, proximal end
	Neck	Below the head
	Greater trochanter	Projection lateral to neck ("measure hips" here)
	Lesser trochanter	Projection on medial surface below neck
	Body	Length of bone between trochanters and distal end
	Linea aspera	Rough ridge on posterior surface of shaft
	Gluteal tuberosity	Rough line below greater trochanter on posterior aspect of bone

male. Also, true and false pelves are described. The true pelvis is the cavity lying below the ilio-pectineal (arcuate) line. The true pelvis is marked by the pelvic outlet, which determines the maximum opening through which childbirth may occur.

The FEMUR (Fig. 6.37) is the longest, heaviest, and strongest bone in the body. It transmits,

Name and number of bones in each appendage	Features on the bone	Description/location of bone/feature
Femur (1) (Continued)	*Medial condyle*	Medial articular surface on distal end
	Lateral condyle	Lateral articular surface on distal end
	Intercondylar fossa	Depression between condyles on posterior surface
Patella (1)		Kneecap
Tibia (1)		Medial bone of leg
	Medial condyle	Medial articular surface on proximal end
	Lateral condyle	Lateral articular surface on proximal end
	Intercondylar eminence	Projection between condyles
	Medial malleolus	Large projection on medial aspect of distal end; commonly called "ankle bone"
	Shaft	Length between proximal and distal ends
Fibula (1)		Lateral bone of leg
	Head	Proximal end
	Shaft	Length between proximal and distal ends
	Lateral malleolus	Lateral projection on distal end of bone; commonly called "ankle bone"
Tarsals (7)		The ankle bones
	Calcaneus	"Heel bone"
	Talus	Articulates with tibia and fibula
	3 cuneiforms	Arranged across "top of foot"; wedge shaped
	Cuboid	Like a cube ⎫ Articulate
	Navicular	Boatlike ⎭ with talus
Metatarsals (5)		Numbered 1–5 beginning with great toe
Phalanges (14)		2 in great toe; 3 in each of the other toes

and bears, the shock of locomotion, and weight of the body.

The PATELLA (Fig. 6.38), a sesamoid bone, protects the knee joint anteriorly and acts as a pulley to increase the contractile efficiency of certain thigh muscles.

The TIBIA and FIBULA (Fig. 6.39) form the leg, that is, the part of the lower appendage between

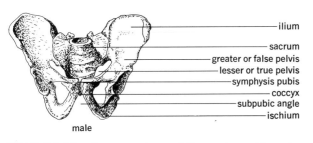

ilium
sacrum
greater or false pelvis
lesser or true pelvis
symphysis pubis
coccyx
subpubic angle
ischium

male

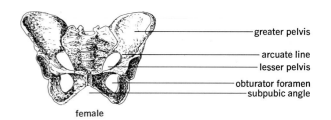

greater pelvis
arcuate line
lesser pelvis
obturator foramen
subpubic angle

female

FIGURE 6.36.   A comparison of the male and female pelves.

FIGURE 6.37.   The right femur. Anterior view (left), posterior view (right).

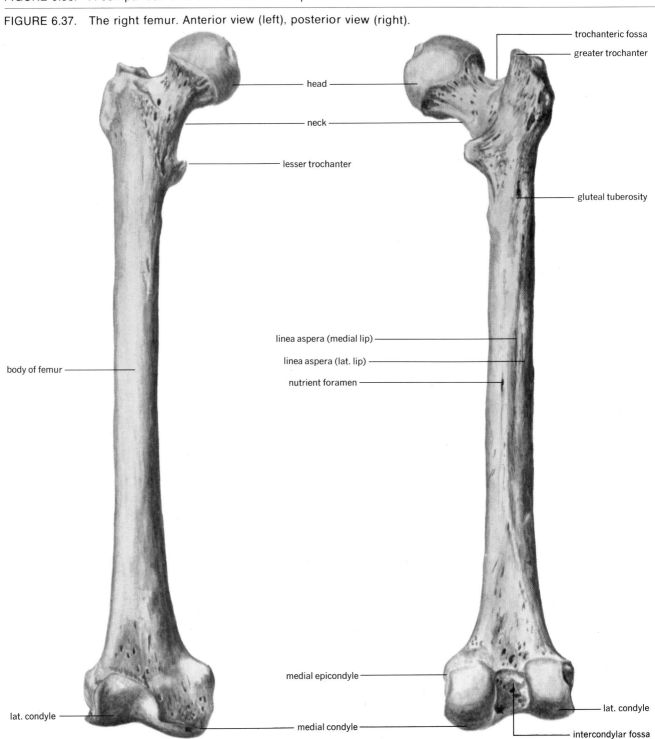

trochanteric fossa
greater trochanter
head
neck
lesser trochanter
gluteal tuberosity
body of femur
linea aspera (medial lip)
linea aspera (lat. lip)
nutrient foramen
medial epicondyle
lat. condyle
medial condyle
lat. condyle
intercondylar fossa

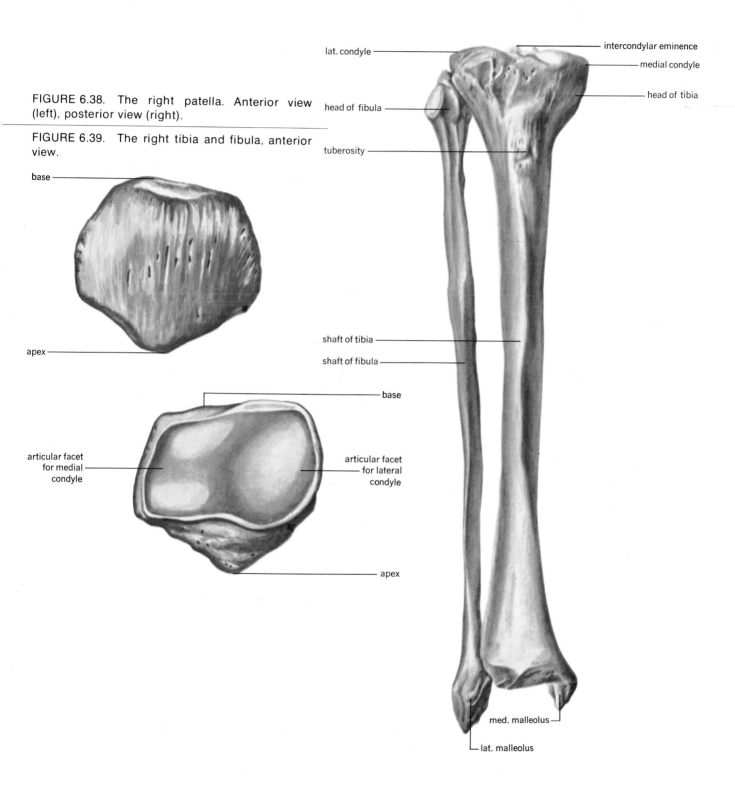

FIGURE 6.38. The right patella. Anterior view (left), posterior view (right).

FIGURE 6.39. The right tibia and fibula, anterior view.

base

apex

articular facet for medial condyle

base

articular facet for lateral condyle

apex

lat. condyle

intercondylar eminence

medial condyle

head of fibula

head of tibia

tuberosity

shaft of tibia

shaft of fibula

med. malleolus

lat. malleolus

the knee and ankle. The tibia is the medial bone of the leg, and the fibula is the lateral bone of the leg. A fracture that splits the lateral malleolus from the fibula is common in trauma that violently twists the foot to the outside; it is called Pott's fracture.

The ankle and foot (Fig. 6.40) Seven TARSAL BONES form the ankle. The high points of a medial longitudinal arch, and a transverse arch meet at the navicular bone; this point also represents the point at which the entire body weight is directed. If the arches are maintained, a good degree of "spring" is given; if not, the individual is "flatfooted."

Five METATARSALS form the center of the foot; these, like those of the hand, are numbered 1–5

FIGURE 6.40.   The right ankle and foot, superior view.

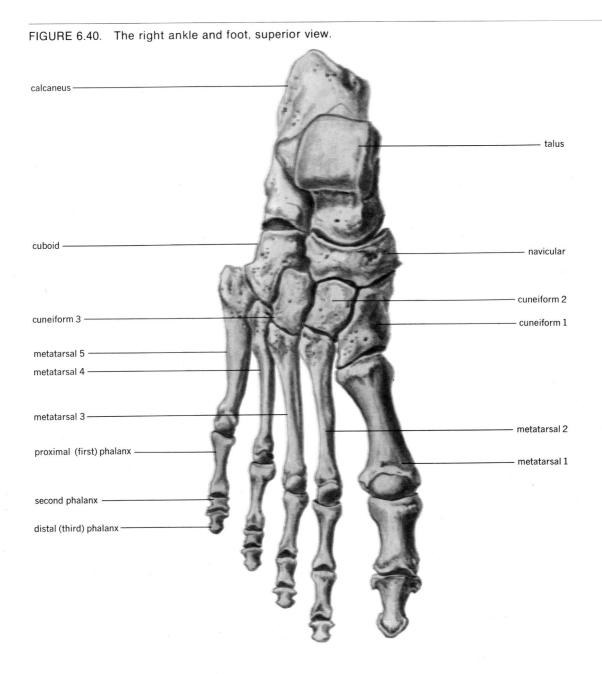

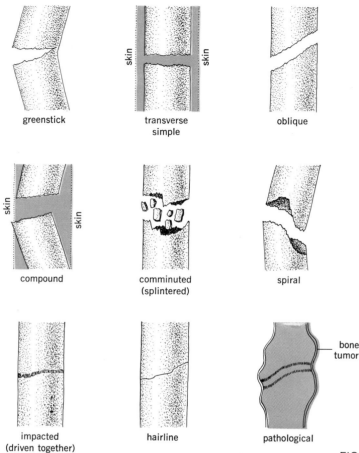

FIGURE 6.41.   Common types of fractures.

beginning with the big toe. Fourteen PHALANGES form the toes, 2 in the big toe, 3 in each of the 4 smaller toes.

Clinical considerations

FRACTURES. A fracture is a loss of continuity of a bone. It is most frequently due to trauma of some sort, such as falls, accidents, or blows. Pathological causes include tumors or loss of inorganic constituents (osteoporosis) which weaken the bone and render it liable to break under "normal use." According to the manner in which the break occurs, several types of fractures are recognized (Fig. 6.41). Treatment of the fracture usually involves immobilizing the broken ends after restoring anatomical relationships, to allow healing to occur. Several steps occur as the bone is repaired.

The bone ends are bound together by fibrin from torn blood vessels, lymph, and exudate.

Granulation tissue is formed. This involves invasion of the area by fibroblasts and vascular elements. At the same time, some fibroblasts form osteoblasts. The periosteum is a main source of both cells and vessels.

Callus formation occurs. This involves formation of bony tissue by the endochondral method between the broken ends.

About 4–6 weeks is required for a broken bone to heal completely.

BONE DISEASES AND DISORDERS. OSTEOMYELI-TIS is a staphylococcal or streptococcal disease of bone. The bacteria settle in the marrow cavity, causing bone destruction, pus formation, and excruciating pain. Antibiotic therapy usually controls the disease.

TUMORS occurring in bone may be benign, malignant, or may arise by metastasis from other tissues and organs. The cause of benign tumors is not known, although some seem to be associated with abnormalities of growth and development of bone. Mild pain or a lump on a bone may signal a benign tumor. Malignant tumors include *osteo-* *sarcomas* (bone tumor), *fibrosarcoma* (fibrous tissue tumor), *chondrosarcoma* (tumor of cartilage), and round cell sarcoma, or Ewing's sarcoma. The latter is a tumor of endothelial origin (vascular) in or on a bone. Metastatic tumors form the most common bone tumors. They occur most often in persons over 40, and consist primarily of metastases from breast, prostate, gut, and lung tumors.

No attempt is made to summarize this chapter since it consists mainly of anatomical terms. The glossary and tables will prove helpful if additional aid is required.

Questions

1. Compare and contrast the two types of bone formation. What is required for calcification to occur?

2. Name and discuss the origin and clinical significance of the fontanels.

3. What bones form the orbit? The nasal cavities?

4. In the disease acromegaly, which occurs in adults, bones grow in width but not in length. Explain how this may occur.

5. What is a "ruptured disc"? What dangers does it present?

6. Compare and contrast spongy and compact bone as to structure, composition, and location within the body.

7. What are the "girdles" of the skeleton? Name the bone(s) composing each.

8. Why is a thumb not a finger, but is a digit?

9. What functions does the skeleton serve?

Readings

Bourne, G. H. (ed). *The Biochemistry and Physiology of Bone*, Vols. 1, 2, and 3. Academic. New York, 1972.

Copenhaver, W. M. et al. *Bailey's Textbook of Histology.* Williams and Wilkins. Baltimore, 1971.

Gray, Henry. *Anatomy of the Human Body.* Edited by C. M. Goss. 29th ed. Lea and Febiger. Philadelphia, 1973.

Hall, M. C. *The Architecture of the Bone.* Thomas Pubs. Springfield, Ill., 1966.

Menczel, J., and A. Harell (eds.) *Calcified Tissue: Structural, Functional and Metabolic Aspects.* Academic. New York, 1971.

Milch, H., and R. Milch. *A Textbook of Common Fractures.* Harper. New York, 1959.

Morris' Human Anatomy. Edited by B. J. Anson. 12th ed. McGraw-Hill. New York, 1966.

Nichols, G., and R. H. Wasserman (eds). *Cellular Mechanisms for Calcium Transfer and Homeostasis*. Academic. New York, 1971.

Rodahl, K. et al. *Bone As a Tissue*. McGraw-Hill. New York, 1960.

chapter 7
Articulations

chapter 7

A joint or articulation is formed where a bone joins another bone, or where a cartilage joins a bone. The structure of the joint depends mainly on the function it must serve. Thus the union may be rigid, or it may permit variable degrees of motion. If permitted, movement may occur in one, two, or three planes of motion, and the joint may be termed uniaxial, biaxial, or triaxial.

Joints depend for their security on closely fitting bony parts, ligaments, or muscles. The closer the fit of the bones, the stronger the joint, but the greater will be the restriction on number of axes of movement.

Classification of joints

Three categories of joints are recognized

FIBROUS JOINTS (*immovable joints*) have no joint cavity, and the bones are held together by fibrous membranes. A SYNDESMOSIS (*syn*, with + *des*, bond) occurs when two bones are held firmly together by collagenous connective tissue as between the radius and ulna (see Fig. 6.33). This type of joint is often called a *ligamentous union*. SUTURES are found only in the skull, and create rigid, boxlike structures for protection of vital body organs. The bones forming the suture are usually of irregular contour for strength, and have minimal amounts of tissue between the two bones. As these joints age, fibrous tissue may disappear entirely, and a bone-to-bone union called a *synostosis* is formed.

CARTILAGENOUS JOINTS (*amphiarthrosis*, slightly movable joints) have no joint cavity, and hyaline or fibrous cartilage joins the two bones. A SYNCHONDROSIS involves hyaline cartilage as the joining material. Synchondroses connect the shaft of long bones with their ends (epiphyseal cartilage) in temporary joints, and connect the upper 10 pairs of ribs to the sternum (costal cartilages) in permanent joints. A SYMPHYSIS utilizes fibrous cartilage as the connecting material. Such joints include the symphysis pubis, and the intervertebral discs between vertebral bodies.

SYNOVIAL JOINTS (diarthroses, freely movable joints) have joint cavities, synovial membranes, synovial fluid, and supporting ligaments (Fig. 7.1). The articulating surfaces are smooth, and are covered with articular cartilage for easy movement. Six types of synovial joints are recognized (Fig. 7.2).

A HINGE (*ginglymus*) JOINT is a uniaxial joint and permits only a back-and-forth type of motion. The knee, elbow, finger joints, and ankle are of this type.

A PIVOT (*trochoid*) JOINT permits a turning (rotation) motion, and the joint is uniaxial. Rotation of the head ("no" movements) and turning the forearm occur in pivot joints.

An OVOID (*condyloid*) JOINT is biaxial, permitting side-to-side and back-and-forth motions. The wrist joint is of this type.

A SADDLE JOINT also is biaxial, permitting side-to-side and back-and-forth motions. The carpal-metacarpal joint of the thumb is of this type.

A GLIDING (*arthrodial*) JOINT is biaxial, as in the intercarpal and intertarsal joints, and permits side-to-side and back-and-forth motions.

The BALL-AND-SOCKET (*spheroidal*) JOINT is triaxial, permitting side-to-side, back-and-forth and rotational movement. The hip and shoulder joints are of this type.

A summary of these articulations is given in Table 7.1.

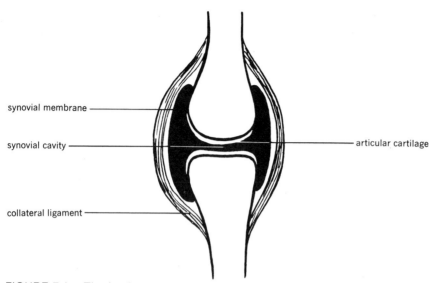

synovial membrane

synovial cavity

collateral ligament

articular cartilage

FIGURE 7.1.   The basic structure of a synovial joint.

FIGURE 7.2.   The six basic types of synovial joints, illustrated by joints of the body.

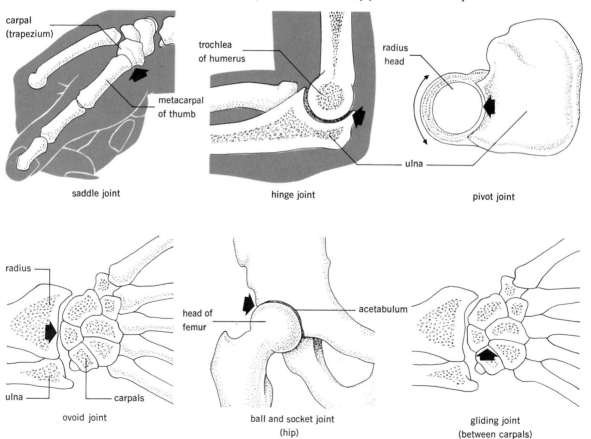

carpal
(trapezium)

metacarpal
of thumb

saddle joint

trochlea
of humerus

ulna

hinge joint

radius
head

ulna

pivot joint

radius

ulna   carpals

ovoid joint

head of
femur

acetabulum

ball and socket joint
(hip)

gliding joint
(between carpals)

TABLE 7.1   Summary of Articulations

Major category	Type	Characteristics	Axes of motion permitted	Examples
Fibrous joints		No cavity, immovable	None	Sutures, syndesmoses
	Syndesmosis	Bones held together by connective tissue	None	Tibia-fibula
	Synostosis	Bone-bone junction	None	Arise by aging of fibrous joints, with loss of fibrous tissue
	Suture	Bone-bone joint, minimal c.t. between	None	Skull; commonly become synostoses
Cartilagenous joints		No cavity; slightly movable; a bone-cartilage-bone joint	Compression only	Synchondroses, symphyses
	Synchondrosis	Bone-hyaline cartilage-bone joint	None	Shaft and ends of long bones; temporary. Between ribs and sternum (costal cartilages); permanent.
	Symphysis	Bone-fibrocartilage-bone joint	Slight compression	Symphysis pubis, vertebral column
Synovial joints		Cavity, freely movable	One to three	Six types
	Hinge	Spool in halfmoon	Uniaxial; back and forth motion	Knee, elbow, ankle, fingers
	Pivot	Cone in depression	Uniaxial; rotation	Radio-humeral, atlas-axis
	Ovoid	Egg in depression	Biaxial; side-to-side and back-and-forth	Wrist
	Gliding	Nearly flat surfaces opposed	Biaxial; side-to-side and back-and-forth	Intercarpal joints, intertarsal joints
	Ball and socket	Ball in cup	Triaxial; side-to-side, back-and-forth and rotation	Hip, shoulder

Discussion of specific joints

Because of their frequent involvement in sports injuries, or because of congenital malformation, the shoulder, sacroiliac, hip, and knee joints will be described in greater detail.

The SCAPULOHUMERAL (*shoulder*) JOINT (Fig. 7.3) is formed by the head of the humerus articulating with the glenoid fossa of the scapula. The fossa has a low rim of cartilage (the glenoid

FIGURE 7.3.   The right shoulder joint. External view (above), frontal section (below).

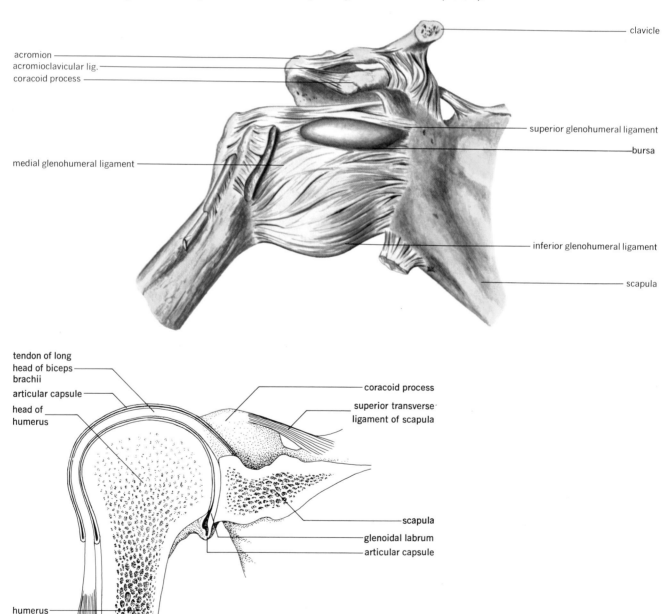

clavicle

acromion
acromioclavicular lig.
coracoid process

superior glenohumeral ligament

bursa

medial glenohumeral ligament

inferior glenohumeral ligament

scapula

tendon of long
head of biceps
brachii
articular capsule
head of
humerus

coracoid process

superior transverse
ligament of scapula

scapula
glenoidal labrum
articular capsule

humerus

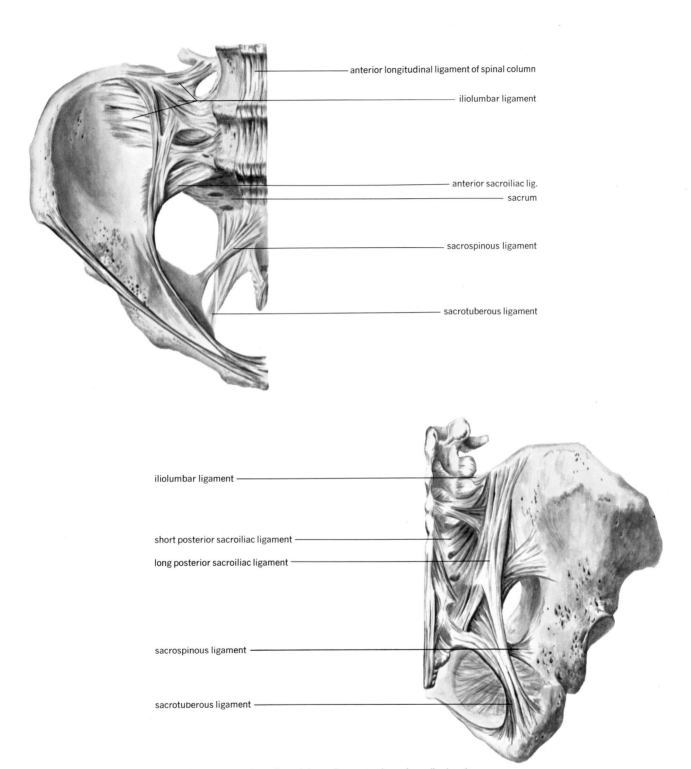

anterior longitudinal ligament of spinal column

iliolumbar ligament

anterior sacroiliac lig.

sacrum

sacrospinous ligament

sacrotuberous ligament

iliolumbar ligament

short posterior sacroiliac ligament

long posterior sacroiliac ligament

sacrospinous ligament

sacrotuberous ligament

FIGURE 7.4. The sacroiliac joint. Anterior view (above), posterior view (below).

FIGURE 7.5.   The right hip joint. External view (above), and frontal section below.

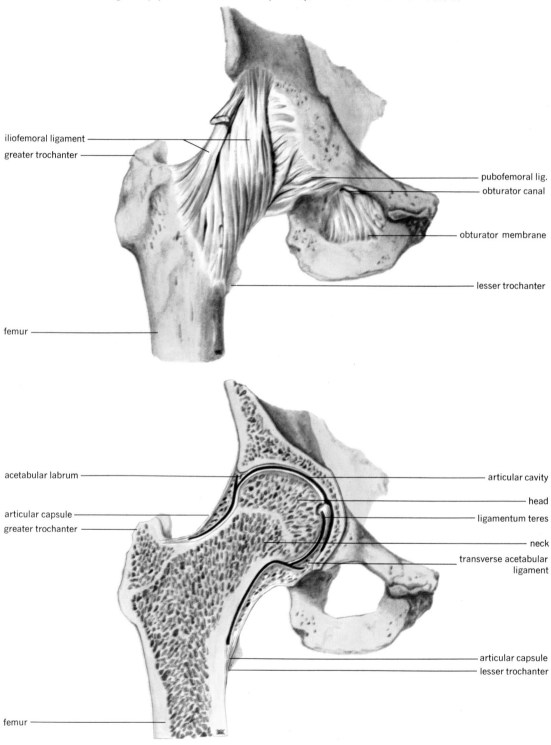

iliofemoral ligament

greater trochanter

pubofemoral lig.

obturator canal

obturator membrane

lesser trochanter

femur

acetabular labrum

articular cavity

head

ligamentum teres

articular capsule

greater trochanter

neck

transverse acetabular
ligament

articular capsule

lesser trochanter

femur

labrum) to deepen it and is surrounded by several ligaments. The muscles around the shoulder also contribute to its security. A "shoulder separation," or dislocation of this joint, occurs commonly as a result of falling on the outstretched arm (football and basketball players). In this disorder, the humerus head is driven out of the glenoid fossa, usually in an upward (superior) direction.

The SACROILIAC JOINT (Fig. 7.4), or union between the sacrum and ilium, is a partially fibrous, partially synovial joint. The joint surfaces are rough and the bones are firmly bound by many ligaments, thus no movement is normally permitted. A "sacroiliac slip" involves stretching of the ligaments holding the joint, and slight movement of the bones may occur. Stretching or tension applied to the sciatic nerve may then result in intense leg pain. "Slips" may occur if heavy loads are carried which put downward pressure on the joint, or in landing on the feet without bending the knees. This tends to push the lower appendages strongly into the hip bones and thus forces upward pressure on the sacroiliac joints.

The HIP JOINT (Fig. 7.5), between the acetabulum and femoral head, is constructed for weight bearing as well as movement. The socket (acetabulum) is deep, and a ligament (ligamentum teres) fixes the head of the femur to the socket. The supporting ligaments are numerous and heavy, and seem to serve to limit motion more than to secure the joint. The joint is rarely dislocated if normal. In congenital malformation of the hip, the acetabular rim is low, and the femoral head slips easily over or out of it. Limited sideward movement of the limb (Fig. 7.6) when hip and knee are flexed (frog position) is usually indicative of congenital malformation. The disorder should be diagnosed in infancy.

The KNEE JOINT (Fig. 7.7) is regarded by many as one of nature's mistakes, because it is basically unstable. Two nearly flat surfaces are held together by a number of ligaments. The CRUCIATE LIGAMENTS tend to limit anterior-posterior movement, while the COLLATERAL LIGAMENTS, with the MENISCI, tend to limit sideward motion. The joint may be disrupted by strong rotation, or by blows to either side, and by blows from the front that tend to overextend it. Such trauma are inherent

FIGURE 7.6. Testing an infant for congenital dislocation of the hip. Sideward movement is limited when knee and hip are flexed 90°. A "click" may be felt at "X," as thigh is abducted.

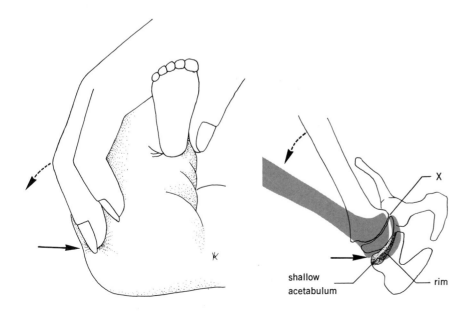

shallow acetabulum

rim

FIGURE 7.7. The right knee joint. Anterior view (above), posterior view (below).

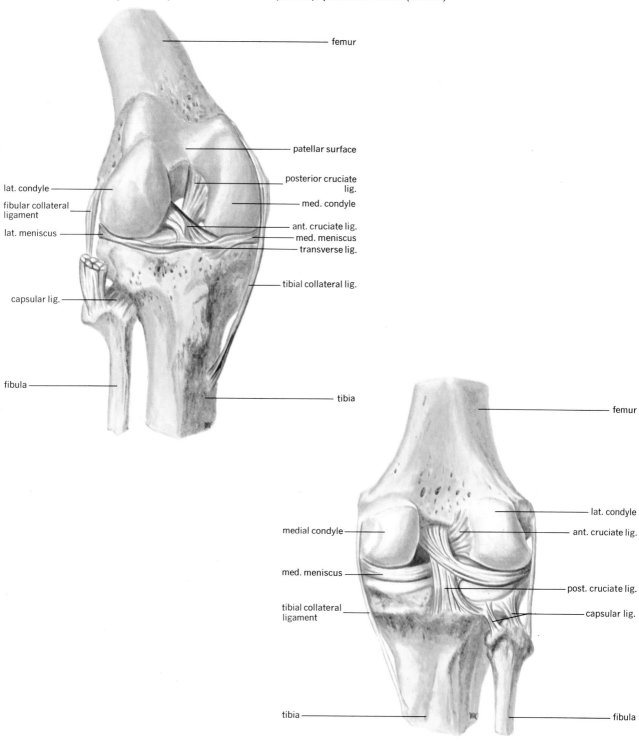

femur

patellar surface

posterior cruciate lig.

lat. condyle

fibular collateral ligament

med. condyle

lat. meniscus

ant. cruciate lig.

med. meniscus

transverse lig.

tibial collateral lig.

capsular lig.

fibula

tibia

medial condyle

femur

lat. condyle

ant. cruciate lig.

med. meniscus

tibial collateral ligament

post. cruciate lig.

capsular lig.

tibia

fibula

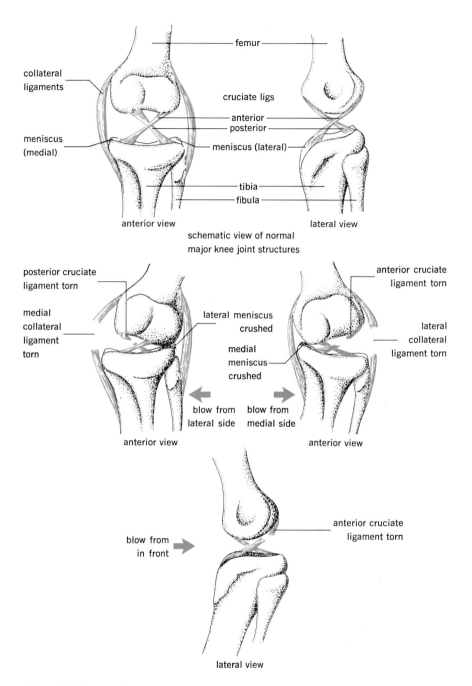

FIGURE 7.8. A diagrammatic representation of the structures of the left knee that will most likely be injured by trauma from various directions.

in the game of football. Crushing of menisci and tearing of ligaments is common, and often requires surgery to repair the damage. Figure 7.8 shows which structures will be disrupted according to the direction of the force imposed on the joint.

Synovial fluid

Synovial fluid lubricates synovial joints. It is secreted by the synovial membrane and, in all synovial joints, has a total estimated volume of 100 milliliters. It functions to lubricate the articular surfaces so as to insure smooth joint action. The fluid contains much mucus and protein-polysaccharides which makes it thicker than other body fluids, and an efficient lubricating material. Excess production of fluid may occur after joint injury, and may limit joint motion (e.g., "water on the knee," inflammation). Joint movement is limited because of the swelling and pressure increase resulting from excessive secretion of synovial fluid.

Clinical considerations

Traumatic damage to a joint

A DISLOCATION is defined as loss of continuity of the joint structures. It may occur as the result of a blow, or because of malformation. A SPRAIN occurs when ligaments and muscles are torn and tendons are stretched as the result of trauma. Pain is common, is often intense, and rapid development of edema (swelling) is usual. Black-blue hemorrhages in the skin (ecchymoses) which change to greenish-brown or yellow with time are also common. Bleeding into the joint produces the common "black-and-blue" mark. As the blood pigments break down, the pigments known as bilirubin (yellow) and biliverdin (green) are produced. Thus the bruise changes color with time. A STRAIN is caused by overstretching or pulling of muscles and tendons, without tearing.

Inflammation of joints

Arthritis is a term used to designate joint inflammation.

RHEUMATOID ARTHRITIS is a disease of connective tissue involving inflammation and destruction of synovial membranes. The joints become enlarged, severely limited in motion, and painful.

OSTEOARTHRITIS is a chronic degenerative disease, especially of weight bearing joints. It is characterized by: destruction of articular cartilage, without great inflammation; overgrowth of bone often with twisting of the affected appendage (e.g., fingers). Pain is minimal, but may increase with hard joint usage.

GOUTY ARTHRITIS is the result of a metabolic error involving the degradation of nitrogenous bases. Uric acid crystals are deposited in certain joints (e.g., ankle, knee, hands), with development of intense pain and swelling.

SEPTIC ARTHRITIS arises as a result of the presence of bacteria or their products in the bloodstream, with involvement of the joints, Gonococcal (gonorrhea), nonhemolytic streptococcal (strep), and staphylococcal (staph) infections are most commonly involved as causative agents.

Summary

1. An articulation is a junction between bones, or between a bone and a cartilage.

   a. Joints may permit no movement, slight movement, or free movement.

   b. Joints are secured by bony parts, ligaments, muscles, or combinations of these.

2. Three (3) categories of joints are recognized:

   a. Fibrous joints are immovable, and are formed by collagenous tissue or nearly bone-to-bone junctions (sutures of skull). Fibrous joints may become bone-to-bone junctions as they age.

   b. Cartilagenous joints are slightly movable and are joined by hyaline cartilage (synchondrosis) or fibrocartilage (symphysis). Synchondroses may be temporary or permanent.

   c. Synovial joints are freely movable and have a joint cavity, with synovial membranes and fluid, and are of six types:

      (1) Hinge. Allows back-and-forth motion. Fingers, knee, elbow.

      (2) Pivot. Permits turning or rotation. Head ("no movement"), radius on humerus.

      (3) Ovoid. Allows side-to-side and back-and-forth. Wrist.

      (4) Saddle. Allows side-to-side and back-and-forth. Thumb.

      (5) Gliding. Allows side-to-side and back-and-forth. Intercarpal, intertarsal.

      (6) Ball and socket. Permits side-to-side, back-and-forth, and rotation. Shoulder, hip.

3. Synovial fluid is produced by the synovial membrane. It is thick and lubricates synovial joints.

4. Specific joints are briefly discussed: shoulder, sacroiliac, hip, knee.

5. Disorders of joints include:

   a. Dislocations. Loss of continuity of joint structures.

   b. Sprains. Tearing of ligaments and muscles with pain and edema.

   c. Strains. Overstretching of ligaments and muscles without tearing.

   d. Arthritis. Inflammation of a joint. It may be due to reactions secondary to infection, may be a disease of aging, or may be the result of metabolic disorders.

Questions

1. Compare fibrous, cartilagenous, and synovial joints as to structure and freedom of movement.

2. Synovial joints may be described as uniaxial, biaxial, or triaxial. Pick six synovial joints in the body, and classify them according to degrees of movement permitted.

3. Does the skull have only sutures? Explain.

4. Compare dislocations, sprains, and strains.

5. Compare the types of arthritis as to cause and symptoms.

Readings

Broer, Marion R. *Efficiency of Human Movement.* Saunders. Philadelphia, 1966.

Committee on Trauma of American College of Surgeons: *An Outline of the Treatment of Fractures,* 8th. ed. Saunders. Philadelphia, 1966.

Gray, Henry. *Anatomy of the Human Body.* Edited by C. M. Goss. 29th ed. Lea and Febiger. Philadelphia, 1973.

Hartung, Edward F. et al (eds). "Arthritis." *Manual for Nurses, Physical Therapists, and Medical Social Workers.* Arthritis and Rheumatism Foundation. New York.

Michele, Arthur. "Principles of Fracture Care." *Amer. J. Orthop.* 9:34-37, 1967.

Morris' Human Anatomy. Edited by B. J. Anson. 12th ed. McGraw-Hill. New York, 1966.

"Rheumatoid Arthritis." *Med. Clin. N. Amer. 52:* entire issue, May 1968.

"Strike Back at Arthritis." USPHS Pub. 747. U. S. Government Printing Office, Washington, D.C.

chapter 8
The Structure and Properties of Muscular Tissue, with Emphasis on Skeletal Muscle

chapter 8

Skeletal muscle

Muscular tissue is the CONTRACTILE tissue of the body. Its cells are elongated and capable of shortening to cause movement, or change in size and shape of an organ in which it is found. Three types of muscular tissue are described:

SKELETAL MUSCLE attaches to and moves the skeleton. It is a voluntary type of muscle, contracting "at will" and is normally stimulated by way of outside nerves.

CARDIAC MUSCLE is found only in the heart, is involuntary, and contracts because of inherent stimulation. It causes the circulation of blood through the body. Its properties are described in Chapter 19.

SMOOTH (visceral) MUSCLE occurs primarily in internal body organs. It is involuntary, and slow acting. It propels materials through the digestive tract, and the reproductive and urinary systems. It also controls the diameter of blood vessels and the tubes of the respiratory system. Its properties are described in Chapter 20.

The present chapter is concerned with skeletal muscle structure and function, with the reminder that the basic mechanisms of contraction and energy sources are shared by all three types of muscle.

Skeletal muscle

Development and growth

Skeletal muscle develops from MESODERM, and the somites (see Fig. 3.5) first give evidence of the formation of the tissue. The cells giving rise to skeletal muscle tissue are termed MYOBLASTS. The mature muscle cell (or fiber) is a multinucleate structure thought to result from embryonic fusion of many uninucleate myoblasts. By about the fifth month of development, the number of fibers is fixed, and any increase in size (growth) of the muscle from this time onward is the result of increase in size of fibers.

Physiological anatomy (Fig. 8.1)

The units of structure and function of skeletal muscle are the FIBERS. They are long threadlike cells and may be 10-100 microns in diameter by 3 millimeters—7.5 centimeters in length. Each fiber has many nuclei (multinucleate) and is anatomically separated from all other fibers by the SARCOLEMMA, or cell membrane. The fiber carries characteristic cross bandings or STRIATIONS. The striations are light and dark stained regions on longitudinally oriented MYOFIBRILS. The striations are designated by letters, and a SARCOMERE is the distance between two Z lines. It is the unit that shortens when the muscle contracts. The myofibrils are suspended in the SARCOPLASM of the fiber (muscle cytoplasm) which contains mitochondria, and a sarcoplasmic reticulum. Tubular structures ("T" or transverse tubules) bring extracellular fluid into the muscle fiber. Individual fibers are bound by connective tissue, known as the ENDOMYSIUM, into groups or fascicles of fibers that are surrounded by PERIMYSIUM. EPIMYSIUM binds fascicles into a whole muscle (Fig. 8.2).

Chemically, skeletal muscle contains about 75 percent water, 20 percent protein, and 5 percent other organic and inorganic substances. The proteins include MYOSIN, a large molecule having enzymelike properties, and clublike endings on its molecule. It forms the thick filaments (see Fig. 8.1). ACTIN is a smaller molecule composed of

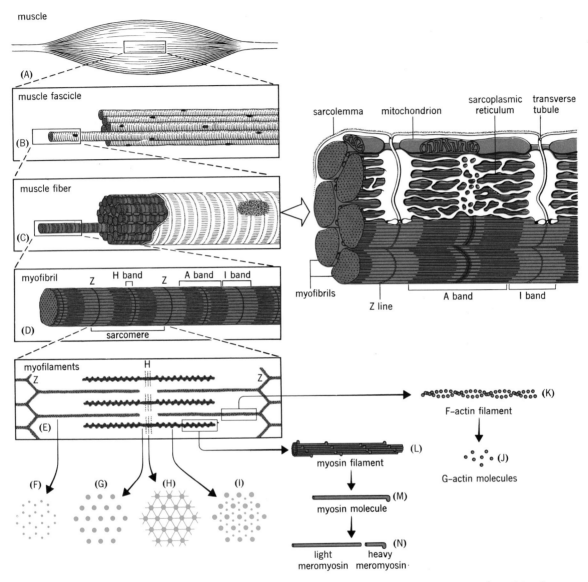

FIGURE 8.1. The organization of a skeletal muscle. (F), (G), (H), and (I) are cross sections of the filament at the levels indicated.

globular actin (G-actin) units, and is found in the thin filaments (see Fig. 8.1). The thin filaments shorten during muscle contraction. CREATINE PHOSPHATE and ATP are high energy phosphate compounds in the muscle. GLUCOSE is a simple sugar in the muscle. The INORGANIC constituents include sodium, potassium, calcium, and chloride ions.

The stimulus that causes contraction of skeletal muscle is normally delivered to the muscle by nerve impulses arriving at a MOTOR END PLATE, a specialized ending of the muscle's motor nerve. A single motor nerve and the muscle fibers it serves constitutes a MOTOR UNIT. A given muscle may contain from a few to several thousand motor units. The end plate (Fig. 8.3) contains vesicles of a chemical (acetylcholine) that are released on arrival of a nerve impulse. The chemical diffuses to the membranes of the fiber and depolarizes them. So that continual depolarization of the membranes does not occur, an enzyme (cholinesterase) rapidly destroys the chemical released by an impulse.

The events of muscular contraction and relaxation

Resting muscle is POLARIZED, that is, the outer surfaces of the cell, T-tubule, and reticulum membranes bear a positive electrical charge when compared to the inner surfaces of these membranes. The electrical difference is due to the separation of ions, chiefly sodium and potassium, by active mechanisms. Calcium appears to be bound to areas inside the sarcoplasmic reticulum; actin and myosin are separate molecules; and ATP molecules are attached to the myosin molecules.

When the muscle is stimulated, through its nerves, acetylcholine is released from the motor end plates. It diffuses to the membranes of the muscle and the polarized state is lost. The fiber membranes become DEPOLARIZED and an elec-

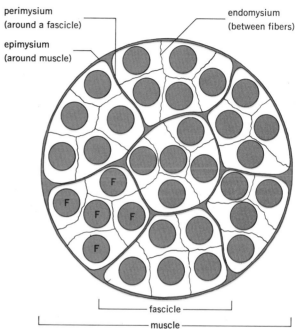

FIGURE 8.2. The connective tissue components of a skeletal muscle. F indicates muscle fibers.

FIGURE 8.3. A diagrammatic representation of a motor end plate. M, mitochondria.

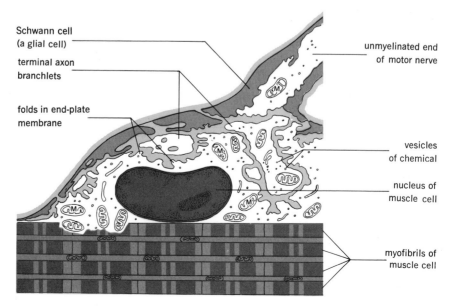

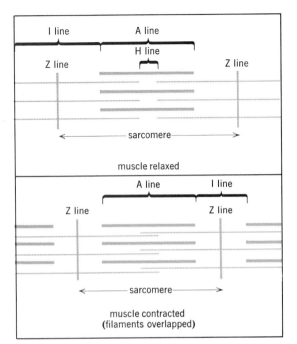

FIGURE 8.4. Changes in lengths of the sarcomere during muscle contraction.

FIGURE 8.5. A length tension curve. Length is measured in arbitrary units, with 100 representing the resting length of the muscle in the body. Note that maximum tension is developed when muscle is of its normal body length.

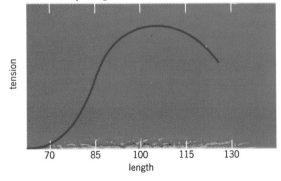

trical disturbance is transmitted over the membrane systems. A result of depolarization is the release of Ca^{++} from its bound state in the sarcoplasmic reticulum. It diffuses from the reticulum to the myosin, activates its enzyme activity, and the myosin splits the end phosphate group from the ATP molecules. Cross bridges are formed, probably via ATP molecules, between myosin and actin. A bridge is then thought to swing or oscillate back and forth, attaching to the actin molecules and drawing them closer together (Fig. 8.4). As the actin filaments are drawn together, they are stretched.

The membranes return to their original condition in the processes of REPOLARIZATION and relaxation. These processes are accomplished by active removal of Ca^{++} to its position in the reticulum, loss of myosin enzyme activity, resynthesis of ATP as the cross-link is broken, and return of the thin filaments to their original length by elastic recoil.

These theories of muscle activity are contained in the "interdigitating filament model" of muscle contraction.

Other events occurring during muscle activity

Depolarization of the membranes of the muscle is associated with the development of electrical disturbances. These disturbances may be recorded from the muscle as an electromyogram (EMG). Absence of an EMG in a resting muscle usually indicates loss of, or damage to, the nerve supply of the muscle.

If a muscle or a fiber at rest is stretched, it shows ELASTICITY much as does a rubber band, and tension increases (to a point) with increasing stretch. If a muscle is stretched slightly, and is then stimulated to contract, it develops more contractile strength than if it is not stretched (Fig. 8.5). Muscles are normally stretched by their attachments in the body, and thus contract more efficiently when stimulated. The muscle may shorten when it contracts and may exhibit an ISOTONIC CONTRACTION. If shortening is prevented while tension is developed, the contraction is called an ISOMETRIC CONTRACTION.

The term isometric is probably familiar to those who have undertaken training programs with a view to increasing muscular strength. Isometric exercises, in which one muscle or group of muscles is pitted against another with no movement being allowed, have been shown to cause a more rapid increase in muscular size and strength than do isotonic exercises (e.g., lifting barbells). While a complete physiological explanation for this phenomenon is lacking, several things occur more rapidly under an isometric program: muscle fibers increase in size more rapidly by adding sarcoplasm; fibers that are smaller or not normally active are brought into play more rapidly; faster increase in capillary networks occurs; motor end plates increase in size and chemical content, bringing stronger stimuli to the muscle.

As a muscle becomes active, it utilizes more energy and produces more HEAT. INITIAL HEAT is produced during stimulation and contraction of the muscle, and appears to be due to the energy required for depolarization, myosin activation, ATP splitting and cross-bridge formation. RELAXATION HEAT is produced during relaxation. RECOVERY HEAT continues after activity is over and is believed to reflect metabolism associated with resynthesis of ATP, and metabolism of lactic acid.

A muscle shows DIFFERENT DEGREES OF CONTRACTION strength, even when stimulated so as to cause all the fibers to shorten. The environment of the muscle appears to be the controlling agent, and includes the following specific factors.

AVAILABILITY OF NUTRIENTS AND OXYGEN, and metabolite removal. Activity and exercise increase the number of capillaries in a muscle. The increased blood supply brings more nutrients to insure stronger contraction and longer activity.

TEMPERATURE. A lower temperature results in a weaker contraction. This is usually not a factor in the more or less constant temperature of the body. In a given active muscle, stronger contractions may occur as activity raises muscle temperature, and this is the basis for "warming up" before strenuous exercise.

Strength of contraction also depends on the following.

LOAD. To a point, a stronger contraction results as load or tension on the muscle is increased. It may be restated that the muscles are normally stretched in the body by their attachments to bones and contract more efficiently when stimulated.

STRENGTH OF STIMULUS. Stronger stimuli activate more nerve fibers in a nerve and, therefore, more motor units, and strength of contraction increases until all muscle fibers are shortening. This factor is the main one responsible for the ability to adjust muscle strength to the task required.

A skeletal muscle also shows INDEPENDENT IRRITABILITY in that it will respond to stimuli delivered directly to the muscle. This property is sometimes utilized to keep muscles from atrophying when their nerves have been damaged. Stimulation is delivered to the muscle through external electrodes placed on the skin.

Energy sources for muscular activity

Resting metabolism in muscle depends primarily on glycolysis and beta-oxidation which utilize glucose and fatty acids, respectively. Studies indicate that glucose breakdown accounts for about 10 percent and fatty acid breakdown 90 percent of the energy released by resting muscle. During exercise, energy is provided by ATP and, to insure continued supplies of ATP, metabolism of several compounds occurs in the muscle to provide the energy necessary for ATP synthesis.

Breakdown of phosphocreatine:
Phosphocreatine + ADP (adenosine diphosphate) $\longrightarrow$ Creatine + ATP

Glycolysis (anaerobic breakdown of glucose):
Glucose + 2 ATP $\longrightarrow$ 2 Pyruvic acid + 4 ATP

Krebs cycle and oxidative phosphorylation:
2 Pyruvic acid + O_2 $\longrightarrow$ 6 CO_2 + 6 H_2O + 36 ATP

Beta oxidation and oxidative phosphorylation:
Fatty acid + O_2 $\longrightarrow$ CO_2 + H_2O + ATP (number depends on length of acid utilized)

The Krebs cycle and oxidative phosphorylation require oxygen, and if it is not available, the pyruvic acid formed in glycolysis is temporarily converted to lactic acid. The muscle will incur an OXYGEN DEBT, that is, the amount of oxygen necessary to combust the lactic acid formed from

FIGURE 8.6.   The events of muscular activity.

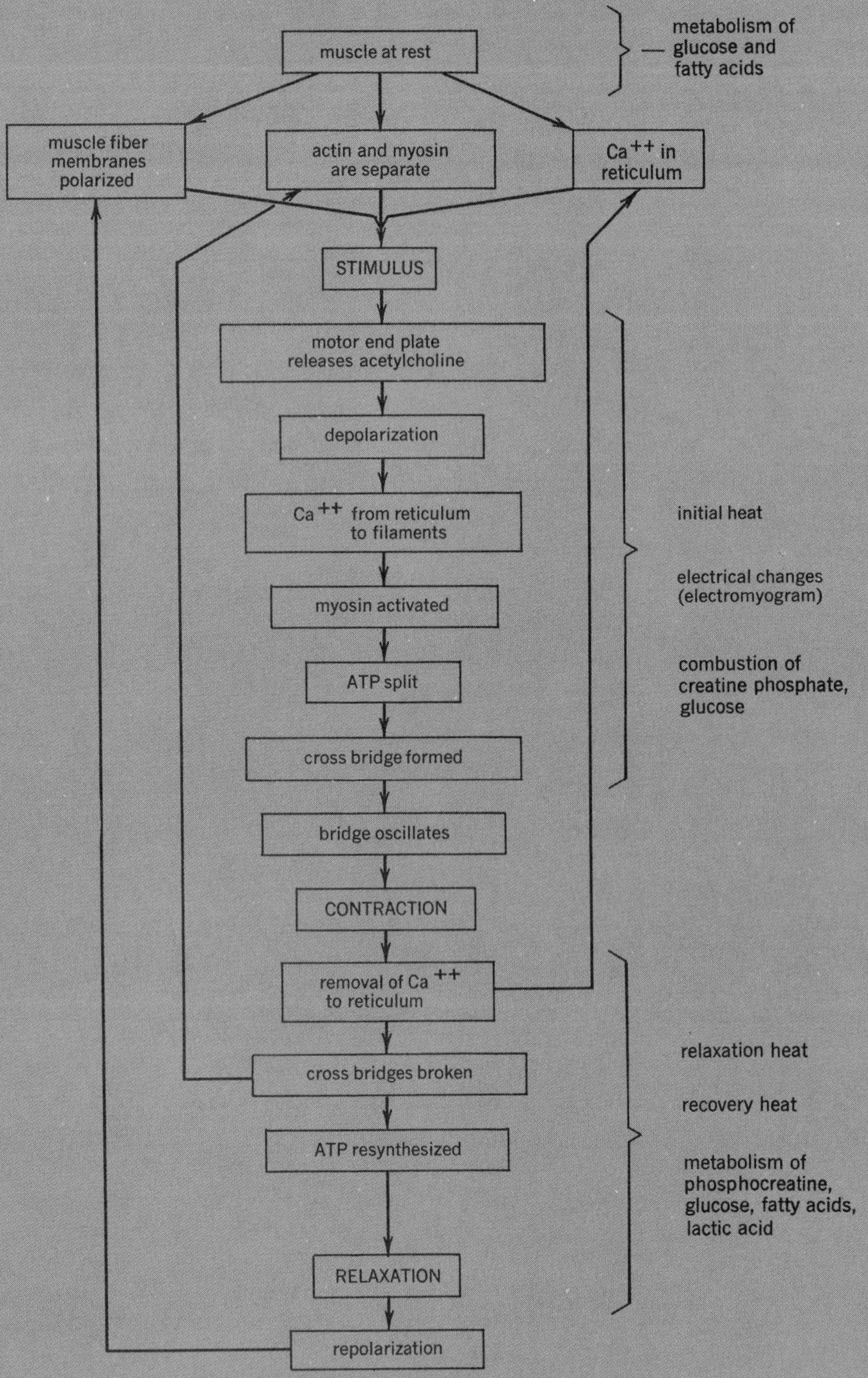

FIGURE 8.7.   A single muscle twitch as recorded on a kymograph.

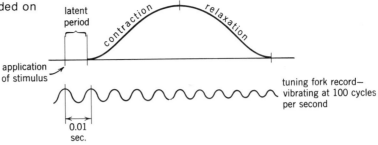

FIGURE 8.8.   The development of tetanus with increasing frequency of stimulation, as recorded on a kymograph.

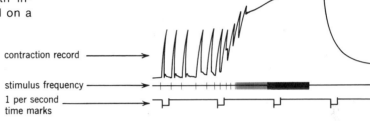

FIGURE 8.9.   Treppe.

pyruvic acid during anaerobic metabolism. When oxygen becomes available, lactic acid is converted back to pyruvic acid and sent through the Krebs cycle. Oxygen debt is part of why we breathe hard after stopping exercise.

A diagram of the interrelationships of the events of contraction is presented in Figure 8.6.

Types of contraction a muscle may show

One or more types of isotonic contractions may occur if a muscle is stimulated.

A TWITCH is normally seen only in the experimental laboratory, and is a single response by a muscle to a single stimulus (Fig. 8.7). It gives information as to the phases of muscular activity and the length of those phases. Muscles on our bodies may "twitch" if they are tired or have suffered nerve damage. This is not what is meant by a muscle twitch as may occur in the laboratory.

Stimuli that are repeated at very close intervals give a series of twitches that fuse into TETANUS (Fig. 8.8), that is, a sustained contraction. Tetanic contractions are used to maintain body posture. Tetanus occurs because the fiber repolarizes very rapidly, and is capable of reacting to a second stimulus before it completes its

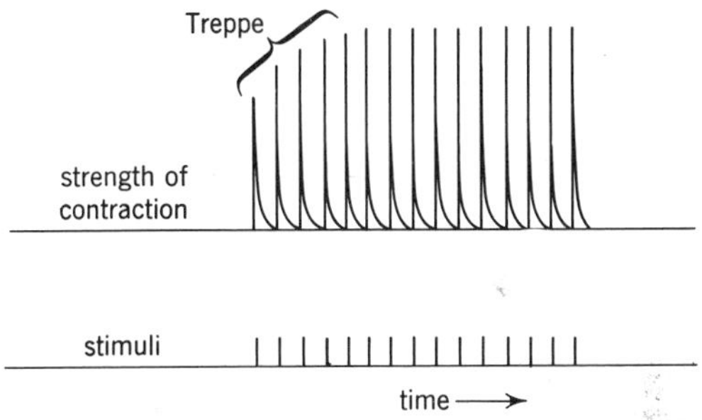

response to the first stimulus. It is said to have a short refractory period, or time where it will not respond to a second stimulus.

TREPPE (Fig. 8.9) results when stimuli strong enough to excite all the muscle fibers are applied in rapid succession, and the first several contractions increase in strength. It is attributed to "warming up" of the muscle and a lowered internal resistance to shortening.

TONE or tonus is said to occur in skeletal muscle as a sustained partial contraction that keeps the muscle in readiness to contract. Because of continual stimulation from the brain to a muscle by way of nerves, a muscle does exhibit a small degree of tension. The tone is maintained without tiring of the muscle, and is due to alternate stimulation of many motor units. Thus a more-or-less constant tension is maintained by using different muscle fibers. If the nerves to a muscle are cut, tone disappears, and the muscle is paralyzed.

A summary of structure and function of skeletal muscle is presented in Table 8.1.

Nerve supply to skeletal muscle

Skeletal muscle is provided with two general groups of nerves. AFFERENT NERVES, from a variety of sensory receptors in the muscle, convey impulses to the spinal cord and brain, and lead to many of the muscular reflexes the body shows. EFFERENT NERVES from the brain and cord to the muscle provide motor impulses that determine contraction, tone, and strength of contraction by the muscle. The nerve fibers to and from the skeletal muscles belong to that part of the nervous system called the SOMATIC NERVOUS SYSTEM. The term somatic, assumes that the muscle contracts only when stimulated through a nerve and "at will" (i.e., it is voluntary).

Clinical considerations

Disorders of skeletal muscle include those associated with its nerve supply, misuse, disuse, or overuse of the muscle, and defects of formation. Most present symptoms of abnormal movement, posture, facial expression, or allow protrusions of visceral organs through muscular walls. Those described here are ones of common occurrence or ones receiving coverage through the various communications media.

MUSCULAR DYSTROPHY is apparently of genetic origin. It involves progressive atrophy of muscles, and weakness. It cripples, and at present, no cure is available.

MYASTHENIA GRAVIS appears to be a disorder of the motor end plate. Either through insufficient production of the transmitting chemical, acetylcholine, or by the presence of excessive enzyme, cholinesterase, which destroys the acetylcholine, passage of impulses across the neuromuscular junction is difficult or impossible. Physostigmine is a chemical that inhibits cholin-

esterase, and can increase muscular contraction in the disorder.

PARALYSIS refers to inability to voluntarily contract a muscle. The muscle may be in a contracted (spastic) state, or in a relaxed (flaccid) state. Spastic paralysis occurs if the motor tracts of the brain and spinal cord are damaged. Flaccid paralysis occurs if the motor nerves from brain or cord to the muscles are damaged. These differences enable determination of where the damage exists when a muscle is paralyzed.

FASCICULATIONS, FIBRILLATIONS, and TREMOR refer to involuntary, repetitive contractions of muscles. The normal cause is nerve damage.

In MUSCLE SPASM, a forcible, often painful contraction of a muscle occurs, usually of appendage muscles. Chemical causes are most common, such as electrolyte imbalances (blood levels of Ca, Na, K) or chemical toxins (tetanus). Massage often relieves the spasm by increasing blood flow to the affected part.

MUSCLE CRAMPS, although similar to spasm, may have a different explanation. Exercise causes intense afferent nerve stimulation, through accumulation of chemicals or other causes. The spinal cord responds to this increased input of signals by an increased output of impulses that cause reflex contraction of the muscle. This may, in turn, lead to greater stimulation of the afferent nerves, and a "vicious cycle" is established. The cramp may often be promptly relieved by voluntary contraction of the opposing muscle group, while preventing movement of the body part. This causes a "reciprocal inhibition" of the contraction. A "charley horse" or "pulled muscle" may be painful and cause spasm and cramps. It usually involves tearing of muscle fibers.

MUSCLE SORENESS, especially after exercise, has not been satisfactorily explained. Metabolities of activity (e.g., lactic acid) may cause edema (swelling), putting tension or pressure on muscle nerves and creating pain; damage to the muscle fibers (rupture) or connective tissue (tearing); or sustained contraction, have all been advanced as causes of soreness.

FATIGUE is a term that is used in several ways. It usually refers to the failure of a muscle to contract when stimulated. It may be due to failure

TABLE 8.1 Summary of Structure and Properties of Skeletal Muscle

Item	Function or structure	Comments
Fiber	The basic unit or "cell" of skeletal muscle	Unit of structure of all named muscles. Size varies according to muscle size.
Sarcolemma	Membrane around fiber	Provides limit to fiber, some control of entry.
Sarcoplasm	The cytoplasm of the fiber	Contains mitochondria, reticulum, myofibrils.
Myofibrils	Longitudinally arranged units of the fiber	Cross banded (striated)
Myofilaments	Protein strands longitudinally arranged inside myofibrils	Contractile units of the muscle. 2 types: thick (contains myosin); thin (contains actin).
ATP, CP, K^+, Ca^{++}, $PO_4^{\equiv}$, glucose	Chemicals, within the fiber	Necessary for contraction and nutrition.
Sarcoplasmic reticulum	The "endoplasmic reticulum" of the fiber	Houses Ca^{++} until required for contraction.
T-tubules	Tubules separate from sarcoplasmic reticulum which communicate to fiber exterior	Avenue for passage of substances into fiber.
Endomysium Perimysium Epimysium	Connective tissue binding fibers into a muscle	Endo—binds fibers together. Peri—surrounds fasciculi. Epi—surrounds muscle.
Interdigitating filament model	Theory of arrangement of filaments in myofibrils	Allows explanation of muscle contraction.
Muscle contraction	Shortening of fibers	Ultimate cause of movement.
Creatine phosphate	Compound in muscle releasing energy for ATP synthesis	Immediate source of energy for ATP synthesis.
Glucose, Lactic acid, Fatty acid.	Combusted to provide energy for ATP synthesis	Provide energy to sustain activity.
Isometric contraction	Tension developed but no shortening	Posture.
Isotonic contraction	Contraction with shortening	Movement.
Twitch	Illustrates phases of muscular activity. 1 response to 1 stimulus.	0.1 second duration (frog muscle). In body, does not normally occur.
Tetanus	Fusion of twitches; sustained maximal contraction	Result of short refractory period allowing twitch fusion.
Tonus	Sustained partial contraction	Depends on intact motor areas in brain, intact upper and lower motor neurons.
Treppe	Increasing strength of contraction with repeated strong stimuli	"Warming up" allows greater contraction.

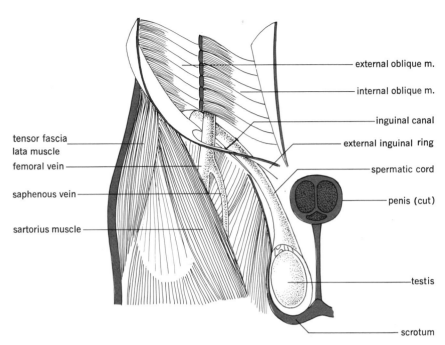

FIGURE 8.10. The anatomy of the inguinal area to show sites of possible hernias.

of a metabolic process to maintain ATP levels, or exhaustion of acetylcholine vesicles in the motor end plate. The term also refers to the subjective sensation of "tiredness" that may occur after exercise, emotional upheaval, and during physical illness.

HERNIA (commonly called "rupture") is a term used to describe the protrusion of abdominal viscera through a defect in the muscle or connective tissue of the abdominal wall, diaphragm, or pelvic floor. Hernias are the result of abdominal weakness, and an intra-abdominal pressure that causes tearing or separation of the components of the abdominal wall. Hernias are more common in obese persons, older individuals, those whose livelihood requires heavy lifting, and in those whose "style of life" results in lack of use and weakening of abdominal muscles.

According to where the hernia occurs, several types are described.

Inguinal hernias account for 80 percent of all hernias. The defect is in the inguinal canal which carries the spermatic cord from the scrotum in the male, and the round ligament of the uterus to the labia majora in the female. Increased intra-abdominal pressure, as in coughing, sneezing, lifting, straining at stool, or in distension caused by fluid or gas, may result in a separation or tearing of the inguinal muscle or abdominal fascia. Abdominal viscera, primarily small intestine, may then protrude into the scrotum (in the male) or labia majora (in the female).

Femoral hernia occurs when there is an enlargement of the femoral ring that normally passes the blood vessels to and from the thigh. A bulging of the skin in the groin is common as intestines push into the area.

Umbilical hernia occurs around the "belly button" (umbilicus) and is more common in infants.

Hiatus hernia is protrusion of the stomach into the chest cavity as a result of a weakness in the opening (hiatus) passing the esophagus through the diaphragm.

Surgical repair is the preferred treatment for those whose activities may lead to difficulties with the hernia. Surgery repairs the defect in the muscle or fascia. After surgery, any type of strenuous activity or lifting should be avoided for several weeks to allow for proper healing and to prevent adhesions from forming.

Common sites of herniation are shown in Figure 8.10.

Summary

1. Muscle is contractile and occurs in three types.
   a. Skeletal muscle attaches to the skeleton. It is voluntary.
   b. Cardiac muscle occurs in the heart and is involuntary.
   c. Smooth or visceral muscle is found in internal organs and is involuntary.

2. Skeletal muscle develops from mesoderm, and after five months *in utero* grows by increase in size of fibers, not by increase in number of fibers.

3. Skeletal muscle shows a characteristic structure.
   a. It consists of multinucleate, striated fibers.
   b. Myofibrils are composed of protein filaments.
   c. The usual cell organelles (mitochondria, reticulum) are present, as is a system of T-tubules; all are present in the sarcoplasm (muscle cytoplasm).

4. Chemically, the muscle:
   a. Is 75 percent water, 20 percent protein, and 5 percent other organic and inorganic substances.
   b. Contains two main proteins, myosin and actin. Myosin acts as an enzyme, actin shortens.
   c. Contains creatine phosphate, which acts as an energy source for ATP production.
   c. Contains inorganic salts for creation of a polarized state and for contraction.

5. The muscle is supplied with nerves that terminate in motor end plates on the fibers. A chemical, acetylcholine, is produced when the nerve is stimulated and depolarizes the muscle fibers. An enzyme destroys the chemical.

6. The events in muscular activity include:
   a. Creation of a resting or polarized state.
   b. Delivery of an impulse to the fiber.
   c. Depolarization of fiber membranes.
   d. Movement of calcium from reticulum to myosin.
   e. Activation of myosin and its splitting of ATP.
   f. Formation of cross bridges between myosin and actin.
   g. Drawing together of actin filaments (contraction).
   h. Active removal of Ca^{++} from sarcoplasm.
   i. Breaking of cross bridges.
   j. Relaxation of muscle and repolarization.

7. Other events occurring during activity include:

   a. Electrical changes.

   b. Development of greater contractile power if muscle is stretched.

   c. Development of tension, whether muscle shortens (isotonic contraction) or not (isometric contraction).

   d. Heat production during initial contraction, relaxation, and recovery.

   e. Strength of contraction is altered according to nutrients, waste removal, temperature, load, and stimulus strength.

8. Skeletal muscle exhibits independent irritability that makes it possible to stimulate it directly.

9. Energy for contraction comes from:

   a. Breakdown of phosphocreatine.

   b. Glycolysis.

   c. Krebs cycle.

   d. Beta oxidation.

   e. Oxidative phosphorylation.

10. Oxygen debt may occur, if a muscle produces lactic acid faster than it can be removed by metabolic aerobic combustion.

11. Skeletal muscles show several types of isotonic contractions.

   a. A single response to a single stimulus is a twitch.

   b. Twitches may be caused to fuse into tetanic contractions.

   c. Treppe occurs as a muscle "warms up" and contracts more strongly with stimulation.

   d. Tonus is a sustained partial contraction due to nervous stimulation.

12. Nerves supplying skeletal muscle include sensory and motor nerves. Both belong to the somatic nervous system.

13. Clinical conditions associated with skeletal muscle include:

   a. Dystrophy, where the muscle atrophies or shrinks.

   b. Myasthenia gravis, a disorder of neuromuscular transmission.

   c. Paralysis is the usual result of motor nerve damage.

   d. A variety of involuntary contractions may follow nerve damage.

   e. Spasm is a painful state of contraction most commonly due to chemical irritants or ionic imbalances.

   f. Soreness is due to nerve irritation as a result of edema, rupture of fibers, or tearing of connective tissue.

   g. Muscle cramps occur when sensory nerve stimulation causes reflex contractions.

   h. Fatigue refers to loss of contraction of muscle on stimulation. It may be caused by loss of energy sources, failure to transmit an impulse across the neuromuscular junction, or emotions, or circulatory failure.

i. Hernia or rupture refers to protrusion of abdominal viscera through a defect in the muscular walls of the abdominopelvic cavity. Inguinal hernias are the most common type. Hernias may be repaired surgically.

Questions

1. How does a skeletal muscle sustain its activity?

2. Describe the structure of a skeletal muscle.

3. Outline the steps occurring in the contraction of a muscle, starting with nerve stimulation.

4. What is the difference between:

    a. Isotonic and isometric contractions?

    b. Treppe and tonus?

    c. Twitch and tetanus?

5. What is meant by:

    a. Independent irritability?

    b. Electromyogram?

    c. Motor unit?

    d. Initial heat?

    e. Oxygen debt?

6. What chemicals are present in a muscle? Give one function for each you name.

7. How does a motor end plate work?

8. What factors determine strength of muscular contraction?

9. Of what value to the body is tone in skeletal muscles?

10. What factors may combine to produce a hernia?

Readings

Bloom, William, and Don W. Fawcett. *A Textbook of Histology.* 9th ed. Saunders. Philadelphia, 1968.

Hoyle, G. "How is Muscle Turned On and Off?" *Sci. Amer.* 222:84 (Apr) 1970.

Huxley, H. E. "The Mechanism of Muscular Contraction." *Sci. Amer.* 213:18 (Dec) 1965.

Merton, P. A. "How We Control the Contraction of Our Muscles." *Sci. Amer.* 226:30 (May) 1972.

Porter, K. R., and C. Franzine-Armstrong. "The Sarcoplasmic Reticulum." *Sci. Amer.* 212:72 (Mar) 1965.

chapter 9
The Skeletal Muscles

chapter 9

The skeletal muscles attach to the bones of the skeleton, and cause body movement by pulling on the skeleton. A joint may allow more than one direction of motion, thus, it follows that there must be separate muscles or groups of muscles to cause each type of motion.

TABLE 9.1 Muscle Actions	
Action	Definition
Flexion	Decrease of angle between two bones
Extension	Increase of angle between two bones
Abduction	Movement away from the midline (of body or part)
Adduction	Movement toward midline (of body or part)
Elevation	Upward or superior movement
Depression	Downward or inferior movement
Rotation	Turning about the longitudinal axis of the bone
Medial	Toward midline of body ("inward")
Lateral	Away from midline of body ("outward")
Supination	To turn the palm up or anterior
Pronation	To turn the palm down or posterior
Inversion	To face the soles of the feet toward each other
Eversion	To face the soles of the feet away from each other
Dorsiflexion (=flexion)	At the ankle, to move the top of the foot toward the shin
Plantar flexion (=extension)	At the ankle, to move the sole of the foot downward, as in standing on the toes

A muscle that causes a given movement is said to be an AGONIST or PRIME MOVER. A muscle causing a movement opposite to that of the agonist is said to be its ANTAGONIST. A SYNERGIST is a muscle that fixes a bone or holds it steady, so as to form a stable base against which movement may occur.

A muscle usually attaches to, at least, two bones. The less movable of those bones forms the muscle's ORIGIN, the more movable bone its INSERTION. The movement that occurs is the muscle's ACTION. A list of actions is presented in Table 9.1, and should be learned. Notice that these actions are presented in antagonistic pairs.

Muscles use the bones as levers to achieve movement. The power or strength the muscle must develop to cause movement depends on the arrangement of the parts of the lever. Any lever has several basic parts.

The *fulcrum* (F) is the point about which the lever turns. It is usually provided by a joint.

The point of *effort* (E) is where the muscle inserts and pull is applied.

The *resistance* (R) is the weight the muscle must move. It is created by the weight of the body part plus any load applied to the body part.

The *effort arm* (EA) is the distance along the lever from F to E.

The *resistance arm* (RA) is the distance along the lever from F to R.

The formula $E \times EA = R \times RA$ enables calculation of any one quantity, given the other three.

Three classes of levers are recognized according to placement of F, E, and R (Fig. 9.1). In a FIRST CLASS lever, F lies between E and R; in a SECOND CLASS lever, R lies between F and E; in a THIRD CLASS lever, E lies between F and R. A moment's reflection will show that the shorter EA is, or the longer RA is, the greater will be the effort required to move a given weight. Everyday examples of the three types of levers include:

Scissors, seesaw, prybars as first class levers.

Wheelbarrow as a second class lever.

Use of tennis racquets or golf clubs as third class levers where

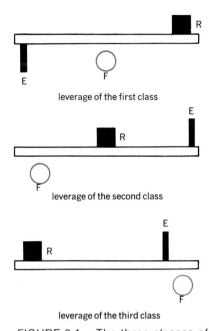

leverage of the first class

leverage of the second class

leverage of the third class

FIGURE 9.1.   The three classes of levers.

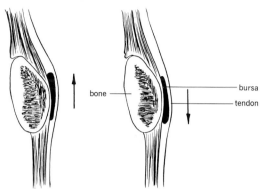

FIGURE 9.2.   The relation of a bursa to a tendon and a bone.

R = the ball
E = the hands on the equipment
F = the shoulder joint

TENDONS of collagenous tissue connect muscles to bones or other connective tissue structures. Tendons are utilized as the connecting structures for several reasons.

The fleshy portion of the muscle may not be long enough to span a distance, or small enough to fit in a restricted area.

Tendons, being tough, may pass over bony prominences without great damage.

Tendons, being small, can pass in groups over a joint, or attach to a smaller area on a bone than could the muscle(s).

As tough as they are, tendons still undergo wear and tear as they rub across bony prominences. BURSAE, small fluid-filled sacs (Fig. 9.2), are often placed between the tendon and bone to cushion the tendon. Bursitis is a painful inflammation of the bursae. "Housemaid's knee" is an inflammation of a bursa of the patella; football players (especially quarterbacks) may get bursitis in the shoulder from the strain of throwing.

The muscles

A general description of the major skeletal muscles follows in the next sections. Muscles are organized regionally into groups that act differently on particular body joints. As an aid to remembering the muscles, it is noted that muscles are generally named by their shape, location, origin and insertion, action, or combinations of these features.

Origins, insertions, actions, and innervations of individual muscles are given in Table 9.2, at the end of the chapter. Figure 9.3 shows the muscular system of the human as it would appear with the skin removed.

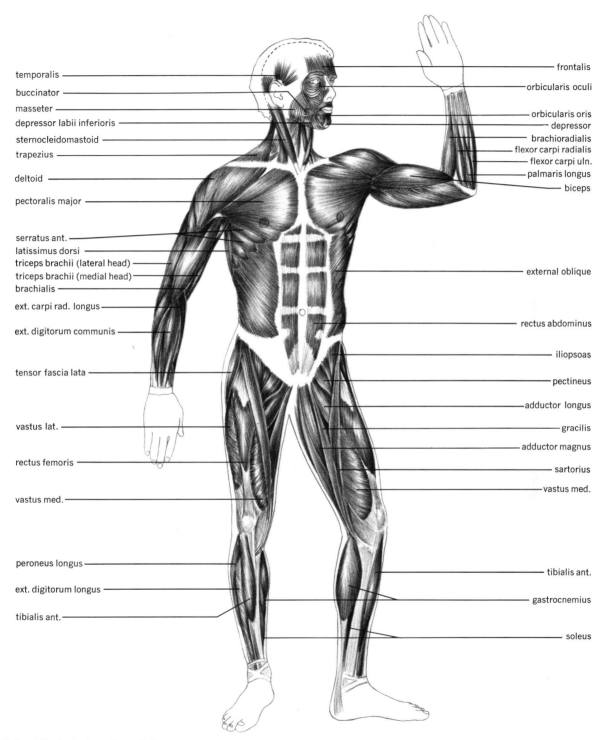

temporalis

buccinator

masseter

depressor labii inferioris

sternocleidomastoid

trapezius

deltoid

pectoralis major

serratus ant.

latissimus dorsi

triceps brachii (lateral head)

triceps brachii (medial head)

brachialis

ext. carpi rad. longus

ext. digitorum communis

tensor fascia lata

vastus lat.

rectus femoris

vastus med.

peroneus longus

ext. digitorum longus

tibialis ant.

frontalis

orbicularis oculi

orbicularis oris

depressor

brachioradialis

flexor carpi radialis

flexor carpi uln.

palmaris longus

biceps

external oblique

rectus abdominus

iliopsoas

pectineus

adductor longus

gracilis

adductor magnus

sartorius

vastus med.

tibialis ant.

gastrocnemius

soleus

FIGURE 9.3. (*A*) Anterior view of the muscular system.

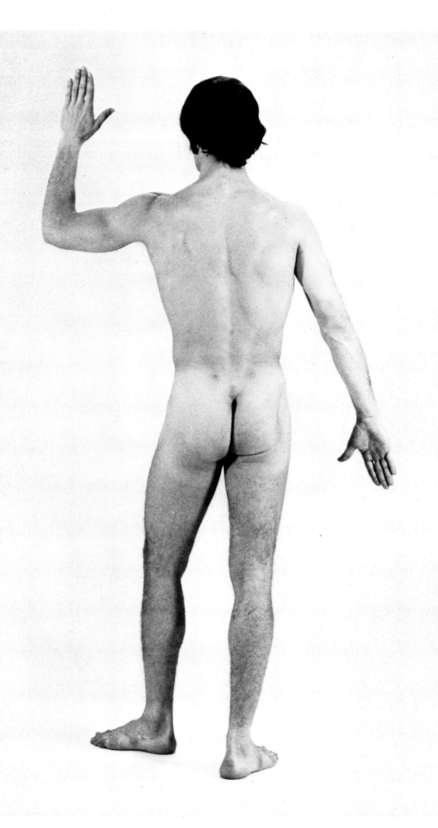

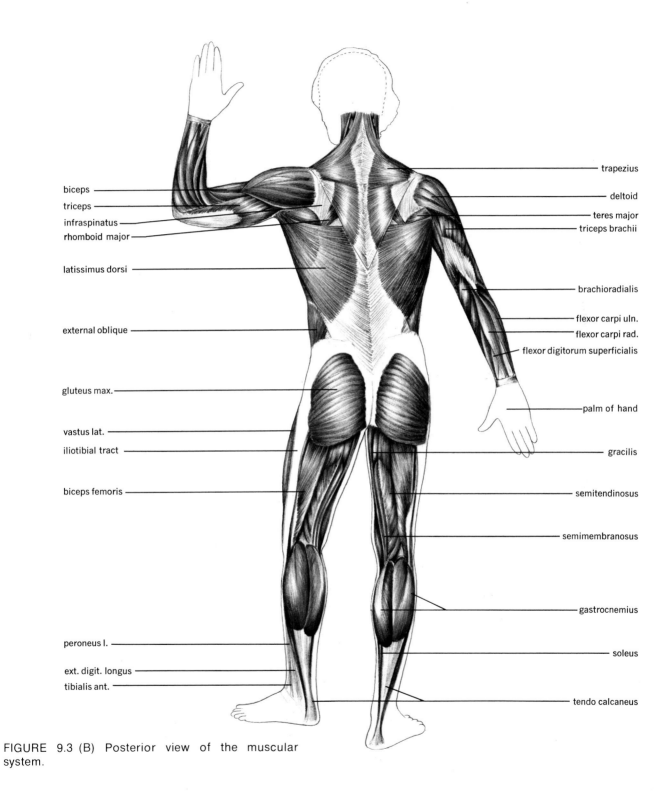

biceps

triceps

infraspinatus

rhomboid major

latissimus dorsi

external oblique

gluteus max.

vastus lat.

iliotibial tract

biceps femoris

peroneus l.

ext. digit. longus

tibialis ant.

trapezius

deltoid

teres major

triceps brachii

brachioradialis

flexor carpi uln.

flexor carpi rad.

flexor digitorum superficialis

palm of hand

gracilis

semitendinosus

semimembranosus

gastrocnemius

soleus

tendo calcaneus

FIGURE 9.3 (B) Posterior view of the muscular system.

Muscles of the head and neck

Facial muscles (Fig. 9.4)

Facial muscles are the muscles of facial expression. Their contractions move the fleshy parts of the face for speech, and cause the various expressions associated with emotions and feelings. The facial muscles include the following.

ORBICULARIS OCULI encircles the eye and closes or winks the eye.

ORBICULARIS ORIS encircles the mouth and closes and protrudes ("puckers") the lips.

LEVATOR LABII SUPERIORIS elevates the upper lip to give an "expression of contempt" (or as in saying "Yech"! or "Yuck"!)

ZYGOMATICUS draws the corners of the mouth up and back, as in smiling.

RISORIUS draws the corners of the mouth directly sideways as in a grimace. (Children often put a finger in each corner of their mouth and pull; this imitates the action of the risorius.)

TRIANGULARIS draws the corners of the mouth down and back to create an expression of sadness ("drooping mouth").

DEPRESSOR LABII INFERIORIS depresses the lower lip.

MENTALIS elevates the lower lip and protrudes it, as in pouting.

BUCCINATOR lies beneath the muscles listed above, and compresses the cheek. It is sometimes called the "trumpeter's muscle," because it is used strongly in playing a brass musical instrument like a trumpet.

CORRUGATOR produces "frown lines" in the center forehead.

Cranial muscles (Fig. 9.5)

The cranial muscles lie on the forehead, back of the head, and around the ears.

The EPICRANIUS is the only cranial muscle. It is divided into an *occipitofrontal group* consisting of an anterior *frontalis* and posterior *occipitalis,* and a *temporoparietal group* consisting of the *auricular muscles* attaching to the pinna (flap) of the ear. The occipitalis and frontalis attach to a broad flat tendon (galea aponeurotica) over the top of the skull. The frontalis pulls the scalp forward and produces transverse wrinkles in the forehead ("expression of surprise"). Occipitalis draws the scalp backward.

FIGURE 9.4.   The facial muscles.

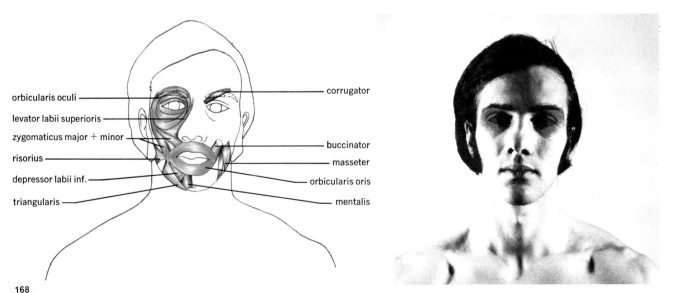

orbicularis oculi

levator labii superioris

zygomaticus + minor

risorius

depressor labii inf.

triangularis

corrugator

buccinator

masseter

orbicularis oris

mentalis

FIGURE 9.5. The cranial muscles.

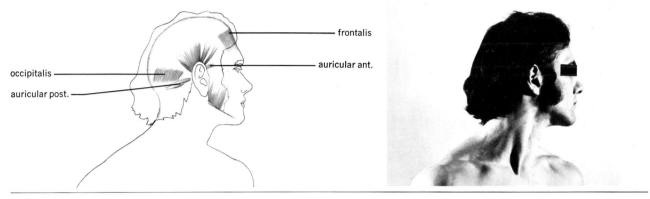

frontalis

auricular ant.

occipitalis

auricular post.

FIGURE 9.6. Superficial muscles of mastication.

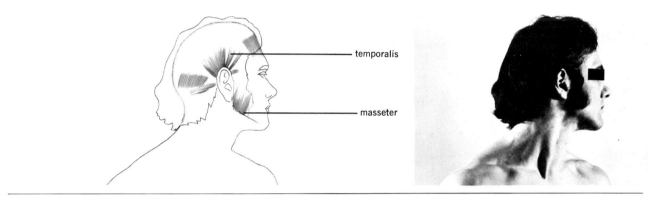

temporalis

masseter

FIGURE 9.7. Deep muscles of mastication.

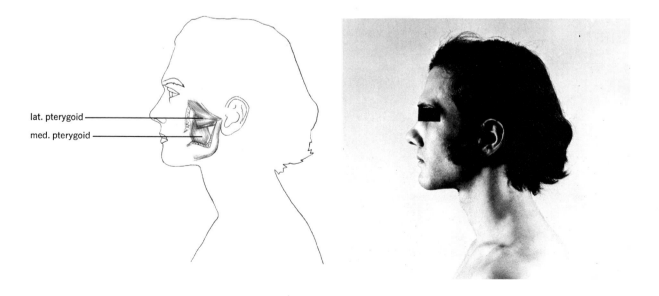

lat. pterygoid

med. pterygoid

The auricular muscles are rudimentary in man. A superior muscle draws the pinna upward; a posterior muscle draws it backward, and an anterior muscle draws it forward.

Muscles of mastication (chewing) (Figs. 9.6 and 9.7)

Four muscles are the most important in mastication.

The TEMPORALIS fills a depression on the side of the skull and elevates the mandible, closing the mouth.

MASSETER, covering the ramus of the mandible, is the most powerful jaw closing muscle.

Both temporalis and masseter can be felt contracting by clenching the jaws and feeling the skull in the appropriate area.

MEDIAL PTERYGOID is internal to the mandibular ramus and elevates the jaw.

LATERAL PTERYGOID lies between the medial

FIGURE 9.8.   The suprahyoid muscles.

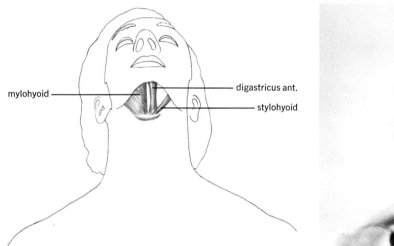

FIGURE 9.9.   The infrahyoid muscles.

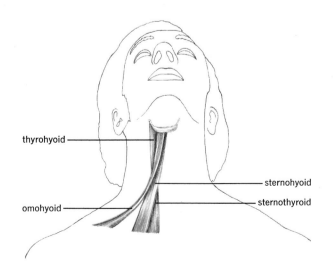

pterygoid and the ramus, and depresses the mandible, opening the jaw (mouth).

The pterygoids also move the mandible laterally.

Muscles associated with the hyoid bone (Figs. 9.8 and 9.9)

These muscles may be grouped according to their position relative to the hyoid bone. Most are named by their origin and insertion, and contribute to the movements that occur when we swallow.

SUPRAHYOID MUSCLES (ABOVE HYOID BONE).
DIGASTRIC consists of two portions or bellies.

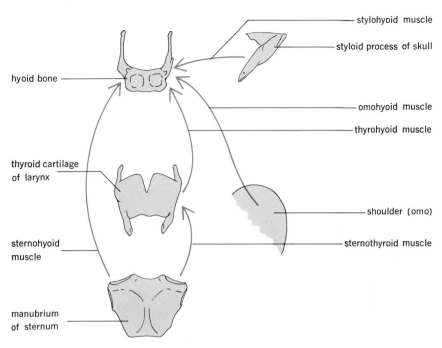

FIGURE 9.10. A diagrammatic representation of how the muscles of the hyoid area are named.

FIGURE 9.11. The anterior neck muscles.

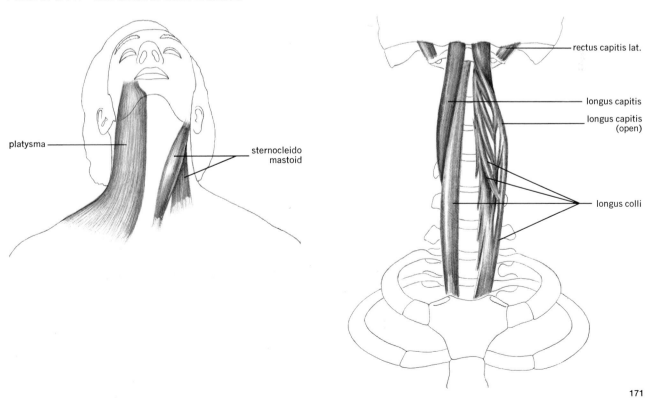

The posterior belly draws the hyoid backward, the anterior belly draws it forward.

STYLOHYOID draws the hyoid bone up and back.

MYLOHYOID forms the floor of the mouth. It lies between the limbs of the mandibular body, and raises the hyoid bone and tongue.

INFRAHYOID MUSCLES (BELOW HYOID BONE).

STERNOHYOID draws the hyoid downward.

STERNOTHYROID draws the larynx downward.

THYROHYOID draws the larnyx upward, if the hyoid is fixed.

OMOHYOID draws the hyoid downward.

Figure 9.10 presents a diagrammatic representation of the positions of these muscles in relation to the skull, neck, and thorax, and shows how they are named.

FIGURE 9.12.   The lateral neck muscles.

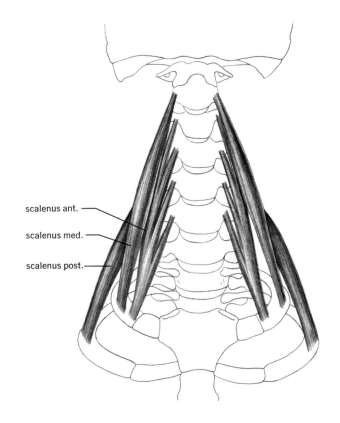

scalenus ant.
scalenus med.
scalenus post.

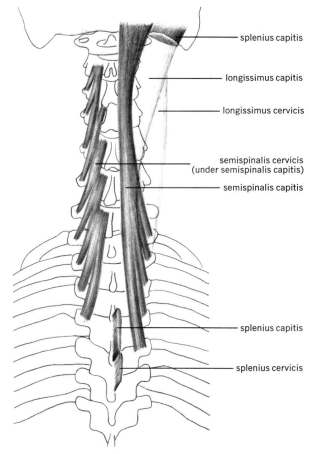

splenius capitis

longissimus capitis

longissimus cervicis

semispinalis cervicis
(under semispinalis capitis)

semispinalis capitis

splenius capitis

splenius cervicis

FIGURE 9.13.   The posterior neck muscles.

Neck muscles (Fig. 9.11 to 9.13)

The muscles of the neck consist of two superficial muscles on the anterior neck, and a number of deep muscles that attach to the vertebrae and skull.

SUPERFICIAL MUSCLES. The PLATYSMA is a broad sheet covering the inner side of the shoulder and lateral neck and mandible. It draws the corners of the mouth down and sideward, as in screaming, or in an "expression of horror."

STERNOCLEIDOMASTOID flexes the neck if the muscles are operating together and, singly, ro-

tates the head to the opposite side while pointing the chin upward.

DEEP MUSCLES. These may be divided into anterior, lateral, and posterior groups.

The anterior group includes LONGUS COLLI, LONGUS CAPITUS, and RECTUS CAPITUS, all of which flex the neck on the chest.

The lateral group includes the three SCALENES, which bend the neck to the side, and help to elevate the ribs if the neck is fixed.

The posterior group includes the SEMISPINALIS, LONGISSIMUS, and SPLENIUS, which extend the neck and raise the chin.

Muscles operating the vertebral column

FIGURE 9.14.   Movements of the spinal column.

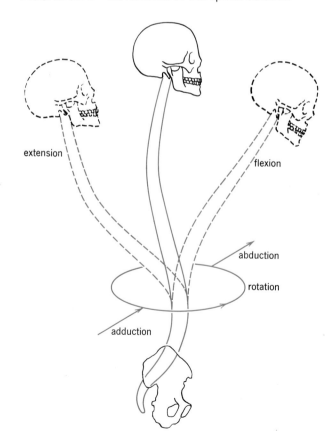

extension

flexion

abduction

rotation

adduction

Movements possible (Fig. 9.14)

The column may be flexed or bent forward, extended or bent backward, abducted or bent to the side, adducted or returned to the midline, and rotated or twisted. Some of the muscles responsible for these actions have other functions, and their names may appear in other sections.

Flexors of the spine (Fig. 9.15)

RECTUS ABDOMINUS is the strongest flexor. It runs down the midline of the belly. One uses it strongly in a "sit-up."

PSOAS and ILIACUS flex the spine, if the thigh is fixed.

Extensors of the spine (Fig. 9.16)

Three muscles of the back, known collectively as the ERECTOR SPINAE or sacrospinalis, are the most important extensors of the spine, and form the major postural muscles holding our backs straight against gravity when we sit or stand.

ILIOCOSTALIS, LONGISSIUMS, and SPINALIS are the three muscle groups, listed laterally to medially.

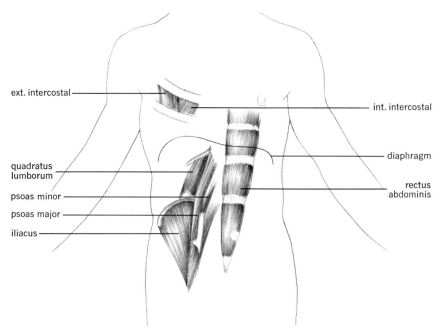

ext. intercostal

int. intercostal

diaphragm

quadratus
lumborum

rectus
abdominis

psoas minor

psoas major

iliacus

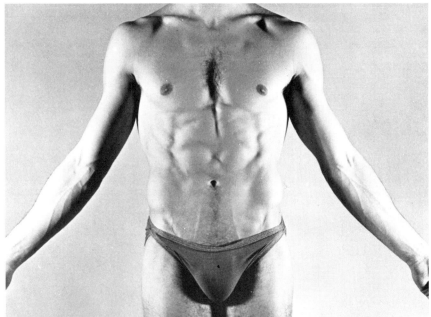

FIGURE 9.15. Flexors of the spinal column and muscles of respiration.

Abductors and adductors of the spine (Fig. 9.17)

ILIOCOSTALIS is lateral enough in placement to bend or to abduct the column. The same muscle on the opposite side of the vertebral column adducts the column. For example, the left iliocostalis abducts the column to the left; the right iliocostalis adducts the column to the midline.

QUADRATUS LUMBORUM pulls the lumbar ver-

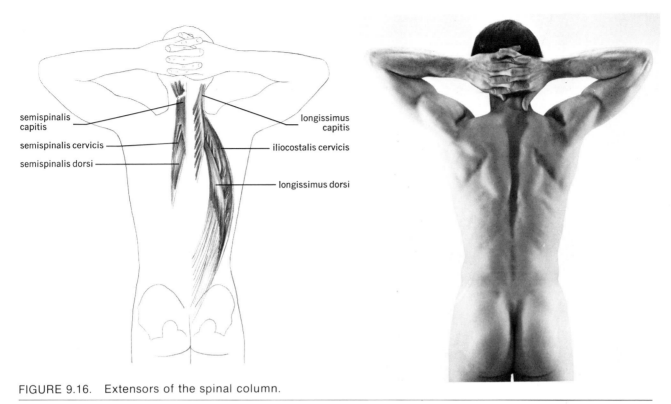

semispinalis capitis

semispinalis cervicis

semispinalis dorsi

longissimus capitis

iliocostalis cervicis

longissimus dorsi

FIGURE 9.16.   Extensors of the spinal column.

FIGURE 9.17.   Abductors and adductors of the spinal column.

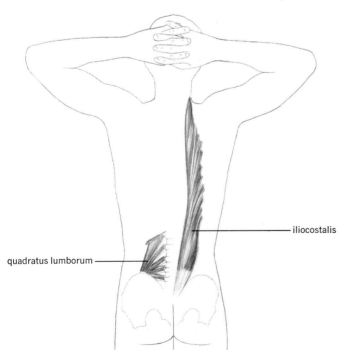

iliocostalis

quadratus lumborum

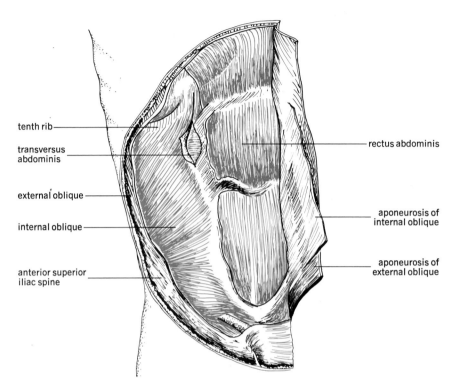

tenth rib

transversus
abdominis

external oblique

internal oblique

anterior superior
iliac spine

rectus abdominis

aponeurosis of
internal oblique

aponeurosis of
external oblique

FIGURE 9.18.   Rotators of the spinal column.

tebrae and bends the column to the same side as the muscle contracting. The same muscle on the opposite side returns the column to the midline position.

Rotators of the spine (Fig. 9.18)

Three abdominal muscles may help to rotate the spine, and also are important in compressing the abdominal contents. Weakness of these muscles allows protrusion of the abdomen (a "pot").

Each EXTERNAL OBLIQUE, acting alone, rotates the spine so as to bring the same shoulder forward. Together, they compress the abdomen.

Each INTERNAL OBLIQUE, acting alone, brings the opposite shoulder forward. Together, they compress the abdomen. (Figure out which muscles, and which sides, would be used to twist the trunk to the left.)

The TRANSVERSE ABDOMINAL is mainly an abdominal compressor, utilized strongly in urination, defecation, vomiting, assisting during childbirth, and forced expiration.

Muscles of the thorax

The muscles of the thorax are those that are associated with the ribs, and are concerned with breathing.

Movements possible (Fig. 9.19)

The ribs may be elevated, which expands the

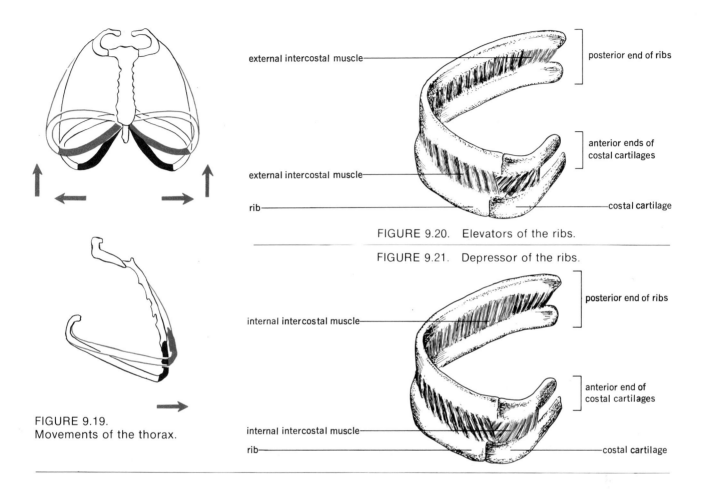

FIGURE 9.20.   Elevators of the ribs.

FIGURE 9.21.   Depressor of the ribs.

FIGURE 9.19.
Movements of the thorax.

thorax in the front-to-back and side-to-side directions, and depressed, which decreases these dimensions. The vertical dimension of the thorax may be increased.

Elevators of the ribs (Fig. 9.20)

The EXTERNAL INTERCOSTALS lie superficially between the ribs. They elevate the ribs, and create what is called "costal breathing." Costal breathing provides about two thirds of the air exchanged during normal breathing.

SERRATUS POSTERIOR INFERIOR also elevates the ribs.

With forced inspiration, the scalenes (see above) and the PECTORALIS MINOR (see below) aid in rib elevation.

Depressor of the ribs (Fig. 9.21)

The INTERNAL INTERCOSTALS lie deep to the externals, and depress the ribs.

Increasing the vertical dimension of the thorax

This task is handled by the DIAPHRAM (Fig. 9.22), a dome-shaped muscle lying between the thoracic and abdominal cavities. Its fibers insert into a central tendon that is drawn downward when the muscle contracts. Its action accounts for about one third of the air exchanged during normal breathing. It becomes the most important breathing muscle when increased depth of breathing is required.

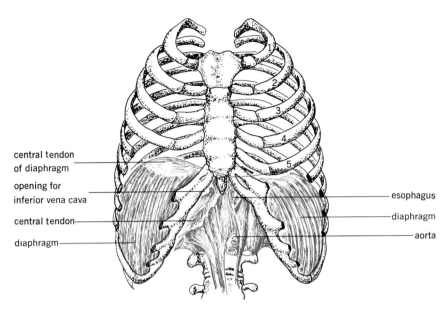

central tendon
of diaphragm

opening for
inferior vena cava

central tendon

diaphragm

esophagus

diaphragm

aorta

FIGURE 9.22.   The diaphragm.

The abdominal muscles (rectus, obliques, transverse) are brought into play during forced expiration. Their contraction presses the abdominal viscera against the diaphragm and hastens its return to its original position.

Muscles of the shoulder (pectoral) girdle

Movements possible (Fig. 9.23)

The scapulae may be elevated depressed, abducted, adducted, and rotated. The clavicles may be elevated and depressed. The muscles involved may be divided into an anterior group, and a posterior group.

Anterior group (Fig. 9.24)

PECTORALIS MINOR draws the scapula forward and down, and if the scapula is fixed, helps to elevate the ribs for breathing.

SERRATUS ANTERIOR abducts the scapula, and acts strongly in pushing movements.

SUBCLAVIUS is a synergistic muscle that depresses and holds the clavicle firmly.

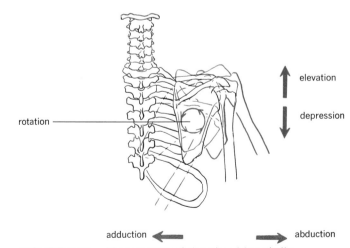

rotation

elevation

depression

adduction   abduction

FIGURE 9.23.   Movements of the shoulder girdle.

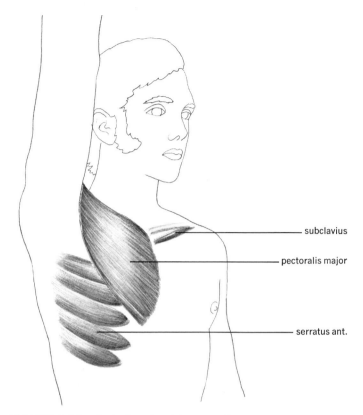

FIGURE 9.24.   Anterior shoulder muscles.

subclavius

pectoralis major

serratus ant.

FIGURE 9.25.   Posterior shoulder muscles.

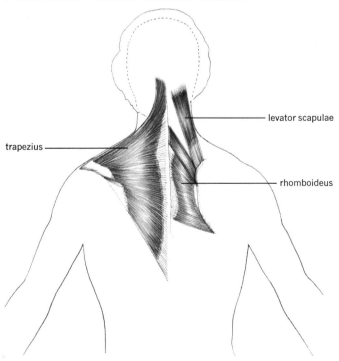

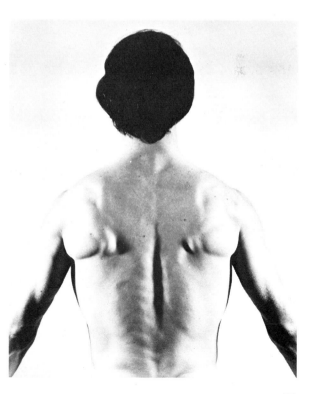

levator scapulae

trapezius

rhomboideus

Posterior group (Fig. 9.25)

TRAPEZIUS has three parts: the upper portion elevates the scapula; the middle part adducts the scapula; the lower part depresses the scapula.

RHOMBOID is a strong adductor of the scapula and also elevates it.

LEVATOR SCAPULAE elevates the scapula, and "shrugs" the shoulder (along with the upper trapezius).

Muscles moving the shoulder joint (humerus)

Movements possible (Fig. 9.26)

As a ball-and-socket joint secured mainly by muscles, the shoulder joint is the most freely movable joint in the body. The humerus thus shows a wide range of motion. It may be flexed, extended, abducted, adducted, and rotated medially or laterally. Circumduction is a circular movement of the entire upper limb, composed of the preceding motions occurring in sequence.

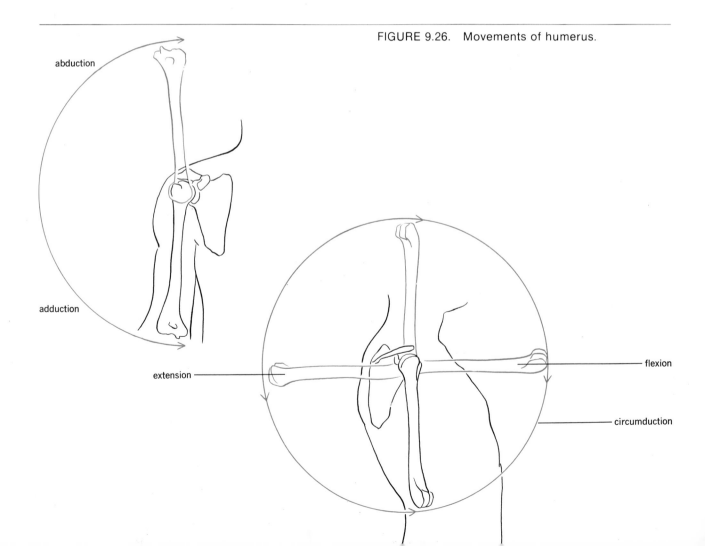

FIGURE 9.26. Movements of humerus.

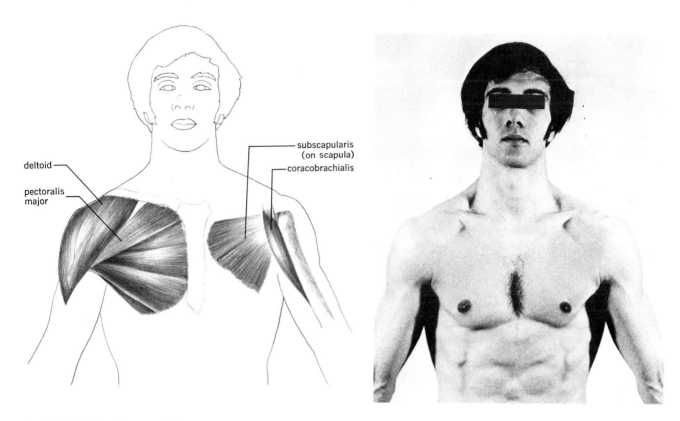

deltoid

pectoralis
major

subscapularis
(on scapula)

coracobrachialis

FIGURE 9.27.   Flexors of humerus.

FIGURE 9.28.   Extensors of humerus.

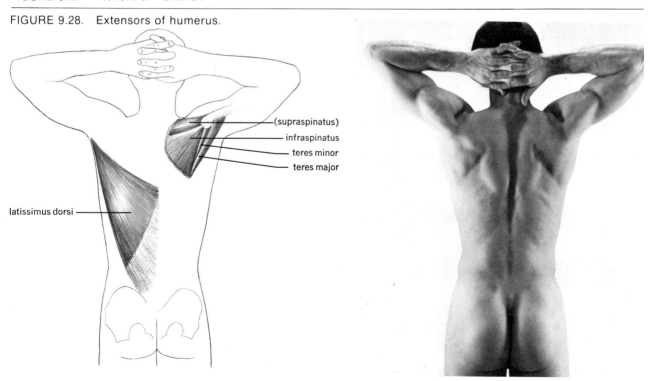

(supraspinatus)

infraspinatus

teres minor

teres major

latissimus dorsi

Flexors of the humerus (Fig. 9.27)

PECTORALIS MAJOR (the "pects" or pectorals) flexes and medially rotates the humerus, as in throwing or delivering a jab.

CORACOBRACHIALIS is a weak flexor.

Extensors of the humerus (Fig. 9.28)

LATISSIMUS DORSI (the "lats") is a powerful extensor of the shoulder, and with the pectoralis major, adducts the humerus.

TERES MAJOR acts with the latissimus.

Abductors of the humerus (see Figs. 9.27 and 9.28)

The DELTOID, acting as a whole, abducts the shoulder as in holding the arms straight out to the side.

SUPRASPINATUS acts with the deltoid.

Rotators of the humerus (see Figs. 9.27 and 9.28)

SUBSCAPULARIS medially rotates the humerus.

INFRASPINATUS and TERES MINOR laterally rotate the humerus.

Muscles moving the forearm

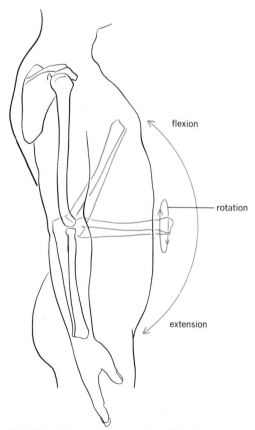

flexion

rotation

extension

FIGURE 9.29. Movements of the forearm.

Movements possible (Fig. 9.29)

The term, elbow joint, refers to three joints between the humerus, ulna, and radius that are enclosed in a common capsule. The three joints are specifically named *humeroulnar, humeroradial,* and *radioulnar.* The humeroulnar joint may only be flexed and extended. The humeroradial joint occurs between the humerus and head of the radius, and may be rotated medially (pronation) or laterally (supination). The radioulnar joint occurs between the proximal ends of radius and ulna and allows the rotation mentioned above.

Flexors of the forearm (Fig. 9.30)

BICEPS BRACHII is a two-headed muscle strongly flexing and laterally rotating the forearm. It is used strongly in a "pull up" (chinning), and is often displayed by small boys to impress others with their "muscle."

BRACHIALIS is a pure flexor, and lies deep to the biceps.

BRACHIORADIALIS lies in line with the thumb on the lateral forearm. It flexes the forearm and returns the forearm to the midpoint position from either full pronation or full supination.

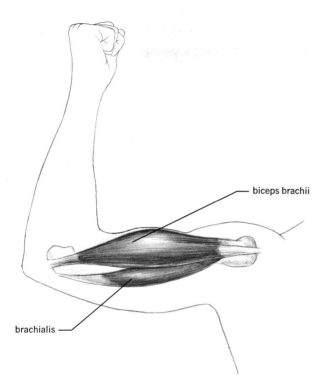

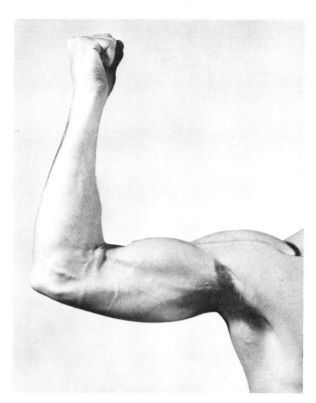

FIGURE 9.30.   Flexors of the elbow.

FIGURE 9.31.   Extensor of the elbow.

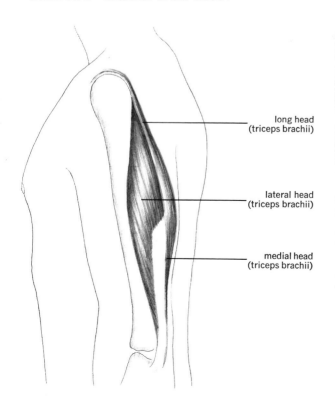

biceps brachii

brachialis

long head
(triceps brachii)

lateral head
(triceps brachii)

medial head
(triceps brachii)

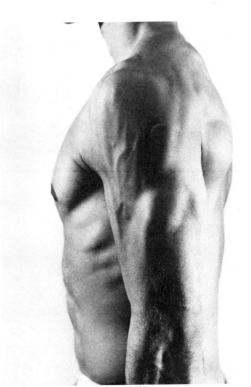

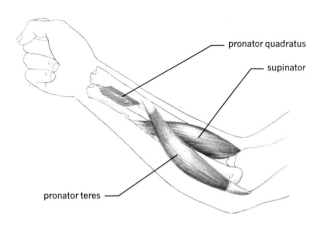

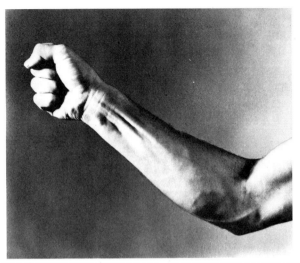

FIGURE 9.32. Rotators of the forearm.

Extensor of the forearm (Fig. 9.31)

The TRICEPS BRACHII is the only extensor of the forearm. It has three heads, designated lateral, long, and medial from lateral to medial on the posterior upper arm. All have a common tendon of insertion on the point of the elbow (olecranon process).

Rotators of the forearm (Fig. 9.32)

In addition to the BRACHIORADIALIS and BICEPS, the SUPINATOR supinates (laterally rotates) the forearm.

The PRONATOR and PRONATOR QUADRATUS pronate (medially rotate) the forearm.

Muscles moving the wrist and fingers

Movements possible (Fig. 9.33)

The wrist and fingers may be flexed, extended, abducted, and adducted.

Flexors of the wrist (Fig. 9.34)

Wrist flexors form a superficial group of muscles on the anterior distal humerus and the anterior forearm. Their tendons cross the anterior aspect of the wrist to attach to the carpal bones. From lateral to medial, there are three muscles.

FLEXOR CARPI RADIALIS flexes the wrist and slightly abducts it.

PALMARIS LONGUS flexes the wrist as its only action.

FLEXOR CARPI ULNARIS flexes the wrist and slightly adducts it.

Flexors of the fingers (see Fig. 9.34)

Beneath the three muscles named above are two layers of muscles that attach to the phalanges of the fingers and flex them. Each muscle has four parts to it, each of which sends a tendon to each finger. Although named as one muscle, the parts operate as single muscles, giving us the ability to move one finger separately from the other fingers.

FLEXOR DIGITORUM SUPERFICIALIS is the first layer, and its tendons reach the second phalanx of each finger.

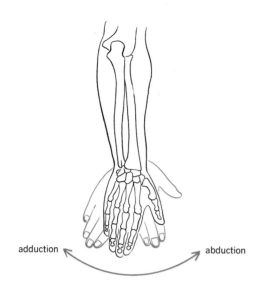

adduction ← → abduction

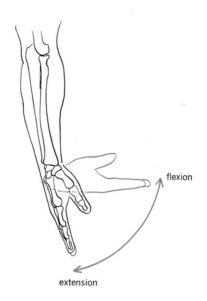

flexion

extension

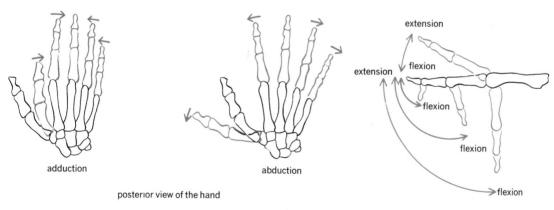

adduction

abduction

posterior view of the hand

extension

extension

flexion

flexion

flexion

flexion

FIGURE 9.33.   Movements of the wrist and fingers.

FIGURE 9.34.   Flexors of the wrist and fingers.

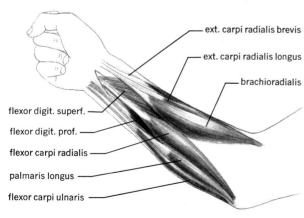

ext. carpi radialis brevis

ext. carpi radialis longus

brachioradialis

flexor digit. superf.

flexor digit. prof.

flexor carpi radialis

palmaris longus

flexor carpi ulnaris

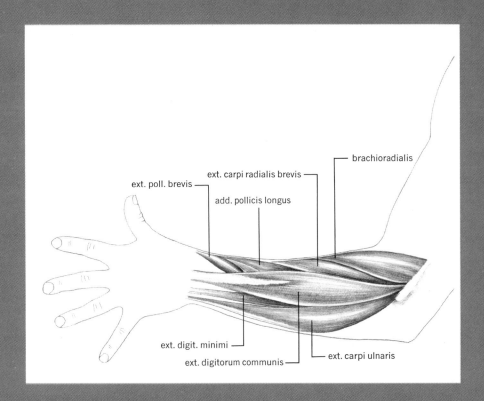

ext. poll. brevis

ext. carpi radialis brevis

add. pollicis longus

brachioradialis

ext. digit. minimi

ext. digitorum communis

ext. carpi ulnaris

FIGURE 9.35. Extensors of the wrist and fingers.

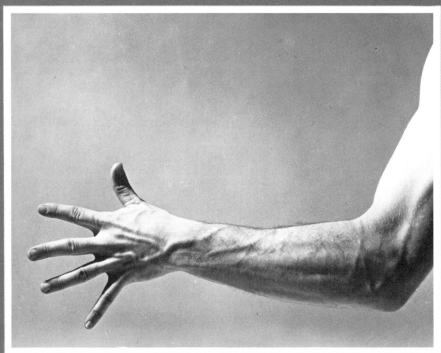

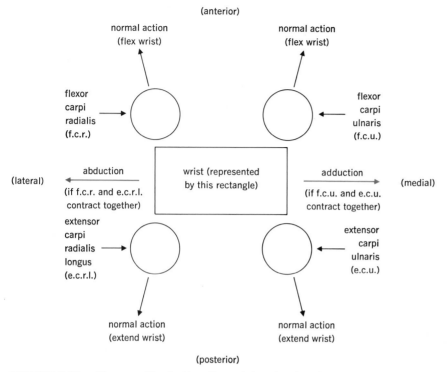

FIGURE 9.36.   Diagram illustrating the origin of wrist abduction and adduction.

FLEXOR DIGITORUM PROFUNDUS is the second layer, and its tendons pass to the distal phalanges.

Both muscles are required to make a fist.

Extensors of the wrist and fingers (Fig. 9.35)

Five muscles in a single layer across the posterior aspect of the forearm, are the extensors of wrist and fingers. From lateral to medial, they are as follows.

EXTENSOR CARPI RADIALIS LONGUS extends and slightly abducts the wrist.

EXTENSOR CARPI RADIALIS BREVIS acts as does the longus.

EXTENSOR DIGITORIUM COMMUNIS is a three-headed muscle sending tendons to all phalanges of the index, middle, and ring fingers, to extend them separately or together.

The EXTENSOR DIGITI MINIMI is a separate muscle sending tendons to all phalanges of the little finger to extend it.

EXTENSOR CARPI ULNARIS extends and slightly adducts the wrist.

Abduction and adduction of the wrist

Strong abduction of the wrist occurs through the cooperation of flexor carpi radialis and extensor carpi radialis longus and brevis.

Strong adduction occurs as the result of co-operation between flexor and extensor carpi ulnaris. Figure 9.36 diagrammatically shows the positions of all the muscles described above with their direction of pull (arrows), and illustrates the manner in which strong adduction and abduction of the wrist is produced by muscle co-operation.

Muscles of the thumb

The thumb has its own supply of muscles independent of those of the fingers. This allows the thumb to be placed in opposition to the fingers for grasping and manipulation of objects, a fact that some authorities credit for man's success on earth. Some of these muscles are shown in Figure 9.37. All thumb muscles carry the action produced, and the word pollicis (L. *pollex,* thumb) in their names.

Miscellaneous hand muscles

Figure 9.37 also shows some of the muscles within the hand that permit abduction and adduction of the fingers. The INTEROSSEI and LUMBRICALES are the two main groups.

FIGURE 9.37.   Muscles of the thumb and hand.

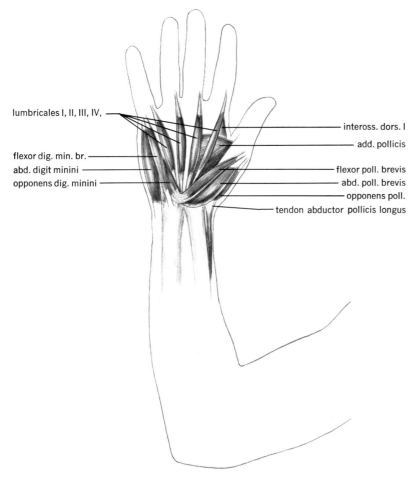

lumbricales I, II, III, IV,

flexor dig. min. br.
abd. digit minini
opponens dig. minini

inteross. dors. I
add. pollicis
flexor poll. brevis
abd. poll. brevis
opponens poll.
tendon abductor pollicis longus

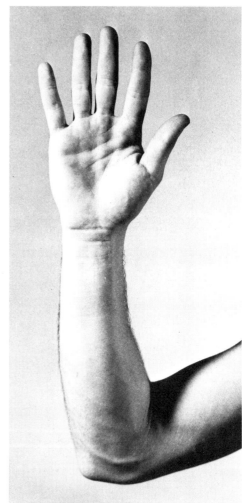

Muscles operating the hip and knee joints

Movements possible (Fig. 9.38)

The hip joint is a ball-and-socket joint that allows the femur to be flexed, extended, abducted, adducted, and medially and laterally rotated. The knee may only be flexed and extended. Some of the muscles operate both joints at once. Locating the muscles by position forms a convenient method of classifying them.

Anterior thigh and pelvic muscles (Fig. 9.39)

These muscles are flexors of the hip and extensors of the knee.

ILIACUS and PSOAS, sometimes considered as a single muscle named *iliopsoas,* flex the femur and medially rotate it. The muscles are strongly used in climbing, kicking, walking, and running.

RECTUS FEMORIS flexes the hip and extends the knee.

VASTUS MEDIALIS, VASTUS INTERMEDIUS, and VASTUS LATERALIS extend the knee as their only action. These three and the rectus femoris form the *quadriceps group,* again, an important muscle in locomotion.

SARTORIUS flexes both hip and knee. It is the main muscle involved in crossing the knees or putting an ankle on the opposite knee.

Posterior thigh and pelvic muscles (Fig. 9.40)

These muscles are extensors of the hip and flexors of the knee.

GLUTEUS MAXIMUS (buttock muscle) extends the hip and rotates it laterally. It is strongly used in bicycling and stair climbing to straighten the thigh on the hip.

BICEPS FEMORIS is the most lateral posterior thigh muscle; it extends the hip and flexes the knee.

SEMIMEMBRANOSUS and SEMITENDINOSUS lie centrally and medially on the posterior thigh, and

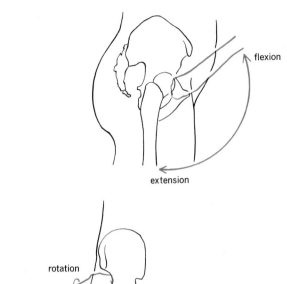

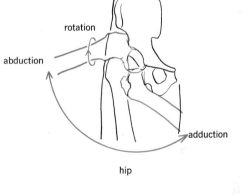

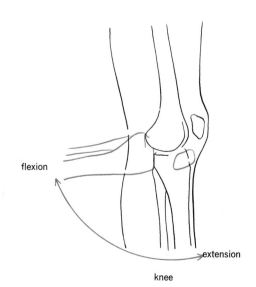

FIGURE 9.38. Movements of the hip and knee.

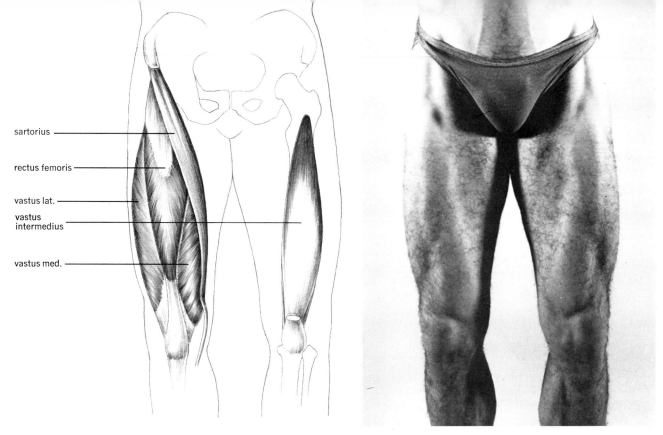

sartorius

rectus femoris

vastus lat.

vastus intermedius

vastus med.

FIGURE 9.39.   Anterior thigh and pelvic muscles.

FIGURE 9.40.   Posterior thigh and pelvic muscles.

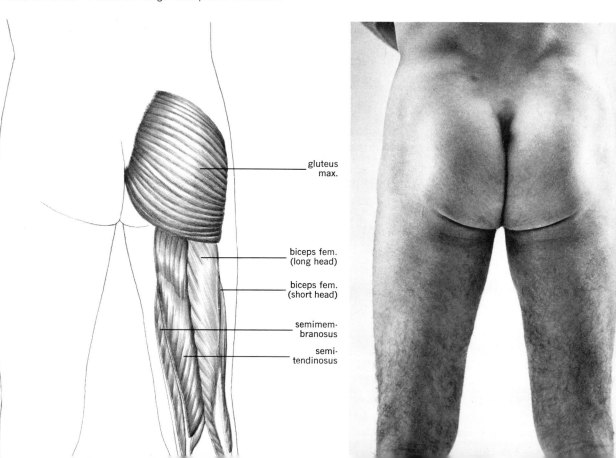

gluteus max.

biceps fem. (long head)

biceps fem. (short head)

semimem- branosus

semi- tendinosus

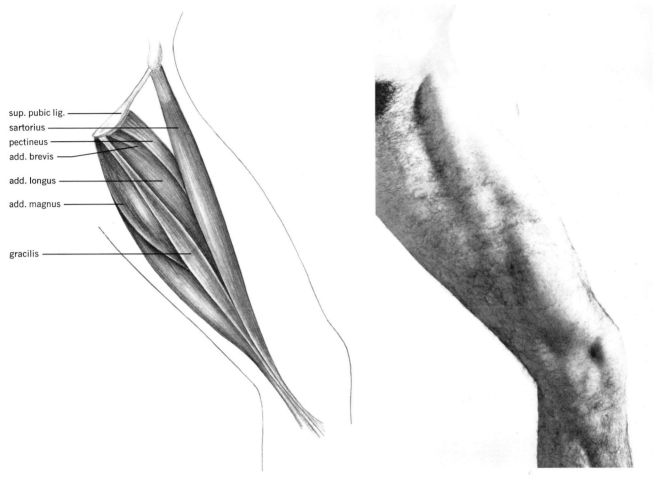

sup. pubic lig.
sartorius
pectineus
add. brevis
add. longus
add. magnus
gracilis

FIGURE 9.41.   Medial thigh muscles.

FIGURE 9.42.   Posterior iliac muscles.

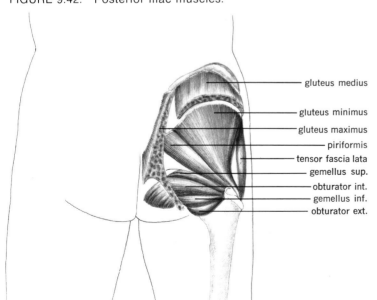

gluteus medius
gluteus minimus
gluteus maximus
piriformis
tensor fascia lata
gemellus sup.
obturator int.
gemellus inf.
obturator ext.

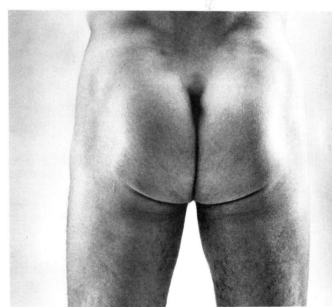

extend the hip and flex the knee. The last three muscles together constitute the "hamstrings."

Medial thigh muscles (Fig. 9.41)

These muscles are adductors of the hip or femur and several carry that action as part of their name.

ADDUCTOR MAGNUS, ADDUCTOR LONGUS, and ADDUCTOR BREVIS are the strongest adductors of the hip, and give the rounded contour to the inner side of the thigh.

PECTINEUS also adducts, and lies above the adductor longus.

GRACILIS is a straplike muscle on the medial side of the thigh, also acting to adduct the hip.

Posterior iliac muscles (Fig. 9.42)

These muscles abduct the thigh.

GLUTEUS MEDIUS and GLUTEUS MINIMUS are the middle and deep muscle layers beneath the gluteus maximus.

TENSOR FASCIA LATA actually lies more on the lateral side of the ilium, and also abducts the thigh.

Posterior sacral muscles

This group includes six muscles in four groups that laterally rotate the thigh. From the superior to inferior direction the muscles are:

Piriformis
Gemellus
       Superior
       Inferior
Obturator
       Internus
       Externus
Quadratus femoris

Muscles operating the ankle and foot

Movements possible (Fig. 9.43)

The ankle may be flexed (dorsiflexion) and extended (plantar flexion). Inversion and eversion occur within the foot. The toes may be flexed, extended, abducted, and adducted. The muscles are grouped into anterior, posterior, and lateral crural (L. *cruralis*, leg) groups.

Anterior crural muscles (Fig. 9.44)

These muscles are on the anterolateral aspect of the leg, and dorsiflex the ankle, invert the foot, and extend the toes.

TIBIALIS ANTERIOR dorsiflexes and inverts the foot. It is the site of a muscle cramp known as "shin splints."

EXTENSOR DIGITORUM LONGUS extends all four outer toes at the same time, because it consists of one muscle that gives rise to four tendons.

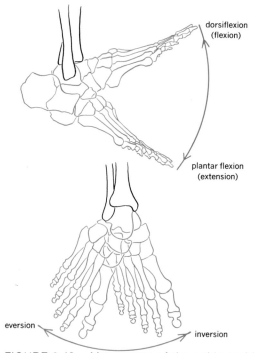

FIGURE 9.43.   Movements of the ankle and foot.

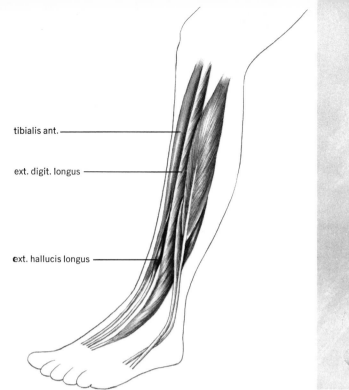

tibialis ant.

ext. digit. longus

ext. hallucis longus

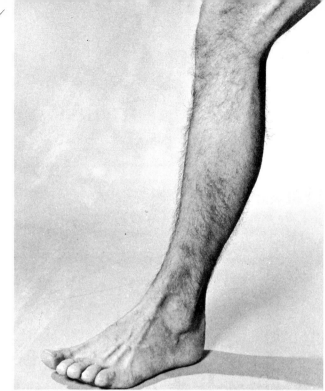

FIGURE 9.44.   Anterior crural muscles.

FIGURE 9.45.   Posterior crural muscles.

gastrocnemius
medial head

gastrocnemius
lateral head

plantaris

flexor digitorum
longus

soleus

tibialis post.

gastrocnemius

flexor
hallucis longus

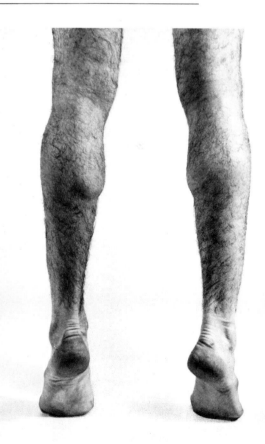

EXTENSOR HALLUCIS LONGUS extends the big toe.

PERONEUS TERTIUS dorsiflexes the foot.

Posterior crural muscles (Fig. 9.45)

These muscles are on the posterior aspect of the leg, and plantar flex the foot and flex the toes.

GASTROCNEMIUS (calf muscle) is an important muscle in walking and running, because it provides the "push" for locomotion. If its tendon, the tendon of Achilles at the back of the ankle, is severed, these activities cannot be performed. It plantar flexes the foot and enables us to "rise on the toes."

SOLEUS lies beneath gastrocnemius and has the same action.

PLANTARIS plantar flexes the foot.

TIBIALIS POSTERIOR plantar flexes the foot.

FLEXOR DIGITORUM LONGUS flexes the four outer toes at the same time.

FLEXOR HALLUCIS LONGUS flexes the big toe.

Lateral crural muscles (Fig. 9.46)

PERONEUS LONGUS AND BREVIS plantar flex and evert the foot.

Miscellaneous foot muscles

Within the foot, as in the hand, there are intrinsic muscles that hold the bones firm so as to provide a stable platform for locomotion. These are shown in Figure 9.47. The INTEROSSEI and LUMBRICALES are the most important groups.

No attempt is made to summarize the skeletal muscles. Refer, instead, to Table 9.2 for details of muscle origins, insertions, and innervations.

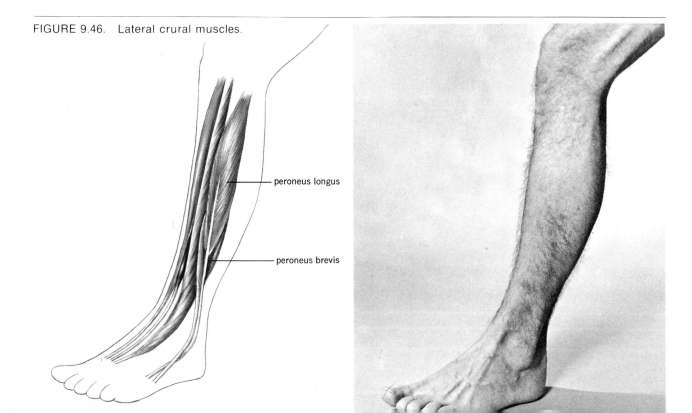

FIGURE 9.46.   Lateral crural muscles.

peroneus longus

peroneus brevis

FIGURE 9.47.   Muscles of the foot.

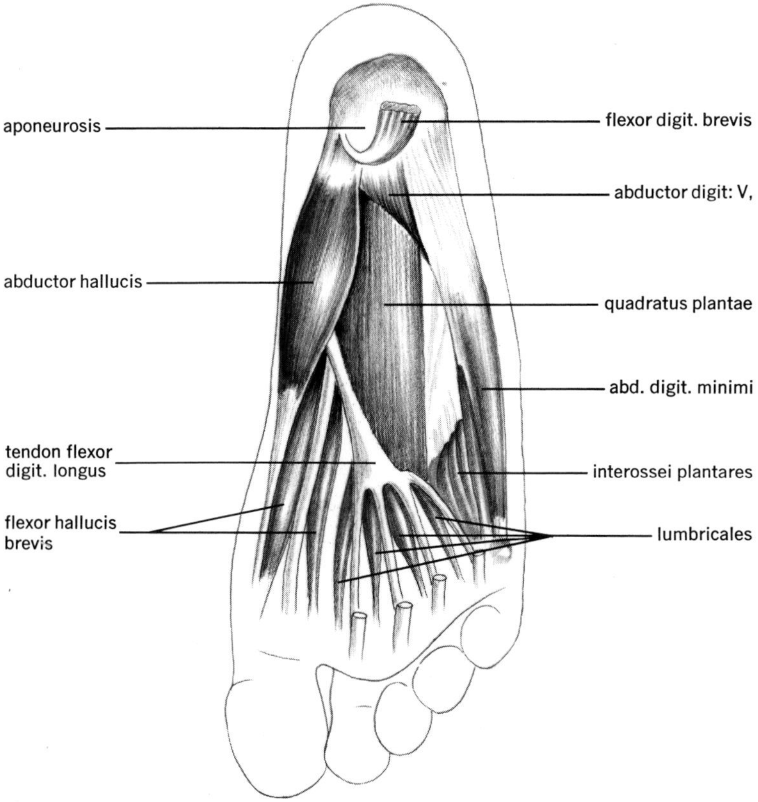

aponeurosis

flexor digit. brevis

abductor digit: V,

abductor hallucis

quadratus plantae

abd. digit. minimi

tendon flexor
digit. longus

interossei plantares

flexor hallucis
brevis

lumbricales

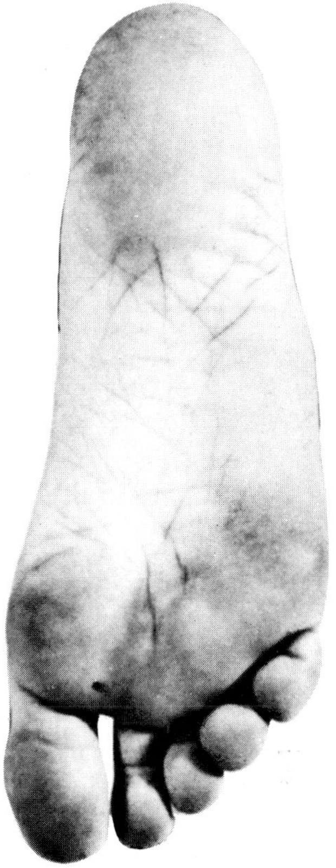

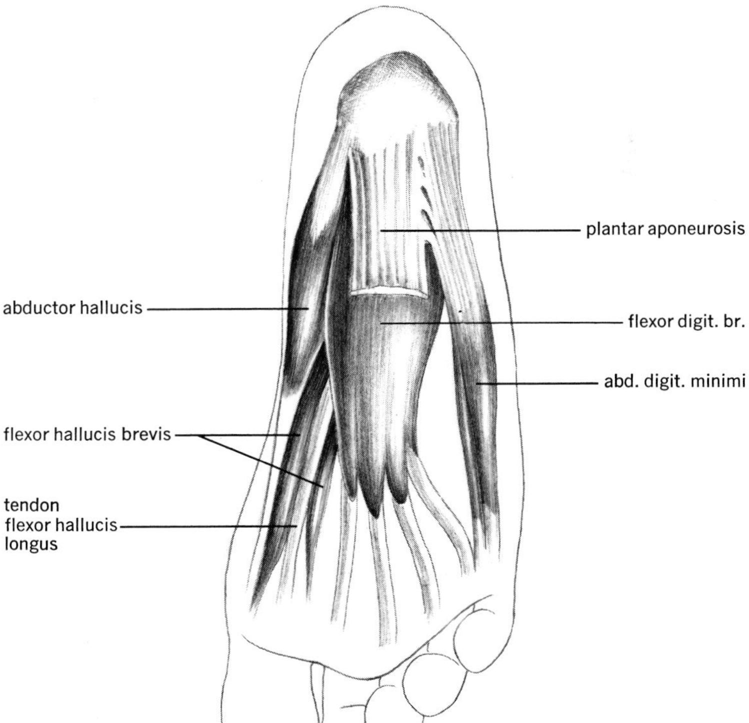

plantar aponeurosis

abductor hallucis

flexor digit. br.

abd. digit. minimi

flexor hallucis brevis

tendon
flexor hallucis
longus

195

TABLE 9.2
Muscles of the Body Arranged Alphabetically
(Includes Those Mentioned or Diagrammed in the Text)   (Italics Indicate Major Muscles)

Muscle	General location	Origin	Insertion	Action	Innervation
Abductor digiti minimi (foot)	Sole of foot	Calcaneus	First phalanx 5th toe	Abducts 5th toe	Greater sciatic
Abductor hallucis	Sole of foot	Calcaneus	First phalanx great toe	Abducts 1st toe	Greater sciatic
Abductor pollicis brevis	Base of thumb on palm	Scaphoid multangular	First phalanx thumb	Abducts thumb	Median nerve
Adductor brevis	Medial thigh	Inferior pubic ramus	Upper third linea aspera	Adducts thigh (some flexion and medial rotation)	Obturator nerve
Adductor hallucis	Sole of foot	Metatarsals	First phalanx great toe	Adduction great toe	Greater sciatic
Adductor longus	Medial thigh	Crest and symphysis of pubis	Middle third linea aspera	Adducts thigh (some flexion and medial rotation)	Obturator nerve
Adductor magnus	Medial thigh	Ramus of pubis ischium	Lower third linea aspera	Adducts, flexes, medially rotates thigh	Obturator and sciatic
Adductor pollicis	Between thumb and index finger palm of hand	Capitate, multangulars, metacarpals 2 to 4	First phalanx thumb	Adducts thumb	Ulnar nerve

Auricularis
  Anterior
  Posterior
  Superior
	
	Anterior to pinna
	Posterior to pinna
	Superior to pinna
		
		Fascia of scalp
		Fascia of scalp
		Fascia of scalp
			
			Pinna
			Pinna
			Pinna
				
				Moves pinna forward
				Moves pinna backward
				Elevates pinna
					7th cranial nerve

Biceps brachii	Anterior upper arm	Long head; above glenoid fossa Short head; coracoid process	Radial tuberosity	Flexes and supinates forearm	Musculo-cutaneous nerve
Biceps femoris	Posterior thigh	Long head; ischial tuberosity Short head; linea aspera	Lateral head fibula	Extends thigh, flexes knee	Sciatic nerve
Brachialis	Anterior upper arm	Lower half anterior humerus	Coronoid process (ulna)	Flexes forearm	Musculo-cutaneous nerve
Brachioradialis	Lateral forearm	Lateral upper ⅔ of humerus	Styloid process (radius)	Flexes forearm (semipronates semisupinates)	Radial nerve
Buccinator	Cheek of face	Alveolar processes of maxillae and mandible	Orbicularis oris fibers	Compresses cheek	7th cranial nerve

Muscle	General location	Origin	Insertion	Action	Innervation
Coraco-brachialis	Medial upper arm	Coracoid process	Midhumerus	Flexes and adducts humerus	Musculo-cutaneous nerve
Corrugator	Forehead	Inner end superciliary ridge	Skin above orbit	Produces frown	7th cranial nerve
Deltoid	Top of shoulder	Clavicle, acromion and spine of scapula	Deltoid tuberosity	Abducts, flexes extends humerus	Axillary nerve
Depressor labii inferioris	Inferior to lower lip	Mental process	Skin of lower lip	Depresses lower lip	7th cranial nerve
Diaphragm	Between thoracic and abdominal cavities	Xiphoid process, last 6 ribs, lumbar vertebrae	Central tendon	Increases vertical dimension of thorax	Phrenic nerve
Digastricus	Runs from mastoid process to hyoid	Posterior belly: mastoid notch of temporal	Greater horn, hyoid	Elevates hyoid, opens mouth	7th cranial nerve
		Anterior belly: mandible	Body of hyoid	Draws hyoid forward	5th cranial nerve
Epicranius *Frontalis*	Forehead	Muscles above orbit	Galea	Raises eyebrows, draws scalp forward	7th cranial nerve
Occipitalis	Base of skull	Superior nuchal line and mastoid temporal	Aponeurotica	Pulls scalp backwards	7th cranial nerve
Extensor carpi radialis brevis	Posterior forearm	Lateral epicondyle humerus	Top of 3rd metacarpal	Extends and abducts wrist	Radial nerve
Extensor carpi radialis longus	Posterior forearm	Lateral epicondyle humerus	Top of 2nd metacarpal	Extends and abducts wrist	Radial nerve
Extensor carpi ulnaris	Posterior forearm	Lateral epicondyle humerus	Base of 5th metacarpal	Extends and adducts wrist	Radial nerve
Extensor digiti minimi	Posterior forearm	Lateral epicondyle, humerus	Top of 1st phalanx, 4th finger	Extends 4th finger	Radial nerve
Extensor digitorium communis	Posterior forearm	Lateral epicondyle, humerus	2nd and 3rd phalanges of fingers	Extends fingers	Radial nerve
Extensor digitorum longus	Anterolateral leg	Lateral tibial condyle, distal fibula	2nd and 3rd phalanges, four outer toes	Extends four outer toes	Common peroneal nerve

(Table 9.2 continued)
Muscles of the Body Arranged Alphabetically
(Includes Those Mentioned or Diagrammed in the Text) (Italics Indicate Major Muscles)

Muscle	General location	Origin	Insertion	Action	Innervation
Extensor hallucis longus	Anterior fibula	Anterior surface of fibula	Top of great toe	Extends great toe	Common peroneal nerve
Extensor pollicis brevis	Posterior radius	Dorsal surface of radius	1st phalanx of thumb	Extends thumb	Radial nerve
Extensor pollicis longus	Posterior ulna	Dorsal surface of ulna	2nd phalanx of thumb	Extends thumb	Radial nerve
External intercostals	Between ribs— superficial	Inferior border of rib	Superior border of rib below	Elevates ribs	Intercostal nerves
External oblique	Abdomen— superficial	Anterior inferior surface, lower 8 ribs	Linea alba, pubis, iliac crest	Flexion, rotation of spine	Intercostal nerves
Flexor carpi radialis	Anterior forearm superficial	Medial epicondyle, humerus	Base 2nd metacarpal	Flexes and abducts wrist	Median nerve
Flexor carpi ulnaris	Anterior forearm superficial	Medial epicondyle of humerus, distal ulna	Base 5th metacarpal pisiform	Flexes and adducts wrist	Ulnar nerve
Flexor digitorum longus	Posterior tibia	Posterior surface tibial shaft	Distal phalanges 4 outer toes	Flexes toes	Tibial nerve
Flexor digitorum profundus	Anterior forearm deep	Proximal three- fourths ulna	Base of distal phalanges	Flexes fingers	Median and ulnar nerves
Flexor digitorum superficialis	Anterior forearm middle	Medial epicondyle, humerus; coronoid process, ulna; radial shaft	Second phalanges of fingers	Flexes fingers	Median nerve
Flexor hallucis longus	Posterior fibula	Lower two-thirds fibula	Undersurface of phalanges of great toe	Flexes great toe	Tibial nerve
Flexor pollicis brevis	Base of thumb on palm	Multangulars	Base of 1st phalanx of thumb	Flexes and adducts thumb	Median nerve
Flexor pollicis longus	Base of thumb on palm	Anterior surface, radius; coronoid process, ulna	Base of distal phalanx of thumb	Flexes and adducts thumb	Median nerve
Gastroc- nemius	Calf of leg, superficial	Posterior surface, condyles of femur	Calcaneus	Plantar flexion flexes knee	Tibial nerve

Muscle	General location	Origin	Insertion	Action	Innervation
Gemellus	Posterior pelvis	Inferior: Ischial tuberosity Superior: Ischial spine	Greater trochanter	Rotates thigh laterally	Obturator nerve
Gluteus maximus	Buttocks, superficial	Gluteal line of ilium	Gluteal tuberosity, femur	Extends and laterally rotates thigh	Sciatic
Gluteus medius	Buttock, inter-mediate	Outer surface of ilium	Greater trochanter	Abducts and medially rotates thigh	Sciatic
Gluteus minimus	Buttock, deep	Outer surface of ilium	Greater trochanter	Abducts and medially rotates thigh	Sciatic
Gracilis	Medial thigh	Symphysis pubis	Medial head tibia	Adducts thigh	Obturator nerve
Iliacus	Iliac fossa	Iliac fossa and crest, sacrum	Lesser trochanter	Flexes and medially rotates thigh	Femoral nerve
Iliocostalis	Midback	Iliac crest, ribs	Ribs and cervical trans-verse processes	Extends spine	Posterior branches of spinal nerves
Infraspinatus	Scapula below spine	Infraspinous fossa	Greater tuber-cle, humerus	Lateral rotator of humerus	Axillary nerve
Internal intercostals	Between ribs, deep	Inner surface of rib	Superior surface of rib below	Depresses ribs	Intercostal nerves
Internal oblique	Abdomen, middle layer	Iliac crest, fascia of back	Lower 3 ribs, linea alba, xiphoid process	Flexes spine, compresses abdomen	Intercostal nerves
Interossei Foot	Between metatarsals	Adjacent sides of metatarsals	Bases 1st phalanges	Abducts toes flexes "knuckles"	Common peroneal nerve
Hand	Between metacarpals	Adjacent sides of metacarpals	Bases 1st phalanges	Abducts fingers flexes "knuckles"	Ulnar nerve
Latissimus dorsi	Lower back, superficial	Spinous processes T6-L5, iliac crest last 3 ribs	Bicipital groove, humerus	Extension, adduction, medial rotation humerus	Radial nerve
Levator ani	Forms part of pelvic floor	Pubis, spine of ischium	Coccyx, central tendon of perineum	Supports and raises pelvic floor	Pudendal nerve
Levator labii superioris	Above upper lip	Lower rim of orbit	Skin of upper lip	Elevates upper lip	7th cranial nerve

(Table 9.2 continued)
Muscles of the Body Arranged Alphabetically
(Includes Those Mentioned or Diagrammed in the Text)   (Italics Indicate Major Muscles)

Muscle	General location	Origin	Insertion	Action	Innervation
Levator scapulae	Posterior neck	Transverse processes C1-4	Superior angle, scapula	Elevates scapula	Cervical nerves
Longissimus	Midback	Transverse processes C5-T5	Mastoid process, skull	Extends spine, bends it to side	Posterior branches of spinal nerves
Longus capitus	Lateral neck	Transverse processes C3-6	Occipital bone	Flexes neck	Cervical nerves
Longus colli	Anterior neck, deep	Transverse processes and bodies C3-T7	Atlas, cervical bodies	Flexes and rotates neck	Cervical nerves
Lumbricales Foot	Plantar surface metatarsals	From flexor digitorum longus	Tendons of extensor digitorum longus	Flexes toes	Common peroneal nerve
Hand	Anterior surface metacarpals	Tendons of flexor digitorum profundus	Tendons of extensor digitorum communis	Flexes knuckles	Median and ulnar nerves
Masseter	Side of mandibular ramus	Zygomatic process of maxilla, zygomatic arch	Ramus, angle and coronoid process of mandible	Elevates mandible	5th cranial nerve
Mentalis	Chin	Mental symphysis	Skin of chin and lower lip	Depresses lip, wrinkles chin	7th cranial nerves
Mylohyoid	Floor of mouth	Mylohyoid lines	Hyoid body	Elevates hyoid and tongue	5th cranial nerve
Obturator	Posterior pelvis	Externus: Rim of obturator foramen	Greater trochanter	Laterally rotates thigh	Obturator nerve
		Internus: Rim of obturator foramen, pubis, ischium			Obturator nerve
Omohyoid	Shoulder to hyoid	Scapula, superior border	Body of hyoid	Depresses hyoid	Cervical nerves
Orbicularis oculi	Around orbit	Medial surface of orbit	Skin of eyelid	Closes eye (wink, blink)	7th cranial nerve
Orbicularis oris	Around mouth	Skin of lips, fibers of other facial muscles	Corners of mouth	Closes and puckers lips	7th cranial nerve

Muscle	General location	Origin	Insertion	Action	Innervation
Palmaris longus	Anterior forearm	Medial epicondyle humerus	Palm of hand	Flexes wrist	Median nerve
Pectineus	Median thigh	Iliopectineal line, pubis	Base of lesser trochanter	Flexes and adducts thigh	Femoral nerve
Pectoralis major	Chest, superficial	Clavicle, sternum, costal cartilages true ribs	Bicipital groove	Flexes, adducts, medially rotates humerus	Pectoral nerve
Pectoralis minor	Chest, deep	Ribs 3 to 5	Coracoid process	Draws scapula forward and down	Pectoral nerve
Peroneus brevis	Lateral leg	Lower two-thirds fibula	Base 5th metatarsal	Plantar flexes and everts foot	Common peroneal
Peroneus longus	Lateral leg	Upper two-thirds fibula	Base 1st metatarsal, 1st cuneiform	Plantar flexes and everts foot	Common peroneal
Peroneus tertius	Anterior ankle	Anterior fibula	Dorsal surface 5th metatarsal	Dorsiflexion of foot	Common peroneal
Piriformis	Posterior pelvis	Anterior sacrum	Greater trochanter	Lateral rotation thigh	Sacral nerves
Plantaris	Posterior lower thigh	Linea aspera	Calcaneus	Plantar flexion	Tibial nerve
Platysma	Anterior neck	Fascia of deltoid and pectoralis major	Skin of lower face	Depresses lower lip	7th cranial nerve
Pronator teres	Anterior upper forearm	Medial epicondyle, humerus; and coronoid of ulna	Midradius	Pronation	Median nerve
Pronator quadratus	Anterior lower forearm	Body of ulna	Lower ¼ radius	Pronation	Median nerve
Psoas	Posterior wall of pelvic cavity	Transverse processes lumbar vertebrae	Lesser trochanter	Flexes thigh	Femoral nerve
Pterygoid Lateralis	Medial to ramus of mandible	Great wing of sphenoid, lateral pterygoid process	Condyloid process	Protrudes and opens jaw, moves jaw side to side	5th cranial nerve
Medialis	Medial to lateralis	Lateral pterygoid process, palatine, maxilla	Ramus and angle of mandible	Closes jaw	5th cranial nerve

(Table 9.2 continued)
Muscles of the Body Arranged Alphabetically
(Includes Those Mentioned or Diagrammed in the Text)   (Italics Indicate Major Muscles)

Muscle	General location	Origin	Insertion	Action	Innervation
Quadratus femoris	Posterior pelvis	Ischial tuberosity	Femur	Laterally rotates thigh	Sacral nerve
Quadratus lumborum	Between last rib and iliac crest	Iliac crest, lower lumbar vertebrae	Rib 12 and upper lumbar vertebrae	Flexes lumbar spine	Lumbar nerves
Quadratus plantae	Sole of foot in front of heel	Calcaneus	Tendons of flexor digitorum longus	Flexes 4 outer toes	Sciatic nerve
Rectus abdominus	Midabdomen	Pubic crest	Cartilages of ribs 5 to 7	Flexes spine	Intercostal nerves
Rectus capitus	Anterior neck, deep	Atlas	Base of occipital bone	Flexes neck	Cervical nerves
Rectus femoris	Anterior thigh, superficial	Anterior inferior iliac spine, acetabulum	Patella	Flexes hips, extends knee	Femoral nerve
Rhomboid	Back, deep	Spinous processes T2–5	Lower third vertebral border of scapula	Adducts and elevates scapula	Cervical nerve
Risorius	Lateral to mouth	Buccinator fascia (masseter)	Skin of corners of mouth	Pulls mouth laterally	7th cranial nerve
Sartorius	Anterior thigh	Anterior superior iliac spine	Medial head, tibia	Flexes hip and knee	Femoral nerve
Scalenes	Lateral neck	Transverse processes cervical vertebrae	1st and 2nd ribs	Elevate ribs and rotates and bends neck	Cervical nerves
Semimembranosus	Posterior thigh	Ischial tuberosity	Medial condyle of tibia	Extends hip, flexes knee	Sciatic nerve
Semispinalis	Midback	Transverse processes lower cervical and thoracic vertebrae	Spinous processes cervical vertebrae and occipital	Extends and rotates spine	Posterior branches of spinal nerves
Semitendinosus	Posterior thigh	Ischial tuberosity	Medial tibial shaft	Extends hip, flexes knee	Sciatic nerve
Serratus anterior	Lateral thorax	Upper 8 or 9 ribs	Vertebral border, scapula	Abducts scapula	Cervical nerves
Serratus posterior	Back, deep	Spinous processes thoracic and lumbar vertebrae	Ribs	Increases lateral dimensions of thorax	Intercostal nerves

Muscle	General location	Origin	Insertion	Action	Innervation
Soleus	Calf, deep	Head of fibula, upper tibia	Calcaneus	Plantar-flexion	Tibial nerve
Spinalis	Posterior neck	Same as semispinalis	Same as semispinalis	Extends spine	Posterior branches spinal nerves
Splenius	Posterior neck	Spinous processes lower cervical and upper thoracic vertebrae	Nuchal lines and cervical transverse processes	Extension of neck, rotation	Cervical nerves
Sternocleido-mastoid	Lateral neck	Sternum and clavicle	Mastoid process of temporal bone	Rotates head, flexes neck	11th cranial and cervical nerves
Sternohyoid	Anterior neck	Sternum	Body of hyoid	Depresses hyoid	Cervical nerves
Sterno-thyroid	Anterior neck	Sternum	Thyroid carti-lage of larynx	Depresses larynx	Cervical nerves
Stylohyoid	Runs from styloid process of skull to hyoid bone	Styloid process	Body of hyoid	Elevates and retracts hyoid	7th cranial nerve
Subclavius	Beneath clavicle	Middle of 1st rib	Clavicle	Depresses clavicle	Cervical nerves
Subscapularis	Scapula, anterior	Subscapular fossa	Lesser tubercle, humerus	Medially rotates humerus	Axillary nerve
Supinator	Anterior upper forearm	Lateral epicondyle humerus	Radial shaft	Supination	Radial nerve
Supra-spinatus	Scapula, above spine	Supraspinous fossa	Greater tubercle, humerus	Abducts humerus	Cervical nerve
Temporalis	Lateral skull	Temporal fossa	Coronoid and ramus, mandible	Closes jaws	5th cranial nerve
Tensor fascia lata	Lateral hip	Anterior superior iliac spine	Fascia lata	Abducts and flexes thigh	Sciatic nerve
Teres major	Inferior angle of scapula to humerus	Inferior angle of scapula	Lesser tubercle, humerus	Adduction, extension, medial rotation humerus	Cervical nerves
Teres minor	Above teres major	Axillary border, scapula	Greater tubercle, humerus	Lateral rotation of humerus	Axillary nerve

(Table 9.2 continued)
Muscles of the Body Arranged Alphabetically
(Includes Those Mentioned or Diagrammed in the Text)   (Italics Indicate Major Muscles)

Muscle	General location	Origin	Insertion	Action	Innervation
Thyrohyoid	Larynx to hyoid	Thyroid cartilage of larynx	Greater horn of hyoid	Elevates larynx	12th cranial and cervical nerves
Tibialis anterior	Anterolateral leg	Upper two-thirds tibia	First cunei-form and 1st metatarsal	Dorsiflex and invert foot	Common peroneal nerve
Tibialis posterior	Posterior leg	Shaft of tibia and fibula	Navicular, calcaneus, all cuneiforms	Plantar flexion and inversion of foot	Tibial nerve
Transverse abdominal	Abdomen, deep	Iliac crest, lumbar fascia, last 6 ribs	Xiphoid process linea alba, pubis	Compresses abdomen	Intercostal nerves
Trapezius	Upper back, superficial	Superior nuchal line, ligamentum nuchae, spines C7-T12	Clavicle, spine and acromion of scapula	Elevates, adducts depresses scapula	11th cranial and cervical nerves
Triangularis	Below corners of mouth	Body of mandible	Skin of lower lip	Depresses lower lip	7th cranial nerve
Triceps	Posterior humerus	Lateral head, humerus Long head, scapula Medial head, humerus	Olecranon process	Extends elbow	Radial nerve
Vastus intermedius	Anterior thigh, deep	Anterior and lateral femur shaft	Patella	Extends knee	Femoral nerve
Vastus lateralis	Lateral thigh	Upper half, lateral linea aspera	Patella	Extends knee	Femoral nerve
Vastus medialis	Medial thigh	Upper half, medial linea aspera	Patella	Extends knee	Femoral nerve
Zygomaticus	Above corners of mouth	Zygomatic bone	Skin of corners of mouth	Draws mouth up and back	7th cranial nerve

Readings

Gray, Henry. *Anatomy of the Human Body.* Edited by C. M. Goss. 29th ed. Lea and Febiger. Philadelphia, 1973.

Morris' Human Anatomy. Edited by B. J. Anson. 12th ed. McGraw-Hill. New York, 1966.

chapter 10
The Basic Organization and Properties of the Nervous System

chapter 10

The nervous system provides a way of DETECTING CHANGES in the internal or external environments of the body, and for PROCESSING the nerve impulses resulting from those changes. It CONTROLS RESPONSE to stimuli so that homeostasis is maintained. Control is rather specific in that particular organs rather than body processes are caused to alter their activity.

Development of the nervous system (Fig. 10.1)

The nervous system is the first system to develop from the ectoderm. It forms as two folds in the embryonic disc, which fuse to form the neural tube. The anterior end of the tube forms three enlargements, which become subdivided as development proceeds (Table 10.1). The cells in the walls of the tube form the cellular layers of the brain and cord. Peripheral nerves arise as cells in the brain and cord send processes to outlying areas of the body.

Organization of the nervous system

The nervous system has several divisions or parts that are distinguished on the basis of location and function. A useful system of classification is presented below.

The CENTRAL NERVOUS SYSTEM (CNS) lies within the skull and vertebral column, and consists of two organs:

Brain.
Spinal cord.

The PERIPHERAL NERVOUS SYSTEM lies outside of

TABLE 10.1 Derivatives of the Primary Neural Tube Enlargements		
Primary enlargement	Subdivision(s)	Structure(s) arising from subdivision
Forebrain (Prosencephalon)	Telencephalon	Cerebrum Basal ganglia
	Diencephalon	Thalamus Hypothalamus
Midbrain (Mesencephalon)	None	Midbrain
Hindbrain (Rhombencephalon)	Metencephalon	Pons Cerebellum
	Myelencephalon	Medulla

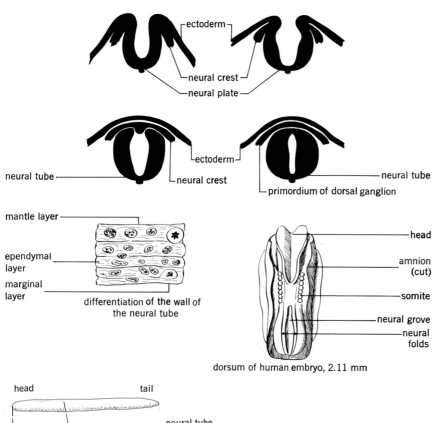

differentiation of the wall of
the neural tube

dorsum of human embryo, 2.11 mm

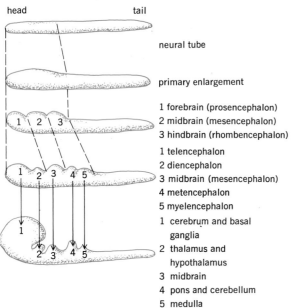

neural tube

primary enlargement

1 forebrain (prosencephalon)
2 midbrain (mesencephalon)
3 hindbrain (rhombencephalon)

1 telencephalon
2 diencephalon
3 midbrain (mesencephalon)
4 metencephalon
5 myelencephalon

1 cerebrum and basal
  ganglia
2 thalamus and
  hypothalamus
3 midbrain
4 pons and cerebellum
5 medulla

FIGURE 10.1.   Early development of the nervous system.

the skull and vertebral column, and includes the cranial and spinal nerves. Its divisions include:

The *somatic system* which supplies skin and skeletal muscles.

The *autonomic system* which supplies smooth muscle, cardiac muscle, and glands of the body. Most organs receive nerves from the two divisions of the autonomic system to control their activity. These divisions are:

The *parasympathetic division* which supplies impulses tending to maintain normal function and conserve body resources.

The *sympathetic division* which supplies impulses tending to cause acceleration of processes that resist stress.

Two other terms may be learned along with the subdivisions of the system. *Sensory* implies carrying impulses from sensory receptors toward the brain or cord; *motor* implies carrying impulses away from the brain or cord to effectors (muscle, glands) that respond by contraction or secretion.

Cells of the nervous system

The cells that form the nervous system are of two types. NEURONS are the excitable and conductile units; GLIA are connective, supportive, and nutritive elements.

Structure of neurons

Neurons are the structural and functional units of the nervous system. They may be quite varied in shape and size (Fig. 10.2), but possess certain features in common, as is illustrated by a motor neuron (Fig. 10.3). A CELL BODY contains a *nucleus,* with one or more prominent *nucleoli,* and finely granular *chromatin.* A nearly invisible *nuclear membrane* separates the nuclear substance or *karyoplasm* from the cell cytoplasm. The *cytoplasm* contains mitochondria, and a Golgi apparatus, but appears to lack centrioles after about 16 years of age. This suggests that mature neurons cannot divide and replace themselves in case of injury or loss. The cytoplasm also contains two structures found only in nerve cells.

FIGURE 10.2. Some different forms of neurons from the human nervous system.

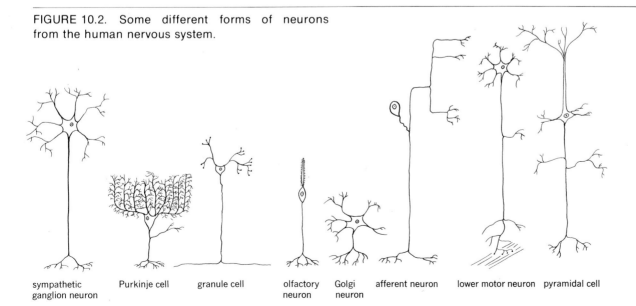

sympathetic ganglion neuron   Purkinje cell   granule cell   olfactory neuron   Golgi neuron   afferent neuron   lower motor neuron   pyramidal cell

Neurofibrils are tiny hollow tubes, running through the entire cytoplasm, which may serve supporting and nutritive functions.

Nissl bodies appear as granular structures and contain RNA. They have ribosomes associated with them, and apparently synthesize the proteins that neurons use for structural and physiological purposes (memory traces for example).

The cell body gives rise to one or more extensions or PROCESSES, which connect the cell body to other neurons, the central nervous system, or an organ. The processes are structurally and functionally of two types.

Dendrites are usually multiple, highly branched, relatively short, and irregular processes, which conduct nerve impulses toward (afferent) the cell body.

Axons are long, usually single, sparsely branched and regular processes, which conduct nerve impulses away from (efferent) the cell body.

Although both types or processes may be called nerve fibers, the term is most commonly applied to axons. Whole neurons may be afferent (as in the sensory portions of the system) and conduct toward the brain or cord, or may be efferent (as in the motor portions of the system) and conduct away from the brain or cord. Internuncial or association neurons may be found between afferent and efferent neurons in the central nervous system.

Both types of processes may develop sheaths,

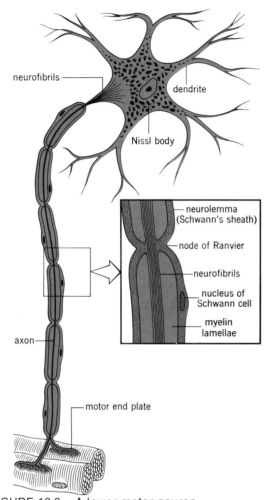

FIGURE 10.3.   A lower motor neuron.

FIGURE 10.4.   The formation of myelin lamellae around an axon. The Schwann cell is believed to "wrap around" the axon.

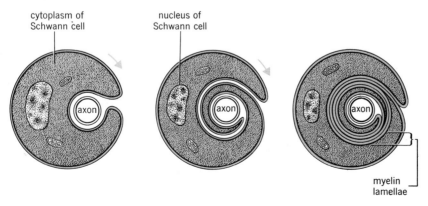

or may remain "naked." A MYELIN SHEATH is a segmented fatty covering formed by glial cells. In peripheral nerves, the glial cells called Schwann cells produce myelin according to the "jelly-roll hypothesis" of myelin formation (Fig. 10.4). Centrally, certain glial cells called oligodendroglia produce myelin. The sheath increases the speed of conduction of impulses along a fiber. A process having a myelin sheath is said to be medullated or myelinated; one lacking it is nonmedullated or nonmyelinated. A NEURILEMMA (Schwann sheath) may also be present. This is a thin membrane external to the myelin, and consists of the cytoplasm and membranes of the Schwann cells. It aids in regeneration and myelination of injured fibers. A generalized scheme of what sheaths are found and where is presented in Figure 10.5.

The endings of an axon or dendrite usually lack sheaths, and are highly branched to form the *telodendria* (terminal arborizations) of the fiber. These form junctions with other neurons, organs, or receptors.

Nerve fibers are microscopic in size. They are bound together to form larger units that may become visible to the naked eye. Such large units

are the NERVES. The organization of a nerve, and the names of its components are shown in Figure 10.6.

Structure of glia (Fig. 10.7)

The glial elements are of several types.

Ependyma line the cavities of the central nervous system. They are wedge-shaped, ciliated cells, and appear to act only as an epithelium for the cavities of the central nervous system.

Astrocytes form supporting networks in the central

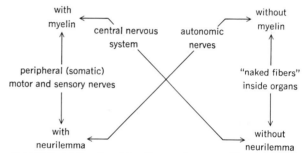

FIGURE 10.5 Combinations of nerve fiber sheaths in the nervous system.

FIGURE 10.6. Diagram showing the connective tissue components of a large peripheral nerve (cross section).

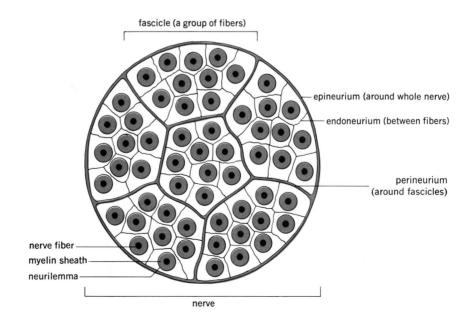

FIGURE 10.7. Some types of glia in the central nervous system. *(a)* Ependyma and neuroglia in the region of the central canal of a child's spinal cord: A, ependymal cells; B and D, fibrous astrocytes; C, protoplasmic astrocytes. Golgi method. *(b)* Interstitial cells of the central nervous system: A, protoplasmic neuroglia; B, fibrous neuroglia; C, microglia; D, oligodendroglia.

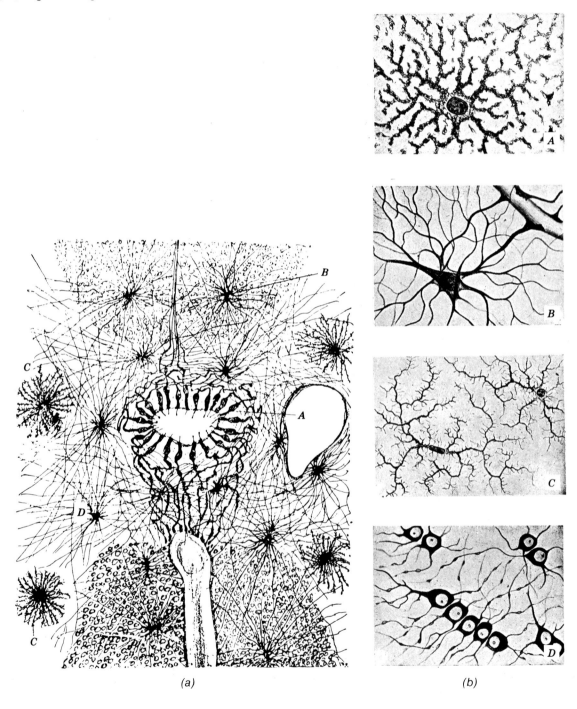

(a)   *(b)*

FIGURE 10.8. The active transport mechanism for sodium and potassium in the membrane of a nerve fiber.

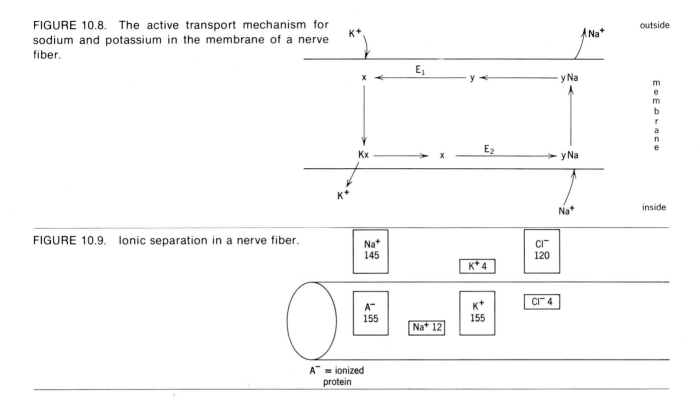

FIGURE 10.9. Ionic separation in a nerve fiber.

A^- = ionized protein

nervous system, and commonly attach to blood vessels. They are believed to transfer nutrients from blood vessels to neurons.

Oligodendrocytes form myelin in the central nervous system.

Microglia are small glial cells in the central nervous system that may become phagocytic under certain conditions. They are stimulated to show their phagocytic properties in the case of injury to the central nervous system.

Schwann cells have already been described.

Properties of neurons; excitability and conductivity

The two basic properties of neurons are excitability, which refers to the ability to respond to stimuli, and conductivity, which refers to the ability to transmit, as a nerve impulse, the disturbance resulting from the reaction to a stimulus.

THE BASIS OF EXCITABILITY. All cells are excitable, and neuron excitability is merely greater than that of other cells. Excitability results from separation of ions (primarily Na^+ and K^+) on two sides of a cell membrane by active processes (Fig. 10.8). As a result, sodium ion (Na^+) tends to accumulate outside the fiber, and potassium ion (K^+) tends to accumulate inside (Fig. 10.9). At rest, the membrane is relatively impermeable to Na^+, and is more permeable to K^+. As a result, there is a slow leakage of K^+ out of the fiber. An excess of positive charges (Na^+ and K^+) thus accumulates on the outside of the fiber, and a potential or electrical charge difference develops across the membrane. In this state the fiber is said to be POLARIZED with the outside of the fiber electrically positive to the inside.

THE BASIS OF NERVE IMPULSE FORMATION. When a stimulus is applied to a fiber, the membrane at that point becomes about 500 times more

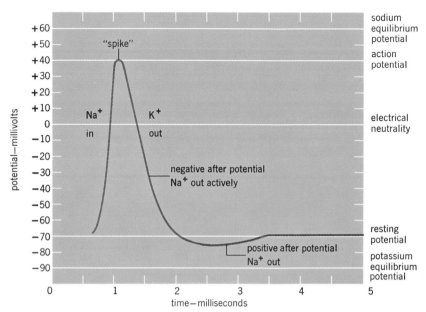

FIGURE 10.10. Changes in membrane potential during development of an action potential.

FIGURE 10.11. Current flow during formation of a nerve impulse.

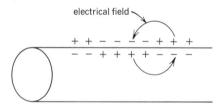

FIGURE 10.12. Saltatory conduction.

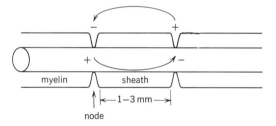

permeable to Na⁺ than it previously was without any great change in permeability to K⁺. Na⁺ diffuses into the fiber according to its concentration difference, while K⁺ flows outward. More Na⁺ enters the fiber than K⁺ leaves, and the potential difference is reversed by the accumulation of positive Na⁺ charges inside the fiber. The fiber is DEPOLARIZED, and a recordable action potential develops (Fig. 10.10) as ions move. Next, flow of electrical current occurs between polarized and depolarized areas in what is called a "battery effect." This current flow occurs through the extracellular fluid, and an electrical field, which is the nerve impulse, is created (Fig. 10.11).

THE BASIS OF CONDUCTIVITY.   The electrical field depolarizes a short segment of the fiber next to the impulse, current flow again occurs, and the disturbance is advanced down the fiber. Behind the advancing impulse, return to the original state, or REPOLARIZATION, is occurring. Repolarization involves active transport removing the Na⁺ that entered the fiber, and recapturing the K⁺ that left the fiber. In myelinated fibers, the only place where current flow and depolarization may occur is at the segments or nodes that are placed several millimeters apart. Thus the impulse "jumps" from segment to segment (*saltatory conduction*) and covers the length of the fiber more rapidly (Fig. 10.12). Speed of conduction is also determined by fiber size, as is shown in Table 10.2.

TABLE 10.2	Classification of Fibers and Conduction Velocity				
Type of fiber	Size (μ)	Velocity of conduction (m/sec)	Myelinated?	Where found	Function
A fiber	1–20	5–120	Yes	Peripheral sensory and motor nerves	Conduct to and from cord and brain
B fiber	1–3	4–14	Usually	Autonomic nervous system	Conduct to and from cord and brain
C fiber	1	0.5	No	Skin, within organs	Ends of fibers to and from cord and brain

The steps in creation of the excitable state, and formation and conduction of an impulse are summarized in Figure 10.13.

It may be noted that a neuron is usually stimulated only at one end, and the impulse normally travels in only one direction. However, if a nerve fiber is stimulated somewhere along its length, conduction may occur in both directions from that point.

Other properties of neurons

In addition to being excitable, nerve cells show numerous additional properties that make them efficient transmitters of nerve impulses.

Neurons follow the all-or-none law. For a given strength of a single stimulus, the fiber either depolarizes or does not. However, repeated weak stimuli, if they occur fast enough, may add or summate their effects, and eventually depolarize the fiber.

Neurons have a very short refractory period. The refractory period is the time when the fiber cannot react to another stimulus, that is, the time when it is depolarized. Nerve fibers repolarize in 3/1000 second (3 millisecond, or 3 msec) or less, and thus can carry many impulses, one after another.

Neurons exhibit a threshold (rheobase) for stimulation. This property indicates that a stimulus must have a certain strength to cause depolarization; that

FIGURE 10.13. The events in formation and conduction of a nerve impulse.

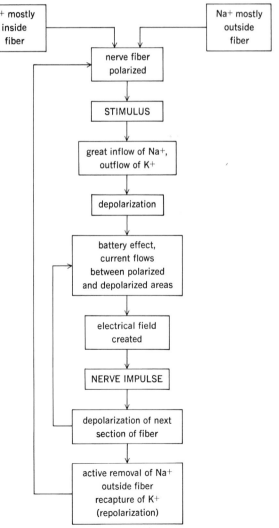

strength is the threshold. Thresholds are different in different fibers, and may be different in the same fiber according to the environmental conditions. This property aids the fibers to be selective of the stimuli which continually bombard them.

Neurons exhibit a chronaxie. A stimulus must act for a certain length of time to be effective; this time is the chronaxie. In general, the stronger the stimulus, the shorter the chronaxie. Strong stimuli may constitute a threat to homeostasis, and are assured of transmission to the central nervous system, even if they only last a short time.

Neurons show accommodation. Threshold rises as a given strength stimulus is applied continuously to a fiber. This enables the system to "ignore" a given stimulus if its intensity does not change, once the body has been notified of its arrival.

Neurons conduct in decrementless fashion. Since the impulse depends on a physicochemical change occurring in the fiber, the strength of the impulse does not change (decrementless) as it travels along a normal fiber. A wire will cause a drop in voltage as current travels along it; a nerve does not.

Neurons require certain substances to sustain their activity. The brain apparently uses only glucose or its metabolites (lactic acid, pyruvic acid) as an energy source. Peripheral nerves utilize glucose and fats. Amino acids are required to synthesize neuron proteins. Oxygen requirements of nervous tissue are 30 times that of an equivalent weight of muscle. Thus, while metabolic activity of nervous tissue does not make itself very visible, it accounts for a large fraction of the energy expended in the resting body.

Synapses

Structure of synapses

Synapses are junctions between neurons, and are areas of functional, but not anatomical, continuity. An impulse crosses a synapse chemically, using what is called a transmitter substance. In the ends of an axon's telodendria are small *vesicles* or granules of chemical substance (Fig.

FIGURE 10.14.   The anatomy of the synapse, as visualized under the electron microscope. M, mitochondria. Notice absence of vesicles in dendrite.

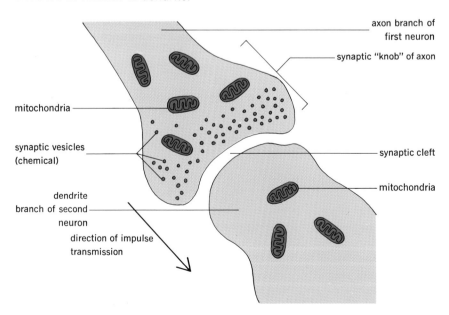

TABLE 10.3 Synaptic Transmitters			
Substance or agent	Enzyme or process destroying the agent	Where found, highest concentration	Comments
Acetylcholine	Cholinesterase	Neuromuscular junctions; peripheral synapses between neurons; brain stem, thalamus, cortex, retina	In CNS, not distributed evenly. Perhaps serves specific functions in each area
Norepinephrine	Monamine oxidase (MAO) or catechol-o-methyl trans-ferase (COMT)	Postganglionic sympathetic neurons; midbrain, pons, medulla	Low concentra-tion in cerebrum
Serotonin	MAO	Hypothalamus, basal ganglia, spinal cord	May be involved in expression of behavior
Histamine	Imidazole-N-methyl transferase, histaminase	Hypothalamus	In mast cells (tissue basophils)
Amino acids, for example GABA (gamma amino-butyric acid)	Oxidation or conversion	Cortex, visual area	GABA is inhibitory to most synapses

FIGURE 10.15.   The events occurring in synaptic activity.

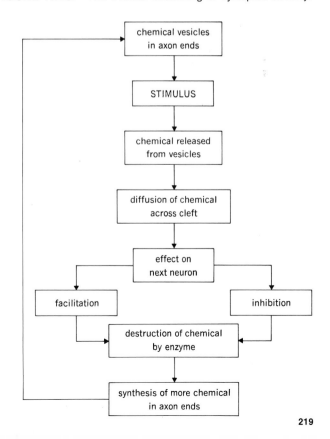

10.14). When an impulse arrives at the fiber end-ing, the chemical is released, it diffuses across the *synaptic cleft* between the two neurons and either depolarizes the next neuron, or inhibits its depolarization. Synapses may thus be excita-tory or inhibitory. In order that the chemical does not continue to exert its effect in the synapse, an *enzyme* is usually present that destroys the chemical, or it is destroyed by chemical means such as oxidation. Some synaptic transmitters and their corresponding methods of destruction are presented in Table 10.3. The steps involved in synaptic transmission are shown in Figure 10.15.

Physiological properties of synapses

Synapses, because of a chemical means of trans-mission, show properties different from the neurons that compose them. Several of these properties enable the synapse to exert a great degree of control over the impulse.

One-way conduction. Only axons contain and can release the chemical transmitters. Thus only axon-dendrite conduction can occur; this prevents "short circuits" in the system by impulses "going the wrong way."

Facilitation. If the chemical lowers the potential of the second neuron (hypopolarization), another impulse will cross more easily.

Inhibition. If the chemical increases the potential of the second neuron (hyperpolarization), another impulse may be blocked. Unimportant information may thus be "filtered out" of the neuronal circuits.

Synaptic delay. A longer time is required for the chemical to be released, to diffuse and to depolarize the second neuron, than to depolarize a nerve fiber. This time is the synaptic delay and it amounts to 0.5–1.0 milliseconds.

Fatigue. If chemical is used up faster than it can be produced, transmission may slow and eventually stop. This constitutes fatigue of the synapse.

Drugs. Many drugs increase the ease of passage of an impulse across a synapse (e.g., caffeine). Others inhibit passage (e.g., aspirin, opium, morphine). The latter group of chemicals are known to relieve pain, probably by influencing transmission.

Summation. Subthreshold stimuli may cause release of a little chemical, but not enough to affect the next neuron. Repeated stimuli, each producing a little chemical, may eventually create enough pooled chemical (summation) to cause an effect on the second neuron. Thus "insistent stimuli," even if weak, may come to the attention of the central nervous system.

Synapses are SENSITIVE TO ENVIRONMENTAL CHANGES. *Alkalosis* (decreased H^+ concentration in body fluids) increases excitability; *acidosis* (increased H^+ concentration in body fluids) depresses excitability. *Hypoxemia* (low blood oxygen levels) causes cessation of synaptic activity. Changes in the environment of the nerve cells during disease processes may result in greatly altered nervous system function.

Reflex arcs (Fig. 10.16)

The reflex arc is the simplest functional unit of the nervous system capable of detecting change and causing a response to that change. Many body activities are controlled reflexly. This term implies an *automatic* adjustment to maintain homeostasis without conscious effort. A reflex arc always has five basic parts to it.

A *receptor* to detect change. Body receptors are sensitive to heat, light, pressure, tension, sound, chemicals, and painful stimuli, so that a wide variety of changes can come to the notice of the CNS.

An *afferent neuron* to conduct the impulse resulting from stimulation of the receptor, to the central nervous system.

A *center or synapse* in the CNS where a junction is made between neurons. An internuncial neuron is commonly found at this point, and it allows the impulse to take a variety of pathways in the CNS. It creates a "side branch" in a neuron circuit for more widespread response.

An *efferent neuron* to conduct impulses to an organ for appropriate response.

An *effector*, or organ that responds and does something to maintain homeostasis.

Activity that results when impulses pass over a reflex arc is called a *reflex act* or, more simply, "the reflex." All reflex acts have several characteristics in common.

Activity is *involuntary* and cannot be started or stopped by an act of will.

Activity is *stereotyped*, that is, stimulation of a receptor in the same way always causes the same response. Reflex activity is thus *predictable*, a property that is useful in determining if reflex pathways are operating normally.

Response is generally *purposeful* in that it serves a valuable function.

Normally the effector will not respond unless the receptor is stimulated. The arc serving the reflex must be whole or complete. If one of the five parts of the arc is not functioning no response will be obtained. This is valuable in determining if part of the nervous system has suffered damage.

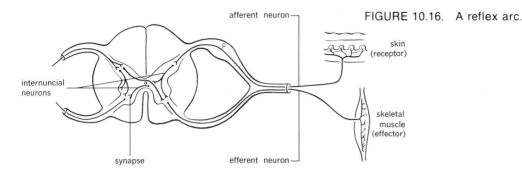

FIGURE 10.16.   A reflex arc.

afferent neuron

skin
(receptor)

internuncial
neurons

skeletal
muscle
(effector)

synapse   efferent neuron

Clinical considerations

Degeneration and regeneration in the nervous system

Injury to neuron cell bodies is usually followed by death of the entire neuron. Damage to an axon results in changes in the cell body (the axon reaction) and in the axon distal to the point of injury (Wallerian degeneration).

In the AXON REACTION, the cell body swells, and the Nissl substance fragments into fine granules. In WALLERIAN DEGENERATION, the distal segment of the axon swells, the myelin sheath degenerates, the axon itself degenerates, and only the outer layer of the neurilemma or glial cells remains. The products of degeneration are removed by glial cells.

In peripheral axons, ameboid, bulbous ends develop on the injured axon, and these grow through the neurilemmal tubes at about 2 millimeters per day. Reestablishment of connections,

if it occurs, is more widespread than originally, and control is less precise than before.

In the central nervous sytem, regeneration is more limited than peripherally. Since central axons lack neurilemmal sheaths, reestablishment of normal connections is haphazard and is complicated by the fact that millions of axons are present in the cord and brain. Thus chances of making proper connections are low.

MULTIPLE SCLEROSIS is a disease of unknown origin, and has no cure. It is characterized by degeneration of myelin sheaths in the brain and spinal cord. Changes in function that commonly occur include: lack of ability to concentrate, speech disturbances, visual changes, and increased vigor of knee and ankle jerk reflexes.

Before the availability of vaccines for POLIOMYELITIS, the viruses often attacked and destroyed the cell bodies of the somatic motor neurons to skeletal muscle. Paralysis followed.

Summary

1. The nervous system provides a means of controlling activities in the body.

2. The system is ectodermal in origin.

   a. The brain and spinal cord develop from the neural tube.

   b. Peripheral nerves are outgrowths of the tube.

   c. The forebrain forms the cerebrum, thalamus, and hypothalamus; the midbrain remains as such; the hindbrain forms pons, cerebellum, and medulla.

3. The system is organized into central and peripheral portions.

    a. The peripheral is subdivided twice more, into somatic and autonomic; the autonomic into parasympathetic and sympathetic.

4. Neurons and glia form the nervous system.

    a. Neurons are excitable and conductile and have a structure similar to other cells.

    b. The characteristic thing about many neurons is their length.

    c. Glia are connective, supportive, and nutritive cells and some also are phagocytic and form myelin.

5. Excitability depends on separation of ions, chiefly Na^+, and K^+.

    a. A fiber is polarized when ions are separated.

    b. A stimulus depolarizes the fiber, allowing Na^+ to enter the fiber, and K^+ to escape.

    c. Repolarization occurs as the ions are returned to their original positions by active transport.

6. Conductivity depends on current flow between polarized and depolarized areas, and movement of that current along the fiber.

    a. Speed of conduction depends on whether the fiber has a myelin sheath (faster) or is larger in diameter (faster).

7. Neurons show other physiological properties.

    a. They depolarize completely or do not at all when stimulated (all-or-none law).

    b. They repolarize rapidly (short refractory period).

    c. They have stimulus thresholds.

    d. They must be stimulated for a certain time before they depolarize.

    e. They require nutrients to live.

8. A synapse is a junction between two neurons.

    a. No anatomical connections exist.

    b. Conduction is chemical.

    c. Synapses control passage across themselves.

9. Synaptic properties differ from those of neurons.

    a. Conduction is one way.

    b. Impulses may pass more easily or be blocked.

    c. It takes longer to cross a synapse.

    d. Drugs affect synapses easily.

10. Neurons form reflex arcs as the simplest multineuron controlling device.

    a. Reflex arcs always have five parts: receptor, afferent nerve, synapse (center), efferent nerve, and effector.

    b. Reflexes operate automatically to control body activities.

    c. Reflex acts are involuntary, stereotyped, predictable, and purposeful.

11. Peripheral nerve fibers may regenerate; central ones show little regeneration. Neuron cell body damage usually causes death of the entire neuron. Demyelinating diseases result in loss of function.

Questions

1. Compare and contrast neuronal and synaptic properties. Account physiologically for as many of the differences as you can.

2. Describe the structure of a neuron.

3. How do synaptic and neuromuscular structure and transmission compare?

4. What is a reflex arc? What importance does it have in the body?

5. Give some characteristics of reflexes, and explain what each means.

6. Can damaged nerve fibers regenerate? Where? What are the chances of normal function being restored?

Readings

Barr, Murray L. *The Human Nervous System, An Anatomical Viewpoint.* Harper & Row. New York, 1972.

Bunge, R.P. "Glial cells and the central myelin sheath." *Physiol. Rev. 48:*197, 1968.

Caldwell, P. C. "Factors governing movement and distribution of inorganic ions in nerve and muscle." *Physiol. Rev. 48:*1, 1968.

DeRobertis, E. "Ultrastructure and cytochemistry of the synaptic region." *Science 156:*907, 1967.

Eccles, John C. *Physiology of Nerve Cells.* Johns Hopkins Press. Baltimore, 1968.

Eccles, John C. "The Synapse." *Sci. Amer. 212:*56 (Jan) 1965.

Science News. "Acupuncture: A Chinese Puzzle." *105,* No. 12. March 23, 1974.

Sidman, R. L., and M. Sidman. *Neuroanatomy: A Programmed Text.* Little, Brown. Boston, 1972.

Wilson, V. J. "Inhibition in the central nervous system." *Sci. Amer. 214:*102 (May) 1966.

chapter 11
The Spinal Cord and Spinal Nerves

chapter 11

The spinal cord serves as the first area of the nervous system where impulses from the periphery are organized and routed, and reflex responses are initiated. The cord also functions as a means of transmitting sensory impulses to the brain for further processing, and as a means for transmitting instructions from the brain to muscles and glands. The spinal nerves serve as the network for carrying messages to and from most peripheral organs.

The spinal cord

Location and structure of the cord (Fig. 11.1)

The spinal cord is located within the VERTEBRAL CANAL formed by the arches of the vertebrae. It is surrounded by three membranes, collectively known as the MENINGES. An outer *dura mater* is a tough, protective membrane of collagenous

FIGURE 11.1.   Three views of the gross anatomy of the spinal cord. *(A)* Lateral view. *(B)* Anterior view. *(C)* Posterior view.

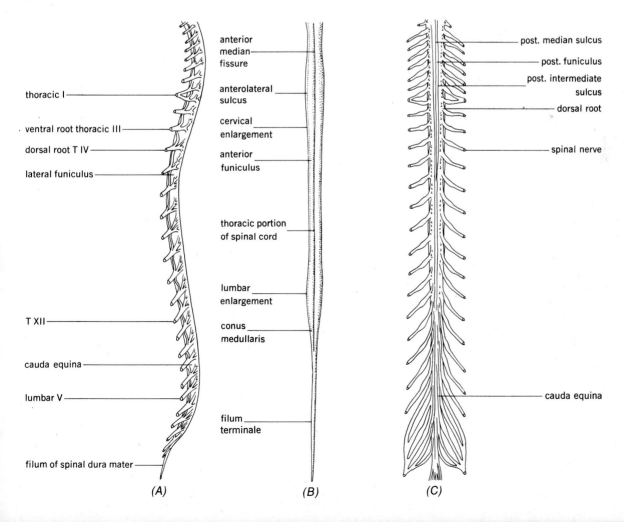

Labels (A):
thoracic I
ventral root thoracic III
dorsal root T IV
lateral funiculus
T XII
cauda equina
lumbar V
filum of spinal dura mater

Labels (B):
anterior median fissure
anterolateral sulcus
cervical enlargement
anterior funiculus
thoracic portion of spinal cord
lumbar enlargement
conus medullaris
filum terminale

Labels (C):
post. median sulcus
post. funiculus
post. intermediate sulcus
dorsal root
spinal nerve
cauda equina

(A)   *(B)*   *(C)*

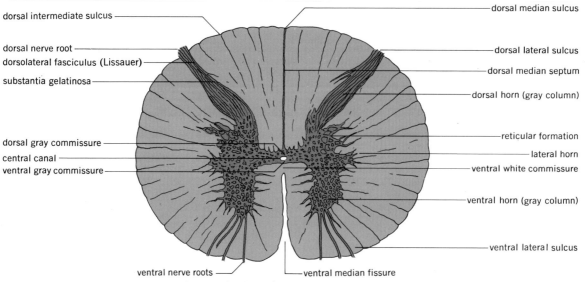

dorsal intermediate sulcus

dorsal nerve root

dorsolateral fasciculus (Lissauer)

substantia gelatinosa

dorsal gray commissure

central canal

ventral gray commissure

dorsal median sulcus

dorsal lateral sulcus

dorsal median septum

dorsal horn (gray column)

reticular formation

lateral horn

ventral white commissure

ventral horn (gray column)

ventral lateral sulcus

ventral nerve roots

ventral median fissure

FIGURE 11.2.   A cross section of the spinal cord.

tissue. A middle *arachnoid* is a delicate membrane enclosing a space (*subarachnoid space*) filled with cerebrospinal fluid. An inner *pia mater* is a vascular membrane carrying blood vessels into the cord. The greatest diameter of the cord itself does not exceed three fourths of an inch, and does not occupy the whole volume of the vertebral canal. The meninges and loose connective tissue fill the remaining space of the canal.

In the adult, the cord has a length of about 18 inches, and stops at the level of the second lumbar vertebra. It is grooved throughout its length by shallow *sulci,* and a deep *anterior fissure,* and is divided into *cervical* (C), *thoracic* (T), *lumbar* (L), and *sacral* (S) *regions.* An upper *cervical enlargement* marks the entry into the cord of the nerves from the upper limb. A lower *lumbar enlargement* marks the entry into the cord of nerves from the lower limb. The tapering end of the cord is the *conus medullaris*, from which the *filum terminale,* a fibrous derivative of the pia mater, extends through the vertebral canal to anchor the cord.

A cross section of the cord (Fig. 11.2) shows an inner H-shaped mass of GRAY MATTER. This consists mainly of neuron cell bodies and synapses. WHITE MATTER, consisting mainly of myelinated nerve fibers traveling up and down the cord, surrounds the gray matter. Each region has several subdivisions as is shown in the figure.

The white matter is divided into functional areas known as TRACTS (Fig. 11.3). Each tract carries particular types of sensory or motor impulses up or down the cord. The tracts receive their fibers from the same or opposite side of the brain or body, and thus a particular pattern of distribution results. The names and functions of the major spinal tracts are given in Table 11.1. The fibers of the descending tracts constitute the *upper* motor neurons of the motor system. Ascending tracts are sensory.

The cord and reflex activity

The spinal cord serves as an area of sensory input, and controls a number of reflexes involving muscle response to sensory stimuli.

Three sensory receptors are found in the skeletal muscles, and the receptors in the skin for touch, pressure, temperature, and pain may also supply input to the cord. The muscle receptors serve several important spinal reflex mechanisms, including the extensor thrust reflex, and the stretch (myotatic) reflexes.

THE EXTENSOR THRUST REFLEX. If pressure is applied to the sole of the foot, a reflex extension of the limb results that tends to maintain posture.

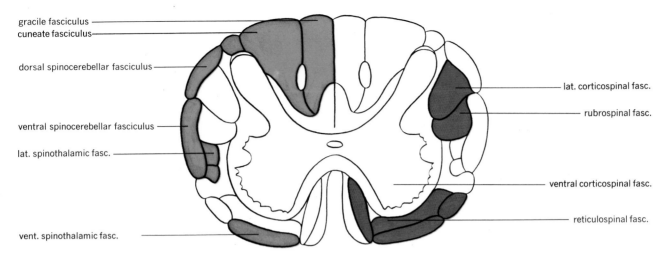

gracile fasciculus
cuneate fasciculus

dorsal spinocerebellar fasciculus

ventral spinocerebellar fasciculus

lat. spinothalamic fasc.

vent. spinothalamic fasc.

lat. corticospinal fasc.

rubrospinal fasc.

ventral corticospinal fasc.

reticulospinal fasc.

FIGURE 11.3 The major spinal tracts or fasciculi. (Red) Sensory or ascending tracts; (blue) motor or descending tracts.

FIGURE 11.4. The segmental distribution of spinal nerves to the body.

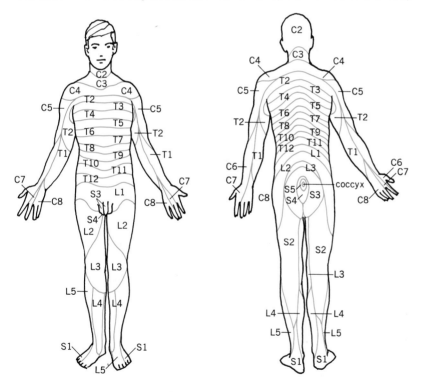

STRETCH (MYOTATIC) REFLEXES. Sudden tapping of the tendon of a muscle momentarily stretches the muscle and stimulates the muscle receptors. An impulse passes to the cord where a synapse is made, and a motor neuron conducts an impulse to the muscle. A reflex contraction of the same muscle results. The knee jerk, ankle jerk, and abdominal reflexes are examples of myotatic reflexes. Such reflexes are usually two neuron reflexes handled by one level of the cord. They are commonly used to assess cord function at various levels, as is shown in Table 11.2.

229

TABLE 11.1   The Major Spinal Tracts

	Name of tract	Area of origin, or region from which tract receives fibers	Area of termination	Crossed or uncrossed	Function or impulse(s) carried
D E S C E N D I N G	Corticospinal Lateral Ventral	Motor areas of cerebral cortex	Synapses in cord	Crossed in medulla, 80% of fibers Uncrossed, 20% of fibers	Carry voluntary impulses to skeletal muscles
	Rubrospinal	Red nucleus of basal ganglia in cerebrum	Synapses in cord	Crossed in brain stem	Involuntary impulses to skeletal muscles concerned with tone, posture.
	Reticulospinal	Reticular formation of brain stem	Synapses in cord	Crossed in brain stem	Increases skeletal muscle tone and motor neuron activity
A S C E N D I N G	Spinothalamic Lateral	Skin	Thalamus, relays to cerebral cortex	Crossed in cord	Pain and temperature
	Ventral	Skin	Thalamus, relays to cerebral cortex	Crossed in cord	Crude touch
	Spinocerebellar Dorsal Ventral	Muscles and tendons	Cerebellum Cerebellum	Uncrossed Uncrossed	Unconscious muscle sense for controlling muscle tone, posture
	Gracile Cuneate	Skin and muscles	Medulla, relays to cerebral cortex	Uncrossed Uncrossed	Touch, pressure, two-point discrimina-tion, con-scious muscle sense concerned with appreciation of body position

WITHDRAWAL REFLEXES. Painful stimulation of skin receptors (e.g., touching a hot object) typically causes withdrawal of the part from the painful stimulus in what is called the *flexion reflex*. It is a protective reflex in that the injured member is usually removed from the painful stimulus. A *crossed extension reflex* is extension of a limb on the opposite side of the body when a flexion reflex occurs in a given limb. For example, stepping on a tack results in a flexion reflex by the limb "a-tacked" and an extension of the opposite limb to maintain posture.

It may be emphasized again that all of these reflex activities are carried out by the spinal cord alone. The cord also monitors visceral reflexes concerned with bladder function, and defecation.

TABLE 11.2 Some Reflexes Used To Assess Cord Function			
Name of reflex	Level of cord involved	How elicited	Interpretation
Jaw jerk	C3	Tap chin with percussion hammer; jaw closes slightly.	Marked reaction indicates upper neuron lesion.
Biceps jerk	C5,6	Tap biceps tendon with percussion hammer; biceps contracts.	If reflex is absent or greatly exaggerated, cord damage, or damage to sensory or motor nerves is probable.
Triceps jerk	C7,8	Tap triceps tendon with percussion hammer; triceps contracts.	
Abdominal reflexes	T9-L2	Draw key or similar object across abdomen at different levels; muscles contract.	
Knee jerk	L2,3	Tap patellar tendon with percussion hammer; quadriceps contract to extend knee.	
Ankle jerk	L5	Tap Achilles tendon; gastrocnemius contracts to plantar flex foot.	
Plantar flexion	S1	Firmly stroke sole of foot from heel to big toe; toes should flex.	If toes fan and big toe extends (Babinski sign), upper neuron lesion is present.

The spinal nerves

Number and location

The cord gives rise to 31 pairs of spinal nerves, divided into *8 cervical* (C), *12 thoracic* (T), *5 lumbar* (L), *5 sacral* (S), and *1 coccygeal.* They are distributed to the skin and muscles of particular body regions (Fig. 11.4) with overlap of about 30 percent by neighboring nerves. Thus, if a given nerve is cut or damaged, complete function in a body area will not usually be lost.

The spinal nerves exit from the vertebral canal through the *intervertebral foramina* between vertebral pedicles. In the upper portions of the cord, the foramina of exit are nearly opposite the cord segment of nerve origin, and the nerves exit immediately to the side. In the lower segments, since the cord is shorter than the spinal column, nerves travel downward within the canal until they find their foramina of exit. Large numbers of these nerves form the *cauda equina* (horse's tail) of the cord (see Fig. 11.1).

Connections with the spinal cord (Fig. 11.5)

Each spinal nerve has a *dorsal root* (*sensory*), which conveys impulses to the cord. Cell bodies of these neurons are located in a *dorsal root ganglion* of each nerve. A *ventral root* (*motor*) conveys impulses from cell bodies in the cord to muscles and glands. *Communicating rami* allow connections for autonomic fibers. Peripherally, *somatic* nerves form anterior and posterior branches to skin and skeletal muscles. The ventral somatic roots to the skeletal muscles constitute the *lower* motor neurons of the motor system.

Organization of spinal nerves into plexuses

In several regions of the body, the spinal nerves

FIGURE 11.5.   The connections of sensory and motor neurons to the spinal cord.

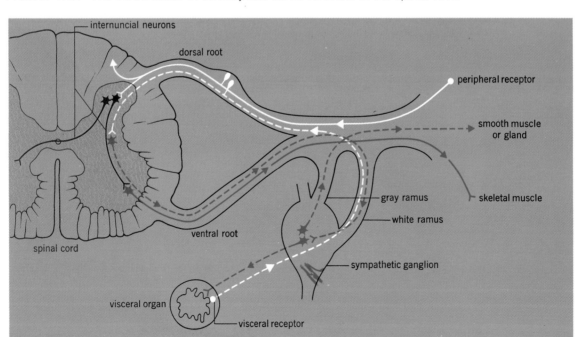

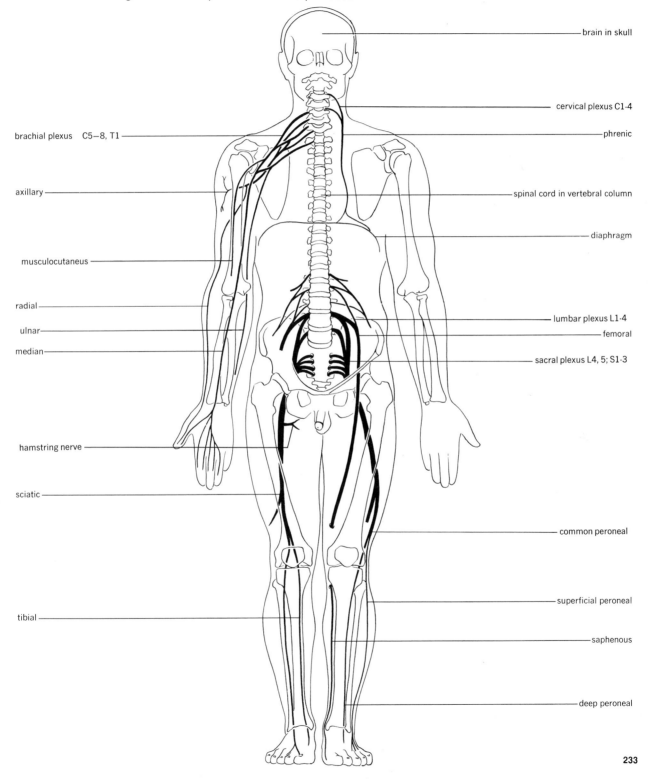

FIGURE 11.6.   The organization of spinal nerves into plexuses.

brain in skull

cervical plexus C1-4

brachial plexus   C5–8, T1

phrenic

axillary

spinal cord in vertebral column

diaphragm

musculocutaneus

radial

lumbar plexus L1-4

ulnar

femoral

median

sacral plexus L4, 5; S1-3

hamstring nerve

sciatic

common peroneal

superficial peroneal

tibial

saphenous

deep peroneal

are "collected" into groups that form a plexus. From these plexuses, the nerves are distributed to muscles and skin.

THE CERVICAL PLEXUS (Fig. 11.7). Cervical nerves 1–4 combine to form the cervical plexus which lies in the upper part of the neck under the sternocleidomastoid muscle. Nerves pass from the plexus to muscles and skin of the scalp and neck, and to the diaphragm by way of the phrenic nerves. A "broken neck" may thus result in respiratory paralysis through injury to the phrenic nerves.

THE BRACHIAL PLEXUS (Fig. 11.8). The anterior roots of spinal nerves C5-T1 combine to form the brachial plexus which lies in the anterior base of the neck just superior to the clavicle. It sends nerves to the skin and muscles of the upper appendage. Five major nerves exit from the plexus.

The *axillary nerve* supplies the shoulder joint and deltoid muscle.

The *musculocutaneous nerve* supplies the structures of the anterior upper arm.

The *median nerve* supplies the anterior forearm.

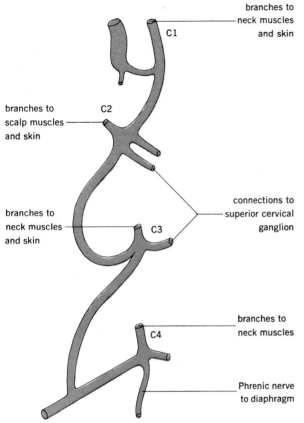

FIGURE 11.7.   The cervical plexus.
C, cervical spinal nerves.

FIGURE 11.8.   The brachial plexus.
C, cervical; T, thoracic spinal nerves.

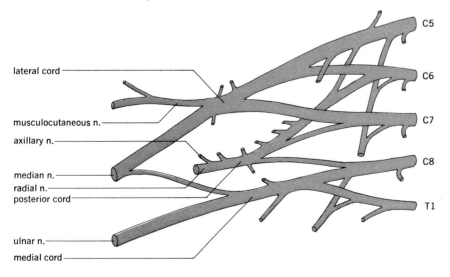

The *ulnar nerve* serves the same areas as the median nerve.

The *radial nerve* supplies the posterior upper arm, forearm, and hand.

The nerves are superficial at the axilla. Injury here may result in paralysis. The "crazy bone" is the ulnar nerve as it passes over the medial epicondyle of the humerus.

THE INTERCOSTAL NERVES. Thoracic nerves 2--12 form the intercostal nerves which supply the intercostal and abdominal muscles. A named plexus is not formed.

THE LUMBAR PLEXUS (Fig. 11.9). Lumbar nerves 1–4 combine to form the lumbar plexus which lies within the posterior part of the psoas muscle. Three main nerves exit from the plexus.

The *femoral nerve* supplies the skin and muscles of the anterior thigh.

The *lateral cutaneous nerve* supplies the skin and muscles of the upper lateral thigh.

The *obturator nerve* supplies the skin and muscles of the medial thigh.

THE SACRAL PLEXUS (Fig. 11.10). Lumbar nerves 4 and 5, and sacral nerves 1–3 combine to form the sacral plexus which lies in the middle portion of the false pelvis. Some anatomists consider the lumbar and sacral plexuses to be one large plexus, which they call the *lumbosacral plexus.* There is some variation as to which nerves contribute to the two plexuses, particularly the lower lumbar nerves. The major nerve of the sacral plexus is the *sciatic nerve,* which forms several important branches.

The *nerve to the hamstring muscles* ("hamstring nerve") arises from the sciatic nerve in the thigh, and supplies the skin and muscles of the posterior thigh.

The *tibial nerve* arises as one branch from the sciatic just above the knee, and supplies the skin and muscles of the knee joint and posterior leg.

The *common peroneal nerve* is the other branch of the sciatic just above the knee. It continues down the leg, giving branches that supply the skin and muscles of the knee, anterior and lateral leg, and foot.

FIGURE 11.9.   The lumbar plexus. T, thoracic; L, lumbar spinal nerves.

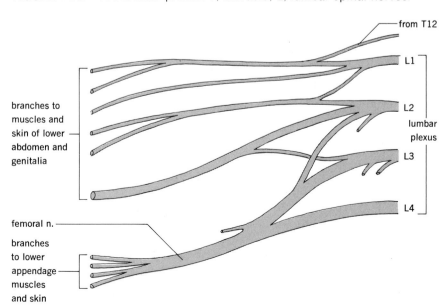

branches to muscles and skin of lower abdomen and genitalia

from T12

L1

L2

lumbar plexus

L3

L4

femoral n.

branches to lower appendage muscles and skin

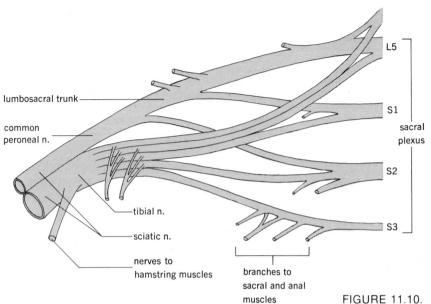

FIGURE 11.10.   The sacral plexus.
L, lumbar; S, sacral spinal nerves.

Clinical considerations

Cord and nerve injury

Dislocation of vertebrae, stab, or bullet wounds, may result in damage to the cord with consequent loss of function. Specific loss depends on what particular cord area and tract(s) are involved. If the cord is severed (complete section) all sensory and motor function will be lost below the cut. Paralysis is of the *spastic* type, characteristic of an upper motor neuron lesion.

Damage to the lower motor neurons results in a *flaccid* paralysis, restricted to the muscles served by the damaged nerve. Regeneration of upper motor neurons occurs, but the chances that normal connections will be reestablished are minimal, thus restoration of function is poor. Peripheral nerves may regenerate, but the new connections established are more widely distributed and less precise than originally, so that function may return but will be less exact than before.

HEMISECTION of the cord (cutting halfway through the cord) produces the *Brown-Sequard syndrome*. This is characterized by:

Loss of voluntary muscle movement, touch, and pressure on the same side of the cut;

Loss of pain, heat, and cold on the opposite side of the cut.

Remembering the spinal tracts and the areas supplied by the spinal nerves, it is easy to determine the spinal level of damage and the side of damage.

Spinal shock

If the cord is completely cut or is severely traumatized (injured) certain losses or alterations of function will result.

Permanent loss of voluntary muscular activity, and sensation will occur in the body segments below the level of the cut.

The development of *spinal shock* will occur. It is a *temporary synaptic depression* that results in loss of reflex activity below the level of the cut. The length of time the state lasts depends on the position of the animal in the phylogenetic scale, and extends from minutes for an animal like the frog, to months in the human. As recovery from shock proceeds, there is a typical pattern of reappearance of reflex activity.

Knee jerk and other simple reflexes reappear.
The flexion and crossed extension reflexes reappear.
Visceral reflexes, including bladder evacuating reflexes, and sexual reflexes reappear (erection may occur but rarely ejaculation).

The human is subject to a postural hypotension (low blood pressure) which results as body position changes from lying, to sitting or standing postures. The individual whose cord has been damaged no longer has connections with the vasoconstrictor centers that are located in the lower part of the brain stem (medulla oblongata). Postural changes are not met by vasoconstriction (narrowing of blood vessels) to maintain blood pressure. Since there is no sensation from the body parts below the level of section, the individual may also develop skin ulcerations from pressure and ischemia, and not be aware of them.

Spinal puncture (tap)

Since the cord stops at about the second lumbar vertebra, it is possible to insert a needle into the subarachnoid space and withdraw spinal fluid. The puncture is usually done at the level of the fourth lumbar vertebra. The fluid may be withdrawn for analysis, or may be replaced with an equivalent volume of anesthetic to create *spinal anesthesia*. Puncture must be done carefully to avoid damaging nerves of the cauda equina.

Intramuscular injections

These injections involve giving medications by insertion of a needle into a muscle. Large nerves may lie in the area into which the injection is to be made, and their path and size must be recalled to avoid injuring them. Branches of the axillary nerve lie beneath the deltoid muscle; the sciatic nerve lies beneath the buttock and lateral thigh muscles; the femoral nerve lies under the anteromedial thigh muscles. Paralysis or sensory loss may result if the nerve is damaged.

Summary

1. The spinal cord integrates reflex activity and transmits impulses to and from the brain. The spinal nerves carry impulses to and from the periphery.

2. The cord is located within the vertebral canal.
   a. It is surrounded by fluid, meninges, and connective tissue.
   b. It is shorter than its canal.
   c. It has four major regions.
   d. A cross section shows gray and white matter.
   e. White matter contains the spinal tracts, areas of particular motor and sensory function.

3. The cord is responsible for much reflex activity.
   a. Stretch, flexion, crossed extension, and withdrawal reflexes are controlled by the cord.
   b. Visceral reflexes such as bladder emptying and defecation are controlled by the cord.

4. Thirty-one pairs of spinal nerves originate from the cord.

a. They connect to the cord by dorsal and ventral roots.

b. They form four major plexuses from which nerves pass to skin and skeletal muscles. These plexuses and the major nerves originating from each area are:

(1) Cervical plexus: nerves to neck and shoulder muscles; phrenic nerves to diaphragm.

(2) Brachial plexus: axillary, musculocutaneous, median, ulnar, and radial nerves to upper appendage.

(3) Lumbar plexus: femoral, lateral cutaneous, and obturator nerves to upper thigh.

(4) Sacral plexus: sciatic: branches to give hamstring nerve; forms tibial nerve, and common peroneal nerve to lower appendage.

5. Thoracic nerves 2–12 form the intercostal nerves which supply the intercostal and abdominal muscles; plexus is not formed.

6. Cord injury produces motor and sensory loss, depending on degree of damage and area of damage.

7. Spinal shock, characterized by loss of function, then return of function in a definite order, is the result of damage to the cord.

8. Spinal puncture is possible because of the fact that the cord does not extend through the entire vertebral canal.

Questions

1. Describe a cross section of the cord, including anatomical features and the locations of the major spinal tracts.

2. If an individual suffers loss of pain, heat, and cold on his right side, muscle paralysis on his left side, and these changes are restricted to his anterior forearm, predict at what level and which side his spinal cord has been injured.

3. Describe some reflexes controlled by the cord.

4. What are/is:

a. Dorsal roots.

b. Plexuses.

c. Conus medullaris.

d. Anterior median fissure.

e. Arachnoid.

5. Describe how it is possible to deliver a spinal anesthetic.

6. What cautions must be observed in giving an intramuscular injection? Relate this to the nerves exiting from the spinal nerve plexuses.

Readings

Anson, Barry J. *Morris' Human Anatomy.* McGraw-Hill. New York, 1966.

Austin, G. *The Spinal Cord.* Thomas Pubs. Springfield, Ill. 1971.

Baker, A. B. *An Outline of Clinical Neurology.* Wm. C. Brown Book Co. Dubuque, Ia. 1965.

Barr, Murray L. *The Human Nervous System; An Anatomical Viewpoint.* Harper & Row. New York, 1972.

chapter 12
The Brain and Cranial Nerves

chapter 12

The term "brain" (Fig. 12.1) refers to the nervous structures enclosed within the cranial cavities. The brain is divided into three major parts, each containing several subdivisions.

The BRAIN STEM supports the other parts and from inferior to superior, consists of the *medulla oblongata* (medulla), *pons, midbrain,* and *diencephalon.* The latter includes the thalamus and hypothalamus. The stem also acts as the point of origin or termination for many of the cranial nerves.

The CEREBELLUM lies posteriorly on the lower part of the stem, and is concerned with involuntary activity.

The CEREBRUM, composed of two cerebral hemispheres, is the human's "crowning glory." It is the largest division of the brain and controls most of the body's mental and physical activity.

The brain stem

The medulla oblongata

STRUCTURE. The medulla is the lowest portion of the stem, and is continuous inferiorly with the spinal cord. It is about 1 inch in length, and consists internally of fiber tracts (white matter) and NUCLEI or groups of cell bodies forming gray matter. The gray and white components are not oriented in any way, but are "all mixed together." On the anterior surface of the medulla are two elongated elevations known as the PYRAMIDS. The OLIVES are rounded elevations on each side of the superior part of the pyramids. The ninth to twelfth cranial nerve nuclei are found in the medulla, and these nerves emerge from the medulla.

FUNCTION. The medulla contains several groups of neurons which form the VITAL CENTERS necessary for survival. Among these are:

The *cardiac centers*, controlling heart rate.
Two of the three *respiratory centers* initiating and regulating rate and depth of breathing.
The *vasomotor centers*, controlling diameter of blood vessels and thus blood pressure.

These centers act primarily in reflexes designed to maintain homeostasis of breathing and blood pressure. The medulla also contains NONVITAL CENTERS for *vomiting, sneezing, coughing,* and *swallowing.*

The major voluntary motor pathways (pyramidal or corticospinal tracts) undergo an 80 percent crossing or decussation in the medulla, and these fibers form the pyramids.

The olives contain synapses of fibers from the equilibrium structures of the inner ear, and send fibers to the cerebellum to insure efficiency of postural muscle adjustments.

The gracile and cuneate tracts of the dorsal white columns undergo synapses in nuclei of the same name in the medulla, and from these nuclei, fibers are relayed to the thalamus, and to the cerebrum.

CLINICAL CONSIDERATIONS. Damage to the medulla, such as may occur in blows to the back of the head, or by infections (e.g., poliomyelitis) are particularly dangerous because they may damage the vital centers. Involvement of the respiratory centers may cause respiratory paralysis, and may make the victim a candidate for a positive or negative pressure breathing apparatus. Involvement of the pyramidal tracts may lead to motor loss, while involvement of the gracile

FIGURE 12.1. The brain. Medial view of the left half of the brain (above). Lateral view of the right side of the brain (below).

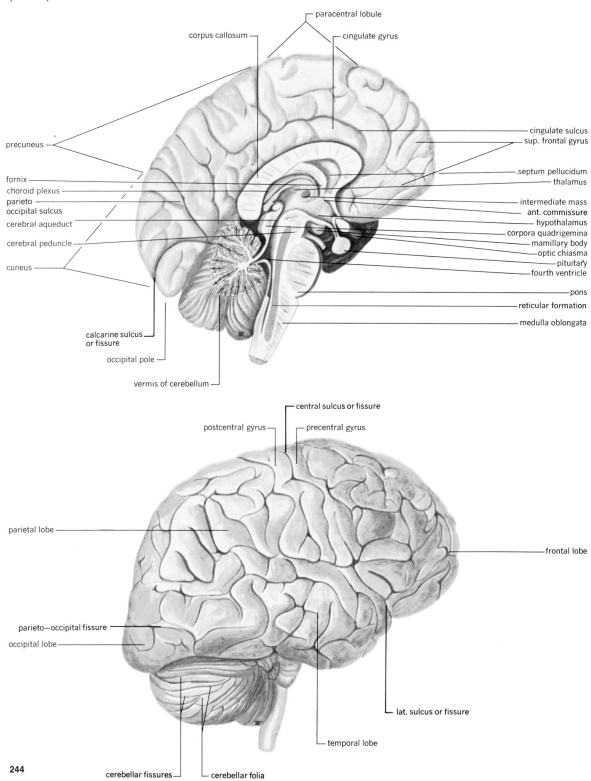

and cuneate nuclei may lead to sensory loss involving touch, pressure, and appreciation of body position.

The pons

STRUCTURE. The pons forms a conspicuous bulge on the anterior surface of the brain stem. It consists mostly of white matter (fiber pathways) passing from medulla to cerebrum, and connecting the pons to the cerebellum via the CEREBELLAR PEDUNCLES. Its NUCLEI include the *pneumotaxic center,* one of several centers concerned with respiration, and the cell bodies of cranial nerves 5–8.

FUNCTION. The pneumotaxic center inhibits sustained inspiration, contributing to expiration during breathing. The peduncles transmit afferent (sensory) impulses from the spinocerebellar tracts to the cerebellum, and carry efferent (motor) impulses from the cerebellum that are concerned with maintenance of balance and equilibrium.

CLINICAL CONSIDERATIONS. Damage to the pons typically produces *motor losses* by interfering with cerebellar afferent and efferent nerves. Voluntary movements are less precise, and posture is difficult to maintain. If the pneumotaxic center is involved, *apneustic breathing* (characterized by long sustained inspiration) may result.

The midbrain

STRUCTURE. The midbrain is a wedge-shaped portion of the brain stem superior to the pons. Its anterior portion is composed of two large bundles of fibers, the CEREBRAL PEDUNCLES. These bundles carry the bulk of the motor fibers from the cerebrum to lower brain-stem regions and the spinal cord. The posterior portion of the midbrain contains the CORPORA QUADRIGEMINA, four small rounded bodies consisting of a pair each of *superior and inferior colliculi.* These bodies are reflex centers that connect higher brain centers with eye muscles and neck muscles, and serve

auditory and visual reflexes. Cranial nerves 3 and 4 originate from the midbrain.

FUNCTION. The midbrain integrates a variety of *visual and auditory reflexes* including: those concerned with objects seen that are approaching the body and that require muscular response to avoid them; and turning the head to achieve the greatest benefit from an auditory stimulus. Certain righting reflexes concerned with maintaining normal posture relative to the eyes and head also appear to be monitored by the midbrain.

CLINICAL CONSIDERATIONS. Damage in the midbrain area will obviously produce disturbances in visual and auditory reflexes and righting reflexes; balance and equilibrium reactions will suffer. Interruption of the peduncles will interfere with the performance of voluntary muscle activity.

The diencephalon

The diencephalon forms the superior end of the brain stem, and is part of the original forebrain. It develops into a superiorly placed thalamus, and a smaller, inferiorly placed hypothalamus.

THE THALAMUS (Fig. 12.2)
   Structure. The thalamus composes about four fifths of the diencephalon and is composed of several nuclei. It occupies the upper lateral wall of the third ventricle.
   Functions. The thalamus is a major RELAY STATION to the cerebral hemispheres for all types of sensory information, except possibly olfactory (smell) impulses. It integrates these impulses by grouping impulses of the same nature from different body areas. These are then sent to particular areas of the cerebral cortex. It can provide "vague awareness" at the conscious level of sensory stimulation without fine discrimination, and may interpret the sense of pain. The thalamus also contains part of the reticular formation (see below), and contributes to the state of wakefulness and alertness of the organism. The affective quality of a sensation, its pleasantness or unpleasant-

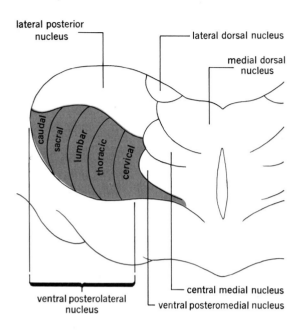

lateral posterior nucleus

lateral dorsal nucleus

medial dorsal nucleus

caudal
sacral
lumbar
thoracic
cervical

ventral posterolateral nucleus

central medial nucleus

ventral posteromedial nucleus

FIGURE 12.2.   The location and internal structure of the thalamus.

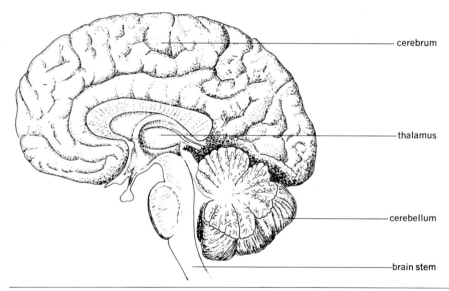

cerebrum

thalamus

cerebellum

brain stem

ness, appears to be determined by the thalamus.

The thalamus also contains areas concerned with motor function, and thus is involved in certain types of movements, especially those involving outward expression of emotions (rage, fear).

Clinical considerations. Damage to the sensory portions of the thalamus produces what is known as the THALAMIC SYNDROME. A lesion here is usually vascular in origin (blood vessel ruptured or blocked), and is associated with diminished awareness of sensation until a certain point is reached. The stimuli may then become intolerable, and very painful.

In older individuals, Parkinson's disease (paralysis agitans or shaking palsy), characterized by tremor and muscle rigidity, may develop. Destruction of the motor regions of the thalamus, or more commonly today, administration of L-DOPA (*l*evorotatory *d*ihydr*oxyp*heny*l*alanine) may relieve or lessen the severity of the disorder.

THE HYPOTHALAMUS

Structure. The hypothalamus forms the lower portion of the diencephalon. It contains many separate NUCLEI (Fig. 12.3) and has connections with the thalamus, cerebral cortex, brain stem, and emotional expression systems (limbic system) of the brain. Its outgoing fibers are concerned primarily with regulating the activity of the pituitary gland and visceral function.

Functions. The cells of the hypothalamic nuclei not only have nerve connections with many body areas but respond to the properties of the blood reaching them. Among the blood properties to which the area responds are pH, osmotic pressure, and glucose levels. Output from the hypothalamus is designed to control many body processes concerned with MAINTENANCE OF HOMEOSTASIS.

Functions of the hypothalamus include:

TEMPERATURE REGULATION

Human body temperature is normally maintained within a degree or two of 98.6°F or 37°C. A proper balance between heat production and loss is maintained by *heat loss and heat gain centers* in the hypothalamus. The heat loss center causes dilation of skin blood vessels, sweating and decreased muscle tone, all of which increase heat loss or reduce heat production. The heat gain center causes skin vessel constriction, shivering, and cessation of marked sweating, all of which conserve body heat or increase heat production.

REGULATION OF WATER BALANCE

Hypothalamic neurons continually monitor blood osmotic pressure and adjust the tonicity of the body fluids by *ADH* (antidiuretic hormone) production. The hormone permits greater water reabsorption from the kidney tubules.

FIGURE 12.3.   The nuclei of the hypothalamus.

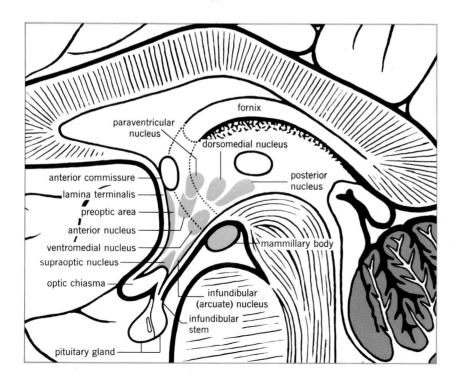

CONTROL OF PITUITARY FUNCTION

At last count, nine chemicals influencing pituitary function have been isolated from the hypothalamus. Properly termed *pituitary regulating factors,* these chemicals are produced in the hypothalamus as a result of blood-borne and nervous stimuli, are passed by blood vessels to the pituitary, and there stimulate or inhibit pituitary production and release of hormones.

CONTROL OF FOOD INTAKE

Initiation and cessation of feeding is controlled by hypothalamic *feeding and satiety centers.*

REGULATION OF GASTRIC SECRETION

The amount of gastric juice produced by the stomach is increased by hypothalamic stimulation. Thus emotions may trigger release of gastric juice when there is no food in the stomach and may lead to ulcer development.

EMOTIONAL EXPRESSION

The hypothalamus is one part of a system necessary for expression of reactions of rage and anger. It also appears necessary for sexual behavior.

Clinical considerations. Depending on where damage occurs in the hypothalamus, and its extent, a wide variety of symptoms may develop. Those associated with disturbances of water balance, temperature regulation, and emotions are the most common. *Diabetes insipidus* is the production of a large volume of very dilute urine as a result of diminished ADH secretion; inability to maintain near constant body temperature indicates hypothalamic damage.

The cerebellum

The cerebellum is the second largest portion of the brain and is an important component of the motor system of the body.

STRUCTURE. The cerebellum (Fig. 12.4) lies posteriorly on the medulla and pons. It has a small centrally placed VERMIS, and two larger CEREBELLAR HEMISPHERES. Both vermis and hemispheres are thrown into small folds known as *folia.* Attaching the cerebellum to the brain stem are three paired bundles of nerve fibers, the CEREBELLAR PEDUNCLES. These bundles carry afferent and efferent impulses to and from the organ. Afferent fibers come from muscles, tendons, skin, inner ear, midbrain, and cerebral cortex. Efferent impulses pass to thalamus, basal ganglia, and spinal cord.

Internally, the cerebellum has an outer CORTEX of gray matter and an inner MEDULLARY BODY of white matter. The latter has a treelike arrangement, and is known as the *arbor vitae.* The characteristic cell of the cerebellum in the Purkinje cell (see Fig. 10.2).

FUNCTION. The cerebellum operates completely at the subconscious level. It coordinates muscular activity, integrates muscular movement, and predicts when to stop movements. Additionally, it coordinates reflexes that serve to maintain posture and equilibrium.

CLINICAL CONSIDERATIONS. Damage to the cerebellum is followed by tremor, asynergia (lack of muscular coordination), "puppetlike" or jerky movements, past pointing (overshooting the end point of a movement), failure to direct movements properly, cerebellar ataxia (a reeling walk), and loss of muscle tone.

The severity of the loss of function depends on the amount of tissue damaged, not the location. If lesions are small, compensation occurs readily, and little evidence of damage may appear.

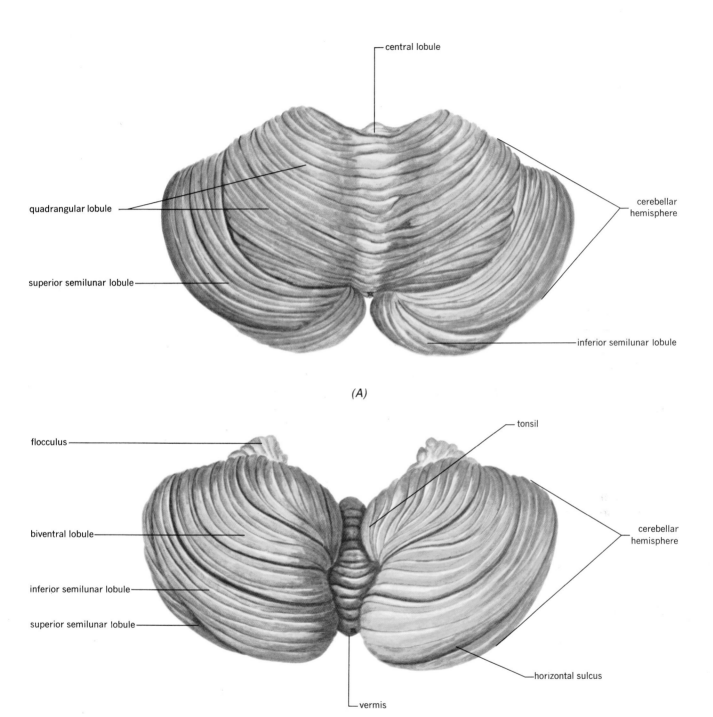

central lobule

cerebellar
hemisphere

quadrangular lobule

superior semilunar lobule

inferior semilunar lobule

(A)

tonsil

flocculus

cerebellar
hemisphere

biventral lobule

inferior semilunar lobule

superior semilunar lobule

horizontal sulcus

vermis

(B)

FIGURE 12.4.   The cerebellum. (A) Posterior surface. (B) Anterior surface.

The reticular formation

The reticular formation is a diffuse network of gray matter placed centrally throughout the brain stem. Its main functions appear to be to ALERT OR AROUSE the organism, and to coordinate reflex and voluntary movements. The reticular formation and its cortical fibers are sometimes designated as the *reticular activating system* (RAS).

Drugs that keep us alert, and those that depress consciousness act, in part, on the RAS.

The cerebrum

The cerebrum develops from the telencephalon, the second subdivision of the original forebrain. It is the largest portion of the brain in humans.

The cerebral hemispheres (Fig. 12.5)

STRUCTURE. The cerebrum is divided into two lateral halves or HEMISPHERES by the longitudinal fissure. Each hemisphere is further separated by fissures, into LOBES that are located and named according to the cranial bones they lie beneath.

A central fissure separates a frontal lobe from a parietal lobe.

The parieto-occipital fissure separates the parietal lobe from an occipital lobe.

The lateral fissure separates a temporal lobe from the frontal and parietal lobes.

FIGURE 12.5.   The fissures and lobes of the cerebrum.

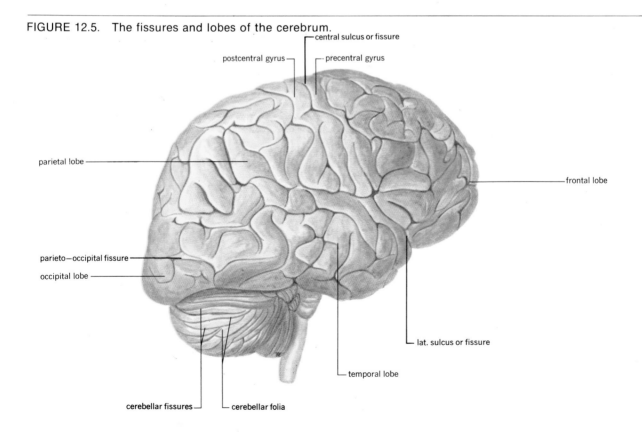

FIGURE 12.6. The basal ganglia. Several major ganglia projected on the cerebral hemisphere (above). Frontal section of the cerebrum (below).

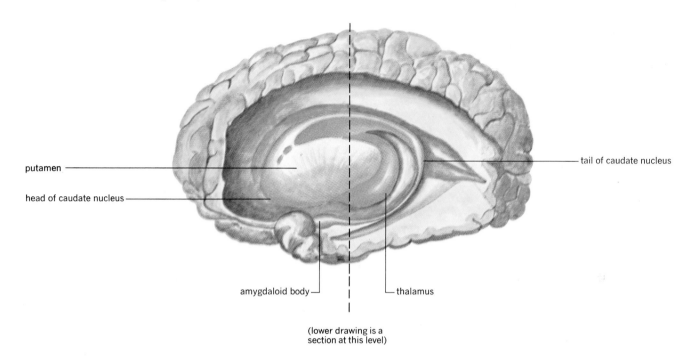

putamen

head of caudate nucleus

tail of caudate nucleus

amygdaloid body

thalamus

(lower drawing is a
section at this level)

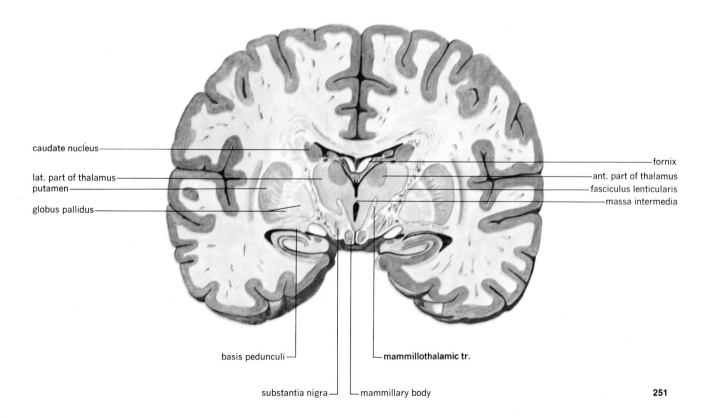

caudate nucleus

lat. part of thalamus
putamen

globus pallidus

fornix

ant. part of thalamus
fasciculus lenticularis
massa intermedia

basis pedunculi

mammillothalamic tr.

substantia nigra

mammillary body

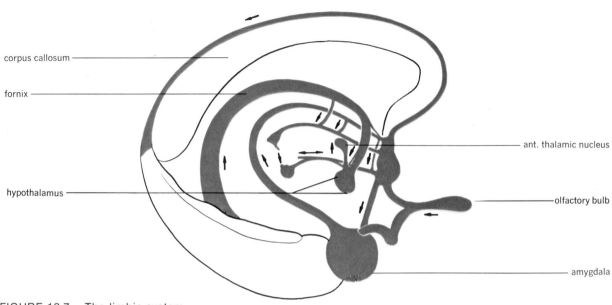

corpus callosum

fornix

hypothalamus

ant. thalamic nucleus

olfactory bulb

amygdala

FIGURE 12.7.   The limbic system.

The cerebrum, throughout its surface, is thrown into upfolds or GYRI, separated by shallow depressions known as SULCI. Each gyrus has a name; two should be emphasized. The *precentral gyrus* lies anterior to the central fissure; the *postcentral gyrus* lies posterior to the central fissure.

The arrangement of gray and white matter in the cerebrum is reversed as compared to that of the cord. An outer layer of gray matter 2–5 millimeters thick forms the CEREBRAL CORTEX. This layer contains an estimated 12–15 billion neuron cell bodies. The inner white matter, or MEDULLARY BODY, contains a large number of fiber tracts connecting the two halves of the cerebrum (*commissural fibers*), tracts that are coming into or leaving the hemispheres (*projection fibers*), and fibers that connect one part of a hemisphere with another (*association fibers*).

The basal ganglia (Fig. 12.6). The basal ganglia are nuclei buried deep within the hemispheres. They include the *caudate* nucleus, *putamen, globus pallidus, subthalamic nuclei, substantia nigra,* and *red nucleus.*

The limbic system (Fig. 12.7). The limbic system is composed of certain cerebral and brain stem structures (see below).

FUNCTION. Specific functional areas within the cerebral cortex exist in the different lobes. A convenient method of designating the areas is to number them according to the scheme devised by Brodmann (Fig. 12.8).

A PRIMARY SOMATIC MOTOR AREA (area 4) lies in the precentral gyrus. The voluntary movements of skeletal muscles are controlled from this region. Representation of body parts is unequal and "upside down" in this region, with more area given to those body parts where skilled or complex movement is required (Fig. 12.9).

A PREMOTOR AREA (areas 6,8) lies anterior to the motor area, and coordinates eye and head movements. The region also creates a postural background (synergistic action) against which skilled movements are performed.

The PRIMARY SOMATIC SENSORY REGION (areas 3,1,2) lies in the postcentral gyrus. As in area 4, the body is represented unequally and upside down in this region (Fig. 12.10). This area interprets most of the skin sensations (heat, cold, touch, pressure, some localization of pain).

The VISUAL AREA (area 17) lies in the occipital lobe and receives impulses from the eyes, by way of the optic nerves, tracts, and radiations.

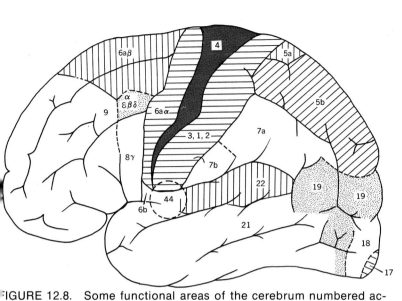

FIGURE 12.8. Some functional areas of the cerebrum numbered according to Brodmann.

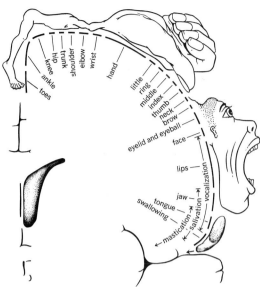

FIGURE 12.9. Cortical location of motor functions; motor homunculus.

FIGURE 12.10. Cortical location of sensory functions; sensory homunculus.

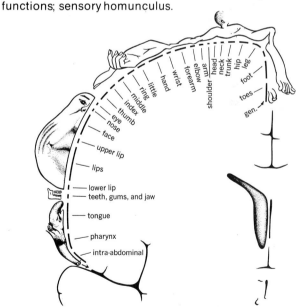

The AUDITORY AREA (areas 41, 42) lies in the temporal lobe, and receives impulses from the cochleas of the inner ear over the auditory pathways.

BROCA'S SPEECH AREA (area 44) includes parts of the motor, sensory, and auditory areas, and is concerned with formation of words.

ASSOCIATION AREAS of the cerebrum are those areas from which specific motor responses or sensations cannot be elicited when they are stimulated.

The *prefrontal areas* (8–12) control behavior, particularly in inhibition of rage reactions.

The *parietal areas* (5, 7, 19, 39, 40) are concerned with interpretation of size, shape, texture, degree of heat, and other qualities of objects touched.

The *temporal areas* (20–22) are concerned with learning and memory of things seen and heard.

The BASAL GANGLIA, as a group, control muscle tone, inhibit movement, and control tremor.

The *caudate nucleus* and *putamen* regulate gross intentional movements of the body, such as movement of a whole limb through space.

The *globus pallidus* provides positioning of the body so that more specific or discrete movements may occur properly.

The *subthalamic nuclei*, substantia nigra and red nucleus are part of the extrapyramidal motor system (see below) concerned primarily with muscle tone and damping of tremor.

The LIMBIC SYSTEM is composed of *hypothalamus, thalamus,* the *amygdaloid nuclei,* and several fibers tracts connecting these areas (see Fig. 12.7). The entire system is concerned with emotions and motivation. Expression of rage, and sexual behavior, are among the functions these structures serve. "Pleasure centers" which create positive drives toward stimuli are also found in these areas.

Functionally, two more groupings of efferent fibers from the brain may be made.

The PYRAMIDAL SYSTEM refers to motor fibers originating from the cerebral cortex and passing through the pyramids. About 40 percent of the fibers in this system come from the primary somatic motor area (area 4); the remainder come from other cortical regions.

The EXTRAPYRAMIDAL SYSTEM consists of motor fibers from basal ganglia, brain stem, cerebellum, and other noncortical areas. Though the pyramidal system has traditionally been considered the "voluntary motor system" of the body, this is not strictly true, since only about 40 percent of its fibers come from area 4. The extrapyramidal system is, however, "involuntary" in the sense that it operates without acts of will, and controls tone and reflex movement.

CLINICAL CONSIDERATIONS. Damage to the primary somatic areas (motor, sensory, visual, auditory) of the cerebrum will produce specific losses of motor or sensory function. Paralysis, partial deafness, or blindness are typical signs of damage to these regions.

Damage in the association areas produces signs correlated with the area involved. Frontal lobe lesions produce behavioral alterations; parietal lobe lesions result in inability to perceive the body correctly as a result of sensory loss; temporal lobe lesions produce *agnosia* (failure to recognize familiar objects), *apraxia* (inability to perform voluntary movements, especially of speech), and *aphasia* (inability to understand written or spoken words).

Damage to the basal ganglia results in loss of muscle tone and in inability or difficulty in mak-

ing precise movements. Also, Parkinson's disease, characterized by involuntary tremors, is most consistently associated with damage to the caudate, pallidus, or substantia nigra.

Damage to limbic structures produces exaggerated responses of rage, and hypersexuality.

Damage to pyramidal or extrapyramidal systems or both produces motor losses characterized by spastic paralysis. If the damage occurs above the level of the brain stem, the opposite side of the body is usually affected; below the brain stem, the same side is usually affected. This occurs because crossing of motor fibers (e.g., lateral corticospinal tract, rubrospinal tract) occurs in the brain stem.

The brain exhibits spontaneous rhythmical electrical activity which may be recorded as an ELECTROENCEPHALOGRAM or EEG. Figure 12.11 shows a normal EEG with several types of wave patterns.

Alpha waves have a frequency of 10–12 per second, and voltages of about 50 microvolts (1 microvolt = 1/1,000,000th volt). They are obtained from an inattentive or "at rest" brain.

Beta waves occur 15–60 times per second and have voltages of 5–10 microvolts.

Delta waves occur during sleep and occur 1–5 per second with voltages of 20–200 microvolts.

Theta waves occur 5–8 times per second with voltages of about 10 microvolts. They are common in children and during times of emotional stress.

CONVULSIVE DISORDERS or seizures are characterized by alterations in the EEG (Fig. 12.12) and alterations in the state of consciousness, muscular activity, and sensory phenomena. The subject may show involuntary contractions of the muscles of the trunk and appendages which constitute the most obvious signs of a seizure.

Fever, infections, lack of oxygen to the brain, tumors, allergic reactions, trauma to the brain, and drugs have all been implicated as causes of convulsive seizures. In other cases, there is no obvious cause of the seizure.

If there is a recurring pattern to the seizures, the subject is usually said to have epilepsy. Epilepsy affects about 0.5 percent of the population, and is divided into two general types:

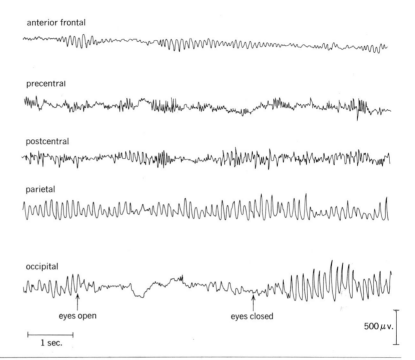

anterior frontal

precentral

postcentral

parietal

occipital

eyes open   eyes closed

500 μv.

1 sec.

FIGURE 12.11. Normal EEG patterns from different regions of the cortex. Alpha waves predominate in parietal and occipital areas, beta waves in precentral area. Alpha waves are blocked when eyes are opened.

FIGURE 12.12. Some abnormal EEGs. Note large spikes and fast rhythm.

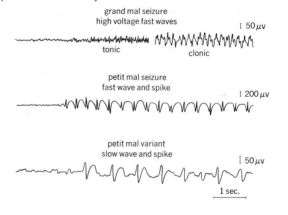

grand mal seizure
high voltage fast waves   I 50 μv

tonic   clonic

petit mal seizure
fast wave and spike   I 200 μv

petit mal variant
slow wave and spike   I 50 μv

1 sec.

Symptomatic (Jacksonian) *epilepsy* may be demonstrated to have a cause, such as a brain lesion.

Idiopathic epilepsy occurs without any demonstrable cause.

According to the type of symptoms the subject shows, there are several kinds of epilepsy described.

Grand mal epilepsy results in loss of consciousness, falling, and spasm of the muscles, often preceeded by a hallucination of a disagreeable odor.

Petit mal epilepsy clouds the consciousness, and is not associated with loss of consciousness or muscular spasms. It produces a "blank" for 1–30 seconds in the subject's behavior.

Psychomotor epilepsy lasts 1–2 minutes and is characterized by loss of contact with the environment, staggering, muttering, and mental confusion for several minutes after the attack is over.

Infantile spasm occurs in the first three years of life, and may be replaced by other types of seizures later in life. It is characterized by flexion of trunk and arms, and extension of the legs.

If a lesion such as a tumor or scar may be shown to cause the seizure, surgery may be indicated. Management of seizures is often achieved by the administration of anticonvulsant drugs alone or in combinations (e.g., phenobarbital, diphenylhydantoin, bromides).

The cranial nerves

The brain gives rise to 12 pairs of cranial nerves which supply motor and sensory fibers to structures in the head, neck, and shoulder regions. These fibers form, with the spinal nerves, the peripheral nervous system.

The relationships of the nerves to the brain are shown in Figure 12.13. A general description of each nerve follows, and a chart of the nerves is shown in Table 12.1.

I. *Olfactory nerve.* The nerve of smell, the olfactory nerve passes from nasal cavities to the olfactory bulb of the cerebrum.

II. *Optic nerve.* The nerve of sight, the optic nerve conveys impulses from retina to brain stem. These impulses are then relayed to the occipital lobe over the optic radiation.

III. *Oculomotor nerve.* Motor fibers in this nerve control four of the six eye muscles which turn the eyeball, cause pupillary size changes, and aid in focusing. A sensory component relays impulses from the muscles, iris, and ciliary body.

IV. *Trochlear nerve.* One of the six eye muscles is supplied by motor and sensory fibers of this nerve.

V. *Trigeminal nerve* (Fig. 12.14). The largest cranial nerve, it supplies sensory fibers to the anterior cranium and face, and motor fibers to the chewing muscles. Ophthalmic, maxillary, and mandibular branches form the nerve, and supply sensory fibers to all structures within their area of distribution.

FIGURE 12.13.   Basal view of the brain showing cranial nerves.

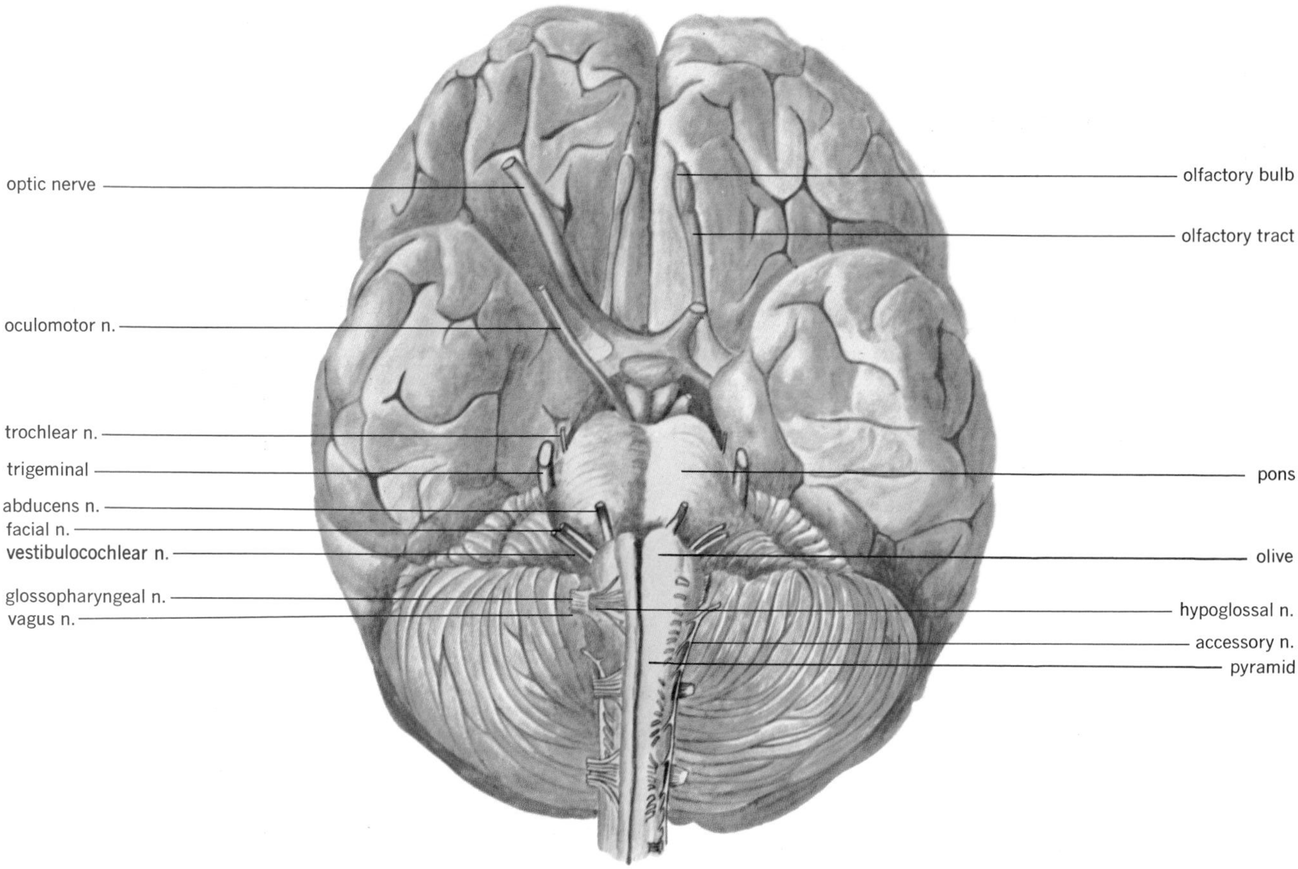

FIGURE 12.14. The distribution of the Vth cranial nerve (trigeminal).

FIGURE 12.15. The distribution of the VIIth cranial nerve (facial).

VI. *Abducent (abducens) nerve.* This nerve supplies motor and sensory fibers to one extrinsic eye muscle.

VII. *Facial nerve* (Fig. 12.15). The facial nerve supplies motor fibers to the muscles of the face, and the salivary glands. Sensory fibers for taste are derived from the anterior two thirds of the tongue.

VIII. *Vestibulochlear* (statoacoustic, acoustic, or auditory) *nerve.* This nerve carries sensory fibers from the cochlea and organs of equilibrium of the inner ear (semicircular canals, utriculus and sacculus), to the temporal lobe.

IX. *Glossopharyngeal nerve.* This nerve supplies sensory fibers to the posterior one third of the tongue for taste, and motor fibers to the throat muscles.

X. *Vagus nerve* (Fig. 12.16). The vagus nerve is a very important component of the autonomic system. It supplies motor and sensory fibers to nearly all thoracic and abdominal viscera.

XI. *Accessory (spinal accessory) nerve.* This nerve supplies motor fibers to muscles of the throat, larynx, soft palate, and to trapezius and sternocleidomastoid muscles.

XII. *Hypoglossal nerve.* This nerve supplies some of the tongue muscles and the infrahyoid muscles.

CLINICAL CONSIDERATIONS. Functions of the cranial nerves are easily assessed by simple procedures, and these form an important part of a neurological examination. The various procedures are presented in Table 12.2.

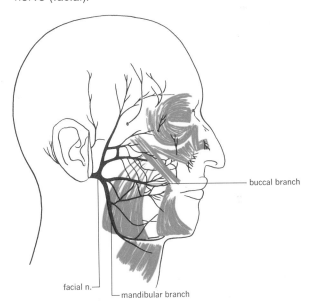

FIGURE 12.16. The distribution of the Xth cranial nerve (vagus).

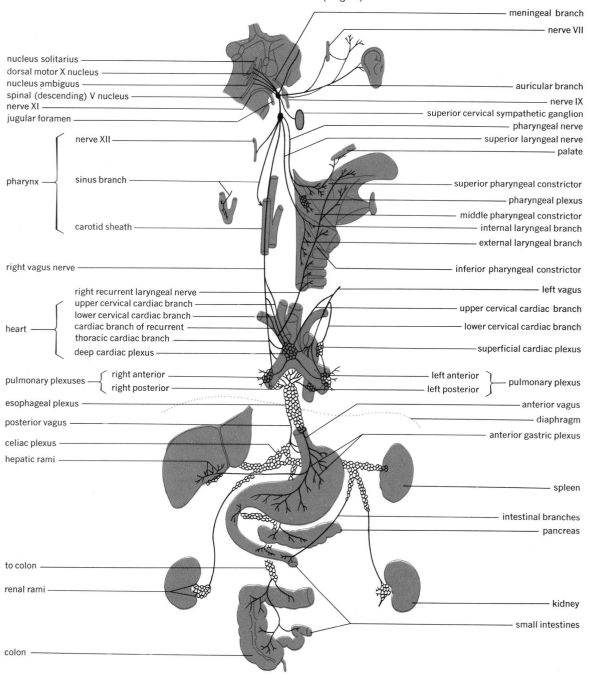

meningeal branch
nerve VII

nucleus solitarius
dorsal motor X nucleus
nucleus ambiguus
spinal (descending) V nucleus
nerve XI
jugular foramen

auricular branch
nerve IX
superior cervical sympathetic ganglion
pharyngeal nerve
superior laryngeal nerve
palate

nerve XII

superior pharyngeal constrictor
pharyngeal plexus

pharynx

sinus branch

middle pharyngeal constrictor
internal laryngeal branch
external laryngeal branch

carotid sheath

inferior pharyngeal constrictor

right vagus nerve

left vagus

right recurrent laryngeal nerve
upper cervical cardiac branch
lower cervical cardiac branch
cardiac branch of recurrent
thoracic cardiac branch
deep cardiac plexus

upper cervical cardiac branch
lower cervical cardiac branch

heart

superficial cardiac plexus

pulmonary plexuses { right anterior
{ right posterior

left anterior }
left posterior } pulmonary plexus

esophageal plexus
posterior vagus
celiac plexus
hepatic rami

anterior vagus
diaphragm
anterior gastric plexus

spleen

intestinal branches
pancreas

to colon
renal rami

kidney
small intestines

colon

TABLE 12.1 Summary of the Cranial Nerves

Nerve	Composition M = motor; S = sensory	Origin	Connection with brain or peripheral distribution	Function
I. Olfactory	S	Nasal olfactory area	Olfactory bulb	Smell
II. Optic	S	Ganglionic layer of retina	Optic tract	Sight
III. Oculomotor	MS	M—Midbrain	4 of 6 extrinsic eye muscles (superior rectus, medial and inferior rectus, inferior oblique)	Eye movement
		S—Ciliary body of eye	Nucleus of nerve in midbrain	Focusing, pupil changes, muscle sense
IV. Trochlear	MS	M—Midbrain	1 extrinsic eye muscle (superior oblique)	Eye movement
		S—Eye muscle	Nucleus of nerve in midbrain	Muscle sense
V. Trigeminal	MS	M—Pons	Muscles of mastication	Chewing
		S—Scalp and face	Nucleus of nerve in pons	Sensation from head
VI. Abducent	MS	M—Nucleus of nerve in pons	1 extrinsic eye muscle (lateral rectus)	Eye movement
		S—1 extrinsic eye muscle	Nucleus of nerve in pons	Muscle sense
VII. Facial	MS	M—Nucleus of nerve in lower pons	Muscles of facial expression	Facial expression
		S—Tongue (ant. 2/3)	Nucleus of nerve in lower pons	Taste
VIII. Vestibulocochlear (Statoacoustic, acoustic, auditory)	S	Internal ear: balance organs, cochlea	Vestibular nucleus, cochlear nucleus	Posture, hearing
IX. Glossopharyngeal	MS	M—Nucleus of nerve in lower pons	Muscles of pharynx	Swallowing
		S—Tongue (post 1/3), pharynx	Nucleus of nerve in lower pons	Taste, general sensation
X. Vagus	MS	M—Nucleus of nerve in medulla	Viscera	Visceral muscle movement
		S—Viscera	Nucleus of nerve in medulla	Visceral sensation
XI. Accessory	M	Nucleus of nerve in medulla	Muscles of throat, larynx, soft palate, sternocleidomastoid, trapezius	Swallowing, head movement
XII. Hypoglossal	M	Nucleus of nerve in medulla	Muscles of tongue and infrahyoid area	Speech, swallowing

TABLE 12.2   Methods of Assessing Cranial Nerve Function

Nerve being tested	Method	Comments
Olfactory	Ask patient to differentiate odors of coffee, bananas, tea, oil of cloves, etc., in each nostril.	Test may be voided if patient has a cold or other nasal irritation.
Optic	Ask patient to read newspaper with each eye; examine fundus with ophthalmoscope.	Perform test with lenses on, if patient wears them.
Oculomotor	Shine light in each eye separately and observe pupillary changes; look for movement by moving finger up, down, right, and left; have subject's eyes follow finger movements.	Look for defects in both eyes at same time. Light in one eye causes other one to constrict to lesser degree.
Trochlear		Requires special devices. If eye movement is normal, nerves are probably all right.
Trigeminal	Motor portion: have patient clench teeth and feel firmness of masseter; have patient open jaw against pressure.	Subject should exhibit firmness and strength in both tests.
	Sensory portion: test sensations over entire face with cotton (light touch) or pin (pain).	Deficits in particular areas indicate problems in one of the three branches.
Abducent	Perform test for lateral movement of eyes.	Must be very careful to observe any deficit in lateral movement.

Summary

1. The brain includes all nerve structures in the cranium. It is divided into:

   a. Brain stem: medulla, pons, midbrain.

   b. Cerebellum.

   c. Diencephalon: thalamus, hypothalamus.

   d. Cerebrum.

2. The medulla is the lowest part of the brain stem.

   a. It contains many important reflex centers for respiration, heart rate, and blood vessel size.

Nerve being tested	Method	Comments
Facial	Have subject wrinkle fore-head, scowl, puff out his cheeks, whistle, smile. Note nasolabial fold (groove from nose to corners of mouth).	Nerve supplies all facial muscles. Fold is maintained by facial muscles and may disappear in nerve damage.
Vestibulocochlear (statoacoustic, acoustic, auditory)	Cochlear portion: test auditory acuity with ticking watch; repeat a whispered sentence; use tuning fork. Compare ears. Examine with otoscope.	Ears may not be equal in acuity; note.
	Vestibular portion: not routinely tested.	If subject appears to walk normally and keep his balance, nerve is usually all right.
Glossopharyngeal and Vagus	Note disturbances in swallowing, talking, movements of palate.	
Accessory	Test trapezius by having subject raise shoulder against resistance; have subject try to turn head against resistance.	Strength is index here.
Hypoglossal	Have subject stick out tongue. Push tongue against tongue blade.	Tongue should protrude straight; deviation indicates same side nerve damage. Pushing indicates strength.

    b. Motor tracts from the cerebrum cross here.

    c. Damage to the medulla may interfere with breathing.

3. The pons lies above the medulla.

    a. It contains one of several respiratory centers (pneumotaxic).

    b. It connects to the cerebellum by way of the cerebellar peduncles.

    c. Damage here produces motor and respiratory deficits.

4. The midbrain lies above the pons.

    a. It shows the cerebral peduncles, receiving fibers from the cerebrum.

    b. It contains the corpora quadrigemina that connect eye and ear with motor fibers for righting and adjustments to visual and auditory stimuli.

    c. Damage here produces deficits in maintenance of balance and equilibrium.

5. The diencephalon includes the thalamus and hypothalamus.

    a. The thalamus has a number of nuclei that are sensory and motor in function. Sensory nuclei provide a means of organizing sensory impulses and crude awareness of sensation. Motor nuclei are concerned with motor activity of emotions. Damage results in the thalamic syndrome.

    b. The hypothalamus contains many nuclei. These nuclei control body temperature, water balance, the pituitary gland, food intake, gastric secretion, and emotional response. Damage produces deficits in the above functions.

6. The cerebellum lies posterior to the medulla and pons.

    a. It has a vermis and two hemispheres, an outer gray cortex and inner white matter.

    b. It coordinates and controls muscular movement and tone.

7. The reticular formation is concerned with arousal and alerting of the organism. It is part of the reticular activating system.

8. The cerebrum is the largest part of the brain.

    a. Each hemisphere contains several lobes, separated by fissures. The hemispheres are folded.

    b. The basal ganglia are nuclei deep within the cerebral hemisphere.

    c. The limbic system is composed of cerebral and lower level structures.

9. The cerebrum contains motor areas, sensory areas, visual and auditory areas, and association areas concerned with sensory interpretation and behavior. Damage interferes with movement, sensory reception, and understanding.

10. The basal ganglia control muscle tone, inhibit movement, and control muscle tremor. Damage produces tremor and loss of tone.

11. The limbic system contains centers for expression of emotion, and pleasure centers. Damage produces exaggerated display of emotions and sexual behavior.

12. The pyramidal system includes nerve fibers causing and modifying voluntary movement; the extrapyramidal system controls muscle reflexes and tone. Damage to either system produces spastic paralysis.

13. Twelve pairs of cranial nerves arise from the brain. They supply head and neck with motor and sensory fibers.

    a. For functions, see Table 12.1.

    b. For testing cranial nerves, see Table 12.2.

Questions

1. Give the functions centered with

  a. The medulla.
  b. Midbrain.
  c. Cerebellum.
  d. Thalamus.

2. What might be expected to be the primary signs of damage to the following areas?

  a. Hypothalamus.
  b. Primary cerebral motor area.
  c. Temporal association area.

3. Compare the contributions of the cerebellum, basal ganglia and cerebrum to movement.

4. What cranial nerves would be involved if the following symptoms were presented?

  a. Inability to smile.
  b. Deviation of the tongue to one side when protruded.
  c. Can't turn eyes upward.
  d. Can't turn eyes laterally.
  e. Can't maintain balance.
  f. No sensation on lower jaw.
  g. Can't swallow properly.

Readings

Baker, A. B. *An Outline of Clincial Neurology.* Wm. C. Brown Book Co. Dubuque, Ia., 1965.

Brazier, Mary A. *Electrical Activity of the Nervous System.* 3rd ed. Williams & Wilkins. New York, 1968.

Ganong, Wm. F. *Review of Medical Physiology.* Lange Medical Pubs. Los Altos, Cal., 1973.

Guyton, Arthur C. Textbook of Medical Physiology. Saunders. Philadelphia, 1971.

Lauria, A. R. "The Functional Organization of the Brain." *Sci. Amer. 222:66* (Mar) 1970.

chapter 13
The Autonomic Nervous System

chapter 13

The autonomic nervous system is usually defined as the part of the peripheral nervous system operating at the reflex level, and providing motor fibers to the visceral organs for control of glandular secretion, and smooth and cardiac muscle activity. Such fibers are termed VISCERAL EFFERENT FIBERS.

Such a definition is too limited, because it is apparent that there are also nerve fibers that convey sensory impulses from the viscera to the central nervous system. These fibers are termed VISCERAL AFFERENT FIBERS.

Connections of the autonomic nerves with the central nervous system (Fig. 13.1)

Afferent pathways

Visceral afferent fibers enter the spinal cord through the posterior or dorsal roots, in company with somatic afferents from the skin. Their cell bodies are located in the dorsal root ganglia, and synapses occur in the dorsal gray columns of the cord. Other visceral afferents enter the brain stem through the vagus nerve.

Efferent pathways

Visceral efferent fibers have their cell bodies in the lateral gray columns of the spinal cord, or in cranial nerve nuclei. Their axons pass away from the brain or cord in cranial nerve efferents, and in the anterior or ventral roots of the spinal nerves. Those that leave via anterior spinal nerve roots separate from the spinal nerves by way of the

FIGURE 13.1. The connections of the autonomic nervous system to the cord.

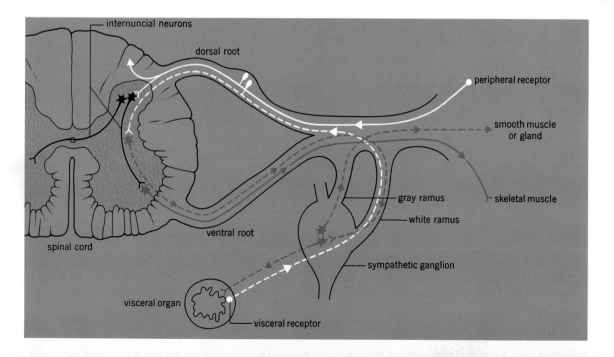

white rami. They may undergo synapses in ganglia close to the cord, or pass through these ganglia without synapsing. The *gray rami* provide a route for fibers that synapse in the ganglia to rejoin the spinal nerves. Two neurons usually extend from the central nervous system to the organ innervated. The first one is termed the *preganglionic neuron*; it leads from the central nervous system to a synapse outside the central nervous system. A *postganglionic neuron* leads from the synapse to the organ. Transmission from pre- to postganglionic neurons is chemical, using acetylcholine. Postganglionic neurons may produce acetylcholine or norepinephrine at their endings. This difference accounts for the different effects exerted by the two parts of the autonomic system.

Autonomic ganglia and plexuses

Synapses between pre- and postganglionic neurons occur in one of three areas.

The LATERAL or VERTEBRAL GANGLIA, also called the *sympathetic ganglia*, form a chain of 22 ganglia that lie alongside the vertebral bodies in the thoracic and abdominal cavities. They receive preganglionic fibers from the thoracic and lumbar spinal nerves.

COLLATERAL or PREVERTEBRAL GANGLIA (*plexuses*) are several groups of nerves and synapses located in association with important body organs or blood vessels. The cardiac, celiac (solar), and mesenteric plexuses are the most important ones. These plexuses receive preganglionic fibers from the cranial, thoracic, lumbar, and sacral portions of the central nervous system.

TERMINAL GANGLIA (*plexuses*) lie close to or in the organ innervated. The Meissner's and Auerbach's plexuses in the walls of the intestine are examples of terminal ganglia. Terminal ganglia receive preganglionic fibers from the cranial and sacral portions of the central nervous system.

Divisions of the autonomic system

Preganglionic autonomic fibers form three outgoing groups. The CRANIAL OUTFLOW consists of the motor fibers of cranial nerves 3, 7, 9, and 10. The THORACOLUMBAR OUTFLOW is composed of the autonomic fibers of all thoracic and lumbar spinal nerves. The SACRAL OUTFLOW consists of the autonomic fibers of sacral spinal nerves 2–4. All of these fibers provide for a dual innervation of most body organs (Fig. 13.2). Functionally, the fibers are grouped into a *parasympathetic (craniosacral) division,* and a *sympathetic (thoracolumbar) division.*

The parasympathetic (*craniosacral*) division

This division is formed from the cranial and sacral outflows. It supplies fibers to all autonomic effector organs except the adrenal medulla, sweat glands, splenic smooth muscle, and smooth muscle of blood vessels in skin and skeletal muscle. The preganglionic fibers are relatively long, synapsing in prevertebral and/or terminal ganglia or both. The postganglionic fibers are shorter, secrete acetylcholine at their endings, and are called CHOLINERGIC fibers. Activity in this division tends to *protect and conserve body resources,* and to preserve normal function.

The sympathetic (*thoracolumbar*) division

This division is formed by the thoracic and lumbar outflows. The preganglionic fibers are relatively short, synapsing in the vertebral and/or prevertebral ganglia. The postganglionic fibers are longer and secrete norepinephrine at their endings. They are thus known as ADRENERGIC

FIGURE 13.2. The autonomic nervous system.

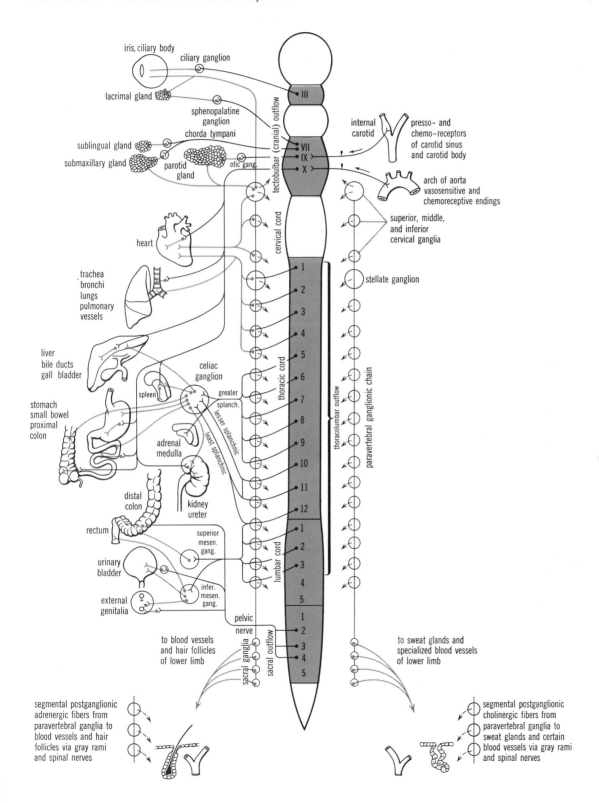

iris, ciliary body
ciliary ganglion
lacrimal gland
sphenopalatine ganglion
chorda tympani
sublingual gland
submaxillary gland
parotid gland
otic gang.

tectobulbar (cranial) outflow

III
VII
IX
X

internal carotid

presso- and chemo-receptors of carotid sinus and carotid body

arch of aorta vasosensitive and chemoreceptive endings

cervical cord

superior, middle, and inferior cervical ganglia

heart

stellate ganglion

trachea
bronchi
lungs
pulmonary
vessels

liver
bile ducts
gall bladder

celiac ganglion

spleen

greater splanch.
lesser splanchnic
least splanchnic

stomach
small bowel
proximal
colon

adrenal medulla

distal colon

kidney
ureter

rectum

superior mesen. gang.

urinary bladder

infer. mesen. gang.

external genitalia

thoracic cord

thoracolumbar outflow

paravertebral ganglionic chain

1
2
3
4
5
6
7
8
9
10
11
12

lumbar cord

1
2
3
4
5

pelvic nerve

sacral ganglia

sacral outflow

1
2
3
4
5

to blood vessels and hair follicles of lower limb

to sweat glands and specialized blood vessels of lower limb

segmental postganglionic adrenergic fibers from paravertebral ganglia to blood vessels and hair follicles via gray rami and spinal nerves

segmental postganglionic cholinergic fibers from paravertebral ganglia to sweat glands and certain blood vessels via gray rami and spinal nerves

TABLE 13.1   Effects of Autonomic Stimulation

Organ affected	Parasympathetic effects	Sympathetic effects
Iris	Contraction of sphincter pupillae; pupil size decreases	Contraction of dilator pupillae; pupil size increases
Ciliary muscle	Contraction; accommodation for near vision	Relaxation; accommodation for distant vision
Lacrimal gland	Secretion	Excessive secretion
Salivary glands	Secretion of watery saliva in copious amounts	Scanty secretion of mucus rich saliva
Respiratory system:		
Conducting division	Contraction of smooth muscle; decreased diameters and volumes	Relaxation of smooth muscle; increased diameter and volumes
Respiratory division	Effect same as on conducting division	Effect same as on conducting division
Blood vessels	Constriction	Dilation
Heart:		
Stroke volume	Decreased	Increased
Stroke rate	Decreased	Increased
Cardiac output and blood pressure	Decreased	Increased
Coronary vessels	Constriction	Dilation
Peripheral blood vessels:		
Skeletal muscle	No innervation	Dilation
Skin	No innervation	Constriction
Visceral organs (except heart and lungs)	Dilation	Constriction

fibers. This division supplies the same organs as does the craniosacral division, and is the only division supplying fibers to adrenal medulla, sweat glands, smooth muscles of the spleen, and blood vessels of skin and skeletal muscles. Thus alteration of activity in these organs occurs by changes in sympathetic activity. Activity of this division tends to *increase utilization of body resources* and to prepare the body to resist stressful situations. Some physiologists refer to the results of massive sympathetic stimulation as preparing the body for "fight-or-flight."

Organ affected	Parasympathetic effects	Sympathetic effects
Stomach:		
Wall	Increased motility	Decreased motility
Sphincters	Inhibited	Stimulated
Glands	Secretion stimulated	Secretion inhibited
Intestines:		
Wall	Increased motility	Decreased motility
Sphincters:		
Pyloric, iliocecal	Inhibited	Stimulated
Internal anal	Inhibited	Stimulated
Liver	Promotes glycogenesis; promotes bile secretion	Promotes glycogenolysis; decreases bile secretion
Pancreas (exocrine and endocrine)	Stimulates secretion	Inhibits secretion
Spleen	No innervation	Contraction and emptying of stored blood into circulation
Adrenal medulla	No innervation	Epinephrine secretion
Urinary bladder	Stimulates wall, inhibits sphincter	Inhibits wall, stimulates sphincter
Uterus	Little effect	Inhibits motility of non-pregnant organ; stimulates pregnant organ
Sweat glands	No innervation	Stimulates secretion (produces "cold sweat" when combined with cutaneous vasoconstriction)

Operation of the two divisions is TONIC or continuous, so that body activity reflects a balance between the actions of both systems. Functioning of the parasympathetic division is more restricted, with individual organs responding; the sympathetic division effects are more widespread. This is because sympathetic preganglionic fibers synapse with more postganglionic neurons supplying a wider body area, and, because norepinephrine is destroyed more slowly than acetylcholine. Table 13.1 summarizes effects of stimulation of the two divisions.

FIGURE 13.3.   Some drugs affecting the autonomic nervous system.

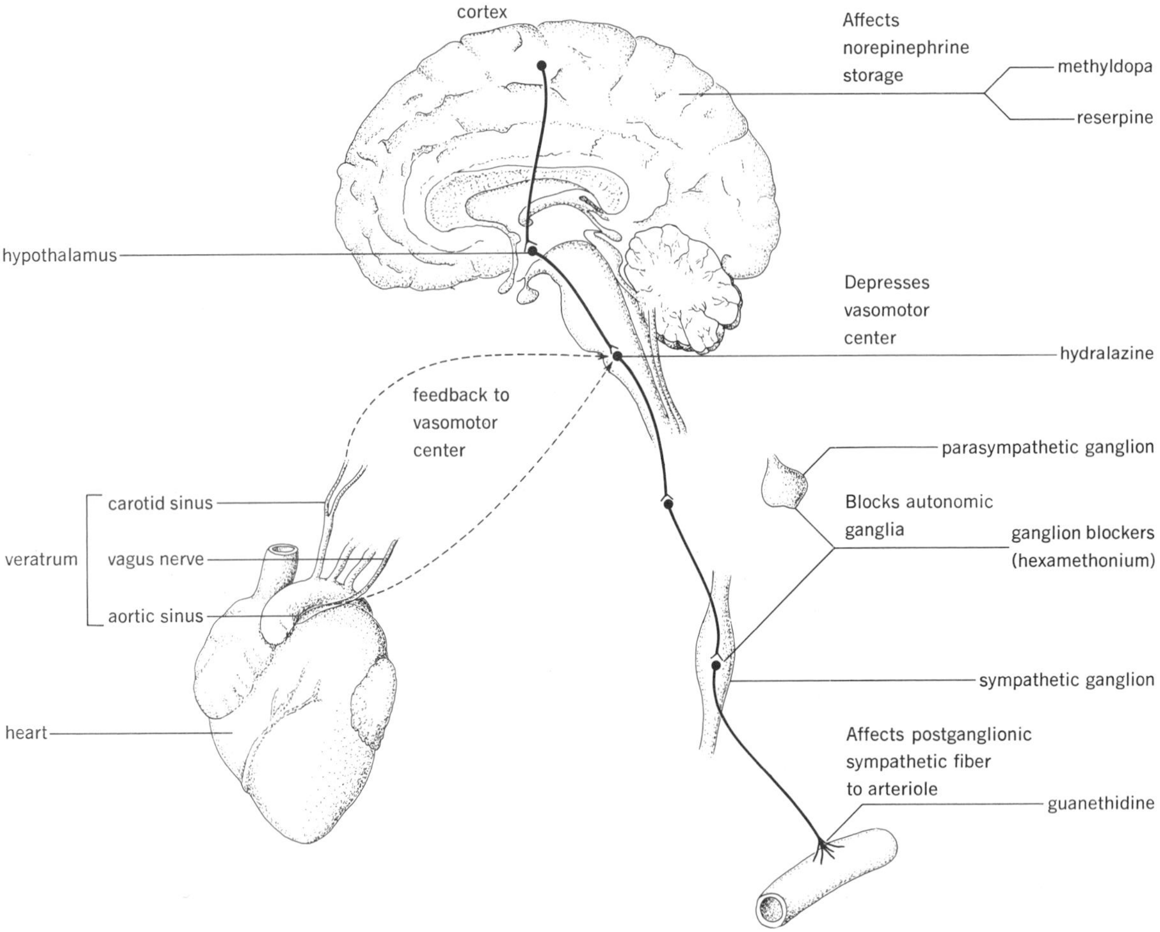

Higher autonomic centers

Several areas in the *brain stem* (other than cranial nerve nuclei), and *cerebral cortex* possess autonomic functions. The medulla contains the cardiac and respiratory centers. Increase in heart and breathing rates are typical sympathetic effects. The opposite effects are parasympathetic. The hypo-thalamus contains, anteriorly, areas that give parasympathetic effects and, posteriorly, areas that give sympathetic effects. The cerebral cortex contains, in the frontal lobes, areas that produce both types of effects.

Clinical considerations

There are some clinical conditions that are life threatening and that are believed to be the result of excessive autonomic activity. One of these is HYPERTENSION (high blood pressure). Surgical or chemical intervention designed to relieve the excessive autonomic activity (if this is the cause) is sometimes employed.

TABLE 13.2	Drugs and the Autonomic System		
Drug	How acts	Use	Comments
Reserpine	Depletes norepinephrine from postganglionic endings by increasing release. Causes decrease in heart action.	To alleviate hypertension	Is one of, and the most potent of a series of alkaloids derived from the Rauwolfia plant. Norepinephrine loss results in vaso-dilation and fall in blood pressure, and decrease in heart action.
Guanethidine	As above, but exerts no effect on heart	As above	Limited side effects make it a ''clinically advantageous drug''
Methyldopa	Lowers brain and heart content of norepinephrine by interfering with synthesis.	As above	May produce toxicity of liver.
Hydralazine	Depress vasoconstrictor center and inhibit sympathetic stimulation.	As above	Many side effects are common.
Veratrum (a series of plant alkaloids)	Slow heart action, stimulate vagus (parasympathetic) activity.	As above	Some side effects.
Hexamethonium	Blocks pre- to post-ganglionic trans-mission, Anticholinergic.	As above	
Acetylcholine	Increase or mimics parasympathetic stimulation. Causes some vasodilation.	As above	Not routinely employed. It is rapidly destroyed and affects many organs other than blood vessels.
Atropine	Inhibits acetylcholine and parasympathetic effects	Eye examinations; before general anaesthesia to dry respiratory secretions.	Sympathomimetic
Pilocarpine	Mimics parasympathetic stimulation.	Treatment of glaucoma.	Parasympatho-mimetic

Surgery and the autonomic system

Since the effect of sympathetic stimulation on most visceral blood vessels is to constrict them, and since constriction of large numbers of blood vessels raises the blood pressure, removal of the sympathetic effect should lower the blood pressure. A *sympathectomy*, which removes the sympathetic ganglia from the levels of T10-L2, may be performed to alleviate hypertension. If the procedure is effective, blood pressure will fall but, since the ganglia are gone, the individual has lost the ability to raise blood pressure in stress or in changes of position.

Drugs and the autonomic system

An alternative to surgical intervention in disorders of autonomic function is to employ drugs that either block or depress the effect of one of the divisions of the system, increase the activity of a division, or mimic the effects of stimulation of a division.

Drugs that mimic effects of stimulation of a given division are said to be *parasympathomimetic* or *sympathomimetic*. Those that prevent synaptic transmission are *ganglionic blocking agents*. Table 13.2 and Figure 13.3 present a summary of some drugs that affect the autonomic system, with notes as to how they act. Most are employed to alleviate hypertension by causing vasodilation.

Summary

1. The autonomic system controls visceral activity, and glandular secretion.

2. Visceral afferent fibers convey sensory impulses from viscera to CNS. Visceral efferent fibers convey motor impulses from CNS to organs.

3. The efferent pathway consists of two neurons and one synapse. The neurons are designated as pre- and postganglionic neurons. Transmission across the synapse is by acetylcholine. Postganglionic fibers may produce either acetylcholine or norepinephrine.

4. Autonomic ganglia and plexuses provide areas for synapse between pre- and postganglionic fibers.

5. Two divisions of the autonomic system are designated.
   a. The parasympathetic (craniosacral) division is composed of cranial and sacral outflows and conserves body resources.
   b. The sympathetic (thoracolumbar) division consists of thoracic and lumbar outflows and increases utilization of body resources.
   c. Most body viscera receive fibers from both divisions. Sympathetic fibers only go to sweat glands, adrenal medulla, spleen, blood vessels of skin and skeletal muscle.
   d. Effects of both systems are tonic (continuous).

6. Higher centers in brain stem, hypothalamus and cerebral cortex are responsible for several autonomic effects.

7. In treatment of hypertension, or high blood pressure, surgical, or chemical intervention is often necessary.
   a. Sympathectomy is a surgical procedure involving removal of sympathetic ganglia to remove constricting nerve impulses to visceral blood vessels.

b. Drugs may block the constricting effect of the sympathetic nervous system on blood vessels.

Questions

1. Compare the effects on the body of stimulation of the two divisions of the autonomic system.

2. Compare the anatomical differences between the two autonomic divisions.

3. How are the differential effects of stimulation of the two autonomic divisions explained?

4. What body organs receive a dual innervation by autonomic fibers? Which receive only a single innervation?

5. How is the activity of an organ that receives only a single autonomic innervation controlled?

6. What areas of the brain have autonomic effects?

7. How do drugs affect the autonomic system?

Readings

DiCara, L. V. "Learning in the Autonomic Nervous System." *Sci. Amer. 222*:30 (Jan) 1970.

Ganong, Wm. F. *Review of Medical Physiology,* 6th ed. Lange Medical Pubs. Los Altos, Cal., 1973.

Mountcastle, Vernon B. *Medical Physiology,* 12th ed. C. V. Mosby Co. St. Louis, Mo., 1968.

Pick, J. *The Autonomic Nervous System.* Lippincott. Philadelphia, 1970.

chapter 14
Blood Supply of the Central Nervous System; Ventricles and Cerebrospinal Fluid

chapter 14

Blood supply

The central nervous system depends on a continual supply of blood perhaps more than any other body organ. The vessels of the brain and cord supply the glucose and oxygen that form the essential nutrients of these organs.

Brain (Fig. 14.1)

A pair of INTERNAL CAROTID ARTERIES, derived from the common carotid arteries of the neck, and the paired VERTEBRAL ARTERIES, arising from the subclavian arteries to the upper appendage, form the arterial supply of the brain. The carotids give rise to three pairs of *cerebral arteries* supplying the cerebrum. The vertebrals join to form the *basilar* artery, and the basilar gives rise to vessels supplying the cerebellum and brain stem. The carotid and basilar arteries are joined by communicating vessels to form the CIRCLE OF WILLIS. The circle allows blood from one set of arteries to flow into the area supplied by the other in case of blockage or diminished flow. The veins draining the brain are largely unnamed except for the dural sinuses (Fig. 14.2).

Spinal cord

A single ANTERIOR SPINAL ARTERY (see Fig. 14.1) arises in a Y-shaped manner from the vertebral arteries. Paired POSTERIOR SPINAL ARTERIES arise separately from the vertebral or posterior cerebellar arteries. These three vessels run the length of the cord. SPINAL ARTERIES arise from the abdominal aorta and also supply blood to the cord.

The blood-brain barrier

Many substances that pass easily through the walls of capillaries elsewhere in the body, are slowed or stopped in their passage through the walls of cerebral capillaries. Protein molecules, antibiotics, urea, chloride, and sucrose are some of the materials whose passage is slowed. Such observations suggest that there is some sort of *blood-brain barrier* that restricts solute passage from the cerebral capillaries. Electron microscope studies of cerebral capillaries show the anatomical features presented in Figure 14.3. Several points may be emphasized.

Cerebral capillaries seem to lack the "pores" found in other capillaries.

Endothelial cells overlap in cerebral capillaries rather than being placed "end-to-end." This creates four membranes that a solute must pass through, rather than two.

A continuous basement membrane runs around each cerebral capillary, not just beneath each endothelial cell.

About 85 percent of the outer cerebral capillary surface is covered by glia. A substance moving from the capillary to the neuron may therefore have to pass through two cells (endothelial and glial).

The barrier may protect the brain neurons from the entry of potentially harmful substances; it also makes antibiotic treatment of brain inflammation very difficult.

Clinical considerations

No attempt will be made to present all possible disorders involving blood vessels of the central nervous system. It may be stated that cerebral vascular lesions ["strokes," cerebral vascular accidents (CVA)] account for more neurological disorders than any other category of pathological processes. "Stroke" is the third leading cause of death in the United States.

FIGURE 14.1. The arteries of the brain.

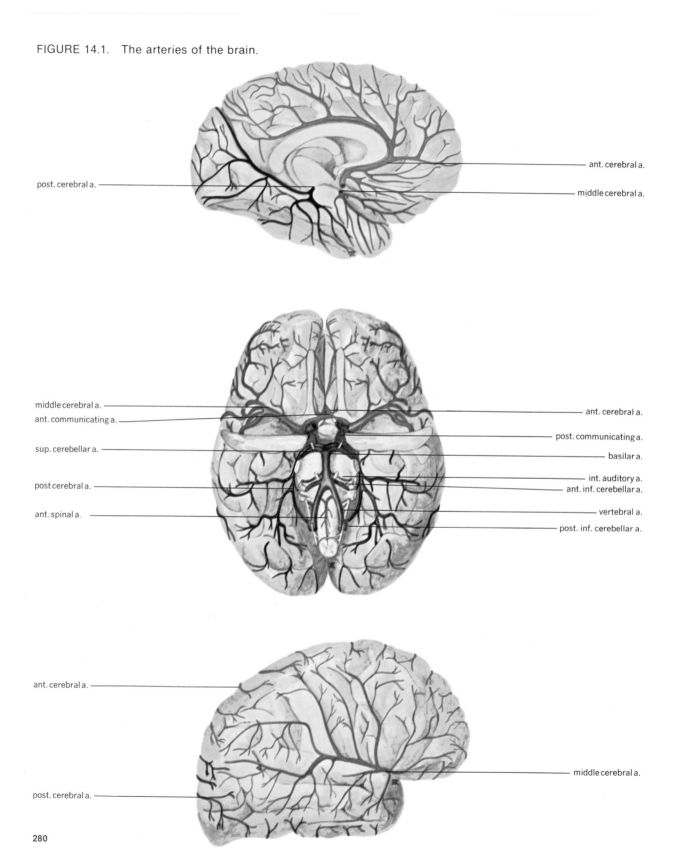

post. cerebral a.

ant. cerebral a.

middle cerebral a.

middle cerebral a.

ant. communicating a.

sup. cerebellar a.

post cerebral a.

ant. spinal a.

ant. cerebral a.

post. communicating a.

basilar a.

int. auditory a.

ant. inf. cerebellar a.

vertebral a.

post. inf. cerebellar a.

ant. cerebral a.

post. cerebral a.

middle cerebral a.

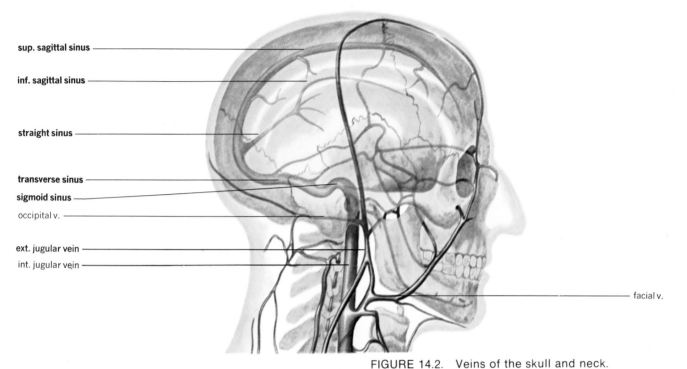

sup. sagittal sinus

inf. sagittal sinus

straight sinus

transverse sinus

sigmoid sinus

occipital v.

ext. jugular vein

int. jugular vein

facial v.

FIGURE 14.2.   Veins of the skull and neck.

FIGURE 14.3.   The anatomy of the blood-brain barrier. (Drawn from an electron micrograph.)

CLASSIFICATION OF VASCULAR DISORDERS OF THE BRAIN.

The vascular disorders of the brain are those in which normal nervous system function is impaired in two ways.

Failure of circulation to all or part of the brain, with death of nerve cells resulting.

Rupture of one or more blood vessels, with development of pressure or dissolution of brain tissue or both.

There are three major categories of cerebral vascular disorders, based on cause.

Narrowing or occlusive disorders can occur in any of the vessels of the brain circulation. Vessel diameter is reduced, flow decreases, and cells starve.

Embolic disorders are those in which brain vessels are blocked by masses of substance, such as, fats, air bubbles, or clots that are floating in the bloodstream.

Weakening of vessel walls, usually followed by dilation (aneurysm) and possible rupture. If the vessel ruptures, an intracranial hemorrhage results.

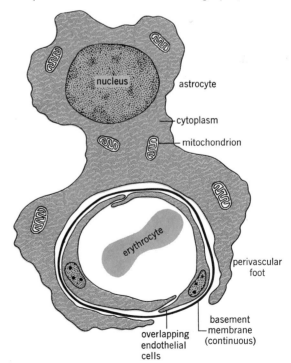

astrocyte

nucleus

cytoplasm

mitochondrion

erythrocyte

perivascular foot

basement membrane (continuous)

overlapping endothelial cells

Narrowing and occlusion usually occur as the result of deposition of fat in the walls of vessels and subsequent incorporation into the vessel wall (atherosclerosis), or hardening of the wall (arteriosclerosis). Infections (syphilis) may thicken the blood vessel walls and reduce diameter.

Embolism may result from sudden decompression and gas bubbles forming in the bloodstream (the "bends"), or by entry of air into the bloodstream after cardiac, pulmonary, or cerebral surgery. Fatty deposits on vessel walls may tear loose and become floating hazards. Clots may form during left heart surgery, and if not removed, may break loose and lodge in the brain.

Weakness is usually the result of structural defects, such as, failure to develop vessel coats, infection, toxic processes (poisons), and is secondary to the mechanical trauma of hypertension.

SYMPTOMS OF VASCULAR DISORDER. Symptoms of any vascular disorder affecting the brain are produced by gradual or sudden reduction of blood supply and are essentially the same. Any differences are due to the area of the brain whose blood supply is diminished. The symptoms common to most vascular disorders of the brain are:

Headache, usually vague and short-lived.
Dizziness and *faintness* which is exaggerated by postural changes.
Nausea and *vomiting*.
Reflexes are unequal; e.g., pupillary light reflex.
Muscular weakness, often one-sided, and on the opposite side from where the lesion has occurred.
Mental confusion and speaking difficulties.

These symptoms suggest that a CVA has occurred, and do not lead to diagnosis of specific disorders.

Ventricles and cerebrospinal fluid

The nervous system develops as a hollow tube. The central cavity of this tube is modified, as the system develops, into a series of brain cavities known as the *ventricles*. Within the ventricles are vascular structures that secrete the *cerebrospinal fluid*. The fluid fills the ventricles and surrounds the spinal cord.

The ventricles (Fig. 14.4)

Each cerebral hemisphere contains a *lateral ventricle* that communicates through an *interventricular foramen* with the *third ventricle*. The third ventricle is a single, narrow and slitlike structure lying between the two halves of the diencephalon. The *cerebral aqueduct* is a small channel running through the midbrain and pons to the *fourth ventricle* located in the medulla. *Three openings* are found in the roof of the fourth ventricle; two lateral foramina of Luschka, and a single midline foramen of Magendie. These foramina communicate with the *subarachnoid space* of the cord and brain.

TABLE 14.1 Plasma and CSF Compared for Some Major Constituents

Constituent or property	Plasma	CSF
Protein	6400–8400 mg%	15–40 mg%
Cholesterol	100–150 mg%	0.06–0.20 mg%
Urea	20–40 mg%	5–40 mg%
Glucose	70–120 mg%	40–80 mg%
NaCl	550–630 mg%	720–750 mg%
Magnesium	1–3 mg%	3–4 mg%
Bicarbonate (as vol % of CO_2)	40–60 mg%	40–60 mg%
pH	7.35–7.4	7.35–7.4
Volume	3–4 liters	200 ml
Pressure	0–130 mm Hg	110–175 mm CSF

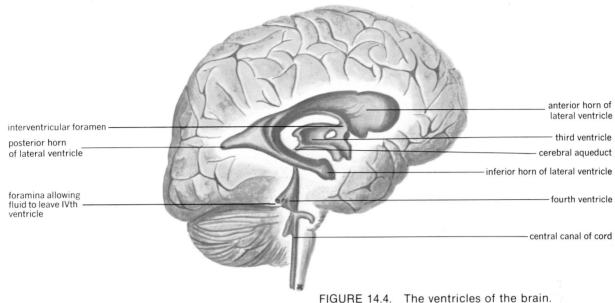

FIGURE 14.4.   The ventricles of the brain.

FIGURE 14.5.   The subarachnoid villi.

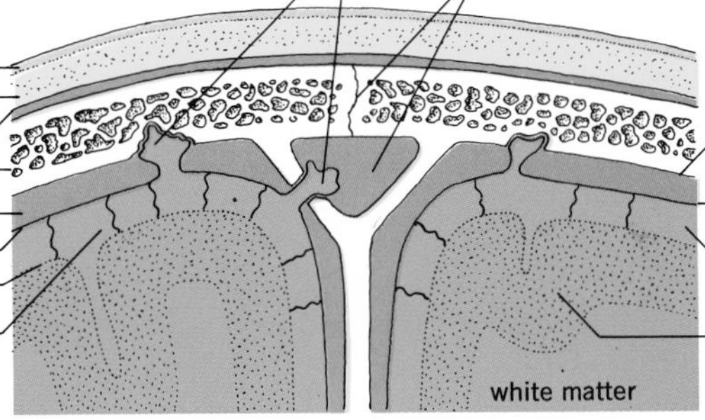

Cerebrospinal fluid

Vascular structures known as *choroid plexuses,* located in the lateral, third, and fourth ventricles, produce the CerebroSpinal Fluid (CSF) from the blood plasma. CSF differs from plasma in enough ways to suggest that it is formed by active processes and not filtration (Table 14.1).

The fluid circulates in a top to bottom direction, that is, from lateral ventricles to third ventricle, to aqueduct, to fourth ventricle, to subarachnoid space. Absorption of the fluid into blood vessels occurs through spinal venous vessels and subarachnoid villi (Fig. 14.5) in the skull.

The fluid acts as a shock absorber for the brain

and cord, is easily removed by spinal puncture for analysis, and compensates for changes in blood volume, keeping cranial volume constant.

Clinical considerations

The term hydrocephalus (G. *ydor,* water, + *kephale,* head) means an increased accumulation of cerebrospinal fluid in the ventricles of the brain. It may be due to increased production of CSF, to decreased absorption, or blockage of the flow of CSF.

Choroid plexus tumor may increase production of

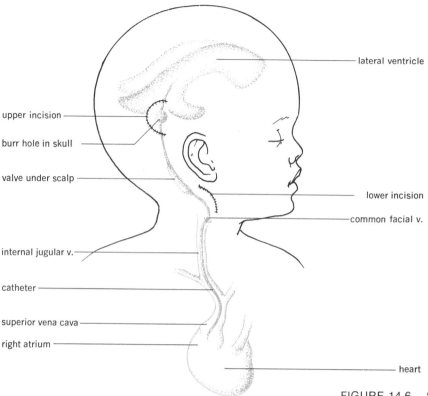

lateral ventricle

upper incision

burr hole in skull

valve under scalp

lower incision

common facial v.

internal jugular v.

catheter

superior vena cava

right atrium

heart

FIGURE 14.6.   Surgical treatment of hydrocephalus.

CSF. A pneumoencephalogram, in which the CSF in the ventricles is replaced by air usually demonstrates the presence of such a tumor.

Decreased absorption is usually due to failure of the subarachnoid villi (see Fig. 14.5) of the meninges to develop.

Blockage, by tumor or failure of the ventricular system to develop, accounts for about one third of the cases. The cerebral aqueduct, because of its small size, is a common site of blockage of CSF flow.

As a result of excessive fluid filling the ven-

tricle(s), ballooning of the ventricles may occur, and the head enlarges because the sutures between the cranial bones have not yet grown together. Diagnosis of the condition is made in infants by repeated measurements of head circumference which demonstrate the abnormal rate of increase of head size.

Treatment is usually surgical, and involves bypassing the obstruction (if present) with a tube and a one-way valve (Fig. 14.6). The CSF may be delivered to the right atrium of the heart or the peritoneal (abdominal) cavity.

Summary

1. The central nervous system has blood vessels designed to provide it with nutrients and to remove its wastes of activity.

  a. The brain is supplied by the internal carotid and vertebral arteries.

b. The spinal cord is supplied by anterior and posterior spinal arteries, and by spinal arteries arising from the aorta.

c. Capillaries in the cerebrum are less permeable to solute passage than elsewhere in the body. A blood-brain barrier exists. The barrier has an anatomical basis.

2. Vascular disorders of the central nervous system deprive nerve cells of their nutrients and lead to cell death. They are of three types.
   a. Occlusive disorders result from narrowing of vessels and deprivation of neurons of oxygen and fuels.
   b. Embolic disorders result when floating masses or bubbles lodge in a cerebral or brain vessel.
   c. Weakening of vessel walls may result in rupture.

3. Symptoms of vascular disorders are basically the same regardless of cause.
   a. Headache, dizziness, nausea, unequal reflexes on both sides of the body, muscular weakness, and mental confusion are the most common symptoms.

4. A series of cavities or ventricles, filled with cerebrospinal fluid, are found within the brain. The fluid also fills the subarachnoid space of brain and cord.
   a. The ventricles are four in number.
   b. The fluid is formed from plasma in choroid plexuses. It is removed by absorption from the subarachnoid spaces.
   c. Excessive cerebrospinal fluid, from whatever cause, produces hydrocephalus.

Questions

1. What would be expected to happen to brain circulation if (as may happen) flow through a carotid artery is reduced? What portions of the cerebral circulation would be most greatly affected?

2. What is implied by the term blood-brain barrier? What is the structural basis of this barrier?

3. What are the major types of vascular disorders that may affect the central nervous system? What are the symptoms that suggest a vascular disorder?

4. Describe the flow of cerebrospinal fluid.

5. How may hydrocephalus develop?

6. What is the clinical importance of the cerebrospinal fluid?

Readings

Baker, A. B. *An Outline of Clinical Neurology.* Wm. C. Brown Book Co. Dubuque, Ia., 1965.

Gillian, L. A. "The Arterial Blood Supply of the Human Spinal Cord." *J. Comp. Neurol.* *110*:75–103, 1958.

Pappas, G. D. "Some Morphological Considerations of the Bloodbrain Barrier." *J. Neurol. Sci. 10*:241–246, 1970.

Stephens, R. B., and D. L. Stillwell. *Arteries and Veins of the Human Brain.* Thomas Pubs. Springfield, Ill., 1969.

chapter 15
Sensation

chapter 15

Homeostatic regulation implies the presence in the body of RECEPTORS that can detect alterations in the internal and/or external environments of the body. AFFERENT NERVES then make connections with the CENTRAL NERVOUS SYSTEM. Spinal tracts and various brain stem or diencephalic structures may be involved in the appreciation of sensations, and the cerebral cortex is involved in the final analysis of sensory input.

Receptors

Definition and organization

A receptor is considered to be the peripheral or outer end of an afferent neuron that is specialized to respond best to a particular type of stimulus. An afferent nerve, its branches, and the attached receptors (if any) constitute a *sensory unit.* The area of the body supplied by that sensory unit is the unit's *peripheral receptive field.* Receptive fields may overlap, and density of receptors (numbers per area) may vary in different regions of the body. Thus *sensitivity* to a particular stimulus is not necessarily the same in all body areas. Typically, three neurons, designated first order, second order, and third order, carry impulses from the periphery to the cerebral cortex.

Basic properties of receptors and their nerve fibers

All receptors follow certain principles of operation.

The *law of adequate stimulus.* A given receptor responds best, but not exclusively, to a particular stimulus, creating specificity on the part of the receptor. This principle is what accounts for the great variety of receptors the body contains, since there has to be a different receptor for each stimulus.

The *law of specific nerve energies.* All impulses in different sensory nerves are alike, regardless of the receptor involved. What we appreciate as a sensation is thus due to the central connections the fiber eventually makes, and the brain becomes the important organ for analysis of sensory input.

All *receptors transduce* (change) a stimulus into a nerve impulse. A particular stimulus produces a permeability change in the receptor, which results in the formation of a *generator potential.* This, in turn, depolarizes the nerve fiber to create the nerve impulse. This property enables a wide variety of different stimuli to be reduced to a form that the brain can interpret, that is, a nerve impulse.

Most *receptors show accommodation.* As a stimulus of a given strength is continued, the frequency of receptor discharge decreases, or may even cease. This "adjustment" constitutes accommodation, and enables familiar or trivial stimuli to be ignored. Thus we do not feel the clothes on our bodies, rings on our fingers, or continue to smell odors in a room.

Classification of receptors

Receptors are most meaningfully classified according to the stimuli that activate them, as is shown in Table 15.1

TABLE 15.1   A Classification of Receptors

Adequate stimulus	Example(s)
Mechanical (mechanoreceptors) pressure, bending, tension	Touch and pressure receptors in skin. Pressure or stretch sensitive receptors (baro-receptors) in blood vessels, lungs, gut. Equilibrium and balance receptors in inner ear (semicircular canals, maculae). Organ of hearing (hair cells of cochlea). Kinesthetic receptors in muscles, tendons, joints.
Chemical (chemoreceptors) substances in solution	Taste Smell
Stimulation by type or concentration of chemical	Carotid and aortic bodies (CO_2 and O_2 sensitive) Osmoreceptors in hypothalamus
Light (photoreceptors)	Eye
Thermal change (thermoreceptors)	Heat Cold
Extremes of most any stimuli (nociceptors)	Pain

Specific sensations: somesthesia

Somesthetic sensations are those resulting from stimulation of receptors other than the ones for sight, hearing, taste, and smell. They are, therefore, sensations originating in the skin, muscles, tendons, joints, and viscera. Sensations from the skin are the *cutaneous sensations*, and include touch, pressure, heat, cold, and some types of pain. Those from muscles, tendons, and joints are the *kinesthetic sensations*. Those from visceral organs are *visceral sensations*.

Touch and pressure

The receptors for touch and pressure are the MEISSNER'S CORPUSCLES of the skin and mucous membranes, and the PACINIAN CORPUSCLES of the skin, mesenteries, and pancreas (Fig. 15.1). Naked nerves, or nerve fibers without corpuscles on their endings around hair follicles, also convey impulses of touch when the hair is bent. The

stimulus that causes a nerve impulse to form is mechanical change of shape of the receptor. Touch receptors are most numerous on the tongue and fingertips, are less numerous on the face, and least numerous in body areas such as the back. The first-order neurons carrying the senses of touch and pressure enter the CNS through the spinal nerves and the fifth cranial nerves. The gracile and cuneate tracts of the cord are the second—order neurons and carry the impulses to the brain stem. A band of fibers goes next to the thalamus. The thalamus sends third-order neuron relay fibers to the cortex (areas 3,1,2).

Thermal sensations: heat and cold

Discrete end organs for heat and cold cannot always be demonstrated in the skin. A group of ill-defined organs called SENSORY END BULBS, and that differ from other receptors in size, shape,

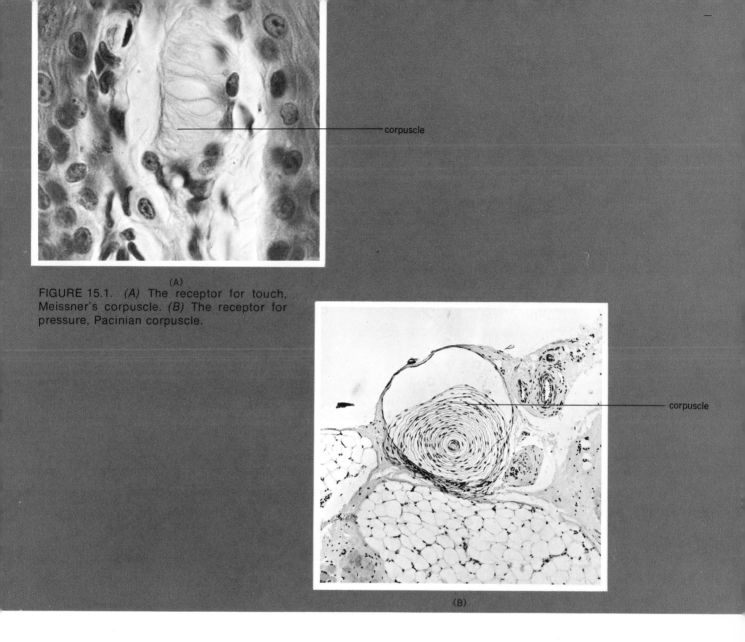

corpuscle

(A)

FIGURE 15.1. *(A)* The receptor for touch, Meissner's corpuscle. *(B)* The receptor for pressure, Pacinian corpuscle.

corpuscle

(B)

and configuration of their nerve fibers have been described. The receptors for heat and cold belong to this group. Sensitivity to thermal change appears "spotlike," with cold spots outnumbering heat spots by 4–10 to 1. Cold thus appears to be a greater threat to body homeostasis than does heat, if numbers of receptors indicate anything.

Warmth receptors respond best between 37 and 40° C (98.6–104° F). Cold receptors respond maximally at 15–20° C (59–68° F) and again strongly at 46–50° C (114–122° F). The latter range of discharge accounts for the *paradoxical cold,* or feelings of coldness that may be experienced when stepping into a hot shower.

The first-order neurons carrying thermal sensations center the CNS through spinal nerves and the fifth cranial nerves. The lateral spinothalamic tracts are the second-order neurons and end in the thalamus. Relay fibers pass from the thalamus to the cerebral cortex (areas 3,2,1), as the third-order neurons.

Pain

Pain is a protective sensation for the body, as it usually indicates damage or potential damage. It may result from overstimulation of any nerve

TABLE 15.2 Referred Pain		
True point of origin of pain	Common cause	Area to which pain is referred
Heart	Ischemia secondary to infarction	Base of neck, shoulders, pectoral area, arms
Esophagus	Spasm, dilation, chemical (acid) irritation	Pharynx, lower neck, arms, over heart ("heart burn")
Stomach	Inflammation, ulceration, chemicals	Below xiphoid process (epigastric area)
Gall bladder	Spasm, distention by stones	Epigastric area
Pancreas	Enzymatic destruction, inflammation	Back
Small intestine	Spasm, distention, chemical irritation, inflammation	Around umbilicus
Large intestine	As above	Between umbilicus and pubis
Kidneys and ureters	Stones, spasm of muscles	Directly behind organ or groin and testicles
Bladder	Stones, inflammation, spasm, distension	Directly over organ
Uterus, uterine tubes	Cramps (spasm)	Lower abdomen or lower back

fiber, and is served by naked nerve endings. There are three varieties of pain.

"*Bright*" or pricking pain is intense, usually short-lived, and is easily localized to the body part affected. Cutting a finger gives bright pain.

Burning pain, such as that experienced in burning a body part, is slow to develop, lasts longer, and is less easily localized.

Aching (visceral) pain is often nauseating, persistent, and may be referred, as shown in Table 15.2, to the skin of body areas other than that of its source.

Pain is referred to a different area than that in which it originates, because impulses from the painful area enter the spinal cord and stimulate neighboring neurons receiving fibers from the skin. Impulses pass to the cortex and are localized as coming from the skin. Referred pain indicates visceral overstimulation and should be of the utmost concern to the physician and allied medical personnel. It may signal life-threatening disorders of visceral organs.

HEADACHE. Headache is probably the most common pain complaint. Several things may contribute to its production.

Tension on blood vessels or meninges, as in brain tumor.

Increased intracranial pressure, from vasodilation. The increased amount of blood entering the rigid skull then puts pressure on the brain and its coverings.

Inflammation of the sinuses or any other pain-sensitive structure in the brain or cranium. Commonly, the pain of an inflamed sinus is felt in the forehead or cheek.

Spasm of cranial muscles. This produces the so-called "tension headaches." This type of headache is most commonly felt in the back of the head, or may be a "pain in the neck."

Eye disorders may result in pain in or behind the eyes as a result of eyestrain.

The conduction pathways for pain include the spinal nerves and the fifth cranial nerves and the spinothalamic tracts. The thalamus may interpret pain, and cortical areas 3,1,2 localize pain and determine its intensity.

Aspirin appears to be effective in treatment of pain by reducing nerve conduction and raising thresholds for pain in the thalamus-hypothalamus areas.

Kinesthetic sense

Sense of body position and movement is provided by receptors in muscles, tendons, and joints. Tension on, or shortening of, the receptor appears to be the adequate stimulus. *Conscious muscle sense,* or appreciation of body position, is carried over afferent spinal and cranial nerves to the gracile and cuneate tracts, to the thalamus, and then to the cerebrum; *unconscious muscle sense,* concerned with muscle tone, is carried over afferent spinal and cranial nerves to the spinocerebellar tracts, and then to the cerebellum.

The MUSCLE SPINDLE is an important sensory receptor of the muscles which provides information on the muscle's length, and the rate of shortening when the muscle contracts. The spindle consists of several tiny skeletal muscle fibers known as the *intrafusal fibers,* around which are wrapped the *annulospiral sensory nerves.* Stretching of the fibers causes stimulation of these spiral endings and provides the information about muscle length. The amount of stretch required to cause

stimulation of those endings is determined by the degree of contraction of the intrafusal fibers themselves. Logically, if the fibers are contracted, it will take more force to lengthen them, that is, a greater stretch on the muscle. The degree of contraction of the intrafusal fibers is determined by so-called *gamma-efferent nerves* that are believed to come from the basal ganglia of the brain. The spindles are important parts of the cross-extensor and other spinal muscular reflexes.

Synthetic senses

Itch, tickle, and *vibrational sense* have no specific nerve endings to serve them. They appear to result from particular modes (methods of how a stimulus is delivered) of stimulation, or simultaneous stimulation of touch, pain, and pressure receptors. Itch appears to result from chemical stimulation of skin pain fibers. If a light stimulus moves across the skin, tickle results, through stimulation of touch receptors. Vibration sense appears to result from rhythmical and repetitive stimulation of pressure receptors; frequencies of 250–300 cycles per second are the most effective.

The reader is reminded of the roles of thalamus and hypothalamus in sensory interpretation. Those sections may be profitably reviewed at this time.

Visceral sensations

Visceral or *organic sensations* result from stimulation of internal organs, and include hunger and thirst. Other sensations (bladder and rectal fullness) will be discussed in appropriate chapters that follow.

Hunger is associated with powerful contractions of the stomach, which may be due to lowering of the blood glucose levels. The compression of the gastric nerves by the powerful contractions may create the unpleasant and often near-painful quality of the sensation.

Thirst occurs when the oral membranes dry as a result of decreased salivary secretion. The decreased secretion is, in turn, due to dehydration, and signals a need for fluid intake.

The somesthetic senses are summarized in Table 15.3.

Clinical considerations

Damage to peripheral nerves produces symptoms that depend on the location of the lesion and the function of the nerve. Sensory disturbances include changes in total sensation (anesthesia) or specific changes in position sense, vibratory sense, or pain, touch, and pressure.

ANESTHESIA. Complete cutting of a nerve produces loss of all sensation in the area the nerve supplies. Injuries from bumps, heat, friction rubs (shoe on the heel) are not felt, and severe damage to the skin may occur. Protection of the anesthesized area becomes the primary consideration.

PARESTHESIA. This term refers to abnormal sensations arising from damage to peripheral nerves or cord tracts. It usually involves pain being referred to the area served by the nerve, although no damage has occurred in that area.

HYPERESTHESIA. This term refers to lowered pain thresholds, and a greater sensitivity to pain.

It is again noted that peripheral nerves may regenerate if the separation or area of damage is not great, and if the neurilemmal sheath is intact. Otherwise, the regenerating fiber may not make its proper connection as before, or it may form a tangled mass known as a *neuroma,* which is both painful and functionless.

Smell and taste

Smell and taste are chemical senses, in that the adequate stimuli are chemicals in solution in air or water. The senses reinforce one another; anyone with a cold and blocked nasal passages knows how "tasteless" his food becomes. Smell has survival value for lower animals, and perhaps also for people, in terms of identification of enemies, territories, and food seeking. Taste may play a part in food selection if an animal is deficient in certain nutritional elements. Children have been shown over the long run to select a diet that is adequate if presented with a variety of foods in separate dishes. Food "fads" should not be a cause for concern unless they continue for long periods of time.

Smell

The receptors for the sense of smell are found in the OLFACTORY EPITHELIUM of the nasal cavities (Fig. 15.2). The olfactory cells are bipolar nerve cells. The outer portion of each cell is a dendrite modified into a cylindrical process that proceeds vertically and then turns parallel to the epithelial surface. On the dendritic process are "receptor sites" into which the molecules to be smelled fit.

Classification of odors appears to depend on molecular shape, and is highly subjective. *Seven basic odors* are described:

Camphoraceous, musky, floral, etheral, pungent, putrid, and pepperminty.

In general, the sense of smell is much more acute than the sense of taste, about 25,000 times more sensitive, as is judged by the concentrations of substances required to create a sensation. Low concentrations of substances cause maximum receptor response, followed by rapid adaptation. This fact suggests that the sense of smell is designed to detect the presence, rather than the amount, of an odor.

The neural pathways for smell (Fig. 15.3) pass from *olfactory cells* to the *olfactory bulb* of the brain, then over the *olfactory tracts* to the *olfactory areas* in the temporal and frontal lobes of the brain. Fibers also pass to autonomic centers in the hypothalamus and brain stem to produce salivation, sexual behavior, and emotional expression as a result of olfactory stimulation.

TABLE 15.3	A Summary of Somesthetic Senses					
Sensation	Receptor	Adequate stimulus	First order neuron	Second order neuron	Third order neuron	Comments
Touch	Meissner's corpuscle	Mechanical pressure to skin	Spinal or cranial afferent	Gracile and cuneate tracts	Thalamus to cortex	Light pressure causes change in receptor shape
Pressure	Pacinian corpuscle	As above	As above	As above	As above	Heavier pressure causes change in receptor shape
Heat	Sensory ending	Rise of temperature, especially between 37–40°C	As above	Spinothalamic tracts	As above	—
Cold	Sensory ending	Fall of temperature, especially between 15–20° C	As above	As above	As above	—
Pain	Naked nerve endings	Overstimulation of any nerve; strong stimulation of naked nerves as by tension, pressure	As above	As above	As above	Three types of pain: bright, burning, visceral
Kinesthesis	Muscle, joint and tendon organs	Tension, stretch or movement	As above	Conscious: gracile and cuneate tracts.	Thalamus to cortex	Conscious is body position
				Unconscious: spino-cerebellar tracts	Cord to cerebellum	Unconscious is tone.
Synthetic (itch, tickle, vibration)	None specific	Chemical and mechanical stimulation	—	—	—	Itch: chemical stimulation of pain fibers Tickle: moving stimulation Vibration: touch organs
Visceral (hunger, thirst)	None specific	Mechanical and drying	—	—	—	Most visceral afferents are carried in the vagus nerve.

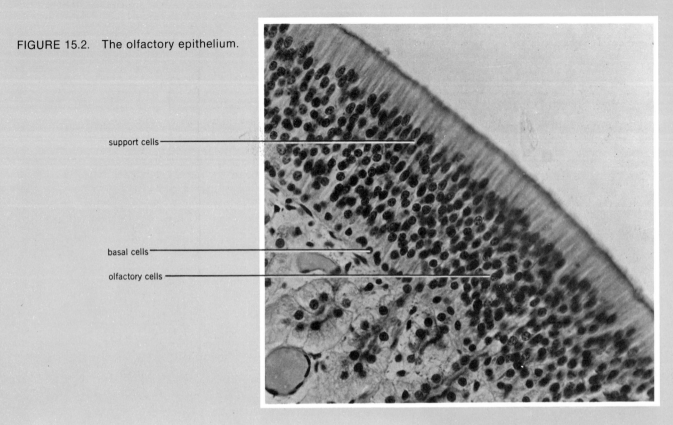

FIGURE 15.2. The olfactory epithelium.

support cells

basal cells

olfactory cells

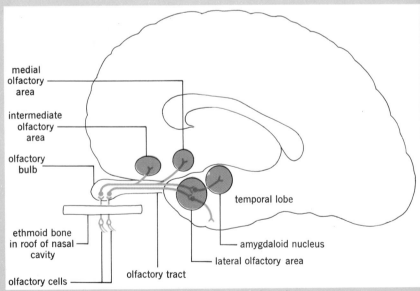

medial
olfactory
area

intermediate
olfactory
area

olfactory
bulb

temporal lobe

ethmoid bone
in roof of nasal
cavity

amygdaloid nucleus

lateral olfactory area

olfactory cells

olfactory tract

FIGURE 15.3. The basic neural pathways for the
sense of smell.

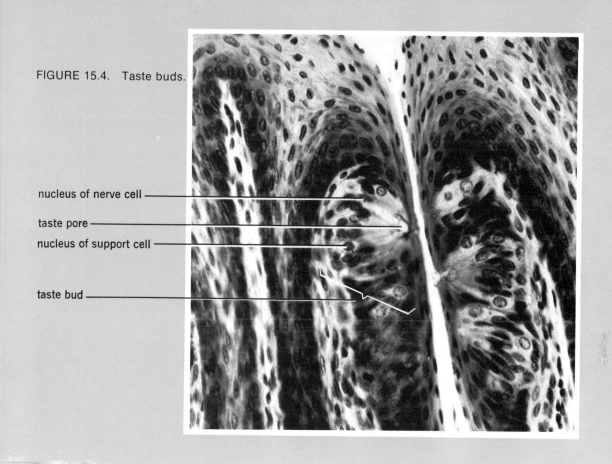

FIGURE 15.4.   Taste buds.

nucleus of nerve cell —

taste pore —

nucleus of support cell —

taste bud —

FIGURE 15.5.   The distribution of the sense of taste
on the tongue.

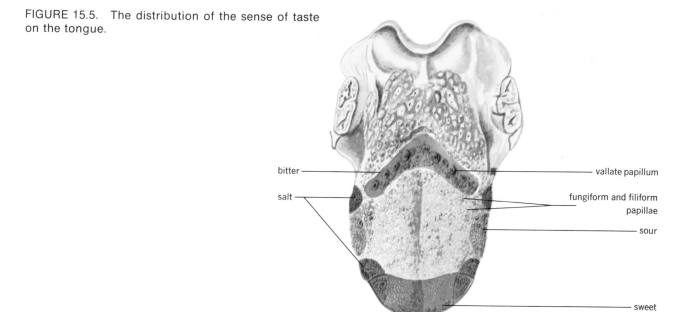

bitter —

salt —

— vallate papillum

fungiform and filiform
papillae

— sour

— sweet

CLINICAL CONSIDERATIONS. Disorders of the olfactory system are rare, but have considerable importance when present. Tumors pressing on or invading any portion of the system may produce loss of smell. Temporal lobe lesions may produce olfactory hallucinations, usually of disagreeable odors. Such odors often precede an epileptic seizure.

Taste

The receptors for taste are the TASTE BUDS (Fig. 15.4), located primarily on the tongue. A few buds may be found on the soft palate, and in the walls of the pharynx and epiglottis. The nerve cells in the buds have a rodlike shape, and have "taste hairs" on their outer ends. The hairs are actually short dendritic processes. The material to be tasted must fit into receptor sites on the taste hairs to depolarize the nerve cell. *Four basic taste sensations* are described:

Sour tastes are produced by H^+ or acids.

Salty tastes are produced by cations of ionized salts, such as Na^+, NH_4^+, Ca^+, Li^+, and K^+.

Sweet tastes are produced by the presence of compounds containing hydroxyl ($-OH$) groups. Alcohols, amino acids, sugars, ketones, and lead salts all taste sweet (children may eat paint for this reason and suffer lead poisoning).

Bitter tastes are produced by alkaloids such as quinine, strychnine, and caffeine and also by many long chain organic molecules.

The buds serving these sensations are not uniformly distributed over the tongue (Fig. 15.5) and do not have the same thresholds. Greatest sensitivity is to bitter, followed by sour, salt, and sweet.

The afferent pathways for taste include the VII (facial) cranial nerve for buds on the anterior one third of the tongue, the IX (glossopharyngeal) nerve for buds on the posterior two thirds of the tongue, and the X (vagus) nerve for those elsewhere. The impulses pass to brain stem, thalamus, and to the parietal cortex (areas 3,1,2). These paths are shown in Figure 15.6.

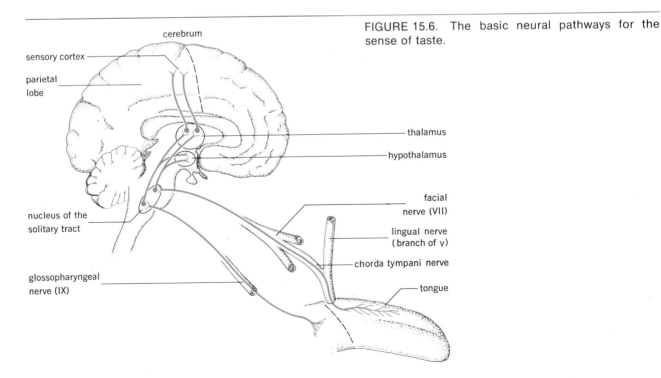

FIGURE 15.6. The basic neural pathways for the sense of taste.

cerebrum

sensory cortex

parietal lobe

thalamus

hypothalamus

facial nerve (VII)

lingual nerve (branch of v)

chorda tympani nerve

tongue

nucleus of the solitary tract

glossopharyngeal nerve (IX)

Vision

The EYES are photoreceptors specialized to respond to light energy. They are located within the orbits of the skull and are moved by six extrinsic muscles (Fig. 15.7 and Table 15.4).

The orbits protect the eyes on all but the anterior one sixth or so, and this latter portion is protected by the conjunctiva and the eye lids (Fig. 15.8). A two-lobed lacrimal gland lies on the upper outer surface of each eye, and produces the tears that cleanse, lubricate, and destroy bacteria on the anterior surface of the eye. Tears are drained into the nose by a system of ducts (Fig. 15.9).

Structure of the eye (see Fig. 15.8)

The adult eyeball is an organ about 25mm (1 inch) in diameter. It is composed of three layers of tissue.

The SCLERA is the outermost coat. Posteriorly, it is opaque and forms the "white of the eye." Anteriorly, it is transparent, and forms the *cornea.*

The UVEA forms the middle coat of the eyeball. Anteriorly, it consists of the *ciliary body,* including the *ciliary muscle, iris, lens,* and *lens ligaments.* Posteriorly, it is composed of the vascular *chorioid* (choroid).

The RETINA is the innermost layer. It extends only as far anteriorly as the posterior border of the iris. It contains the *rods* and *cones* which are the visual receptors of the eye. Its nerve fibers form the *optic nerves,* which exit from the eye.

The portion of the eye anterior to the lens is

FIGURE 15.7. The six extrinsic muscles of the left eye, viewed from the lateral side. Notice how the pulley converts the direction of pull of the superior oblique.

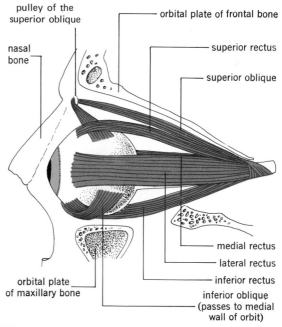

TABLE 15.4 The Extrinsic Eye Muscles		
Name	Innervation (cranial nerve)	Action on eyeball
Lateral rectus	VI Abducent	Out
Medial rectus	III Oculomotor	In
Superior rectus	III Oculomotor	Up and in
Inferior rectus	III Oculomotor	Down and in
Inferior oblique	III Oculomotor	Up and out
Superior oblique	IV Trochlear	Down and out

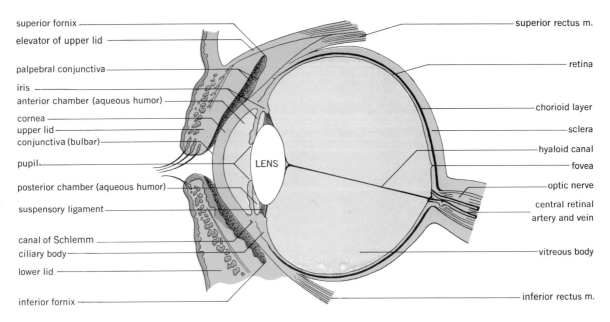

FIGURE 15.8. The eye and eyelids in section.

filled with the watery *aqueous humor,* secreted from the ciliary body. A ring of veins known as the *canal of Schlemm,* lies at the junction of sclera and cornea (see Fig. 15.8), and drains the humor from the anterior chamber. Blockage of the canal raises intraocular pressure higher than its normal 20 mm Hg (millimeters of mercury), and *glaucoma* results. Glaucoma causes compression of blood vessels in the eye and compression of the optic nerve. Retinal cells die and the optic nerve may atrophy, producing blindness. Glaucoma is the leading cause of blindness in the world.

The remaining portion of the eye is filled with the gel-like *vitreous humor,* which serves mainly to maintain the shape of the eyeball.

The blood vessels that are responsible for most of the nourishment of the eye enter the eyeball from behind at the *optic disc.* The disc is where the optic nerve leaves the eye, and is lacking in visual receptors, hence the name "blind spot." Ophthalmoscopic examination of the eye is often performed to determine the status of the eye structures and vessels. What is seen is termed the fundus, and it normally appears as is shown in Figure 15.10.

Functioning of the eye

IMAGE FORMATION. Light waves entering the eye must be bent or *refracted* so that they form a sharp image on the retina. Ability to bend light waves depends on the curvature of the refracting surface and the media (air or fluid) on either side of the surface. Most of the refraction of light entering the eye occurs at the anterior corneal surface which might be said to act like a coarse adjustment on a microscope. The refracting power of the cornea is fixed by its anatomical curvature and cannot change. Final focusing occurs by changes in curvature of the lens to create sharp retinal images.

CLINICAL CONSIDERATIONS. *Errors of refraction* include *myopia, hypermetropia* (or hyperopia) *astigmatism,* and *presbyopia.*

The normal or *emmetropic eye* focuses an inverted or reversed image directly on the retina. The *myopic* (nearsighted) eye focuses images in front of the retina, because the eyeball is too long for the refracting power of the eye. The *hypermetropic* (farsighted) eye focuses images behind the retina because the eyeball is too short for the re-fracting power of the eye. These abnormalities are easily corrected by fitting lenses ahead of the cornea. These conditions and their correction are shown in Figure 15.11. *Astigmatism* results when the cornea or lens, instead of having the same curvature vertically and horizontally, is shaped like a teaspoon. The unequal curvatures cause two separate focal points and double vision. *Presbyopia* ("old eye") results from hardening of the lens with age, and causes an inability to accurately focus near objects.

ACCOMMODATION. A series of three changes occur in the eye as images are formed: change in lens curvature for focusing, pupillary constriction, and convergence of the eyes.

Change in lens curvature occurs as the ciliary muscle contracts, tension is relaxed on the lens liga-

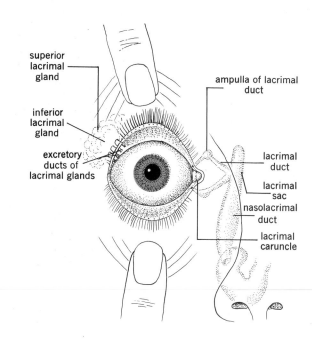

FIGURE 15.9.   The lacrimal apparatus.

FIGURE 15.10.   The fundus of the right eye as seen through an ophthalmoscope.

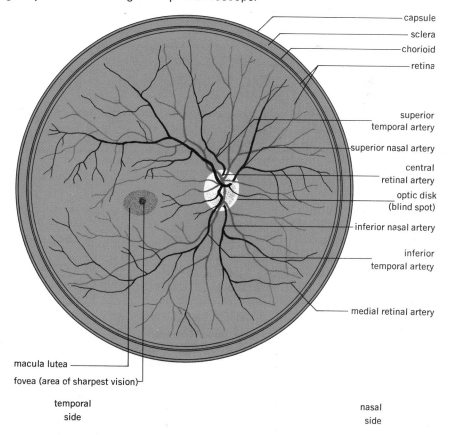

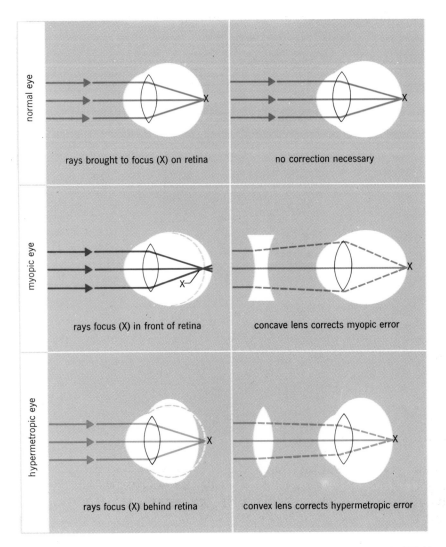

normal eye

rays brought to focus (X) on retina

no correction necessary

myopic eye

rays focus (X) in front of retina

concave lens corrects myopic error

hypermetropic eye

rays focus (X) behind retina

convex lens corrects hypermetropic error

FIGURE 15.11.   *(A)* Emmetropic, normal. *(B)* Myopic, nearsighted. *(C)* Hypermetropic, farsighted eyes. Diagram indicates type of lens required to correct abnormal eyes.

ments, and the lens becomes more rounded due to its own elastic capsule. Curvature is adjusted so that images are normally brought to a sharp focus on the retina.

Pupillary constriction blocks light rays that would pass through the edges of the lens and which cannot be brought to sharp focus. Pupillary constriction also occurs in response to light shown into the eye, and is a protective response.

Convergence directs both eyes at the object being viewed. If one of the extrinsic eye muscles contracts more forcibly, or is shorter than the others, accurate directing of the eyes on the object may not occur, and double vision may result.

CLINICAL CONSIDERATIONS. If the oculomotor nerve, or its central connections are damaged, pupillary constriction to light may not occur. An Argyll-Robertson pupil results. Pupil size also changes with anesthesia, being dilated in light and deep stages, and constricted in the intermediate stage.

Strabismus (cross-eye) is the result of improper

convergence. It is a condition that, if uncorrected, can lead to the brain ignoring one of the two images it receives from the eyes, and the subject may become functionally blind in one eye.

Cataract is lens opacity, and may result from aging, infection, or trauma of the eye, or as the result of diabetes mellitus. It results in failure to transmit light rays to the retina. The lens may be surgically removed if it is too opaque to transmit enough light to see by; glass lenses then correct the vision.

FIGURE 15.12.   The organization of the retina.

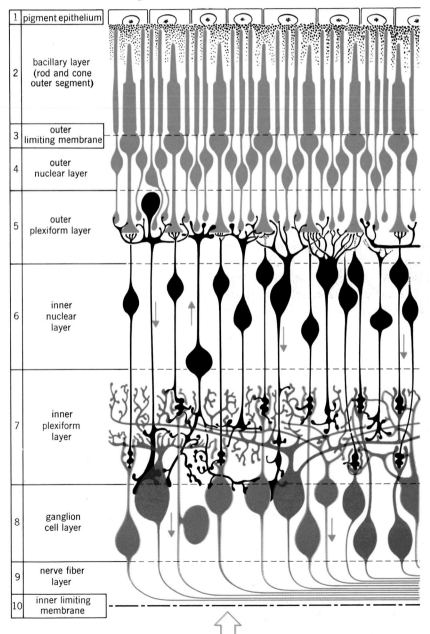

1	pigment epithelium
2	bacillary layer (rod and cone outer segment)
3	outer limiting membrane
4	outer nuclear layer
5	outer plexiform layer
6	inner nuclear layer
7	inner plexiform layer
8	ganglion cell layer
9	nerve fiber layer
10	inner limiting membrane

light

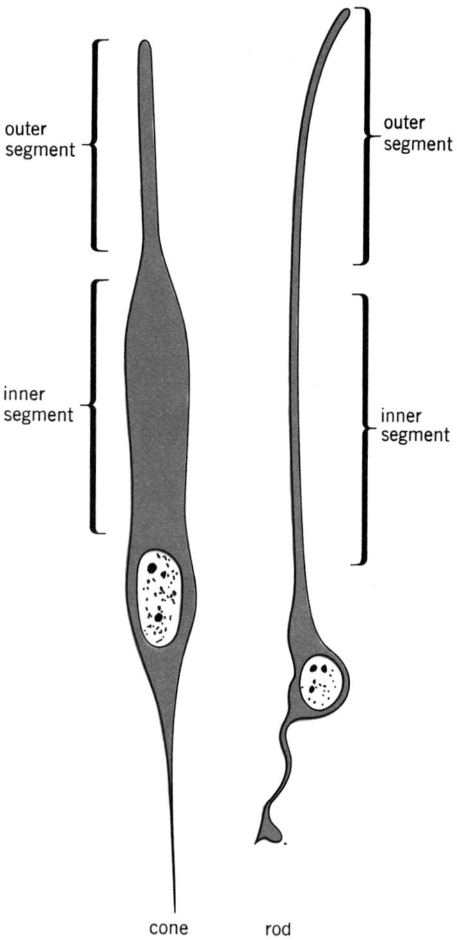
outer segment
inner segment
outer segment
inner segment
cone
rod

FIGURE 15.13. The structure of the rods and cones.

Corneal transplant may be employed in corneal damage. The tissue is not vascular, and rarely is rejected when transplanted.

RETINAL FUNCTION. The retina consists of 10 layers of neurons and fibers (Fig. 15.12). The second layer contains the rods and cones (Fig. 15.13).

The rods. Rods are sensitive to low-intensity light and are responsible for *scotopic vision* (night vision). Rods contain a visual pigment called *rhodopsin.* The pigment is thought to be split by light, and the products of the splitting cause depolarization of the rod cell. A nerve impulse is formed. Resynthesis of the pigment requires a derivative of *vitamin A.* Thus, if the body is deficient in vitamin A, "night blindness" may result. Yellow vegetables are sources of vitamin A.

The cones. There appear to be three classes of cones, each with a different visual pigment in it. There are cones that respond maximally to red, green, or blue light. It is known that all colors may be created by mixing these three basic colors. Differential (unequal) stimulation of the three cone types may be the basis for *color vision,* one of the functions of the cones. *Photopic vision* (daylight vision) is the other function of the cones. The basic mechanism of cone function is presumed to be the same as that for rods, that is, light causes pigment breakdown and depolarization of the cone.

CLINICAL CONSIDERATIONS. *Color blindness* is easily explained on the basis of the three cone types, by presuming that either a given type of cone is missing, or that the pigment is not synthesized. An individual is said to have *protanopia* (red blindness), *deuteranopia* (green blindness), or *tritanopia* (blue blindness). Most color blindness is inherited, and is sex-linked (defective gene is carried on sex chromosomes), so that males are more commonly affected than females.

OTHER RETINAL FUNCTIONS. *Visual acuity* is a function of the retina. This refers to sharpness of vision, and depends on the "density of packing" of cones in the fovea. The fovea contains only cones that are narrower than in other parts of the retina. Acuity depends on the presence of an unstimulated cone between two stimulated ones. If the cones are narrower, an object may be moved farther away before its image falls on neighboring cones. How "far" we can see clearly is the most common measure of acuity.

The ability to *fuse* separate images into a continuous sequence (as in viewing a movie) depends on *visual persistence* in the retina. The products of destruction of visual pigments are not removed immediately and the "image lasts" a while. The next image then fuses with the first, and we see "continually," and not in "bits and pieces."

After images also result from persistance of visual pigment breakdown, and one may "see something" for several seconds after the stimulus has disappeared.

Visual pathways (Fig. 15.14)

Impulses originating in the retina pass over the *optic nerves* to the *optic chiasma*, and from there over the *optic tracts* to the *thalamus*. From the thalamus, fibers pass over the *optic radiation* to the *occipital lobes* (area 17) of the cerebrum. Further interpretation of visual impulses is provided by the visual association areas (18,19).

CLINICAL CONSIDERATIONS. Figure 15.15 presents the types of visual loss that would result from damage to the visual pathways. One should note that there is a definite spatial relationship of the various retinal fibers, so that loss is characterized for the site of injury.

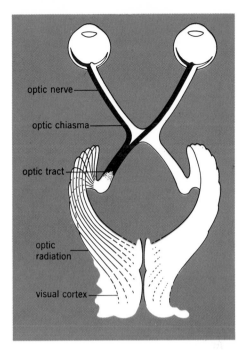

FIGURE 15.14. The neural pathways for the sense of sight.

FIGURE 15.15. The types of visual loss resulting from lesions in various parts of the visual pathways.

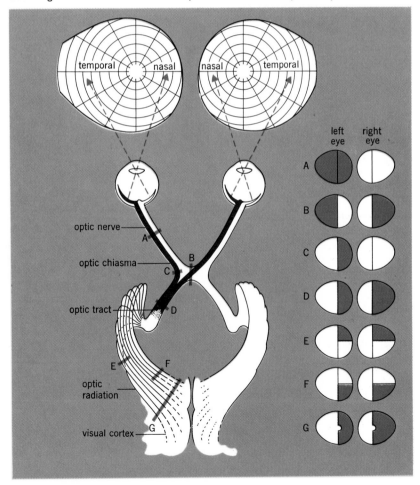

Hearing

The auditory mechanism responds to sound waves, and is divided into three portions; the *outer, middle, and inner ears* (Fig. 15.16).

The outer ear

The outer ear consists of the AURICLE or *pinna,* the fleshy flap attached to the side of the skull, and the EXTERNAL AUDITORY MEATUS or ear canal. The pinna may aid in collection and directing of sound waves to the middle ear. The canal is slightly S-shaped, and directs sound waves to the eardrum. The canal is guarded externally by large hairs, and by glands (ceruminous glands) which produce the ear wax (cerumen). The wax is bitter to the taste, and is supposed to aid in keeping insects out of the ear canals.

CLINICAL CONSIDERATIONS. Accumulation of cerumen may block transmission of sound waves down the canal, and result in one type of *transmission deafness.* This cause of the condition is more common in children.

The middle ear

The middle ear consists of the MIDDLE EAR CAVITY, and its contents. The cavity is an irregular, air-filled space which communicates with the throat by way of the *pharyngo-tympanic (Eustachian) tube.* The tube allows equalization of air pressure on the two sides of the eardrum. The tympanum or eardrum forms the lateral wall of the cavity. Sound waves cause vibration of the drum. Three EAR OSSICLES: the *malleus, incus,* and *stapes,* are caused to move by the vibrations of the drum. The ossicles form a system of levers that increase the pressure, and decrease the movement of the stapes on the oval window of the cochlea. Since the drum vibrates in air and the cochlea in fluid, more energy is required to vibrate the fluid. Two small muscles in the middle ear (tensor tympani and stapedius) contract when sounds are very loud, and prevent excessive movement of the ossicles which could damage the inner ear. The contraction of the muscles is reflexly controlled, and is called the *auditory reflex.*

FIGURE 15.16. The ear.

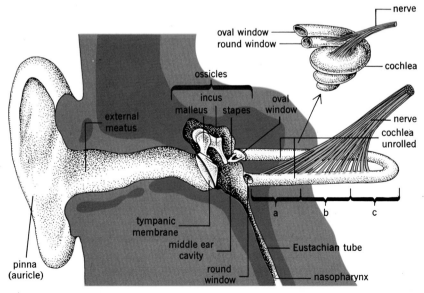

pinna (auricle) · external meatus · tympanic membrane · middle ear cavity · round window · ossicles · incus · malleus · stapes · oval window · Eustachian tube · nasopharynx · nerve · oval window · round window · cochlea · nerve · cochlea unrolled · a · b · c

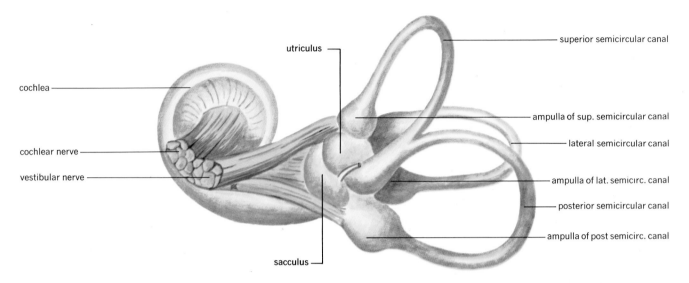

utriculus

superior semicircular canal

cochlea

ampulla of sup. semicircular canal

lateral semicircular canal

cochlear nerve

vestibular nerve

ampulla of lat. semicirc. canal

posterior semicircular canal

ampulla of post semicirc. canal

sacculus

FIGURE 15.17. The membranous labyrinth of the inner ear.

FIGURE 15.18. Pitch analysis by the basilar membrane of the cochlea. Numbers indicate cycles per second (cps). Membrane averages 32 millimeters long, 0.04 millimeters wide at base, 0.5 millimeters wide at apex.

CLINICAL CONSIDERATIONS. Transmission (conduction) deafness occurs when sound waves are prevented from reaching the eardrum or by failure of the ossicles to transmit the sound waves to the cochlea. *Inflammation* of the middle ear (otitis media), *fusion* of the ossicles or *fixation* of the stapes in the oval window (otosclerosis) may all result in loss of hearing. Bone conduction may be employed if the ossicles cannot be freed. This utilizes the fact that vibrations may reach the fluid of the cochlea by transmission through the skull bones, rather than by way of the ossicles. This forms the basis for many types of hearing aids.

The inner ear

The inner ear (Fig. 15.17) contains the cochlea, the organ of hearing, and organs concerned with balance and equilibrium (maculae and semicircular canals).

THE COCHLEA AND HEARING. As the stapes moves in the oval window of the cochlea, shock waves are created that cause vibration of the BASILAR MEMBRANE (Fig. 15.18) of the ORGAN OF

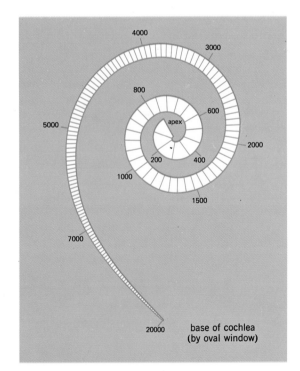

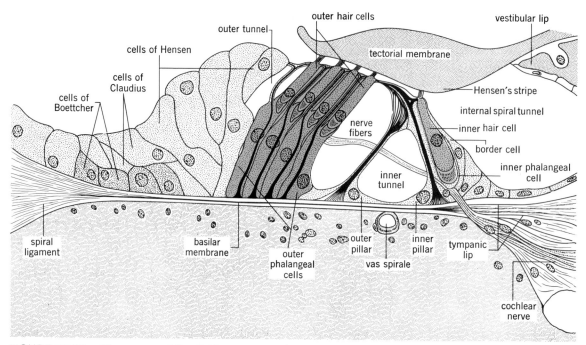

FIGURE 15.19.   The organ of Corti. The sensory elements are the hair cells.

CORTI (Fig. 15.19). The basilar membrane is composed of 25,000–30,000 strands of tissue that are "tuned" to vibrate to particular pitches of sound (place or selective theory of hearing). The motion of the basilar membrane is translated into nerve impulses through bending of the HAIR CELLS that form the receptors of the organ of Corti. The hair cells are bent as the basilar membrane vibrates; this creates generator potentials that depolarize the hair cells to create a nerve impulse. So that the shock waves do not continue to "bounce around" in the cochlea and cause continuous nerve impulse formation, the vibrations are damped by movement of the round window at the base of the cochlea. As the stapes moves inward at the oval window, the round window bulges outward and tends to cushion the shock wave.

AUDITORY PATHWAYS. Impulses pass from the hair cells over the cochlear portion of the *eighth cranial nerve* to the *cochlear nuclei* in the brain stem. From the brain stem, impulses pass to the *thalamus,* then to the *auditory cortex* in the temporal lobes (area 41,42). Both cochleas are represented in both temporal lobes, with high tones anteriorly, and low tones posteriorly. The temporal lobes determine localization of sounds, the quality of the sound, and its intensity. The temporal association areas are concerned with understanding of auditory input.

CLINICAL CONSIDERATIONS. *Nerve deafness* results from damage to the cochlea and/or the auditory pathways to the brain. The antibiotic streptomycin, if given to treat bacterial infections, may damage the eighth cranial nerve and may cause deafness and disturbances of equilibrium.

The inner ear and equilibrium and balance

The SEMICIRCULAR CANALS are three fluid-filled channels in each ear (see Fig. 15.17). An enlarged ampulla contains a CRISTA that responds, by being bent, to changes in acceleration and direction of motion. Bending occurs as the head moves and causes the fluid in the canals to move toward or away from the cristae. The three canals are placed in three mutually perpendicular planes to one another so that any head movement will cause excitation of one or more cristae. The MACULAE are organs of position, not movement. They consist of HAIR CELLS whose processes are set in a gelatinous mass containing tiny granules called otoliths (ear stones). The entire macula rests in an air-filled cavity in the vestibule of the inner ear. The maculae are activated by gravity pulling or pushing on the otoliths of the organ and bending the hair cells. Excitation of semicircular canals or maculae results in reflex responses (labyrinthine reflexes) tending to maintain the body properly in space. LABYRINTHINE REFLEXES are categorized into the two following groups.

Acceleratory reflexes, arising from stimulation of the semicircular canals, produce responses of eyes, neck, trunk, and appendages to "starting," "stopping," or "turning" motions. *Nystagmus,* a horizontal, vertical, or rotary movement of the eyes, occurs on angular acceleration. Limb responses are those of extension on the side away from an angular acceleration, and toward the force if linear. Thus a person "leans into" a curve, leans forward on acceleration, and backward on deceleration, maintaining his balance.

Positional reflexes, arising from stimulation of the maculae, involve righting reflexes and notification of head position. The classical statement that "a cat always lands on its feet" illustrates the operating of the righting reflexes. In this example the head is first turned into a proper relationship to the ground, and the body is then brought into relationship to the head. The pathway for communication of information from the canals and maculae is via the *vestibular portion* (vestibular nerve) of the eighth cranial nerve. The fibers pass to brain stem *vestibular nuclei,* located in the lower lateral pons, and thence to the *cerebellum,* where appropriate muscular response is integrated.

CLINICAL CONSIDERATIONS. Although not strictly a disorder, *motion sickness* involves repetitive changes in angular and linear acceleration affecting the receptors. *Nausea* and *vomiting* may result. Vomiting results from motion sickness as follows.

There is, in the medulla of the brain stem, a "trigger zone" that works with the vomiting center to cause vomiting. The zone is larger than the actual center. Impulses that arrive at this trigger zone from the semicircular canals, excite the zone; impulses pass from the zone to the vomiting center and cause the nausea and vomiting associated with motion sickness.

Infections of the vestibular nerve may result in loss of equilibrium as may trauma that injure the nerve. Indeed, any basic loss of equilibrium without change of muscle tone is indicative of altered inner ear equilibrium function. Testing equilibrium function typically involves spinning the subject in a Barany chair with his head in various positions to excite the different canals. If nystagmus, dizziness and loss of equilibrium are not produced, a lesion is presumed to exist somewhere in the system.

Summary

1. Receptors detect changes in internal and external environments, and lead to homeostasis-maintaining responses.

2. Receptors:
   a. Follow the law of adequate stimulus and respond best to one type of stimulus.
   b. Follow the law of specific nerve energies which states that all impulses in different nerves are the same, and the sensation we appreciate depends on where the nerve pathways terminate in the brain.
   c. Cause a stimulus to be changed into a nerve impulse.
   d. Show accommodation (decrease of discharge with continued stimulation).

3. Receptors are best classified by adequate stimulus (Table 15.1).

4. Touch and pressure sensations:
   a. Are initiated by a change in shape of the receptor.
   b. Are served by Meissner's and Pacinian corpuscles, respectively.
   c. Are conducted over gracile and cuneate tracts to the brain stem. Fibers then pass to the thalamus and then to the cerebrum.

5. Heat and cold receptors:
   a. Respond to changes in temperature.
   b. Is served by specific receptors.
   c. Send impulses over spinothalamic tracts to the thalamus and then fibers go to the cerebrum.

6. Pain:
   a. Results from overstimulation of any nerve.
   b. Is protective in nature.
   c. Is served by naked nerve endings.
   d. Is transmitted over spinothalamic tracts to thalamus and then to the cerebrum.
   e. Is of several types: bright pain, burning pain, and aching pain; the latter may be referred to a different body area than its point of origin. Referred pain indicates involvement of visceral organs.
   f. Headache results from tension on cranial structures, increased intracranial pressure, or inflammation, spasm, and eyestrain.

7. Kinesthetic sense originates in muscles, joints, and tendons.
   a. It maintains awareness of body position.
   b. It maintains muscle tone.
   c. Results from stimulation of specific end organs.

8. Synthetic senses (itch, tickle, vibration):
   a. Have no specific end organs.
   b. Result from stimulation of other end organs.

9. Visceral sensations (hunger, thirst):

   a. Originate in visceral organs.

   b. Are important in regulation of nutrition and water balance.

10. Loss or alteration of the above sensations is significant in determining what nerves have been damaged. Anesthesia, paresthesia, or hyperesthesia may occur when sensory nerves are damaged.

11. Smell:

   a. Has receptors in the nasal cavities.

   b. Has seven basic odors.

   c. Is more sensitive than taste.

   d. Is connected by olfactory bulbs and tracts to temporal and frontal lobes of the brain.

12. Taste:

   a. Has the taste bud as receptor.

   b. Has four basic sensations: sour, salt, sweet, and bitter.

   c. Uses VII and IX cranial nerves as pathways, and terminates in the parietal lobes.

13. Vision is served by the eyes.

   a. The eyes are located in the orbits.

   b. Each eye is moved by six muscles.

   c. Three coats form each eye.

      (1) The sclera and cornea are outermost.

      (2) The uvea is central and includes: chorioid, ciliary muscle, iris, lens, and lens ligaments.

      (3) The retina, which responds to light.

   d. The eyes contain aqueous and vitreous humors.

      (1) Blockage of aqueous humor drainage creates glaucoma.

14. Image formation by the eye:

   a. Requires action of cornea, lens, and iris.

   b. If defective, causes myopia, hyperopia, and astigmatism.

   c. Involves accommodation, which includes three things:

      (1) Lens shape changes for focusing.

      (2) Pupil constriction for correction of aberrations and light control.

      (3) Convergence for directing eyes to object.

15. The retina:

   a. Contains rods and cones.

   b. Is responsive to low light levels (rods).

   c. Is responsive to high light levels and colors (cones).

   d. Is responsible for visual acuity.

e. Fuses separate images into a continuum.

f. Creates after images.

16. The visual pathways include the optic nerves, chiasm, tract, thalamus, and occipital lobes.

17. The ear is divided into outer, middle, and inner ears.

18. The outer ear:

a. Gathers sound waves.

b. Conducts sound waves to the eardrum.

19. The middle ear:

a. Consists of middle ear cavity, ossicles, and Eustachian tube.

b. Transmits eardrum vibrations to the cochlea.

c. Equalizes pressure on the drum.

d. Controls ossicle movement according to intensity of sound.

20. The inner ear:

a. Contains the cochlea, with its organ of Corti and hair cells for hearing. Vibrations (sound waves) cause selective vibration of Corti's organ and hearing results. Auditory pathways pass from cochlea to brain stem to thalamus to temporal lobes.

b. Contains organs for equilibrium. The maculae signal held position of the head. The semicircular canals detect motion and initiate motor responses to maintain posture.

Questions

1. What is a receptor? What is a peripheral receptive field?

2. What basic properties do all receptors share?

3. Give examples of chemoreceptors, baroreceptors, and thermoreceptors, and describe what the adequate stimulus is for each that is listed.

4. What is somesthesia?

5. Name the receptor, spinal pathway, and interpretive area for the senses of touch, pain, and unconscious muscle sense.

6. What are the varieties of pain? How do they differ?

7. What are paresthesia and hyperesthesia? How may each be brought about?

8. What are the roles of the thalamus in sensation?

9. What structures are involved in image formation by the eye? What does each contribute to the formation of the image?

10. What changes occur during accommodation?

11. Describe the defects present that result in myopia. How is the condition corrected?

12. What evidence exists to suggest the presence of two types of receptors in the retina?

13. What functions are served by the structures located in or communicating with the middle ear?

14. Describe the structure and function of the cristae ampullaris.

15. What are the basic taste sensations and what are the adequate stimuli for each?

16. What are some basic odors? Compare sensitivity of the olfactory organ to that of taste.

Readings

Ammore, J. E. *Molecular Basis of Odor.* Thomas Pubs. Springfield, Ill., 1970.

Bekesy, G. von. "The Ear." *Sci. Amer.* (Aug) 1970.

Botelho, S. Y. "Tears and the lacrimal gland." *Sci. Amer.* 211:78 (Oct) 1964.

Cagan, R. H. "Chemostimulatory Protein: A new type of taste stimulus." *Science 181*:32, 1973.

Holvey, David N. (ed). Merck Manual. "Glaucoma," p. 1024. Merck & Co., Inc. Rahway, N. J., 1972.

Jones, R. Clark. "How Images are Detected." *Sci. Amer.* 219:110 (Sept) 1968.

Neisser, U. "The Processes of Vision." *Sci. Amer.* 219:204 (Sept) 1968.

Rosensweig, M. R. "Auditory Localization." *Sci. Amer.* 205:132 (Oct) 1961.

Science News. "Narcotics: How they kill pain." *105,* No. 11:174 (Mar 16) 1974.

Young, Richard W. "Visual cells." *Sci. Amer.* 223:801 (Oct) 1970.

chapter 16
Body Fluids and Acid-base Balance

chapter 16

The fluids of the body constitute an internal environment which supplies substances necessary for cellular activity, and which removes the wastes of that activity. Maintenance of homeostasis within these fluids is vital to survival of the cell and the entire organism. This chapter examines the nature of these fluids and of some of the mechanisms responsible for regulation of volume and composition of these fluids.

Fluid compartments

While there is considerable individual variation, total fluid in the adult body amounts to 55-60 percent of the body weight. Two major compartments or divisions in total body water are recognized (Fig. 16.1).

Extracellular fluid (ECF)

ECF refers to all fluid not contained within cells. It accounts for about 35 percent of the body water. It has several subdivisions.

PLASMA water is the fluid portion of the blood, and constitutes 7 percent of the body water.

INTERSTITIAL OR TISSUE FLUID is the immediate environment of the cell; LYMPH is the tissue fluid that has entered lymph vessels. Together these two constitute 18 percent of the body water.

FLUIDS OF DENSE CONNECTIVE TISSUE and bone are actually continuous with interstitial fluid, but water in these tissues exchanges very slowly due to the density of the tissue. It behaves like a separate fluid compartment. These fluids form 10 percent of the body water.

TRANSCELLULAR FLUIDS are all fluids separated from other fluids by an epithelial membrane. This compartment includes the fluids of the eye, joints, nervous system, body cavities, and those within

FIGURE 16.1. Relationships of body water compartments. Numbers represent percent each compartment comprises of total body water in an adult.

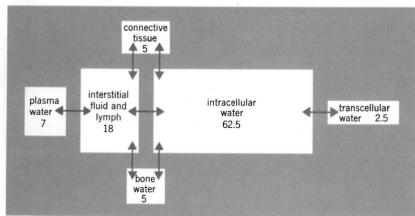

TABLE 16.1 Compartments of the Body Water		
Compartment	Approximate percentage of adult body weight	Approximate percentage of adult body fluid
Extracellular fluid	21.4	37.5
Plasma	5	7
Interstitial fluid	15	28
Transcellular fluid	1.4	2.5
Intracellular fluid	38.6	62.5
Total body water	60	100

hollow organs other than blood vessels. Collectively, transcellular fluids constitute 2.5 percent of body water.

Intracellular fluid (ICF)

ICF is fluid contained within cells, and amounts to 62.5 percent of the body water. The body water compartments are summarized in Table 16.1.

Water requirements

The amount of water required to maintain water balance depends on water loss by various routes, and on the metabolism of the individual. The adult requires about 2500 milliliters of water per day to maintain normal body water levels. Of this 2500 milliliters per day requirement, 2300 milliliters is taken in food and drink, and 200 milliliters is added by metabolism. This intake balances a daily loss of water of 2500 milliliters through the lungs (300 ml/day), perspiration (500 ml/day),

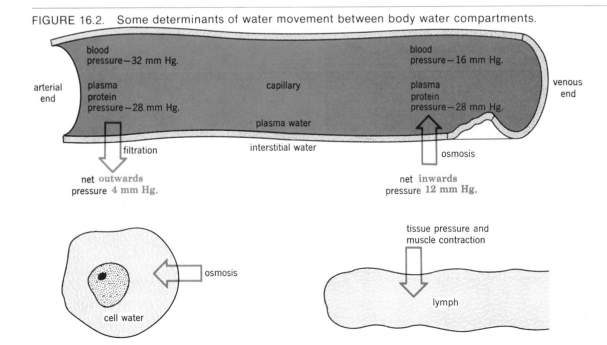

FIGURE 16.2. Some determinants of water movement between body water compartments.

feces and evaporation from the mouth (200 ml/day), and via the urine (1500 ml/day).

Water supplied by food and drink is absorbed by osmosis from the alimentary tract and passes to the plasma. From here it is filtered from capillaries into the tissue spaces and is removed by osmosis back into the capillaries, or passes into the lymph vessels of the tissues. Some fluid may also move into cells. Figure 16.2 shows these interrelationships. A "circulation" of water is thus created, with filtered water ultimately being returned back to the circulatory system from which it came.

FIGURE 16.3.   A comparison of the constituents of the three major compartments of the body water. (Reproduced, with permission, from J. L. Gamble, Jr. Chemical Anatomy, Physiology, and Pathology of Extracellular Fluid. 6th ed. Harvard University Press. Boston, 1954.)

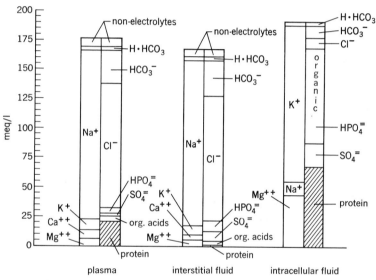

| Constituents and Properties | Extracellular fluid | | Intracellular fluid |
	Plasma	Interstitial fluid	
Sodium	142 meq/L	145 meq/L	10 meq/L
Potassium	4 meq/L	4 meq/L	160 meq/L
Calcium	5 meq/L	5 meq/L	2 meq/L
Magnesium	2 meq/L	2 meq/L	26 meq/L
Chloride	101 meq/L	114 meq/L	3 meq/L
Sulfate	1 meq/L	1 meq/L	20 meq/L
Bicarbonate	27 meq/L	31 meq/L	10 meq/L
Phosphate	2 meq/L	2 meq/L	100 meq/L
Organic acids	6 meq/L	7 meq/L	—
Proteins	16 meq/L	1 meq/L	65 meq/L
Glucose (av)	90 mg %	90 mg%	0-20 mg%
Lipids (av)	0.5 gm %	—	—
pH	7.4	7.4	6.7-7.0*

TABLE 16.2   Composition of Extracellular and Intracellular Fluid Compartments

*Average value; difficult to measure.

Composition of the compartments

In general, the subdivisions of the ECF (transcellular fluids excepted), are nearly identical in electrolyte composition, and differ mainly in protein concentration. Intracellular fluid differs greatly from ECF. These differences are summarized in Table 16.2, and are shown in Figure 16.3. Sodium and chloride are seen to be the chief extracellular substances, while potassium, phosphate, and proteins are the chief intracellular materials.

It becomes obvious that if solute concentrations change greatly between ICF and ECF, changes will occur in water movement (osmosis) between compartments; *regulation* of both solute and solvent concentrations thus becomes of utmost importance.

Homeostasis of volume and osmotic pressure of ECF

Control of the volume of the ECF is established by governing both intake and output of fluid.

The *thirst mechanism* is activated by increased loss of water, or retention of sodium. Both of these cause the ECF to become more concentrated in solutes. Secretion of fluid by the salivary glands diminishes, the mouth linings dry, and a sensation of thirst is created. Intake restores the normal balance.

Cells (*osmoreceptors*) sensitive to the osmotic

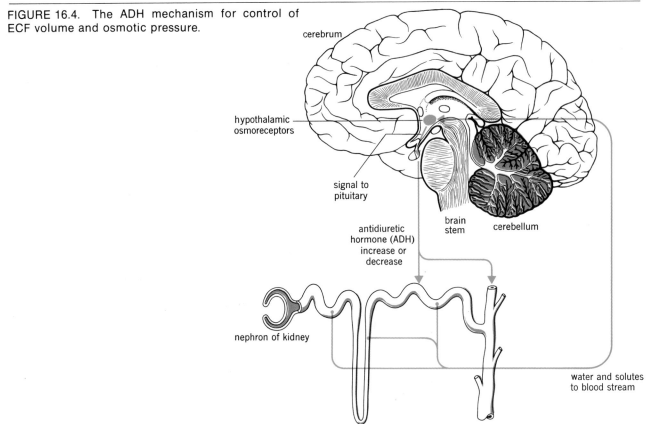

FIGURE 16.4. The ADH mechanism for control of ECF volume and osmotic pressure.

cerebrum

hypothalamic osmoreceptors

signal to pituitary

antidiuretic hormone (ADH) increase or decrease

brain stem

cerebellum

nephron of kidney

water and solutes to blood stream

pressure of the ECF are found in the hypothalamus (Fig. 16.4). These cells continually measure the osmotic pressure of the ECF, primarily the blood, and govern the release of a hormone called ANTIDIURETIC HORMONE (ADH) from the pituitary gland. ADH in turn controls passage of water from the kidney tubules into the bloodstream. The more ADH released, the greater the reabsorption of water from kidney to blood.

Cells in the kidney sensitive to blood volume and pressure produce a substance called *renin* whenever there is a low blood pressure in the kidney. Renin is an enzyme that converts a plasma substance to an active material known as angiotensin. Angiotensin raises blood pressure by causing vasoconstriction, and also causes the adrenal glands to secrete a hormone, ALDOSTERONE. The hormone increases sodium reabsorption by the kidney, which causes an in-

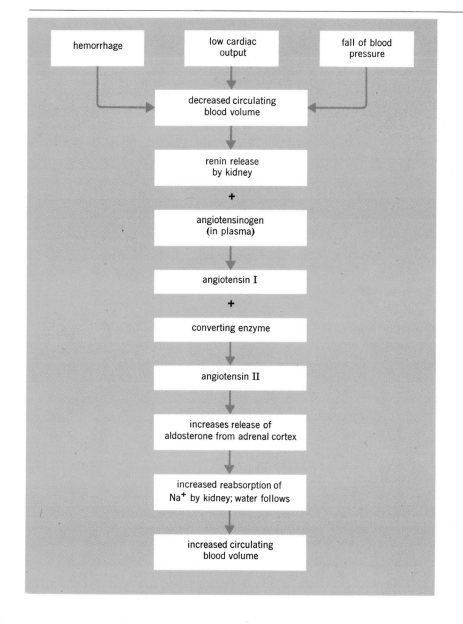

FIGURE 16.5. The relationships of angiotensin and aldosterone to the regulation of ECF volume.

creased osmotic flow of water from tubules to bloodstream. Thus both solute and solvent con-

centrations of the ECF are increased. This sequence of events is depicted in Figure 16.5.

Clinical aspects of fluid and electrolyte balance

Dehydration

Isotonic dehydration (equal loss of fluid and electrolytes) occurs during loss of whole blood from the body (hemorrhage) or during short-term vomiting or diarrhea. The total volume of the ECF changes, but composition and osmotic pressure remain within normal limits.

Hypertonic dehydration (excess fluid loss compared to electrolyte loss) occurs when there is a lack of water intake while loss continues through skin, lungs, and urine. It may also occur during excessive sweating, in burned patients, and during diabetes mellitus (sugar diabetes). The loss of relatively more water than solutes tends to raise the concentrations of sodium, potassium, and calcium in the body fluids, singly or in combination, and makes the ECF hypertonic to the cell. The cell tends to lose water. Excessive sodium levels tend to cause water retention by the kidney, and this is a beneficial effect. Excessive potassium levels lead to muscular weakness and eventual paralysis, to cardiac and respiratory irregularities, and to diminished urine formation. Excessive calcium levels cause relaxation of muscles, flank pain, and loss of reflexes.

Hypotonic dehydration (excess electrolyte loss compared to fluid loss) occurs in deficiency of certain adrenal cortical hormones (aldosterone), in kidney disease, during encephalitis (ADH secretion may be abnormal), during prolonged diarrhea or vomiting, and on a low sodium diet. The ECF tends to become hypotonic relative to the cell, and the cell tends to gain water. Loss of sodium leads to muscle cramps, convulsions, and a feeling of apprehension. Loss of potassium leads to relaxed muscles (the muscles feel like half-filled hot water bags), paralysis, weakness, and irregularities of heart beat. Lowered calcium levels lead to muscle cramps and tetany (muscle in a sustained contraction), and

tingling of the fingers. Loss of magnesium causes disoriented behavior, muscle tremor, and exaggerated reflexes.

Edema

Edema is a condition caused by accumulation of excess fluid in the interstitial compartment. It results in swelling and "waterlogging" of the tissues. Some factors involved in the production of edema are the following.

Increased filtration of fluid and solutes from capillaries. If capillary permeability permits protein loss through the capillary wall, the osmotic force tending to draw water from tissues to capillary is decreased. Thus water remains in the tissues.

Decreased heart action blocks blood flow through veins, and causes engorgement of tissues with blood. Venous pressure rises and opposes osmotic return of water from the tissues to the capillaries. Therefore more water remains in the tissues.

Salt retention or high salt diets cause water to be retained as well, and ECF volume increases. If ECF volume rises, cell volume may also increase by increased water movement into cells. The nervous system shows such diluting effects first, with the development of disoriented behavior, convulsions, and coma.

Although it is usually not possible to determine which particular solute has been altered and in which direction, without blood analysis, one can get an idea of whether he is dealing with an altered state of hydration or an alteration of solutes in general. Several physiological observations should be made on a patient suspected of having a hydration disorder, or on one who is under treatment for such a disorder.

Temperature. In the absence of an obviously fever-producing condition, an increase in temperature indicates decreased water or increased sodium levels in the body.

Pulse. Rapid weak but regular pulses indicate hemorrhage or sodium loss. Weak and irregular pulses suggest potassium deficiency.

Respirations (breathing). Difficult breathing suggests excessive fluid volume and possible pulmonary edema. Shallow breathing may indicate potassium deficiency, while harsh, high-pitched sounds (stridor) during breathing suggests severe calcium deficiency.

Blood pressure. Increased pressures suggest excessive fluid volumes; decreased pressures suggest sodium and/or potassium deficit, or loss of fluid volume.

Skin condition. Does the skin appear dry, does it stay "tented" when pinched, or does it retain a depression when pushed? Such observations give a clue as to the state of skin hydration (water content) and thus of the body in general.

General behavior. Is the patient thirsty, does he talk and walk well, is he apprehensive or anxious?

Keep in mind that these symptoms are not exclusively the result of alterations of water and solute levels; they do, however, strongly suggest an alteration in fluid and/or electrolyte balance in the body.

Acid-base balance of body fluids

Metabolism produces large quantities of substances that release free hydrogen ions. Free hydrogen ions are potentially dangerous because they change the acidity of the body fluids and thus threaten many enzyme controlled chemical reactions in the body. The body normally maintains a pH value of 7.4 ± 0.02 in the ECF to insure normal body function.

Acids and bases

Acids are materials that, in water, are capable of losing or giving up hydrogen ions to create free H^+; bases are substances capable of accepting free hydrogen ions. Strong acids release hydrogens easily, or may be said to have a high degree of dissociation ("come apart" in solution). A weak acid prefers to remain associated, and gives up hydrogen ions grudgingly. It has a low degree of dissociation. Concentrations of hydrogen ion are usually described by the symbol pH. The pH values range from 0 to 14 (Fig. 16.6), or from strongly acid to weakly acid solutions. At a pH of 7, a solution is neither acidic (values below 7), nor basic (values above 7), and is said to be neutral, because acid and base concentrations are equal. A change of one pH unit represents a ten fold change in hydrogen ion concentration, as is shown in Table 16.3.

Because of the production of large quantities of acids during metabolism, most clinically important deviations in pH occur between pH 7.00 and 7.25, that is, to the acid side of normal (pH 7.4). In general, living processes cannot be maintained if pH falls below 6.8, or rises above 7.8.

Sources of hydrogen ion in the body

The COMPLETE COMBUSTION of carbon containing compounds, such as carbohydrates and fatty acids, produces CO_2 which reacts with water to

TABLE 16.3 Relationships Between pH and Concentration of Hydrogen ion	
pH	Concentration of H^+ (Gm/L)
1	0.1
2	0.01
3	0.001
4	0.0001
5	0.00001

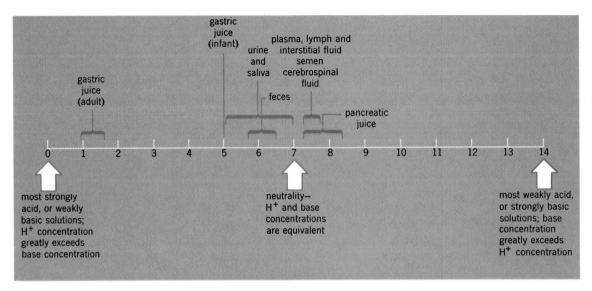

FIGURE 16.6.   The range of pH values.

form carbonic acid. The carbonic acid then dissociates to liberate H^+. These reactions are shown by the equation:

$$CO_2 + H_2O \rightleftharpoons \underset{\substack{\text{carbonic} \\ \text{acid}}}{H_2CO_3} \rightleftharpoons \underset{\substack{\text{hydrogen} \\ \text{ion}}}{H^+} + \underset{\substack{\text{bicarbonate} \\ \text{ion}}}{HCO_3^-}$$

The reactions are reversible, and while carbonic acid is a weak acid, so much of it is produced from CO_2 that it forms a major source of H^+ for the body fluids.

The PRODUCTION OF ORGANIC ACIDS results from the incomplete combustion of carbon containing compounds in metabolic cycles such as glycolysis and the Krebs cycle. The hydrogens then dissociate from the carboxyl (—COOH) groups of the acids. The equation below shows dissociation of H^+ from pyruvic acid as an example.

$$\underset{\text{pyruvic acid}}{CH_3-\overset{\overset{\displaystyle O}{\|}}{C}-COOH} \rightleftharpoons H^+ + CH_3-\overset{\overset{\displaystyle O}{\|}}{C}-COO^-$$

The INGESTION OF ACID-producing substances may add H^+ to the body fluids. While one does not normally ingest acid, it is of interest to note that substances such as aspirin contain organic acids (salicylic acid) capable of dissociating H^+.

Methods of minimizing the effects of free hydrogen ion

The threat posed to body homeostasis by the production of free H^+ from such processes as described above, is that pH will fall and inhibit vital enzyme-controlled reactions that are pH dependent. The pH of either intracellular or extracellular fluid may be altered with detrimental effects on body function. Several methods are employed to keep hydrogen ion concentrations within normal limits. Most are aimed at maintaining normal ratios between acidic and basic substances in the body fluids.

BUFFERING. Buffering is a process that utilizes certain chemicals in the body fluids to react with and thus "trap" hydrogen ion in a more or less nondissociable compound. The reaction requires a weak acid, and the "conjugate base" (completely ionizable salt) of that acid. Both compounds will share a common anion (negatively charged ion), but will have different cations (positively charged ions). The anion of the base reacts with excess H^+ to form the weak acid. The combination of the weak acid and its conjugate base is termed a *buffer pair* or *buffer system*. The system of *carbonic acid and sodium bicarbonate* is the major buffer of the extracellular fluid.

weak acid: $H_2CO_3 \rightleftharpoons H^+ \quad + HCO_3^-$
carbonic   hydrogen   bicarbonate
acid   ion   ion

conjugate base: $NaHCO_3 \rightarrow Na^+ \quad + HCO_3^-$
sodium   sodium   bicarbonate
bicarbonate   ion   ion

The conjugate base provides a reserve of anion (in this case, bicarbonate ion) which reacts with H ion, from whatever source, to form the weak acid. For example,

$H^+ \quad + HCO_3^- \rightleftharpoons H_2CO_3$
hydrogen   bicarbonate   carbonic acid,
ion   ion   a weak acid

Normal body mechanisms maintain a ratio of one molecule of H_2CO_3 to 20 molecules of $NaHCO_3$. The extra HCO_3^- provides an alkali reserve, or base excess, to react with H^+ produced under most normal activity.

As H^+ reacts with HCO_3^- to form the weak acid, it would appear that the base cation Na^+ would tend to increase its concentration in the body fluids. What does the body do to prevent accumulation of excess Na^+ which can cause disturbances in excitability? Keeping in mind that the body's primary task is to maintain nearly constant ratios of acids and bases in the body regardless of absolute concentrations of these substances, it should become clear that if there is

an increased amount of H_2CO_3 and Na^+ in the fluids, the body must add HCO_3^- to the fluids to keep ratios constant. This job may be taken care of by the kidney. The kidney reabsorbs HCO_3^- from the filtrate (fluid formed by the kidney). Alternatively, the breathing rate may be stimulated by a slight fall in pH, which results as excessive H^+ is released from the carbonic acid. This response causes a greater liberation of CO_2 from H_2CO_3 according to the equation:

$$H_2CO_3 \xrightleftharpoons{\text{enzyme}} H_2O + CO_2$$

The reaction is aided by an enzyme called carbonic anhydrase. Thus acidic and basic substances are again balanced. The excess water is then eliminated by the kidney. The responses the body makes to keep acid and base substances in proper ratio constitute compensation.

While the H_2CO_3 : $NaHCO_3$ system is probably the most important buffer system of the extracellular fluid, it is not the only buffer system operating in the body. Other ECF buffer systems include

Phosphate systems
weak acid: $NaH_2PO_4 \rightleftharpoons Na^+ + H_2PO_4^-$
conjugate base: $Na_2HPO_4 \longrightarrow Na^+ + NaHPO_4^-$

Plasma proteins
weak acid: H Proteinate (Pr) $\rightleftharpoons H^+ + Pr^-$
conjugate base: Na Proteinate $\longrightarrow Na^+ + Pr^-$

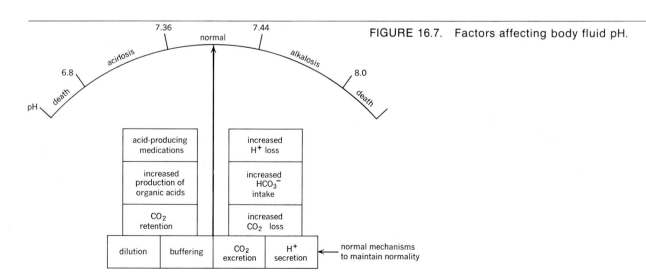

FIGURE 16.7.   Factors affecting body fluid pH.

Intracellular buffer systems include

Hemoglobin (Hgb) in red cells

weak acid: $HHgb \rightleftharpoons H^+ + Hgb^-$

conjugate base: $KHgb \longrightarrow K^+ + Hgb$

Phosphate systems

weak acid: $KH_2PO_4 \rightleftharpoons K^+ + H_2PO_4^-$

conjugate base: $K_2HPO_4 \longrightarrow K^+ + KHPO_4^-$

Other methods of minimizing free H^+ effects are

DILUTION. Some hydrogen ion, as it is produced, is set free in the body fluids. It is diluted by these fluids and thus local accumulation is prevented. This method does not remove the H^+; it only prevents excessive local accumulation.

UPTAKE BY HYDROGEN ACCEPTORS. Associated with the basic metabolic cycles, are molecules that take up hydrogen ions as they are released by those cycles. These acceptors then carry the hydrogens to oxidative phosphorylation or to a synthetic scheme that uses them. Three acceptors are recognized:

Nicotinamide adenine dinucleotide (NAD)

Nicotinamide adenine dinucleotide phosphate (NADP)

Flavine adenine dinucleotide (FAD)

As long as the hydrogens are on the acceptor or in a cycle, they are not free to cause pH change. It is critical to the understanding of acid-base disturbances to recognize that both acids and bases are present in the body, and that pH can change either as a result of too much of one substance or too little of the other. A "scalelike" balance is thus achieved, which can be upset from either side (Fig. 16.7).

Clinical aspects of acid-base balance

Normally, mechanisms for maintaining ECF pH regulate it between 7.38 and 7.42. Alkalosis (a pH value greater than 7.42), or acidosis (a pH value less than 7.38) occur beyond these limits. Four basic types of disturbances may occur.

Respiratory acidosis

Respiratory acidosis occurs when elimination of CO_2 by the lungs is diminished. CO_2 is retained, more carbonic acid and hydrogen ion is formed, and pH falls. This disturbance occurs in many types of lung conditions. In mild respiratory acidosis, excess H^+ resulting from CO_2 retention stimulates breathing, causing an increased CO_2 elimination; the kidney increases H^+ secretion, and the condition is compensated.

Metabolic acidosis

Metabolic acidosis results from loss of base, as in diarrhea, which causes loss of HCO_3^-; by impaired secretion of hydrogen ion by the kidney; by excessive production of organic acids by abnormal or exaggerated metabolic processes (e.g., diabetes mellitus). Excessive liberation of hydrogen ion from these acids causes pH to fall. Fall of pH again stimulates breathing and eliminates CO_2 to reduce the source of H^+, and achieve compensation.

Respiratory alkalosis

Respiratory alkalosis results from excessive loss of CO_2, as in hyperventilation; excessive alkali accumulates, and pH rises. Slowing of breathing occurs when pH rises, and decreased kidney secretion of H^+ raises the concentration of H^+ in the body fluids. This compensates for the alkalosis.

Metabolic alkalosis

Metabolic alkalosis results from excessive alkali intake (e.g., antacids) and by vomiting, which causes loss of hydrogen ion in stomach fluids.

FIGURE 16.8. Production and compensation of respiratory and metabolic acidosis.

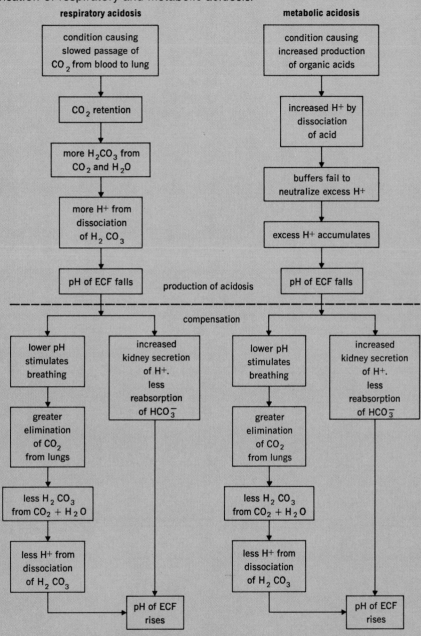

FIGURE 16.9. Production and compensation of respiratory and metabolic alkalosis.

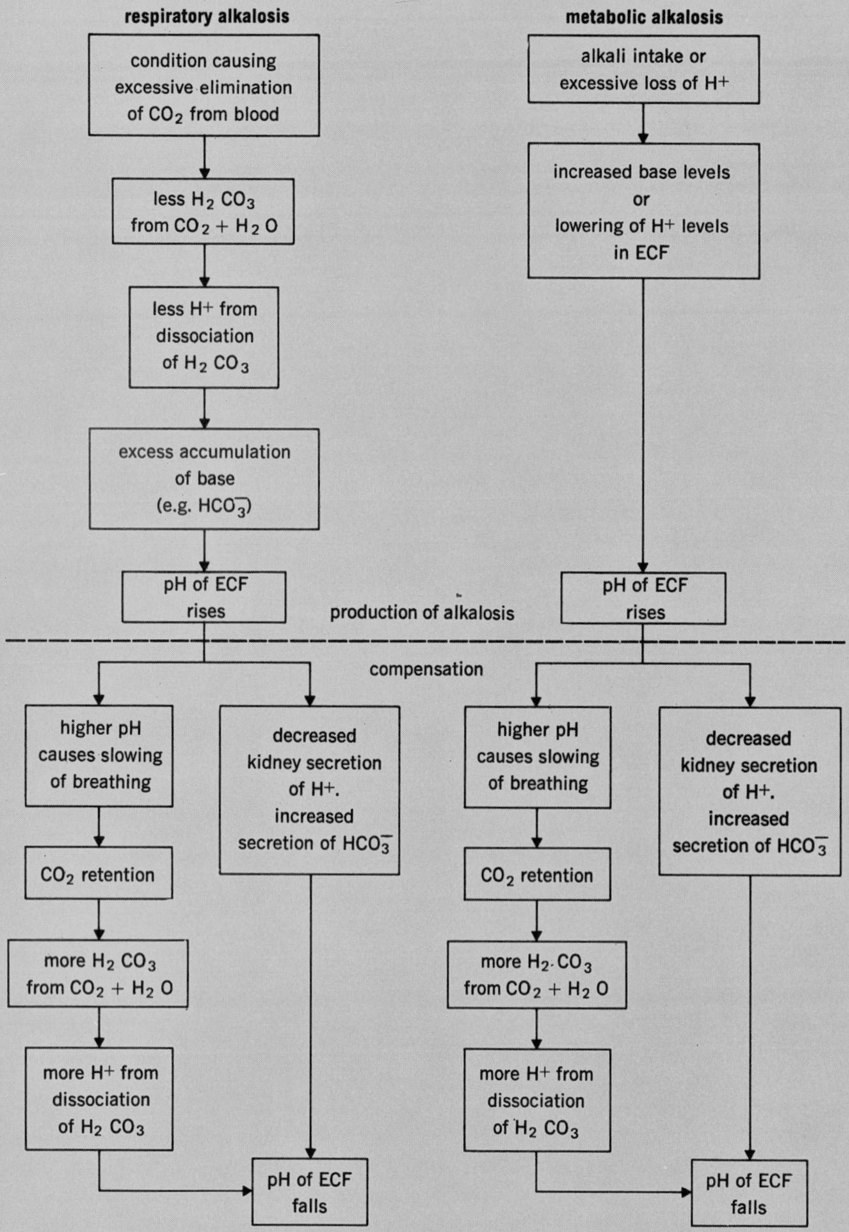

respiratory alkalosis

condition causing excessive elimination of CO_2 from blood

less H_2CO_3 from $CO_2 + H_2O$

less H^+ from dissociation of H_2CO_3

excess accumulation of base (e.g. HCO_3^-)

pH of ECF rises

metabolic alkalosis

alkali intake or excessive loss of H^+

increased base levels or lowering of H^+ levels in ECF

pH of ECF rises

production of alkalosis

compensation

higher pH causes slowing of breathing

CO_2 retention

more H_2CO_3 from $CO_2 + H_2O$

more H^+ from dissociation of H_2CO_3

decreased kidney secretion of H^+. increased secretion of HCO_3^-

pH of ECF falls

higher pH causes slowing of breathing

CO_2 retention

more H_2CO_3 from $CO_2 + H_2O$

more H^+ from dissociation of H_2CO_3

decreased kidney secretion of H^+. increased secretion of HCO_3^-

pH of ECF falls

Again excessive alkali accumulates and pH rises. Compensation occurs as in respiratory alkalosis. These conditions and their compensations are summarized in the "flow sheets" below (Figs. 16.8 and 16.9).

In achieving recovery from acid-base disturbances, CO_2 elimination and secretion of hydrogen ion by the body are considered to be of critical importance. Metabolic acidosis and alkalosis can normally be compensated over several days to the extent of 50–75 percent by alteration of breathing. Respiratory acidosis and alkalosis can be nearly completely compensated, also in several days, by alteration of kidney secretion of hydrogen ion. If it seems paradoxical to alter a given condition by using the opposite organ (e.g., kidney to compensate for a respiratory acidosis), remember that initially something was wrong with the organ that caused the condition, and that the body will seek another route to correct it.

Summary

1. The body fluids constitute an internal environment for the body cells, and compose 55–60 percent of the body weight.

    a. The fluids supply necessary nutrients for cellular activity and serve as the route for removal of wastes.

    b. Regulation of composition and characteristics of the body fluids is essential to life.

2. Two major divisions and several minor subdivisions may be made in total body fluids.

    a. Extracellular fluids (37.5%), fluid outside cells:

    (1) Blood plasma (7%).
    (2) Interstitial fluid (tissue fluid) and lymph (18%).
    (3) Connective tissue fluids (10%).
    (4) Transcellular fluids (2.5%): fluids in hollow organs (eye, digestive organs, lungs, joints, cavities of and surrounding the central nervous system).

    b. Intracellular fluid (62.5%), fluid within cells.

3. Maintenance of water balance requires intake equivalent to output.

    a. Requirements for water are increased with increase of metabolism.

    b. Sources of water include:

    (1) Ingestion of food and drink.
    (2) Production by metabolism.

    c. Routes of loss of water include:

    (1) "Fixed loss"; lungs, perspiration, feces.
    (2) Variable loss, kidneys.

4. Most absorbed water enters the plasma.

    a. Plasma water is filtered into the interstitial fluid.

5. Water is returned to the plasma by osmosis due to plasma proteins.

6. Interstitial water passes osmotically into the plasma, is removed by lymph vessels, or enters cells.

7. All extracellular fluids are nearly identical in electrolyte types and concentrations, but differ in protein concentration. Sodium and chloride are the primary extracellular electrolytes. Protein concentration is highest in the plasma compartment.

8. Intracellular fluid contains much more protein than extracellular fluid and contains potassium and phosphate as the primary electrolytes.

9. Regulation of volume and osmotic pressure of extracellular fluid is by:

   a. Thirst which increases intake.

   b. Osmorceptors in the brain (hypothalamus) which determine release of antidiuretic hormone.

   c. The adrenal hormone, aldosterone, which deals with sodium reabsorption.

10. Dehydration results when fluid and electrolyte loss exceeds intake.

    a. "Isotonic dehydration" involves equivalent loss of fluid and electrolytes.

    b. "Hypertonic dehydration" involves greater fluid than electrolyte loss. Cell water is lost.

    c. "Hypotonic dehydration" involves greater electrolyte than fluid loss. Water enters cells.

11. Edema results from accumulation of fluid in the interstitial compartment. It occurs when:

    a. Filtration from blood vessels exceeds osmotic return.

    b. The heart fails to pump blood effectively and venous pressure increases.

    c. There is salt and water retention.

    d. Lymph vessels are blocked.

12. Regulation of pH of body fluids insures continuance of enzymatic reactions in the body.

    a. Regulation of pH maintains pH at 7.4 ± 0.02.

13. Acids are substances that release hydrogen ions.

    a. Strong acids release much hydrogen ion.

    b. Weak acids release little hydrogen ion.

14. Bases are substances that accept free hydrogen ions and neutralize their effects.

15. Sources of hydrogen ion include:

    a. Reaction of CO_2 with H_2O to produce carbonic acid with release of hydrogen ion.

    b. Release of hydrogen ion from organic acids.

    c. Ingestion of acidic substances.

16. Methods of minimizing the effect of free H ion is by:

    a. Buffering. A weak acid and its corresponding completely ionized salt are a buffer pair. Hydrogen ion added to the salt forms the weak acid. Several buffer systems in ECF (bicarbonate, phosphate, protein) and in ICF (hemoglobin, phosphate) are described. Compensation is discussed as a method by which breathing and kidney secretion of H^+ are adjusted to maintain proper H^+ and base ratios.

b. Dilution. Prevents local accumulation.

c. Uptake by hydrogen acceptors.

17. The body has base in excess of normal need to neutralize hydrogen ion. The extra base is the alkali reserve or base excess.

18. Acid-base disturbances include acidosis (pH less than 7.38), and alkalosis (pH greater then 7.42).

a. Respiratory acidosis results from CO_2 retention and is compensated by increased kidney secretion of H^+.

b. Metabolic acidosis results from loss of alkali, excessive intake of acids, or accumulation of organic acids, and is compensated by increased breathing, and increased kidney secretion of H^+.

c. Respiratory alkalosis results from excessive CO_2 loss, and is compensated by decreased breathing and decreased kidney secretion of H^+.

d. Metabolic alkalosis results from loss of hydrogen ion, and is compensated in the same manner as respiratory alkalosis.

Questions

1. What are the three largest body water compartments? Give the appropriate percentage of body weight for each in the adult.

2. Which of the body water compartments does the kidneys regulate and, thus, ultimately regulate all of the fluid compartments of the body?

3. What are the sources and routes of loss of body water?

4. What forces are responsible for water movement into and out of each fluid compartment?

5. Compare the compositions of intracellular and extracellular fluid for the predominant ions.

6. How is the volume of the ECF regulated?

7. What is the ADH mechanism for control of fluid osmotic pressure?

8. What is a weak acid?

9. Where does hydrogen ion come from in the body?

10. How does the body minimize the effect of hydrogen ion?

11. Discuss the "automatic mechanisms" that go into operation, when acidosis or alkalosis arises, to compensate for the excessive or deficient H^+ concentration.

12. Compare a metabolic and a respiratory acidosis as to cause and method of compensation.

Readings

Abbott Laboratories. *Fluids and Electrolytes.* North Chicago, Ill., 1960.

Kee, Joyce LeFever. *Fluids and Electrolytes with Clinical Applications.* Wiley. New York, 1971.

Maxwell, Morton H. and Charles R. Kleeman (eds). *Clinical Disorders of Fluid and Electrolyte Metabolism.* 2nd ed. McGraw-Hill. New York, 1972.

Pitts, Robert F. *Physiology of Kidney and Body Fluids.* 2nd ed. Yearbook Medical Pubs. Chicago, 1968.

Taylor, Ann et al. ''Vasopressin: Possible Role of Microtubules and Microfilaments in Its Action.'' *Science* 181:370, 1973.

Winters, Robert W. et al. *Acid-base Physiology in Medicine, A Self Instruction Program.* 2nd ed. The London Co. Westlake, O., 1969.

chapter 17
The Blood and Lymph

chapter 17

Blood is a connective tissue containing a liquid intercellular material in which cells or cell-like structures are found. It is the extracellular fluid contained within the arteries, capillaries, and veins of the blood vascular system. Blood makes up about 7 percent of the body weight and, in the 70-kilogram (154-lb) male adult, amounts to 5–6 liters in volume. It is easily obtained from the body, and since it circulates to and from cells, it has a composition that depends on cell activity. Thus analysis of the blood gives a good idea of the status of body function.

Lymph is formed by filtration from the blood and was originally tissue or interstitial fluid. When it enters lymphatic vessels, it is called lymph. It is ultimately returned to the blood vessels from the tissue spaces.

Development of blood (Fig. 17.1)

Since growth of the embryo depends on adequate supplies of oxygen and nutrients, blood and blood vessels are among the earliest structures to develop. At about 2½ weeks of age, when the embryo is only 1.5 millimeters long, dense masses of cells called BLOOD ISLANDS appear in the mesoderm of the embryo and its surrounding tissues. The peripheral cells of the islands flatten and form the linings (endothelium) of the embryonic blood vessels, and the cells inside differentiate into

FIGURE 17.1. The development of blood islands.

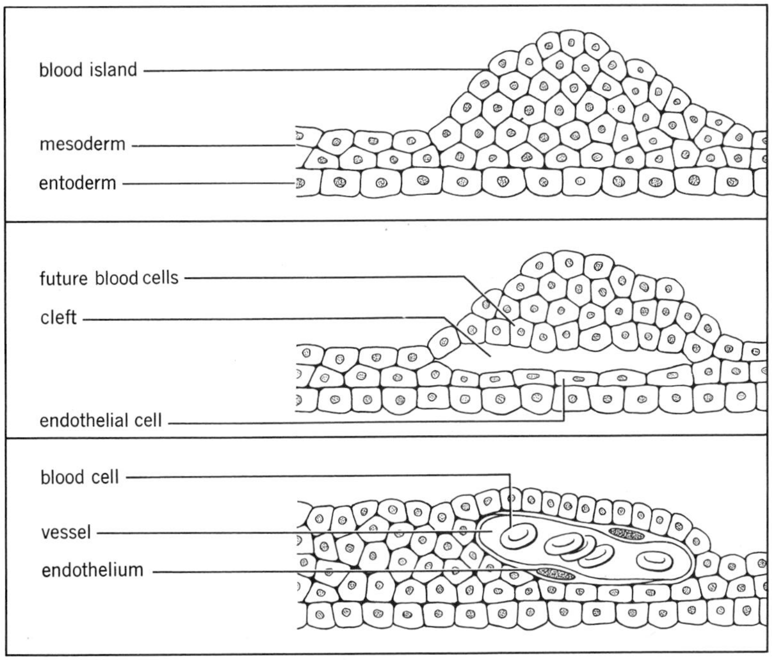

primitive blood cells. The first cells formed are nucleated red blood cells, capable of transporting oxygen and carbon dioxide. Later, the yolk sac, liver, spleen, lymph glands, and bone marrow form blood cells. At birth, the bone marrow and lymph glands remain as the only important sites of blood cell formation.

Functions of the blood

Functions of the blood center about two main activities: TRANSPORT and REGULATION OF BODY HOMEOSTASIS, the latter including regulation of water content, pH, temperature, and protection against invasion by chemicals and microorganisms.

Transport

Because of its liquid intercellular material, the blood is capable of simultaneously dissolving and suspending many different substances and transporting them to and from cells as the blood is circulated by heart action. *Nutrients,* including amino acids, vitamins, glucose, salts, and lipids, are picked up from the gut and carried to cells. Oxygen is picked up in the lungs. *Metabolic wastes,* such as heat, CO_2, and nitrogen containing compounds (urea, uric acid) are carried to their appropriate organs of excretion. *Regulatory substances,* including enzymes and hormones, are picked up from their organs of production and are carried to other cells to exert their effects.

Regulation of homeostasis

Among the processes essential to homeostatic regulation in which the blood is involved are:

Regulation of tissue water content. Since fluids in other body water compartments (Chapter 16) are derived from the blood, the osmotic and hydrostatic pressures of the blood will determine the passive movement of water between compartments.

Regulation of pH. The blood contains a number of buffer systems (Chapter 16) capable of resisting changes in pH. The pH of arterial blood is thus normally maintained at a pH of 7.4; venous blood is maintained at a pH of 7.36. The difference is due to the H^+ produced from the reaction of CO_2 and H_2O.

Regulation of body temperature. The water of the blood absorbs much heat of metabolic reactions, and carries it to the skin and lungs, where it is radiated and conducted from the body if the vessels are open to permit flow of blood and radiation of heat. Conversely, narrowing of the vessels prevents circulation through these areas and conserves heat.

Protection. The blood has the ability to coagulate or clot, and thus protects the body from blood loss. Several of the formed elements are capable of phagocytosis, or the engulfing and digesting of particles (bacteria, debris of injury); certain of the formed elements are capable of producing chemicals (antibodies) that afford protection against disease, and the products of disease processes.

Components of the blood

The liquid portion of the blood is the PLASMA. The cells or cell-like structures floating in the plasma are the FORMED ELEMENTS.

Plasma

Plasma is the straw-colored, liquid portion of the blood composing 55–57 percent of its volume. Plasma is an extremely complex solution and suspension of chemical substances. Table 17.1 presents some of the important plasma constituents.

SOURCES OF PLASMA COMPONENTS. Water, electrolytes, carbohydrates, lipids, and vitamins are derived almost entirely by absorption from

TABLE 17.1 Some Important Plasma Constituents		
Constituent/type	Amount/concentration	Functions/comments
Water	90% of plasma	Dissolves, suspends, ionizes electrolytes, carries heat.
Electrolytes	About 1% of plasma . (Examples: Na^+, K^+, Ca^{++}, Mg^{++}, Cl^-, HCO_3^-, PO_4	They buffer, establish osmotic gradients, are responsible for excitability of cells.
Proteins	6-8% of plasma	General functions: give viscosity to blood, clotting, antibodies, reserve of amino acids, establish colloid osmotic pressure of plasma.
Albumins	50-65% of total proteins	Are responsible for most of the plasma colloidal osmotic pressure.
Globulins (alpha, beta, gamma)	14.5-27% of total proteins	α and β serve general functions; γ are major source of antibodies.
Fibrinogen	2.5-5% of total proteins	Clotting factor
Other substances		
Glucose	about 0.1% of plasma	Nutrients
Gases (O_2, CO_2)	variable (O_2 av 20 ml/100 ml blood) (CO_2 av 9 ml 100 ml blood)	or
		Wastes
Lipids	variable	
Vitamins	variable	
Nitrogenous materials (urea)	variable	

the gut. Gases are derived from the lungs (O_2) or tissues (CO_2). The proteins are manufactured mostly by the liver, with some coming from cell disintegration in the bloodstream and from plasma cell (γ globulins).

The formed elements (Fig. 17.2)

Three categories of formed elements occur in normal human peripheral blood. They are: erythrocytes or red blood cells, leucocytes or white blood cells, and platelets.

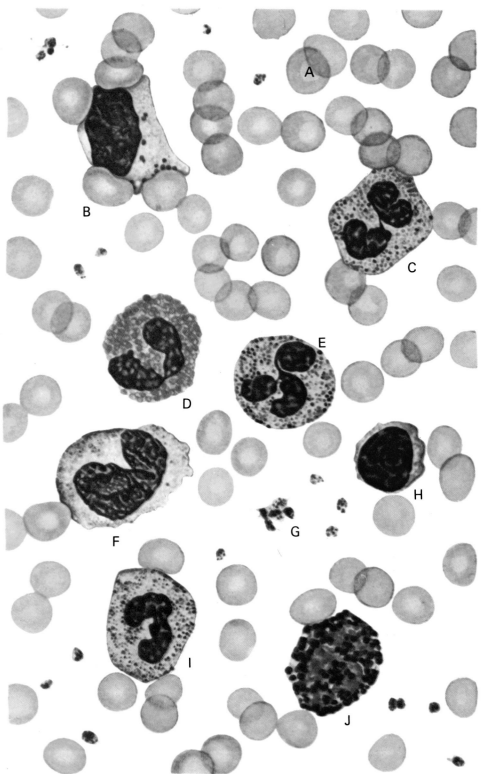

Legend key.
CELL TYPES FOUND IN SMEARS OF
PERIPHERAL BLOOD FROM NORMAL
INDIVIDUALS. The arrangement is
arbitrary and the number of leukocytes
in relation to erythrocytes and
thrombocytes is greater than would
occur in an actual microscopic field.

A   Erythrocytes
B   Large lymphocyte with azurophilic
    granules and deeply indented by
    adjacent erythrocytes
C   Neutrophilic segmented
D   Eosinophil
E   Neutrophilic segmented
F   Monocyte with blue gray cytoplasm,
    coarse linear chromatin and blunt
    pseudopods
G   Thrombocytes
H   Lymphocyte
I   Neutrophilic band
J   Basophil

FIGURE 17.2.   The morphology of the blood elements.

FIGURE 17.3. The formation of blood cells.
The exact morphology of the stem cell is still open to question. Wright stain. 1200X. (Courtesy of American Society of Hematology National Slide Bank and Health Sciences Learning Resources Center, University of Washington. Used with permission.)

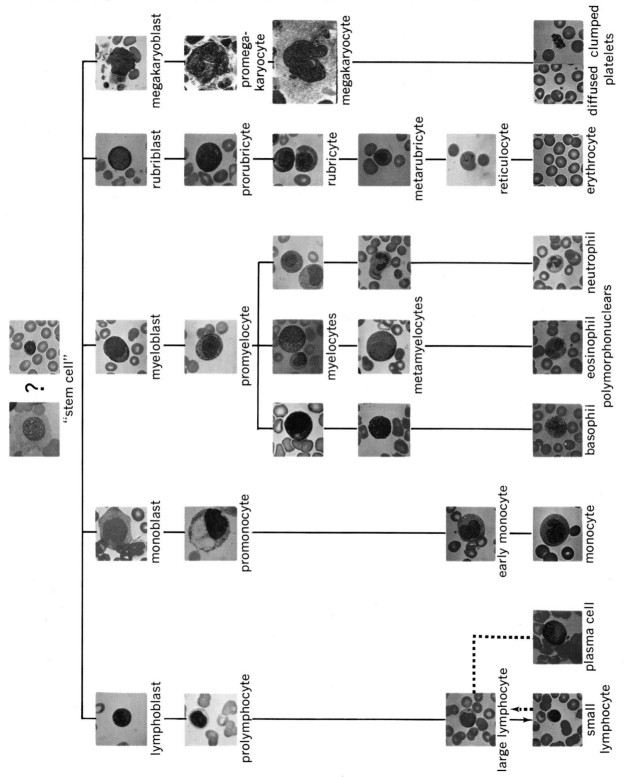

ERYTHROCYTES (see Fig. 17-2). Mature erythrocytes are tiny (8μ), flexible, biconcave discs with no nuclei. They consume small amounts of O_2, glucose, and ATP, and liberate CO_2, indicating a low level of metabolic activity. Most of this activity is concerned with the operation of active transport systems in the cell membrane.

Numbers. In the adult, erythrocytes range between 4.8 and 5.4 million per cubic millimeter of blood. This number reflects a balance between production and destruction of cells; production and destruction are estimated to occur at rates of 5 to 10 million cells per second.

TABLE 17.2 Requirements for Erythrocyte Production	
Substance	Description and/or use
Lipid	Cholesterol and phospholipids; incorporated in membrane and stroma
Protein	Incorporated into cell membrane
Iron	Incorporated into hemoglobin
Amino acids	Incorporated into hemoglobin
Erythropoietin	A glycoprotein, it is released from the kidney with hypoxia, hemorrhage, and excessive androgen secretion, and stimulates production of erythrocytes
Vitamin B_{12}	Used in the formation of DNA in nuclear maturation
Intrinsic factor	A mucopolysaccharide produced by the stomach. It combines with vitamin B_{12} and insures absorption of the vitamin from the gut
Pyridoxin	Increases the rate of cell division
Copper	Catalyst for hemoglobin formation
Cobalt	Aids synthesis of hemoglobin
Folic Acid	Promotes DNA synthesis

Life history. Production of red cells occurs in the bone marrow in a series of stages (Fig. 17.3). Production begins with a cell known as a *hemocytoblast*. The cell develops into a *rubriblast* with a very dark staining nucleus and cytoplasm. Next a *prorubricyte* develops, and then a *rubricyte*. The rubricyte shows the first stages in the formation of hemoglobin as tiny droplets known as hemoglobin islands. The cells become smaller and accumulate more hemoglobin, and form *metarubricytes*. The nucleus condenses, fragments, and is lost from the cell to form the *mature erythrocyte*. Red cells carry out their functions in the bloodstream for 90–120 days, and are then removed from the bloodstream by phagocytic cells in the liver, spleen, and bone marrow. The components of the cells are largely conserved and recycled into the production of new cells. The yellow to brown coloring matter in urine and feces comes from the breakdown of hemoglobin.

Many substances are required for the production of red cells. Some of the most important are shown in Table 17.2.

Of interest is erythropoietin. It is a substance that is released by the kidney under any condition which lowers O_2 levels of the blood reaching the kidney. In 3–4 days after its release from the kidney, erythropoietin stimulates the production of red cells to a peak of about 2 percent of the total circulating mass of red cells. The homeostatic mechanism controlling red cell production is shown in Figure 17.4.

Hemoglobin. Hemoglobin is a respiratory pigment located in the red cells. It consists of an iron containing pigment known as HEME, and a protein portion named GLOBIN. Each molecule possesses the property of combining loosely with eight atoms of oxygen. Oxygen is taken up by the hemoglobin in the lungs to form OXYHEMOGLOBIN; release of oxygen to the tissues results in the formation of REDUCED HEMOGLOBIN. The relationships between these compounds may be summarized by the equation:

$$\underset{\substack{\text{reduced hemoglobin}\\\text{(bluish in color)}}}{\text{Hgb}} + \underset{\text{oxygen}}{O_2} \rightleftharpoons \underset{\substack{\text{oxyhemoglobin}\\\text{(red in color)}}}{\text{Hgb } O_2}$$

Hemoglobin may also combine with carbon dioxide to give CARBAMINOHEMOGLOBIN. Normal amounts of hemoglobin in the blood are 13.5–16 grams per 100 milliliters of blood.

Functions of the erythrocytes. Red cells transport oxygen and carbon dioxide through the body. Hemoglobin also forms part of the buffer system of H · Hgb (weak acid) and K · Hgb (salt of the acid).

Clinical considerations. Conditions associated with erythrocytes may result in low or high numbers of cells, abnormal hemoglobin, and alterations in size (anisocytosis) or shape (poikilocytosis).

ANEMIA results when there is insufficient hemoglobin in the bloodstream to carry the oxygen needed for body chemical reactions. Anemia may result from low production of cells, excessive destruction of cells, or abnormal hemoglobin types or concentrations. Among the symptoms that commonly develop during anemia are shortness of breath, cyanosis (blue color of skin), and feeling of chronic fatigue. Table 17.3 summarizes some of the anemias.

POLYCYTHEMIA refers to excessive numbers of red cells, and is usually the result of hypoxemia or low blood oxygen levels, which stimulate red cell production.

A HEMOGLOBINOPATHY results when there is an abnormality in the hemoglobin of the red cells. Defects may result in abnormal heme or globin portions of the molecule. Since enzymes are required in the synthesis of the hemoglobin molecule, a gene mutation can result in the synthesis of abnormal fractions of the molecule. More than a dozen hemoglobinopathies are known. SICKLE CELL DISEASE is the result of the production of an abnormal hemoglobin that precipitates into spindle-shaped units when exposed to lowered oxygen concentrations. Sickled cells are phagocytosed more rapidly, and the sufferer usually exhibits anemia and weakness.

LEUCOCYTES (see Fig. 17.2). The white blood cells differ from red cells in that they are all nucleated and contain no hemoglobin. They range in size from about 8 to 25 microns.

Numbers. White cells are the least numerous of the formed elements. In the adult, they normally range from 5000–9000 per cubic millimeter, with an average of 7500 per cubic millimeter.

Life history (see Fig. 17.3). There are two categories of white cells: GRANULAR white cells (granulocytes) are produced in the bone marrow

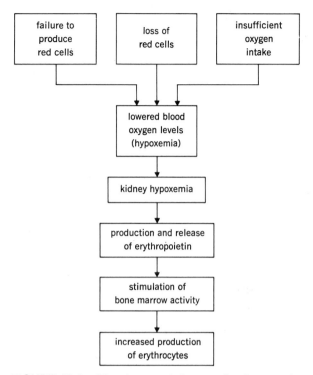

FIGURE 17.4. The homeostatic mechanism maintaining red cell production.

(myeloid origin) and have lobed nuclei and characteristic granules in their cytoplasm; NON-GRANULAR white cells (agranulocytes) are produced in the lymphatic tissue (lymphoid origin) and in the bone marrow.

Granular leucocytes remain functional for 10 hours to about 3 days; nongranular leucocytes remain functional for 100–300 days. Leucocytes are lost by migration through epithelial surfaces, by disintegration in the bloodstream, and by tissue destruction during inflammation and infections.

Production of leucocytes is influenced primarily by blood levels of adrenal cortical hormones. Products of infection and a *leucocyte mobilizing factor* are released when cells are damaged. Both increase the production of white cells.

Functions. All white cells are capable of ameboid motion and phagocytosis. All are chemotaxic, responding to chemicals, and capable of DIAPEDESIS, or the ability to pass through capillary walls to enter the tissues. The cells fight disease organisms or their chemical products or

TABLE 17.3   Some Common Anemias

	Type of anemia	Causes	Characteristics	Symptoms	Treatment
I N C R E A S E D L O S S	*Hemorrhagic* Acute	Trauma Stomach ulcers Bleeding from wounds	Cells normal	Shock	Transfusion
	Chronic (iron deficiency)	Stomach ulcers Excessive menses	Cells small (microcytes), deficient in hemoglobin content	None or fatigue	Iron administration
	Hemolytic	Defective cells Destruction by parasites, toxins, antibodies	Young cells (reticulocytes) very prominent serum haptoglobin* reduced. Morphology of cells may be abnormal when detected.	None or fatigue	Dependent on cause
D E C R E A S E D F O R M A T I O N	*Deficiency states* Folic acid	Nutritional deficiency	Cells large (macrocytes), with normal hemoglobin content	None or fatigue	Folic acid
	Vitamin B_{12}	Lack of intrinsic factor in stomach (pernicious anemia)	Cells large with normal hemoglobin content		Administration of vitamin B_{12}
	Hypoplastic or *Aplastic*	Radiation Chemicals Medications	Bone marrow hypoplastic	None or fatigue	Transfusion Androgens Cortisone

*A protein strongly binding hemoglobin in plasma.

both by phagocytosis, or the production of antibodies. Additional functions are carried out by specific white cells as is shown in Table 17.4.

Types of cells. There are five types of white cells, the structure and characteristics of which are shown in Figure 17.2, and are described in Table 17.4. Neutrophils, monocytes, and lymphocytes are the best phagocytic cells. Eosinophils increase in allergy.

LYMPHOCYTES have been recently grouped as B-LYMPHOCYTES, produced in bone marrow, and T-LYMPHOCYTES, possibly produced in the thymus gland. B-lymphocytes form the cells of the lymph nodes, spleen, tonsils, and the nodules in the digestive and respiratory systems. They develop the ability to transform into plasma cells and form antibodies. T-lymphocytes assume greater importance in delayed reactions, and in

TABLE 17.4 Summary of Formed Elements

Element	Normal numbers	Origin (area of)	Diameter (micron)	Morphology	Function(s)
Erythrocytes	4.5-5.5 million /mm	Myeloid (marrow)	8.5 (fresh) 7.5 (dry smear)	Biconcave, non-nucleated disc; flexible	Transports O_2 and CO_2 by presence of hemoglobin; buffering
Leucocytes	6000-9000/mm^3		9-25		
Neutrophil	60-70% of total	Myeloid	12-14	Lobed nucleus, fine heterophilic specific granules	Phagocytosis of particles, wound healing. Granules contain peroxidase for destruction of microorganisms.
Eosinophil	2-4% of total	Myeloid	12	Lobed nucleus; large, shiny red or yellow specific granules	Detoxification of foreign proteins? Granules contain peroxidases, oxidases, trypsin, phosphatases. Numbers increase in autoimmune states, allergy, and in parasitic infection (schistosomiasis, trichinosis, strongyloidiasis).
Basophil	0.15% of total	Myeloid	9	Obscure nucleus; large, dull, purple specific granules	Control viscosity of connective tissue ground substance? Granules contain heparin (liquefies ground substance) serotonin (vasoconstrictor), histamine (vasodilator)
Lymphocyte	20-25% of total	Lymphoid			Phagocytosis of particles, globulin production
Small			9	Nearly round nucleus filling cell, cytoplasm clear staining	
Large			12-14	Nucleus nearly round, more cytoplasm	
Monocyte	3-8% of total	Lymphoid	20-25	Nucleus kidney or horseshoe-shaped, cytoplasm looks dirty	Phagocytosis, globulin production
Platelets (Thrombocytes)	250,000-350,000 mm^3	Myeloid	2-4	Chromomere and hyalomere	Clotting

combating viruses, bacteria, and fungi that enter the body.

Clinical donsiderations. Leucopenia refers to a decrease in the numbers of leucocytes. It may result from increased destruction of cells such as occurs during viral infections, or from treatment with certain drugs (sulfonamides, anticonvulsives). Leucocytosis refers to an increase in numbers of leucocytes. A physiological leucocytosis (temporary increase) results from exercise, as cells are "washed out" of normally closed capillaries. Sustained elevations of white cell numbers suggest pathological processes in the body, such as leukemia.

Leukemia is a neoplastic disorder of the blood-forming tissues, primarily those giving rise to the white blood cells. Viruses have been suggested to be the cause of the cell proliferation, but no proof exists for this cause in the human. The major symptoms that appear are the production of great numbers of white cells, and diminished production of erythrocytes. According to the type of cell that predominates in the bloodstream, and the maturity of those cells, four general types of leukemia are distinguished.

Acute lymphoblastic leukemia (ALL) is a disease primarily of children, and is characterized by increased numbers of immature lymphocytes. Anemia, due to decreased production of red cells is found in 90 percent of patients with ALL.

Acute myeloblastic leukemia (AML) can occur at any age, and is characterized by increase in all of the granulocytes (neutrophil, eosinophil, basophil) and by increased numbers of immature granulocytes.

Chronic lymphocytic leukemia (CLL) occurs most commonly in the middle aged and elderly, and is characterized by greatly increased numbers of small, mature lymphocytes. Anemia is milder than in the acute leukemias.

Chronic myelocytic leukemia (CML) occurs more commonly in young adults and is characterized by increased numbers of all granular leucocytes.

Radiation and chemotherapy are the measures most frequently employed to treat leukemia. The acute forms of the disease are more difficult to arrest than the chronic forms; remissions of as much as 15 years have been achieved with adequate therapy that has been instituted early.

Hodgkin's disease is a chronic, ultimately fatal lymphoma of unknown case, characterized by enlargement of the lymph nodes, spleen, and liver. White cell numbers do not increase greatly in this disorder, but a giant cell with many-lobed or multi-nuclei (Reed-Sternberg cells) appear in the nodes. There is no certain cure for the disease, but radiation and chemotherapy are employed to lengthen life.

Mononucleosis is a disease thought to be due to a virus. It causes elevation of nongranular white cells, fatigue, swollen lymph nodes, fever, and sore throat. It lasts 10 days to 2 weeks, and is specifically diagnosed by the presence of abnormal lymphocytes and a specific antibody (Paul-Bunnel antibody) in the bloodstream.

PLATELETS (see Fig. 17.2). Platelets are nonnucleated structures 2−4 microns in diameter. They may have one or more dark-staining granules (chromomere) in a clear light-staining area (hyalomere) and are found singly or in groups among the other formed elements of a peripheral blood smear.

Numbers. Platelet numbers normally lie between 250,000 and 300,000 per cubic millimeter.

Life history. In the bone marrow are giant cells known as *megakaryocytes*. As these cells mature, they form granules in their cytoplasm. They then send cellular projections (pseudopods) through capillary walls, and pinch off a portion of the cytoplasm as a platelet. Thus platelets are actually fragments of the cytoplasm of the giant cell. Platelets remain in the bloodstream for 10−12 days, and are believed to be removed by phagocytosis when they have aged.

Function. Platelets contain one of many factors necessary for blood clotting. They are very adhesive and aid in plugging holes in damaged blood vessels. They also aid in maintaining the strength of capillary walls. Once a clot has formed, platelets aid in its shrinkage (syneresis) to form an even tighter bond across a damaged vessel.

Clinical considerations. Because of their involvement in the clotting process, platelet disorders are discussed in the next section. A summary of the formed elements is given in Table 17.4.

Hemostasis

When blood vessels are damaged, a series of reactions occurs that aids in preventing blood loss through the wound. These reactions constitute hemostasis (G. *amia*, blood, + *statikos*, standing). Three types of reactions occur.

There is NARROWING OF THE VESSELS (vasoconstriction) in the area of the wound, which presumably causes some reduction of the blood flow in the injured area.

A PLATELET PLUG is formed across the tear in the vessel.

The blood changes from its normal fluid consistency to a gelated or semisolid state, in the process of COAGULATION or clotting.

Vascular responses and the platelet plug

VASOCONSTRICTION occurring when blood vessels are traumatized is first due to nerve impulses arriving at the vessels because of the pain associated with the injury. A reflex involving the cord is responsible, but this response lasts only a few minutes. Second, the vessel muscle undergoes an intense spasm because of the direct mechanical damage to the vessels. This reaction lasts for 20-30 minutes.

Injury to the vessels also creates a roughened surface to which the PLATELETS ADHERE. Several layers of platelets accumulate across a tear in a vessel and may completely seal the injury. This platelet mechanism operates continually to seal the ruptures in the small vessels of the body that occur as a normal consequence of living. If platelets are low in numbers, many tiny pinpoint hemorrhages appear on the body surface as capillaries rupture and are not sealed by platelet plugs.

Coagulation

At least 12 factors or chemical substances are required for the blood to clot. These are present in the platelets and plasma, or are produced during the clotting reactions. They are shown in Table 17.5 (see page 346).

In general outline, blood clots in a definite sequence of events as shown below:

Thromboplastin + prothrombin $\longrightarrow$ thrombin
Thrombin + fibrinogen $\longrightarrow$ fibrin (clot)

Production of thromboplastin requires the interaction of several other substances, and certain substances speed the reactions producing thromboplastin. Additionally, thromboplastin may be produced by two methods: an EXTRINSIC SCHEME occurs when cells are damaged; an INTRINSIC SCHEME occurs within the blood itself. One theory of how these chemicals and schemes fit together is presented below.

These reactions are autocatalytic, that is, each step speeds the production of materials in the next step, so that, once begun, the clotting process proceeds rapidly. The clot is eventually removed

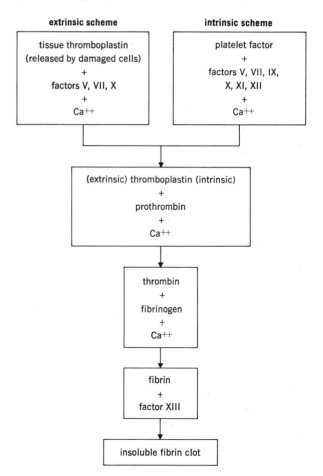

International committee designation	Synonyms	Origin	Location
Factor I	Fibrinogen	Liver	A plasma protein
Factor II	Prothrombin	Liver	A plasma protein
Factor III	Thromboplastin	By series of reactions in blood; also found, as such, in cells	Produced in the clotting process or released into fluids by injured cells
Factor IV	Calcium	Food and drink; from bones	As Ca^{++} in plasma
Factor V	Labile factor (accelerator globulin)	Liver	Plasma protein
Factor VI*			
Factor VII	SPCA (serum prothrombin conversion accelerator)	Liver	Plasma
Factor VIII	AHF (antihemophilic factor) AHG (anti-hemophilic globulin)	Liver	Plasma
Factor IX	PTC (plasma thrombo-plastin component)	Liver	Plasma
Factor X	Stuart-Prower factor; develops full factor III power	Liver	Plasma
Factor XI	PTA (plasma thrombo-plastin antecedent)	Liver	Plasma
Factor XII	Hageman factor; contact factor; initiates reaction	?	Plasma
Factor XIII	Fibrin stabilizing factor; renders fibrin insoluble in urea. (Laki-Lorand factor)	?	Plasma
Platelet Factor	Cephalin	Marrow	Platelets

TABLE 17.5 Factors definitely implicated in blood coagulation

*No longer considered a separate entity; considered to be identical to Factor V.

Disorder or deficiency	Synonyms	How inherited	Treatment
Factor VIII deficiency (80% of all hemophilias)	Classical hemophilia; Hemophilia A	Sex linked (on X chromosome) recessive; males show disorder	Transfusion of *fresh* blood containing Factor VIII (VIII disappears with aging of blood)
Factor IX deficiency (15% of hemophilias)	PTC deficiency; Christmas disease; Hemophilia B	Sex linked recessive; males show disorder	Transfusion of stored blood is effective
Factor XI deficiency	PTA deficiency; Hemophilia C	Rare; autosomal dominant	Transfusion of stored blood is effective
Hypofibrino-genemia	Afibrinogenemia	Autosomal recessive	Transfusion of whole blood, or plasma

TABLE 17.6 Some genetically determined disorders of clotting

by an enzyme, FIBRINOLYSIN (plasmin), produced by activation of a plasma precursor.

Procoagulation and anticoagulation

Clot formation may be speeded by heat or by spraying a wound with a clotting factor (e.g., fibrin). Prevention of clotting is essential when drawing blood for tests, transfusions, or storage, and is most easily accomplished by removing one of the materials in the clotting process; Ca^{++} is the one most easily removed, by simply mixing the blood with substances (e.g., oxalates) that exchange their cation for Ca^{++}. The blood is then said to be *decalcified. Heparin* is an organic substance that decreases thromboplastin production; *dicoumarol* is an organic substance that interferes with liver synthesis of prothrombin, and factors V and VII. Dicumarol requires 36-48 hours for its effect to be manifested.

Clinical considerations

Since many of the major clotting factors are proteins (e.g., Factors I, II, VIII, IX, and XI), and since proteins are produced under the direction of DNA, alterations in genes (mutations) may result in defective production of clotting factors. Table 17.6 presents several genetically determined disorders of clotting, all of which are characterized by failure of the blood to clot when the body is wounded, bruised, or traumatized.

Disorders of platelet production leads to the development of THROMBOCYTIC PURPURA, characterized by easy bruising and bleeding from the nose, gums, gastrointestinal tract, and kidneys. Low platelet levels result in fragile capillaries and slowed clotting.

SPONTANEOUS INTRAVASCULAR CLOTTING (clotting in vessels in the absence of trauma) may create a THROMBUS (stationary clot) that blocks a vessel and causes tissue death beyond the block (e.g., coronary thrombosis).

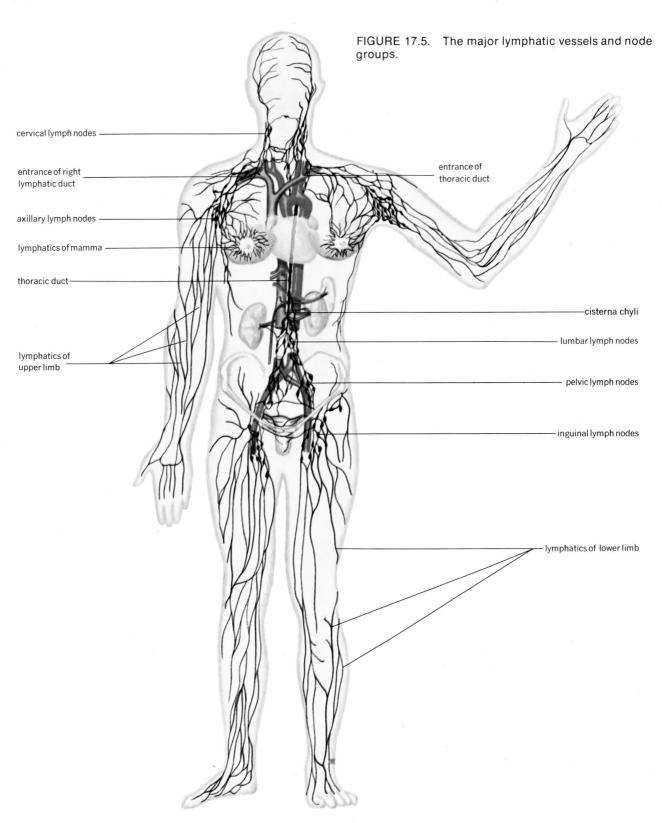

FIGURE 17.5.   The major lymphatic vessels and node groups.

cervical lymph nodes

entrance of right lymphatic duct

axillary lymph nodes

lymphatics of mamma

thoracic duct

lymphatics of upper limb

entrance of thoracic duct

cisterna chyli

lumbar lymph nodes

pelvic lymph nodes

inguinal lymph nodes

lymphatics of lower limb

Lymph and lymph organs

Lymph formation, composition and flow

Lymph is tissue or interstitial fluid that has entered the lymph vessels (Fig. 17.5) of the body. It arises by a process of filtration of the blood from blood capillaries. Thus it has the same composition as cell free, low protein plasma.

Tissue fluid enters lymph vessels when they are expanded as muscle becomes active in the body ("suction effect"), and because of a small pressure gradient that exists from blood to tissue space to lymph vessel. Protein enters the lymph vessels easily because they are much more permeable than blood capillaries. The lymph moves through the vessels because of the massaging action of muscles as they contract (muscle pump), and because of breathing movements which alternately compress and release the lymphatics (respiratory pump). Valves assure flow toward the veins beneath the clavicles, where the large lymph ducts connect to the blood vascular system. Rate of flow is about 1.5 milliliters per minute.

FUNCTIONS. The lymph returns, to the blood vascular system, filtered protein and excess tissue water that is not osmotically returned to blood capillaries. The lymph of the intestine contains much absorbed end products of fat digestion, because these end products enter the lymph vessels of the intestinal villi and not the blood vessels.

Lymph nodes

FORM AND FUNCTIONS. Lymph nodes (Fig. 17.6) are ovoid or rounded structures varying in size from 1 to 25 millimeters. They are situated along most medium-sized lymph vessels and form four major groups (see Fig. 17.5) in the *axilla, groin, neck,* and *abdomen.* The nodes are sites of production of nongranular white cells. Within the nodes are many fixed phagocytic cells (macrophages). Lymph passing through the nodes is FILTERED or cleansed of particulate matter (bacteria, dirt, and cell debris). The nodes may also trap cancer cells

that have entered the lymph vessels from a neoplasm.

CLINICAL CONSIDERATIONS. Enlargement of the nodes is common in infections, due to an increase in the number of cells to trap the infectious agents. Nodes are generally removed in conditions where it is believed that cancer cells have metastasized. The status of the nodes is commonly determined by feeling (palpation) in individuals presenting themselves for physical examination.

Other lymph organs

The TONSILS (Fig. 17.7) are three pairs of lymphoid organs located in the oral cavity and the upper

FIGURE 17.6. A diagrammatic representation of a lymph node. Cortical and medullary areas contain both lymphocytes and macrophages. Arrows indicate direction of lymph flow.

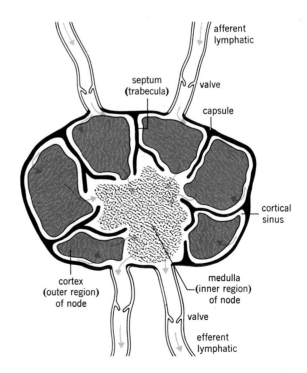

afferent lymphatic

septum (trabecula)

valve

capsule

cortical sinus

cortex (outer region) of node

medulla (inner region) of node

valve

efferent lymphatic

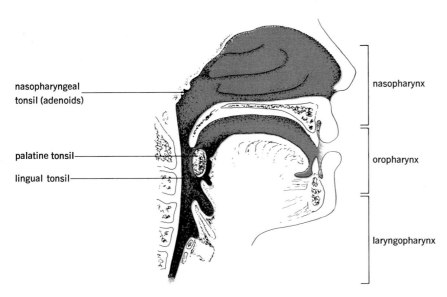

nasopharyngeal
tonsil (adenoids)

palatine tonsil

lingual tonsil

nasopharynx

oropharynx

laryngopharynx

FIGURE 17.7.   The location of the tonsils.

part of the throat. The *palatine tonsils* are found in the side walls of the posterior part of the oral cavity, and are the ones referred to when one says he is "having the tonsils out." The *pharyngeal tonsils,* also called the adenoids, are located in the upper part of the throat or pharynx just behind the hard palate. The *lingual tonsils* are found embedded in the base of the tongue. All the tonsils consist of masses of lymphocytes and germinal centers in which lymphocytes are produced. The lymphocytes are shed into outgoing lymph vessels and eventually reach the bloodstream. The tonsils do not filter or cleanse the lymph because they do not have afferent or incoming lymph vessels.

The SPLEEN (Fig. 17.8) is a soft reddish-colored organ located behind the left side of the stomach. It lies between arteries and veins and contains areas where lymphocytes are produced which are then placed into outgoing lymphatic vessels. These areas are known as the *white pulp* of the spleen. The spleen is also an area where old red blood cells are phagocytosed and removed from the circulation, and where up to 200 milliliters of blood can be "stored" and injected into the circulation when the spleen contracts. The spleen does

not cleanse the lymph, because it, like the tonsils, has only outgoing lymphatic vessels.

The functions of the THYMUS are considered in Chapter 18. Masses of lymphoid cells form NODULES in the walls of the digestive and respiratory systems and are considered in those chapters.

CLINICAL CONSIDERATIONS. The palatine, pharyngeal, and lingual tonsils form a ring (Waldeyer's ring) of lymphoid tissue that encircles the entrance of the digestive and respiratory systems into the body. The ring offers a means of combating microorganisms that may penetrate the mucous membranes of the upper respiratory and digestive systems. Removal of any part of the ring may thus interfere with the protective function the ring serves.

The components of Waldeyer's ring normally enlarge as the child grows, reaching peak development at puberty. It grows because it is exposed to a variety of new infectious agents. Mere size of one of the components should not be used as the primary reason for its surgical removal. Only if one of the tonsils is chronically infected, or has enlarged to the point where it obstructs the

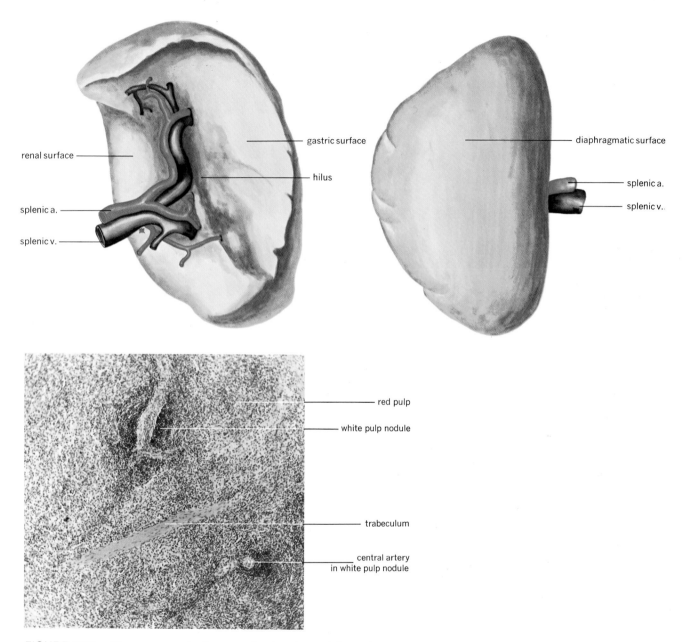

renal surface

gastric surface

hilus

splenic a.

splenic v.

diaphragmatic surface

splenic a.

splenic v..

red pulp

white pulp nodule

trabeculum

central artery
in white pulp nodule

FIGURE 17.8.   The gross and microscopic structure of the spleen.

breathing or digestive passageways, should it be removed.

The spleen is not an organ essential to life, since it may be removed without harmful effects to the body. It may enlarge during infections (typhoid fever, malaria, syphilis, tuberculosis), or during conditions that stress its red-cell-destroying capacity (e.g., sickle cell anemia). Trauma may cause the spleen to rupture, and this usually requires surgical removal of the organ to avoid "bleeding to death."

Summary

1. Blood is a connective tissue that flows through the blood vessels. It constitutes about 7 percent of the body weight.

2. Blood develops from mesodermal blood islands at about 2½ weeks of embryonic life.

3. Blood functions in transport, and in regulation of several aspects of homeostasis (heat control, water balance, pH regulation). Its cells and antibodies protect against bacteria and their products.

4. Two main components compose blood.
   a. The liquid plasma exceeds one half the blood volume, and contains organic and inorganic components. It dissolves, suspends, and transports many materials.
   b. The formed elements are cells or cell-like structures, and constitute a little less than one half the blood volume.

5. Erythrocytes or red cells:
   a. Are the most numerous of the formed elements (4.5-5 million/mm^3); and are produced in the bone marrow.
   b. Are of small (8μ) size.
   c. Contain hemoglobin for O_2 transport.
   d. Last 90–120 days in the circulation.

6. Hemoglobin:
   a. Contains a pigment (heme) portion.
   b. Contains a protein (globin) portion.
   c. Requires many substances for its production.
   d. Acts as a buffer.
   e. Forms a loose combination with O_2.

7. Anemia:
   a. Results when hemoglobin levels are decreased.
   b. May be due to decreased production of cells or hemoglobin, to increased loss, or to shortage of building blocks.
   c. May be due to formation of abnormal hemoglobins due to genetic causes.

8. Leucocytes or white cells:
   a. Are all nucleated and contain no hemoglobin.
   b. Are the least numerous of the formed elements (7500/mm^3).
   c. Are produced in marrow and lymph organs.
   d. Last from hours to a year in the body.
   e. Afford protection against disease organisms and their products.
   f. Alter their numbers in various types of diseases, such as leukemia and mononucleosis.

9. Platelets:
   a. Are fragments of large bone marrow cells.
   b. Are intermediate in number (250,000/mm^3).
   c. Are important in blood clotting.

10. Blood flow and characteristics change when vessels are injured. The vessels constrict, a platelet plug is formed, and the blood clots.
   a. Many factors are required to form a clot.
   b. Thromboplastin, prothrombin, fibrinogen and, Ca^{++} are essential for clotting.

11. Failure of the blood to clot involves lack or removal of an essential clotting factor.
   a. Removal of calcium may stop clotting.
   b. Hemophilias result from failure of DNA to cause production of essential clotting proteins.

12. Lymph:
   a. Is formed by filtration of blood.
   b. Is tissue fluid in lymph vessels.
   c. Flows by "pumping action" of muscles and breathing.
   d. Serves to return protein and water to the blood vessels and as a route of fat absorption from the gut.

13. Lymph nodes:
   a. Produce nongranular white cells.
   b. Filter and cleanse lymph.
   c. Trap cancer cells that have metastasized.
   d. Enlarge during infections.
   e. Are 1-25 millimeters in size, and are disposed in four main areas of the body.

14. Tonsils and spleen are accessory lymphoid organs and produce lymphocytes and monocytes, but do not filter the lymph. The tonsils form Waldeyer's ring around the entrance of the respiratory and digestive systems.

Questions

1. What are the functions of the blood?

2. What are the components of the blood and their percents?

3. What are the major components of plasma?

4. What are the subdivisions of the plasma proteins and what are their functions?

5. Describe erythrocytes as to appearance, size, numbers, and function.

6. What is the importance of hemoglobin to the body?

7. What are the two major causes of anemia?

8. What are the properties and functions of leucocytes?

9. Outline the phases of the clotting of the blood.

10. Compare the extrinsic and intrinsic schemes of clotting as to materials required and results.

11. Compare the composition of plasma and lymph and account for any significant differences.

12. Describe the functions of the lymph nodes.

13. Compare the composition of plasma, lymph, and intracellular fluid.

14. What functions do lymph nodes and tonsils have?

Readings

Arehart-Treichel, J. "Out for Blood." *Science News,* vol. 105, no. 13 (March 30) 1974, pp. 210-211.

Alper, Chester A. "Beta Lymphocyte Malignancy." *N. Eng. J. Med.* 289:154, 1973.

Diggs, L. W. et al. *The Morphology of Blood Cells.* Abbott Laboratories. North Chicago, 1954.

Krantz, S. B., and L. O. Jacobson. *Erythropoietin and the Regulation of Erythropoiesis.* U. Chicago Press. Chicago, 1970.

Linman, J. W. "Physiologic and Pathophysiologic Effects of Anemia." *N. Eng. J. Med.* 279:819, 1968.

Lerner, Richard A., and Frank J. Dixon. "The Human Lymphocyte as an Experimental Animal." *Sci. Amer.* 228:82 (June) 1973.

Mayerson, H. S. "The Lymphatic System." *Sci. Amer.* 208:80 (June) 1963.

Ratnoff, O. D., and B. Bennet. "Genetics of Hereditary Disorders of Blood Coagulation." *Science* 179:1291, 1973.

Yoffey, J. M., and F. C. Courtice. *Lymphatics, Lymph and Lymphomyeloid Complex.* Academic Press. New York, 1970.

chapter 18
Tissue Response to Injury; Antigen-Antibody Reactions; Immunity; Blood Groups

chapter 18

The body has a number of devices designed to protect it from injury or invasion by microorganisms and chemicals, and to react if the first lines of defense have been penetrated. The skin provides a mechanical barrier to the entry of foreign organisms and substances. Mucous membranes trap bacteria, pollens, and other substances in their mucus coatings. Blood and tissue cells phagocytose particles, and secretions dissolve or destroy infectious agents. Thus a variety of mechanisms are employed when the body is threatened. The reaction to threats may be a local one, or may mobilize the whole body. In some instances, the defensive response confers protection against that particular agent for the duration of the individual's life.

Tissue response to injury

The inflammatory response

Inflammation is defined as a tissue response to injury whose purpose it is to destroy or neutralize the injurious agent and hasten repair of the injury. The tissues respond with similar reactions regardless of the nature of the agent.

NARROWING of blood vessels is followed by opening, to allow a greater blood flow to the injured area.

CELL DEATH occurs as lysosome enzymes are released by injured cells.

CAPILLARY and VENULE increase in permeability is due to release of chemicals such as histamine from injured cells. This creates redness and an exudate or fluid.

LEUCOCYTES INVADE the area and take up the tasks of removing cell debris, phagocytosing bacteria, and healing, as cell reproduction replaces destroyed cells. How long the process takes and its final outcome depends on the extent of tissue damage, the tissue damaged (more vascular tissues recover faster) and effectiveness of the response of white cells and tissue cells in walling off the area and producing antibodies.

Cellular response to challenge

The reticuloendothelial (RE) system

The more or less fixed phagocytic or MACROPHAGE CELLS of the body, the PLASMA CELLS, and the monocytes comprise the RE system. The macrophages are found in the spleen, lymph nodes, liver, bone marrow, and loose connective tissues of the body; plasma cells may be found anywhere in the body, but are most numerous in the walls of the digestive and respiratory systems; monocytes are found in the tissues and bloodstream. The plasma cells are believed to be derived from lymphocytes, and produce antibodies (gamma globulins or immunoglobulins; Ig) against foreign substances entering the body.

Two types of lymphocytes are believed to be present in the body. "T-cells" originate from the thymus and bone marrow and require a chemical

FIGURE 18.1. Relationships of T- and B-lymphocytes.

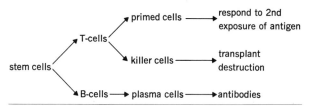

from the thymus to make them capable of responding to antigens (see below) produced within the body or those introduced by tissue transplantation. Some T-cells become "killer cells" that reject tissue transplants; others become "primed cells," capable of responding to a second exposure of a given antigen. With primed cells, a first exposure to an antigen merely sensitizes the cell and makes it capable of a reaction.

"B-cells" originate from the bone marrow and form the lymphoid tissue of the spleen, lymph nodes, and lymph nodules of the intestine. They do not require a thymus chemical, and are capable of transforming into antibody-producing plasma cells. The relationship of these two types of lymphocytes is shown in Figure 18.1.

The two types of cells thus appear to create protection against all possible means of challenge by foreign cells, chemicals, or both.

Antigens and antibodies

ANTIGENS are agents which, if introduced into the body, stimulate the production of antibodies by RE cells to neutralize or remove the antigen as a threat. They are typically proteins, polysaccharides, or lipid-carbohydrate compounds of high molecular weight.

ANTIBODIES are specific gamma globulins that are produced by plasma cells as a result of antigenic stimulation. They react with the antigen that caused their production.

TABLE 18.1 Methods of Achieving Immunity		
Method	Example(s)	Comments
Live, weakened organism is given	Polio, measles	Organism stimulates antibody production, but is too weak to cause disease (usually).
Use of organism similar to disease organism	Use of cowpox for smallpox	Use is based on similarity of antigens on organism surface.
Use of killed organism	Polio, typhoid	Cannot produce disease, but stimulate antibody production.
Use of toxoid	Diphtheria, tetanus	Toxoid is a toxic substance of the infectious organism which has been treated to reduce its toxic properties but not its antibody stimulating properties.
Injection of gamma globulin		A "shotgun" approach to any disease; may decrease severity of the disease in the exposed subject.
Antitoxins	Used against snake venoms	Is antibody to that specific antigen.

Two major theories are advanced to explain how antibodies are produced.

The **template theory** assumes that all plasma cells can respond to any antigen, and that the antigen "takes over" the DNA replicating processes of the cell to cause production of an antibody that is the mirror-image of the antigen. The antigen then reacts with the antibody that caused its production, to neutralize it.

The **selective theory** (modified clonal theory) suggests that cell mutation during embryonic development has produced many groups of cells or clones, each capable of responding to *one* antigen the body may contact (over 100,000 possible!). When challenged, only the clone capable of responding to a given antigen will produce antibodies to that antigen. Other clones remain capable of responding to other antigens.

An antigen-antibody reaction occurs when an antigen meets its corresponding antibody, and the antigen is destroyed, precipitated, or otherwise removed as a threat to the body.

Immunity

Antibody production may result in a permanent resistance to a given antigen. If this occurs, IMMUNITY is said to have developed to that antigen or organism. Table 18.1 summarizes some methods by which immunity may be achieved, other than by actually acquiring the disease.

Several "types" of immunity are recognized, as summarized in Table 18.2.

VACCINATION provides a convenient method of introducing the antigen into the body. One to two weeks is usually required to produce immunity.

ALLERGY. In exposure to some antigens, body

TABLE 18.2 Types of Immunity		
Type of immunity	Examples	Comments
Tissue (cellular)	Measles	Antibody stays in cell; does not go to bloodstream, thus, less liable to destruction.
Humoral (chemical)	Diphtheria, Tetanus	Antibody enters bloodstream and is liable to destruction, thus, "booster" required.
Active	Any viral or bacterial disease	Subject gets disease and acquires immunity to it.
Passive	Tetanus, poisonous snakebite	Antibody is given directly.
Passive-active	?	Antibody to protect; antigen to stimulate further antibody production.
Active-passive	Measles	Antibodies given to reduce disease severity; disease itself causes antibody production.

TABLE 18.3   Methods of Minimizing Tissue Transplant Rejection

Method	Comments
Culture of tissue (e.g., skin, adrenal glands, corneas)	Cells assume embryonic characteristics, and (presumably) do not produce antigens.
Radiation of the lymphoid tissues	Kills plasma cells; also kills other cells.
Tissue matching	Uses lymphocytes in a typing or matching test to get as close a genetic match as possible. Reduces severity of rejection reaction, but does not eliminate it.
Drug therapy	Reduces inflammatory response to transplant.
Antilymphocyte serum (ALS)	Take subject's lymphocytes, inject into animal, isolate antibodies and give to subject. "Kills" subject's lymphocytes; also reduces resistance to other diseases.
Remove lymphocytes from lymph	Lessens numbers of cells that can become plasma cells.

reaction is incomplete or not great enough to confer an immunity to the antigen. The subject thus continues to respond, with hives, runny nose, or edema, to repeated contacts with the antigen. This constitutes an allergy. Some allergic reactions are very severe and lead to *anaphylactic shock.* In this condition there is circulatory collapse as the result of the reaction between the antigen and the antibodies present in the bloodstream of the patient. Death may occur (e.g., penicillin reaction) if the circulatory problems such as low blood pressure and weak heart action deprive vital organs of their blood supply, or are not treated adequately.

TRANSPLANTATION. Surgical replacement of diseased or worn-out organs is a relatively new field of medicine. Replacement organs (except in identical twins) constitute foreign antigens and call forth reactions to reject the transplant. Transplants of tissues or organs from one part of a person's body to another are called AUTOLOGOUS TRANSPLANTS. Skin, bone, and cartilage grafts are examples of autologous transplants or *autografts* (G. *autos,* self, + L. *graphium,* grafting knife). A

transplant from one species to a member of the same species is called a HOMOLOGOUS TRANSPLANT. If the exchange of organs is between gentically identical individuals, for example, between identical twins, it is called an *isograft* or *homograft* (G. *isos,* equal, or G. *homo,* likeness). The term *allograft* (G. *allos,* other) describes an organ transplant between gentically different members of the same species and results in the introduction of foreign antigens into the recipient. The transplantation of hearts, kidneys, lungs, and livers between humans are examples of homologous transplants. HETEROLOGOUS TRANSPLANTS, also known as *heterografts* or *xenografts* (G. *eteros,* other, or G. *Xeno,* foreign), involve transplantation of organs between different species, and are the least successful of all types of grafts because they introduce nonhuman or dissimilar antigens into the recipient's body. The use of chimpanzee livers to remove toxic wastes from the bloodstream of humans is an example of a heterologous transplant. Several methods may be employed to reduce the tendency of the body to reject the transplant, as is shown in Table 18.3.

Blood groups

The blood groups are examples of genetically determined antigens which have corresponding antibodies. Two main groups are recognized, the ABO SYSTEM of antigens, and the RH SYSTEM of antigens. In both groups, the antigens are designated as ISOANTIGENS (isoagglutinogens, agglutinogens), and the antibodies are designated as ISOANTIBODIES (isoagglutinins, agglutinins).

The ABO system

Two basic isoantigens on the red cell surface are involved, called A and B. With two antigens, four possible combinations may exist: the cell may have one, the other, both, or neither antigen. The particular antigen involved determines the blood group. The four basic blood groups are of variable occurrence, and are summarized in Table 18.4.

Genetically, A and B inheritance is dominant to O, so that a given individual may be homozygous (have two like genes) or heterozygous (have two unlike genes) for his blood group.

Blood group	Genotype (genes present)
A	AA, Ai
B	BB, Bi
AB	AB
O	ii

where

      A = group A ⎫ codominant, and
      B = group B ⎭ dominant to group O
    AB = group AB
      i =   group O, recessive to both A and B

A child thus may have a blood group different from that of his parents, unless they are both group O.

The plasma contains isoantibodies corresponding to the isoantigens listed above. These are designated as **a**, *Anti-A,* or *alpha,* and **b**, *Anti-B,* or *beta. Isoantibodies are not present at birth.* They are gamma globulins that appear between 2 and 8 months after birth in response to A and/or B antigens ingested in foods (meats). In the blood of any one given individual, *the antibody present is always the reciprocal of the antigen.* This avoids an antigen-antibody reaction between corresponding materials. Thus, the setup of antigens and antibodies in the basic blood groups would be as shown below.

Blood group	Dominant Antigen(s)	Antibody
A	A	b
B	B	a
AB	AB	None
O	None	ab

The Rh system

This blood system is composed of three allelic genes for each group. The dominant genes are usually designated **CDE**, the recessives as **cde**. Eight genotypes, grouped into two major categories are recognized. These are shown below.

TABLE 18.4 Blood Groups and Frequency of Occurrence			
Antigen present	Blood group (same as antigen)	Frequency of population	
		Percentage of white	Percentage of black
A	A	40.8	27.2
B	B	10.0	19.8
AB	AB	3.7	7.2
Neither A nor B	O	45.5	45.8

TABLE 18.5   Transfusion Fluids Other Than Blood

Fluid	Use	Comments
Plasma	To preserve blood volume and pressure	Used when fluid and protein has been lost (e.g., burns). "Pulls" water from tissues since proteins are still present. May contain antibodies.
Colloidal solutions:		
Albumin (a protein)	As above	As above; also, may be antigenic.
Dextran (a poly-saccharide)	As above	A "plasma expander," it filters slowly and is not antigenic. Draws water from tissues by osmosis.
Crystalloid solutions (isotonic salt solutions)	As above	Particles are very small and filter rapidly. Short-lived effect.

CDE		Cde	
	Rh positive		Rh negative
cDE	(85% of white Americans,	cDe	●% of white Americans,
CdE	88% of black Americans)	cdE	12% of black Americans)
CDe		cde	

These antigens (CDE) are inherited independently of the ABO antigens, and any combination between the two systems may exist. The Rh positive individual has Rh antigen(s) on his red cells, but has no antibodies in his plasma; the Rh negative individual has no antigen(s) on the red cells, *and* no antibody in the plasma. *The Rh negative individual can produce antibodies to Rh positive cells if they ever enter his bloodstream.*

Typing the blood

The blood may be typed for ABO and Rh antigens (and therefore blood group) by mixing commercially available antibody on a slide with the blood to be typed. The antibodies will then visibly react with antigen-laden red cells to CLUMP or AGGLUTINATE them. As the slide is inspected, one

fact and one question should be kept in mind: fact—a reaction will occur only if corresponding antigens and antibodies are mixed; question—what had to be present in the unknown blood to give a reaction with known antibody? The chart below summarizes the possible reactions that may result (+ indicates a reaction; − indicates no reaction).

Antibody		Group and antigen present is:
a	b	
+	−	A
−	+	B
+	+	AB
−	−	O
Anti Rh		
+		Rh positive
−		Rh negative

Transfusions

The significance of the blood group antigens is apparent when it becomes necessary to transfuse blood from a DONOR into a RECIPIENT. If the mixed bloods contain corresponding antigens and antibodies, a reaction may occur with life-threatening results. It thus becomes necessary to perform tests to determine the COMPATIBILITY of the two bloods.

The blood is TYPED (as described above), and CROSS MATCHED. A cross match mixes donor's red cells with recipient's serum and vice versa. If no reaction occurs in each combination, the bloods are usually considered compatible and capable of being transfused without the development of reactions.

Whole blood is obviously the best transfusion fluid, for it provides all necessary substances and cells. If blood is not readily available, other substances may be employed to preserve blood volume and pressure. Table 18.5 describes some alternative substances.

Clinical considerations

Other than the transfusion or agglutination reactions mentioned above, the major clinical consideration concerned with the blood groups is HEMOLYTIC DISEASE OF THE NEWBORN (HDN) or, as it is sometimes still called, erythroblastosis.

If a child inherits from his father an antigen *not* possessed by his mother, and if that antigen crosses the placenta into the mother's bloodstream before birth, the mother will make an antibody to that antigen. The antibody may then recross the placenta to the child and damage his red cells. ABO and Rh antigens may cause this series of events, with Rh difficulties the more severe of the two. If Rh antibodies are discovered in the mother's bloodstream or if the infant is born with hemolytic disease, the antibodies may be neutralized by giving the mother, within 72 hours after birth of her child, massive doses of anti-Rh gamma globulin (RhoGAM* or Rho-Imune†). A product that may be excreted by the kidney is formed, and the mother suffers no harm. The infant may have to have replacement of his/her blood supply in an exchange transfusion if the hemolytic disease is severe. If it is not severe, transfusion or no treatment is employed.

*Ortho Chemical Co.
†Lederle Laboratories

Summary

1. The inflammatory response occurs when tissues respond to injury, and aids in destruction of the injurious agent, and in repair of the injury. Several steps occur in the response:

  a. Vasoconstriction.

  b. Cell death.

  c. Histamine release and vasodilation.

  d. Increased capillary and venule permeability, with formation of exudate.

  e. Leucocyte invasion of the injury and commencement of healing.

2. Intensity of injury, duration of irritation, type of tissue injured, and individual response to injury determine the course of the response. Tissues of younger people heal more rapidly.

3. The reticuloendothelial system (RE system) is important in body defense against disease.

  a. Definition of RE system: all fixed phagocytic cells of body plus those cells capable of producing antibodies.

4. Antibodies, modified globulins, are produced in response to antigens or foreign chemicals that enter the body.

   a. Plasma cells, derived from lymphocytes, are the cells producing most of the antibodies.

   b. Plasma cells are caused to produce antibodies only when the antigen enters an unspecialized plasma cell and directs synthesis of antibodies specific to that antigen (template theory), or are genetically produced by mutation, are specialized as to the antigen they may react to, and possess the ability to produce antibodies to any antigen that may enter the body and stimulate only a certain cell to activity (selective hypothesis).

   c. Antigen-antibody reactions occur if an antigen and its corresponding antibody are brought together. The reaction is usually protective in nature and removes the antigen as a threat to the body.

5. Antibodies may create immunity to many disease agents that act as antigens.

   a. Immunity may be achieved by exposing the body to killed or live organisms or injection of chemicals (toxoids, antitoxins, or gamma globulins).

   b. Several types of immunity exist, involving antibodies in cells (cellular immunity), in the bloodstream (humoral immunity), and various combinations of active immunity (involving the disease agent and production of antibody by the individual), and passive immunity (involving administration of antibody from another person or animal).

6. Allergy is an incomplete response to an antigen and lasting immunity is not achieved.

7. Transplantation of organs causes antigen-antibody reactions with the donor's organ(s) acting as the antigen(s) and the recipient's body producing the antibodies. Transplants may be autologous (from one area of the body to another in the same person), homologous (from one human to another), or heterologous (from one species to another).

8. The blood groups are examples of blood antigens and antibodies.

   a. The ABO group consists of two antigens (A and B) and two antibodies (a and b) usually present in reciprocal relationship. For example: Ab, Ba, AB, and Oab.

   b. The Rh group consists of one antigen (Rh) and it may be present (Rh positive) or absent (Rh negative) in the blood.

   c. Transfusion of blood requires that the antigens are the same between donor and recipient bloods to avoid an antigen—antibody reaction.

   d. Erythroblastosis may result if a mother produces antibodies to her child's blood Rh antigens while the child is still *in utero,* and the antibodies recross to the child through the placenta.

9. Replacement of lost blood may be made by transfusion of whole blood, plasma, colloids, or crystalloids. Whole blood provides all missing elements; the other solutions provide volume.

Questions

1. What are the steps in the development of the inflammatory response?

2. What factors determine the outcome and course of inflammation?

3. Describe the components and functions of the reticuloendothelial system.

4. Define antibody; antigen.

5. How can immunity to disease be achieved?

6. How does an allergy differ from the immune response?

7. Why are transplants subject to the "rejection phenomenon"?

8. What are some of the methods employed to reduce the rejection of a transplant?

9. What determines a person's blood type?

10. Using the blood antigens and antibodies as examples, describe some of the results of mismatching the bloods in a transfusion.

Readings

Berczi, I. et al. "Rejection of Tumor Cells *in vitro*." *Science* 180:1289, 1973.

Clarke, C. A. "The Prevention of Rhesus Babies." *Sci. Amer.* 219:46 (Nov) 1968.

Edelman G. M. "The Structure and Function of Antibodies." *Sci. Amer.* (Aug) 1970.

Edelman, G. M. "Antibody Structure and Molecular Immunology." *Science* 180:830, 1973.

Hilleman, M. R., and A. A. Tytell. "The Induction of Interferon." *Sci. Amer.* 225:26 (July) 1971.

Jerne, N. K. "The Immune System." *Sci. Amer.* 229:52 (July) 1973.

Notkins, A. L., and H. Koprowski. "How the Immune Response to a Virus Can Cause Disease." *Sci. Amer.* 228:22 (Jan) 1973.

Reisfeld, R. A., and B. D. Kahan. "Makers of Biological Individuality." *Sci. Amer.* (June) 1972.

Rocha, M. et al. *Chemical Mediators of Acute Inflammatory Reaction.* Vol. 37. Pergamon Press. New York, 1972.

Ross, Russell. "Wound Healing." *Sci. Amer.* 220:40 (June) 1969.

Rowley, D. A. et al. "Specific Suppression of Immune Responses." *Science* 181:1133, 1973.

Science News. "A Major Advance in Preventing Rejection of Transplanted Tissue." Vol. 104, No. 1, p. 4. Science Service Inc., Washington, D.C. (July 7) 1973.

Science News. "Chemical for Immunity." Vol. 105, No. 6, p. 86. Science Service Inc., Washington, D.C. (Feb 9) 1974.

Weiss, L. *The Cells and Tissues of the Immune System.* Prentice-Hall. Englewood Cliffs, N.J., 1972.

Zweifach, B. *The Inflammatory Process.* Academic Press. New York, 1965.

chapter 19
The Heart

chapter 19

The heart provides a source of pressure to cause the circulation of blood through the body. It is a hollow muscular pump whose contractions normally insure circulation of enough blood to meet the varying demands of the body for nutrients, oxygen, and waste removal.

The development of the heart (Fig. 19.1)

As the embryo grows, cells tend to become removed from immediate environmental sources of nutrients.

The development of vessels to carry blood to these cells, and a source of pressure to move the blood thus become early priorities in embryonic development.

At about 2 weeks of age, a mass of mesoderm called the CARDIOGENIC PLATE develops in the anterior part of the embryo. A cavity forms around the plate as the future pericardial cavity. By 3 weeks of age, paired HEART TUBES, actually blood vessels, have formed in the plate. By 4 weeks of age, the paired tubes have fused from anterior to

FIGURE 19.1. Stages in the development of the heart. (Numbers 5 to 8 occur between 5 and 8 weeks.)

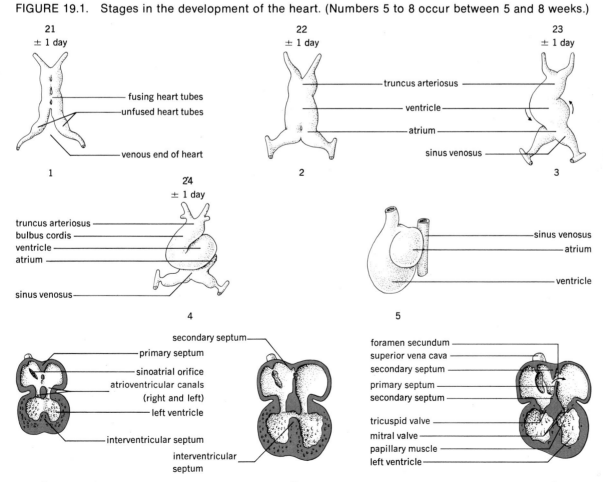

369

posterior to create a single tube. Constrictions from anterior to posterior in the tube mark the positions of the BULBUS ARTERIOSUS, VENTRICLE, ATRIUM, and SINUS VENOSUS. It should be noted that, at this time, the chambers are reversed in position from the fetal or adult position. Rapid growth of the heart tube occurs next, and the ventricle and sinus venosus fold over the atrium and come to lie posterior to the atrium, which is the normal fetal position. Between 5 and 7 weeks, a series of changes occur that result in the formation of the typical fetal structures of the heart. These changes include the following.

The common atrium is divided into two chambers by a primary septum that is perforated almost immediately by an opening called an ostium. Later, a secondary septum develops just to the right of the first one, but it remains incomplete. The opening between the two portions of the secondary septum is the FORAMEN OVALE. The foramen allows a large portion of the blood entering the right atrium of the heart to go to the left atrium and bypass the nonfunctional lungs of the fetus. The lower part of the primary septum overlaps the foramen, and seals it at birth to establish the adult type of flow through the heart.

The ventricle is partitioned into two chambers by an interventricular septum, which grows upward from the apex of the heart.

The bulbus arteriosus is subdivided into two ves-

FIGURE 19.2. The plan of the fetal circulation. Colors indicate state of oxygenation of blood. (Blue, lowest; red, highest; gray, intermediate.)

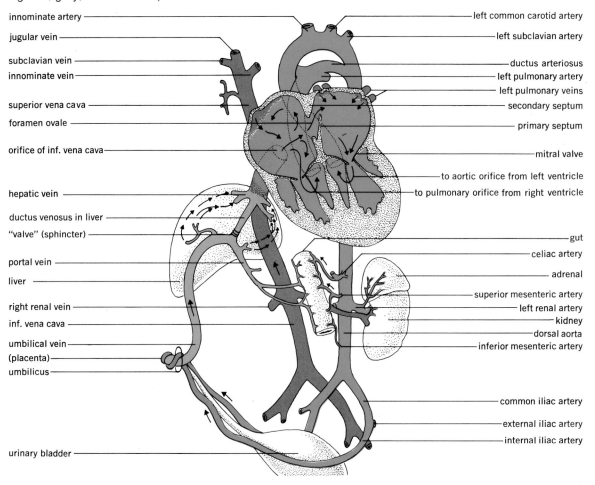

innominate artery

jugular vein

subclavian vein

innominate vein

superior vena cava

foramen ovale

orifice of inf. vena cava

hepatic vein

ductus venosus in liver

"valve" (sphincter)

portal vein

liver

right renal vein

inf. vena cava

umbilical vein

(placenta)

umbilicus

urinary bladder

left common carotid artery

left subclavian artery

ductus arteriosus

left pulmonary artery

left pulmonary veins

secondary septum

primary septum

mitral valve

to aortic orifice from left ventricle

to pulmonary orifice from right ventricle

gut

celiac artery

adrenal

superior mesenteric artery

left renal artery

kidney

dorsal aorta

inferior mesenteric artery

common iliac artery

external iliac artery

internal iliac artery

sels, the aorta and pulmonary arteries, which leave the left and right ventricles, respectively.

The three layers of the heart wall are formed, as is the special nodal tissue that causes the heart to beat.

Simultaneously, connections with the placenta, for exchange of nutrients and wastes, are established. The heart and circulation show the features depicted in Figure 19.2.

A critical time period in these changes occurs when the atria and ventricles are being divided.

Interference with these processes, as by rubella or rubeola (measles) virus, may lead to ventricular septal defects (VSD), atrial septal defects (ASD), or defects in the arteries. At birth, the shunts in the fetal circulation are closed and the normal adult pattern of circulation is established. Two shunts are closed: the foramen ovale; and the ductus arteriosus, which allows any blood entering the pulmonary artery to go to the aorta, bypassing the lungs.

Basic relationships and structure of the heart

Size

The human heart approximates the size of the clenched fist of its owner. At birth, it measures about an inch (25 mm) in length, width, and depth, and weighs about 20 grams. In the adult, the heart is about 5 inches (12.5 cm) long, 2 inches (5 cm) deep, and 3½ inches (9 cm) wide, and weighs about 300 grams. There are two periods of greatest growth; between 8 and 12 years of age, and between 18 and 25 years of age.

Location and description (Fig. 19.3)

The adult human heart is located in the central

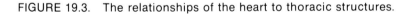

FIGURE 19.3.   The relationships of the heart to thoracic structures.

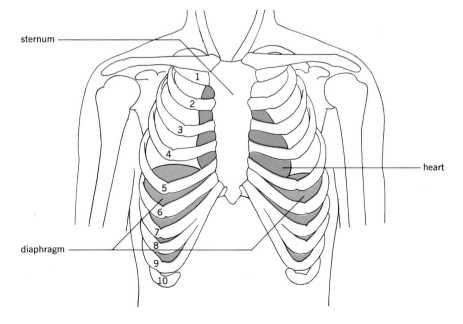

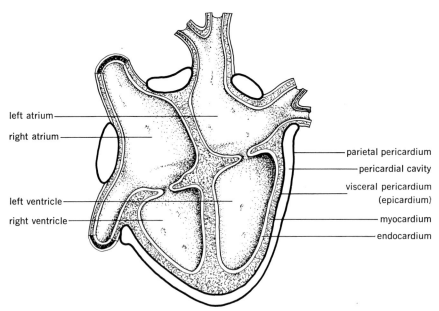

left atrium

right atrium

left ventricle

right ventricle

parietal pericardium

pericardial cavity

visceral pericardium
(epicardium)

myocardium

endocardium

FIGURE 19.4.   The relationships of the pericardial sac to the heart wall.

FIGURE 19.5.   The structure of the heart wall.

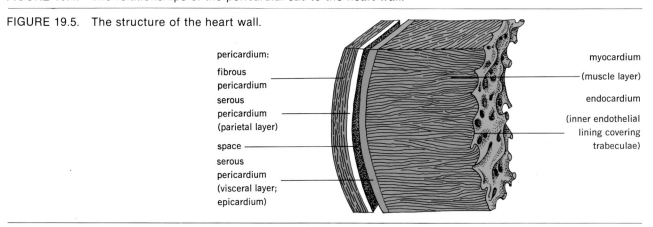

pericardium:

fibrous
pericardium

serous
pericardium
(parietal layer)

space

serous
pericardium
(visceral layer;
epicardium)

myocardium

(muscle layer)

endocardium

(inner endothelial
lining covering
trabeculae)

portion of the thorax known as the MEDIASTINUM. The broad upper or atrial end of the heart is known as the BASE, and is directed toward the right shoulder. The tapering lower end of the heart, or APEX, is directed toward the left hip. Thus, the AXIS of the heart (a line drawn from the center of the base to the apex) does not lie vertically within the chest. The heart is not centered within the chest; about one-third lies to the right of the midsternal line, and two-thirds lies to the left.

The heart is enclosed within a double-walled sac known as the PERICARDIUM (Fig. 19.4), which

consists of two layers of tissue. An outer tough FIBROUS PERICARDIUM attaches to the bases of the pulmonary artery and aorta, the diaphragm, and sternum. It aids in maintaining the heart's position, and protects it. The SEROUS PERICARDIUM is a thin delicate membrane that lines the fibrous sac and is reflected onto the surface of the heart at the bases of the arteries. The part of the serous pericardium lining the fibrous sac is termed the PARIETAL PERICARDIUM, and the portion on the heart is the VISCERAL PERICARDIUM or epicardium. A space exists between the visceral and parietal layers, known as the PERICARDIAL CAVITY. A thin film of pericardial fluid lubricates the surfaces of the visceral and parietal layers and allows nearly frictionless movement of the heart within the pericardial sac. In PERICARDITIS, the serous membrane becomes inflamed and *adhesions* (sticking together of the membranes) may develop. In CARDIAC TAMPONADE, excessive pericardial fluid secretion or blood accumulation in the cavity may hinder the heart's action.

The heart wall (Fig. 19.5)

As development and differentiation of the heart occurs, its wall develops three basic layers of tissue. An inner ENDOCARDIUM is composed of a lining endothelial layer, and of several layers of connective tissue and smooth muscle. The endocardium also forms the valves of the heart. In some parts of the heart, a *subendocardial layer* is present and carries the specialized nodal tissue that causes the contraction of the heart muscle. A central MYOCARDIUM, composed of cardiac muscle, makes up about 75 percent of the thickness of the heart wall. Many capillaries and small arteries and veins are found between the cardiac muscle fibers of the myocardium. The fibers are organized in circular or spiral fashion around the chambers of the heart, so as the heart muscle contracts, the chamber "wrings" the blood from it,

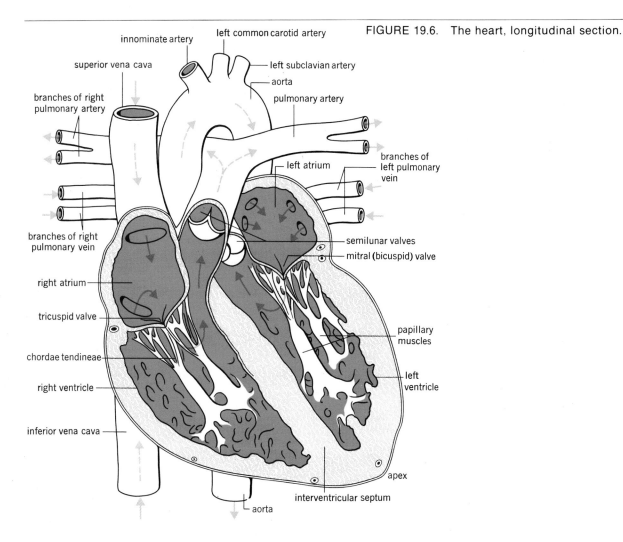

FIGURE 19.6.   The heart, longitudinal section.

FIGURE 19.7. Some of the most common congenital anomalies of the heart.

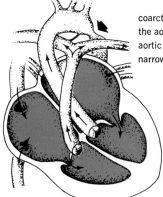

coarctation of
the aorta. the
aortic lumen is
narrowed.

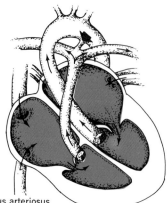

patent ductus arteriosus.
the shunt between the pulmonary
artery and aorta is open.
blood bypasses the lungs.

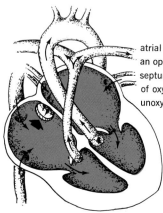

atrial septal defect.
an opening in the atrial
septum permits mixing
of oxygenated and
unoxygenated blood

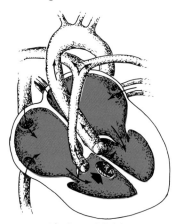

ventricular septal defect.
an abnormal opening between
the ventricles. blood passes
from left to right ventricle.

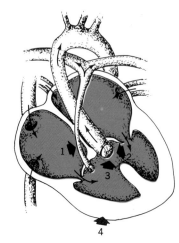

tetralogy of Fallot. four defects
occur simultaneously. (1) narrowing
of pulmonary artery or valve (2) ventricular septal defect, (3) aorta (overriding)
receiving blood from both ventricles
(4) hypertrophy of wall of right
ventricle.

rather than pushing it out. The apex twists as the beat occurs, and taps the chest wall to give the apex beat, or apical pulse. This pulse is often used to determine the effects of medication on the heart rate. The outer layer of the heart wall is the EPI-CARDIUM, and consists of the visceral layer of the serous pericardium described above.

Chambers and valves (Fig. 19.6)

The heart is actually two pumps in one. The right side of the heart acts as a pump to receive blood from the body generally and to send it to the lungs for oxygenation and carbon dioxide elimination. This circuit constitutes the PULMONARY CIRCULA-TION. The left side of the heart receives blood from the lungs and pumps it to the body generally through the SYSTEMIC CIRCULATION.

The upper receiving chambers of the heart are called the ATRIA, and the lower pumping chambers are the VENTRICLES. The RIGHT ATRIUM is a thin-walled chamber that receives blood from the "great veins" or venae cavae. It pumps its blood through the right atrioventricular orifice into the RIGHT VENTRICLE. The orifice is guarded by the RIGHT ATRIOVENTRICULAR or TRICUSPID VALVE. This valve consists of three flaps (cusps) of tissue attached by their bases to a fibrous ring around the orifice. The free edges are attached to strands of fibrous tissue called CHORDAE TENDINAE. The chordae attach to PAPILLARY MUSCLES of the ventricular wall and prevent the valve from reversing as the ventricle contracts. The right ventricle has a thicker muscular wall than the right atrium and pumps blood into the pulmonary artery for oxygenation in the lungs. In the base of the pulmonary artery is the PULMONARY VALVE. It is composed of three pockets of tissue, without supporting chordae; the valve prevents return of blood to the right ventricle. After being oxygenated in the lungs, the blood returns to the thin-walled LEFT ATRIUM through four pulmonary veins. No valves are present in these vessels. The left atrium sends blood through the left atrioventricular orifice into the LEFT VENTRICLE. This orifice is guarded by the LEFT ATRIOVENTRICULAR VALVE, also known as the MITRAL, or BICUSPID

VALVE. It has a structure similar to that of the tricuspid valve, except it has only two flaps of tissue. The very thick-walled left ventricle pumps blood into the aorta for distribution to the entire body. Again, a valve, the AORTIC VALVE, is found in the base of the aorta. It is constructed like the pulmonary valve and prevents return of blood to the left ventricle.

Thickness of chamber wall reflects the amount of work each chamber must do. Atria receive blood and move it to the adjacent ventricles; thus, little force is required, and the atria have thin walls. The right ventricle pumps to the lungs, a task requiring more force; thus, it has a thicker wall. The left ventricle pumps to the entire body and has a wall about three times thicker than the wall of the right ventricle.

CLINICAL CONSIDERATIONS. Congenital anomalies of the heart include: failure of the ductus arteriosus to close at birth (patent ductus arteriosus); failure of the foramen ovale to close; and a variety of atrial and ventricular septal defects that may occur in various combinations. The most common of these congenital defects are shown in Figure 19.7. Short descriptions of each condition are provided under each picture. Valves may also not develop properly, or may be damaged by disease and trauma (as in hypertension). If a valve fails to open completely, it is said to be stenosed; the condition is called STENOSIS. If the fibrous rings supporting the valves are stretched, the orifice may not seal properly when the valve closes. Such a valve is said to be INCOMPETENT, and permits REGURGITATION of blood in a wrong-way flow. Rheumatic fever, the result of streptococcal bacterial infection, may result in stiffening and erosion of the mitral cusps causing stenosis and incompetence.

Blood supply of the heart (Fig. 19.8)

Two CORONARY ARTERIES, a right and a left, arise from the aorta just behind the pockets of the aortic valve. This origin permits the heart to receive blood with the highest oxygen level and under the highest possible pressure. The right

FIGURE 19.8.   The coronary vessels. *(A)* Anterior view. *(B)* Posterior view.

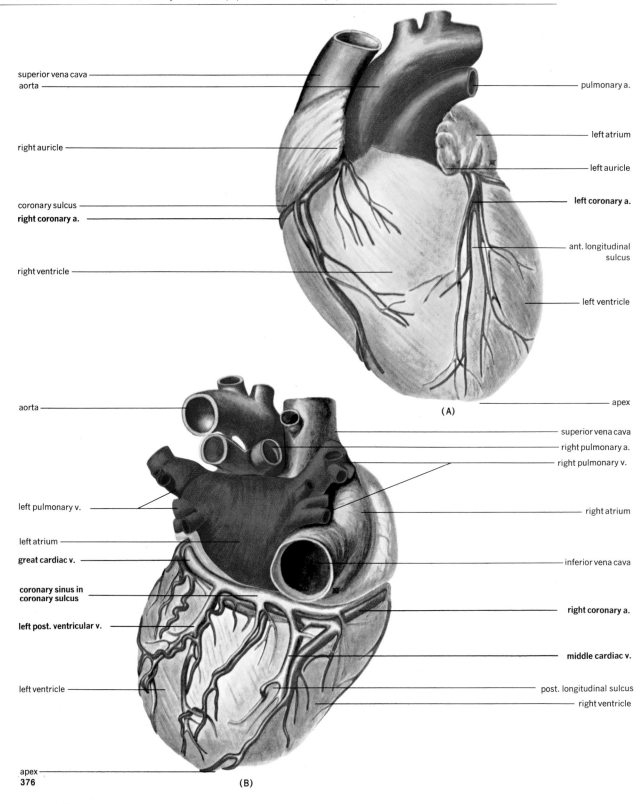

superior vena cava
aorta

right auricle

coronary sulcus
right coronary a.

right ventricle

pulmonary a.

left atrium

left auricle

left coronary a.

ant. longitudinal
sulcus

left ventricle

apex

(A)

aorta

left pulmonary v.

left atrium
great cardiac v.

**coronary sinus in
coronary sulcus**

left post. ventricular v.

left ventricle

apex

superior vena cava
right pulmonary a.
right pulmonary v.

right atrium

inferior vena cava

right coronary a.

middle cardiac v.

post. longitudinal sulcus
right ventricle

(B)

coronary artery passes to the posterior side of the heart, giving off branches to the left ventricle, right ventricle, and right atrium. The left coronary artery supplies the left ventricle and left atrium. *Anastomoses*, or communications between the smaller branches of both coronary arteries, are common. About 48 percent of the United States population has what is called a right coronary predominance, in which the right coronary artery supplies more than just the right side of the heart. About 18 percent have left coronary predominance. The remainder, about 33 percent, have a balanced coronary supply.

The arteries form capillary and sinusoidal (large irregular vessels) beds in the myocardium, and substances diffuse to and from these beds to the muscle and other layers of the heart wall. About 30 percent of the blood reaching the heart wall passes through the sinusoids directly into the ventricular cavities. The remaining 70 percent is collected by the coronary veins which, by way of the coronary sinus, empty into the right atrium.

Blood flows through the system of coronary arteries, capillaries and veins, only when the myocardium is *not* contracted. Contraction compresses the small vessels of the heart and stops flow. Also, when the ventricles are ejecting blood, the open semilunar valve in the aorta blocks the openings of the coronary arteries. Thus, the only time when nutrients may be supplied to the heart is when the myocardium is relaxing or at rest, and when the aortic valve is closed.

The coronary arteries bring nutrients and oxygen to the heart to sustain its activity. About 60 percent of the energy for heart activity is derived from the metabolism of fatty acids. About 35 percent comes from the metabolism of carbohydrates, and 5 percent from the metabolism of amino acids. Inorganic salts are also extremely important in the regulation of the beat and depolarization of the heart muscle. Low blood Na levels (hyponatremia) interfere with the depolarization of the heart muscle, and the heartbeat ceases. Increase or decrease of potassium levels in the blood (hyper- and hypokalemia) cause difficult depolarization and repolarization, respectively. Increased blood calcium level (hypercalcemia) results in rapid incoordinated contraction or fibrillation of the cardiac muscle. Low blood magnesium levels lengthen a beat, and high blood magnesium levels cause cardiac arrest (stoppage).

CLINICAL CONSIDERATIONS. Blockage of the coronary vessels, as by a blood clot, results in coronary thrombosis, the myocardium is deprived of blood flow (ischemia), and death of tissues beyond the block (infarction) may occur. If the thrombosis results in incomplete blockage of the vessel, a reduced blood flow to the area supplied (ischemia) may be the only result. Decreased flow of blood to the heart is usually associated with the development of the pain of ANGINA. In this condition, pain associated with myocardial ischemia is referred to the left chest and arm. If the vessel plugged is relatively small, the anastomoses between coronary arteries may permit sufficient blood flow to compensate for the blockage. If a large vessel is plugged, the amount of tissue that dies may be too great to permit continued heart action. The *s*erum *g*lutamic *o*xaloacetic *t*ransaminase levels or SGOT (an enzyme involved in heart metabolism) bears a relationship to severity of cardiac muscle death following a thrombosis. Normally found in a concentration of about 10–40 units per milliliter of serum, the enzyme increases in amount according to the severity of cardiac damage. Thus an SGOT of 90–100 units indicates mild heart muscle damage, and 200 or more units indicates severe damage. The SGOT may be used to indicate the severity of the initial attack, and to follow recovery from the attack.

Nerves of the heart

The heart receives MOTOR FIBERS from both divisions of the autonomic nervous system. Parasympathetic fibers are carried in the vagus nerves, and act primarily to slow the beat. Sympathetic fibers, carried in the cardiac nerves, increase both heart rate and strength of beat. SENSORY FIBERS arise within the heart and are carried in the vagus nerves to the brain stem. These fibers carry impulses for the reflex control of heart rate and carry pain impulses to the central nervous system.

The physiology of the heart. The tissues composing the heart

Two types of tissue are of utmost importance in the contraction of the heart muscle. The NODAL TISSUE initiates and distributes nerve impulses that cause heart muscle contraction. The CARDIAC MUSCLE itself contracts and creates the pressures necessary to circulate the blood.

Nodal tissue

Nodal tissue develops from cardiac muscle. It has largely lost its power of contractility, but has developed the abilities of spontaneous depolarization and conductivity to a high degree. The nodal tissue generates and transmits the electrical impulses that cause contraction of the cardiac muscle. The nodal tissue is organized into several discrete masses and bundles (Fig. 19.9).

The SINOATRIAL (SA) NODE is imbedded in the wall of the right atrium at its junction with the superior vena cava. It is about ¾ inch in length by ¼ inch in width (18 × 6 mm). While all parts of the nodal tissue can depolarize spontaneously, the SA node does the task most rapidly and strongly. Its activity thus tends to dominate that of the other parts of the nodal tissue and, for this reason, the SA node acts as the PACEMAKER for the heartbeat. The SA node sets a basic rate of beat at about 75–80 per minute, which may be changed by nervous influences.

The ATRIOVENTRICULAR (AV) NODE lies in the right atrium near the base of the interatrial septum. It receives the SA node impulses which reach it through the atrial muscle, delays the impulses slightly to allow completion of atrial contraction, then passes the impulses rapidly to the ventricles.

The AV BUNDLE, BUNDLE BRANCHES, and PURKINJE FIBERS distribute impulses to the ventricular myocardium nearly simultaneously, because conduction is very rapid in these portions. Thus nearly simultaneous contraction of the muscle assures maximum pressure development to circulate blood.

CLINICAL CONSIDERATIONS. ARRHYTHMIA is a

FIGURE 19.9.   The locations of the nodal tissue in the human heart.

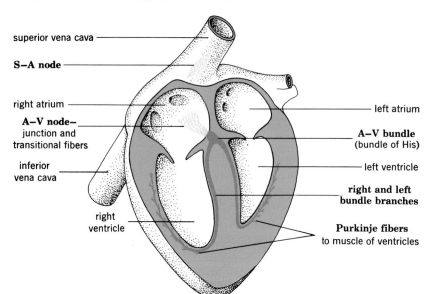

superior vena cava

S–A node

right atrium

A–V node–
junction and
transitional fibers

inferior
vena cava

right
ventricle

left atrium

A–V bundle
(bundle of His)

left ventricle

right and left
bundle branches

Purkinje fibers
to muscle of ventricles

term applied to variations from normal rhythm and/or sequence of events in heart excitation. Arrhythmia may result from the following conditions.

Alterations of SA node activity. Drugs may slow or speed SA node activity. Rate increase is caused by atropine, caffeine, epinephrine; rate decrease is caused by amylnitrite, nitroglycerine, and alcohol in small quantities.

Interference with conduction. HEART BLOCK is the result of interference with conduction in the nodal tissue. Normally, atria and ventricles beat at the same rate. In heart block, interference may cause slowing of beat (first degree block); missed beats by the ventricle, for example, atrial rate 75, ventricular rate 60 (second degree block); or complete separation of atrial and ventricular activity, for example, atrial rate 80, ventricular rate 40 (third degree block). In third degree block, ventricular rate is usually too low to circulate enough blood for body needs, and an *artificial pacemaker* may be required. The pacemaker gives electrical stimulation to the ventricles by wires, and may provide a constant rate of stimulation, a variable rate of stimulation depending on activity levels, or may become active only when natural stimulation falls below some preset rate.

Cardiac muscle (Fig. 19.10)

Cardiac muscle fibers form a branching system or syncytium of fibers disposed in two main masses. One mass forms the atria, the second forms the ventricles. There are cell membranes between the cardiac fibers, but they do not hinder the passage of impulses along the fibers. Thus each entire mass behaves as though it was one large cell. Cardiac muscle is EXTENSIBLE, in that it may be stretched to allow greater filling of the heart chambers. After the chambers are filled, and tension is put on the muscle fibers, they contract more forcibly than they would if they had not been stretched. This property is called the "LAW OF THE HEART," or STARLING'S LAW. It enables adjustment of blood pumped to match blood received by a chamber. The muscle is INVOLUNTARY, in that it does not require stimulation by outside nerves to contract. Because the nodal tissue is entirely within the heart, the heart is said to possess INHERENT RHYTHMICITY. In part, this property permits activity in transplanted hearts. The cardiac fibers follow the ALL-OR-NONE LAW, which assures that any stimulus strong enough to cause contraction will cause the strong-

FIGURE 19.10.   *(A)* Diagram showing the structure of cardiac muscle. *(B)* The courses of the major groups of ventricular muscle fibers.

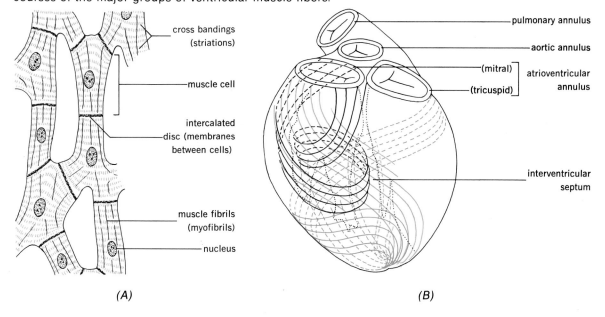

(A)   *(B)*

est one possible. Therefore, the greatest possible pressure is created to move the blood. The fibers usually cannot be caused to undergo a sustained contraction if the blood's ionic balance is normal, and thus pumping continues rhythmically to insure blood circulation.

Nodal and cardiac tissue combine to give the heart a rhythmical, continuous beat whose force is adjustable according to body requirements.

The cardiac cycle

A cardiac cycle is defined as one complete series of events occurring within the heart, and usually is said to begin with depolarization of the SA node. Each cycle requires a certain length of time for its completion, and pressure, volume, electrical, and sound changes occur during each cycle.

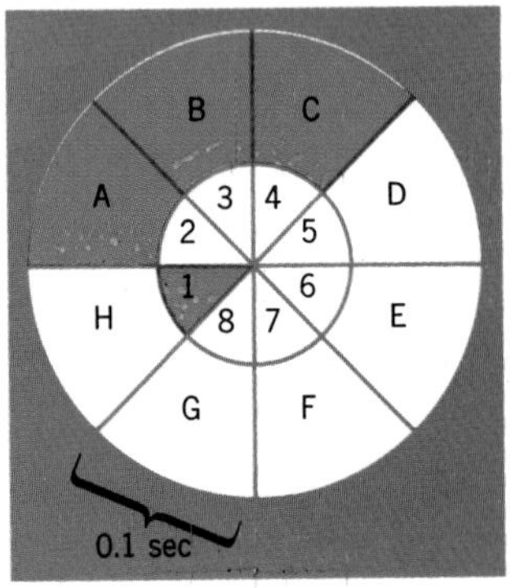

Timing of the cycle (Fig. 19.11)

At a rate of 75 beats per minute, considered to be a normal or resting rate, it takes 0.8 seconds for the completion of one cardiac cycle. Within this time span, atrial SYSTOLE (contraction) takes about 0.1 seconds, atrial DIASTOLE (relaxation) about 0.7 seconds. Ventricular systole takes about 0.3 seconds, and ventricular diastole takes 0.5 seconds. Each chamber, therefore, spends more time relaxing or resting than contracting. If heart rate increases, the periods of relaxation and rest become shorter. Relaxation and rest permit blood flow through the coronary vessels to occur and allow filling of the heart chambers with the blood to be pumped during the next cycle.

Electrical changes during the cycle

Formation and transmission of impulses through the nodal tissue and depolarization of the cardiac muscle, is associated with electrical changes that may be recorded as an ELECTROCARDIOGRAM or ECG. A normal ECG is shown in Figure 19.12. The parts of the recording are explained as follows.

The *P wave* indicates atrial excitation.

The *P-R segment* or *interval* is the time required for the impulses to reach the ventricles.

The *QRS wave* represents ventricular excitation.

The *ST segment* or *interval* is the time between the end of ventricular depolarization and the start of repolarization.

FIGURE 19.11. The timing of a cardiac cycle. Rate of beating is 70-72 beats per minute. 1, atrial systole; 2, atrial diastole; 3-8, atrial diastasis. A,B,C, ventricular systole; D, ventricular diastole; E,F,G,H, ventricular diastasis.

FIGURE 19.12. The normal electrocardiogram.

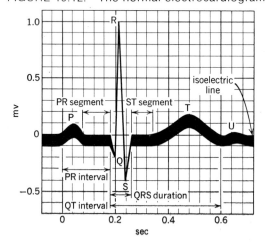

The *T wave* indicates completion of ventricular repolarization.

The *U wave* is not present in all ECG's. If it shows up, it indicates either slow papillary muscle repolarization, or hypokalemia.

CLINICAL CONSIDERATIONS. The value of an ECG lies in the fact, that any changes in the nodal tissue or cardiac muscle will be seen as alterations of the normal ECG. Alterations of function are reflected as changes in direction of the waves, in their height, or in the length of the intervals. Damage of the nodal tissue or cardiac muscle following "heart attacks" may be determined; recovery may be followed; the severity of the damage may be assessed. Some abnormal ECG's are presented in Figure 19.13. Each should be examined to determine the specific changes that occur in the ECG in each type of abnormality.

Pressure and volume changes during the cycle (Fig. 19.14)

Since blood only moves through the body from an area of higher pressure to one of lower pressure, the creation of pressure "highs" by the heart is essential to blood circulation. Pressure changes also operate the valves in the heart and the valves in the arteries leaving the heart. Pressure is usually measured in mm Hg (millimeters of mercury).

Blood enters the right atrium in a steady stream from the venae cavae. Right atrium contraction creates a pressure of about 7 mm Hg which aids in moving blood to the right ventricle. Right ventricular contraction raises the pressure in that chamber above that of the right atrium. The blood trying to return to the right atrium shuts the tricuspid valve. Pressure rises in the

FIGURE 19.13. Abnormal electrocardiograms in various conditions.

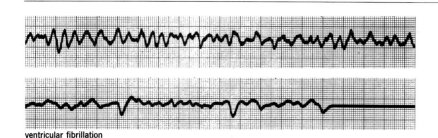
ventricular fibrillation

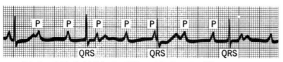
complete heart block (atrial rate, 107; ventricular rate, 43)

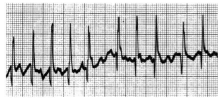
atrial fibrillation

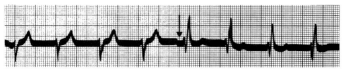
intermittent right bundle branch block

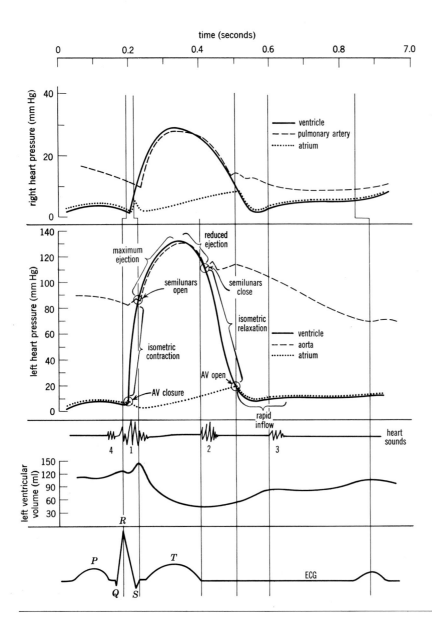

FIGURE 19.14. Composite diagram of heart activity.

right ventricle until it reaches about 18 mm Hg, at which time the pulmonary valve is forced open and blood is ejected into the pulmonary circulation. Pressures rise to a maximum of about 30 mm Hg in the right ventricle. Relaxation of the right ventricle causes pressure to fall to less than that in the pulmonary artery and blood tries to return to the ventricle. This closes the pulmonary

valve. Continued fall of right ventricular pressure causes, eventually, a fall of pressure to less than that in the right atrium, the tricuspid valve opens, and blood again fills the ventricle. The same series of events occurs in the left side of the heart as blood passes from left atrium, to left ventricle, to aorta. Several important differences may be noted in the action of the left side of the heart.

TABLE 19.1 Causes of the Heart Sounds		
Sound (S)	Cause(s)	Comments
First (S_1)	Closure of AV valves	Sound is low pitched and drawn out ("Lubb")
	Tensing of valves	
	Vibration of chordae	
	Contraction of ventricular muscle	
Second (S_2)	Closure of arterial valves	Sound is higher pitched and short duration ("Dup")
Third (S_3)	Rush of blood from atrium to ventricle	Occurs during or after atrial contraction; a sound of turbulent flow
Fourth (S_4)	Muscular contraction of atria	— —

The left atrium achieves a peak pressure of about 10 mm Hg, as compared to 7 mm Hg in the right atrium.

The left ventricle must create a pressure of 80 mm Hg to open the aortic valve as compared to 18 mm Hg by the right ventricle, and a peak pressure of about 130 mm Hg is achieved by the left ventricle, as compared to 30 mm Hg by the right ventricle.

As the ventricles contract, their volume obviously decreases as blood is pumped into the pulmonary and systemic circulations. Each ventricle has a normal volume of 100–120 milliliters. At normal or resting rates and volumes of beat, each ventricle empties about two thirds of its contained blood into the two circulations. Thus about 65–80 milliliters of blood enters the circulation at each heartbeat. Increased volumes of blood may be pumped into the vessels by either increasing the heart rate, or the volume emptied per beat. Normally, each ventricle pumps an equal amount of blood, so that no "accumulation" of blood occurs in any part of the system.

Heart sounds during the cycle

As one listens (auscultation) to the heart with a stethoscope placed on the chest, it may be determined that the heart creates sounds as it beats. Most of these sounds occur as valves open or close, and as the myocardium contracts. With a stethoscope, two heart sounds are heard, designated the FIRST HEART SOUND, and the SECOND HEART SOUND. Sensitive microphones may pick up a third and fourth heart sound. The times of occurrence of each sound are shown in Figure 19.14, and their causes are presented in Table 19.1.

CLINICAL CONSIDERATIONS. Heart MURMURS are abnormal sounds produced during heart activity, or are modifications of normal sounds. They usually indicate abnormal valve action (stenosis or incompetence). FUNCTIONAL MURMURS are usually of no concern, are produced only when the heart is beating forcibly, and are usually outgrown. RESTING MURMURS are a cause for concern, for they usually indicate valve damage.

The heart in creation and maintenance of blood pressure

The discussion of heart action has emphasized the role of the heart in creating a pressure to eject blood from the heart into the pulmonary and systemic circulations. In order that blood flow be continuous in spite of the fact that heart action is intermittent, and to insure that the heart can develop a pressure on the blood, factors other than heart action enter the picture. The five factors responsible for the origin and maintenance of arterial blood pressure are presented here, so that an introduction to the concepts of blood pressure may be made; then the specific role of the heart in the creation and maintenance of arterial pressure is discussed in this chapter, and that of the

FIGURE 19.15. Cardiac reflex pathways.

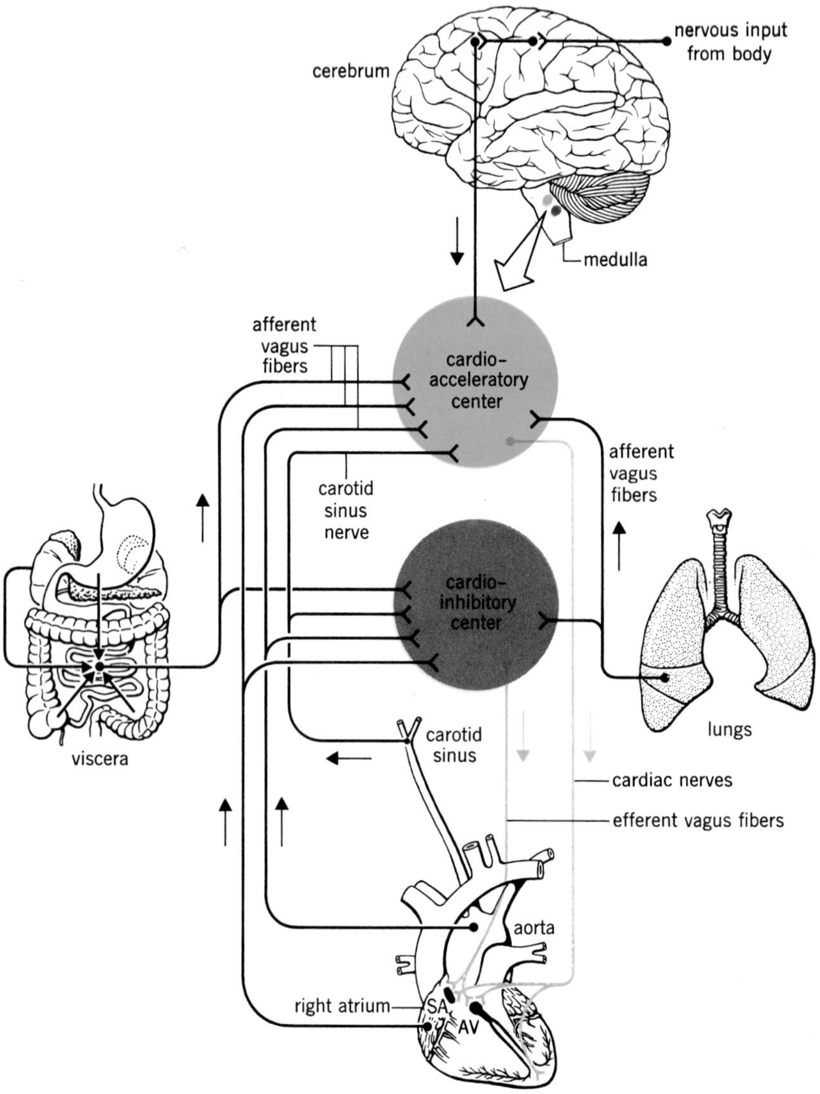

TABLE 19.2 Cardiac Reflexes						
Name of reflex	Input from	Stimulus triggering	*Effect on		*Effect on	
			CIC	CAC	SR	SV
Aortic depressor	Baroreceptor (sensitive to stretch) in aorta	Stretch of aorta	+	−	↓	0
Carotid sinus	Baroreceptor in carotid sinus	Stretch of sinus	+	−	↓	0
−	Baroreceptors in pulmonary arteries	Stretch of arteries	+	−	↓	0
Bainbridge reflex (existence disputed)	Vena cava or right atrium	Increased filling of atrium (stretching)	−	0/+	↑	0
−	Cerebrum— anger/fear	Stress, thoughts	−	+	↑	0
−	Strong stimu- lation of any nerve	Pain	−	+	↑	↑

*+ (stimulate); − (inhibit); 0 (no change); ↑ (increase); ↓ (decrease).

blood vessels is considered in Chapter 20. The five factors responsible for creating and maintaining arterial blood pressure are as follows.

1. *The pumping action of the heart.* Ventricular contraction injects into a vascular system of a given volume, variable quantities of blood per minute. The greater the amount of blood injected into the arteries, the higher is the pressure in the system. Without heart action, *no* pressure would exist. The term *cardiac output* refers to the volume of blood pumped by one ventricle per minute.

2. *Peripheral resistance.* Mainly a function of the diameter of blood vessels, this factor refers to the *ease of outflow* of blood through the small muscular arteries of the body. The smaller the vessel, the greater will be the pressure required to force blood through it.

3. *Elasticity of arteries* near the heart. Ventricular

activity injects a quantity of blood into the large arteries near the heart. They expand to contain this volume and, in so doing, prevent excessive rise of systolic blood pressure. During ventricular relaxation, elastic recoil of these arteries prevents diastolic pressure from falling to low values, by moving the blood onward through the arteries.

4. *Viscosity of the blood.* A thicker (viscous) fluid such as the blood requires more pressure to circulate it; therefore, pressures are higher than if a fluid such as water was being pumped.

5. *Volume of blood* in the system. In a system of pipes with a given volume or capacity, the placing of more fluid into that system must raise the pressure.

Of these factors, *heart action* and *peripheral resistance* are the most important in creating and maintaining arterial blood pressure, and are the

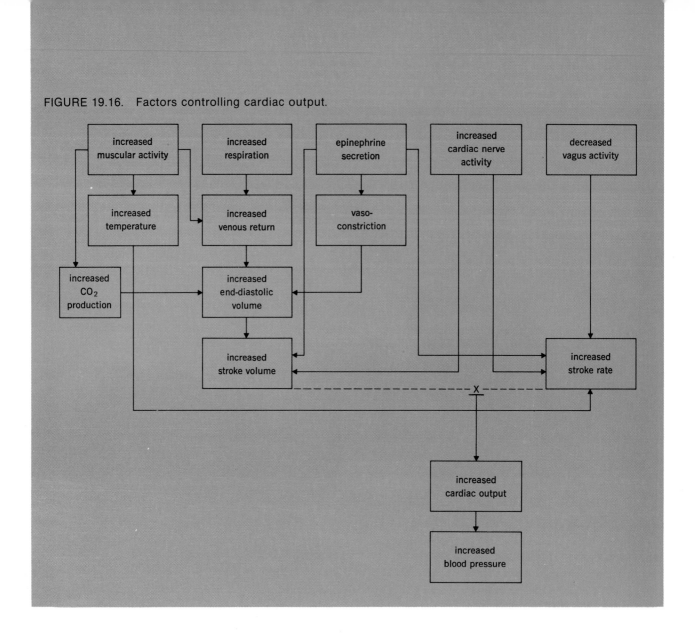

FIGURE 19.16. Factors controlling cardiac output.

only ones that can be changed to meet the varying demands of the body for blood flow.

Heart action and blood pressure; cardiac output

Cardiac output is defined as the amount of blood pumped per minute by one ventricle and is the product of rate of beat (stroke rate) and milliliters of blood ejected per beat (stroke volume) according to the following equation:

Cardiac output (CO) in ml/min = stroke rate (SR) in beats/min × stroke volume (SV) in mls/beat

or

$$\textbf{CO} = \textbf{SR} \text{ (beats/min)} \times \textbf{SV} \text{ (ml/beat)} = \text{ml/min}$$

The maximum pressure created when the ventricles contract is termed the SYSTOLIC PRESSURE; the low value to which the pressure falls between beats when the ventricles relax is termed the DIASTOLIC PRESSURE. The difference between the two is the PULSE PRESSURE. Mechanical, chemical, thermal, and nervous factors control the cardiac output by altering stroke rate and stroke volume. Therefore, blood pressure is altered, and the amount of blood delivered to the tissues may be changed.

CONTROL OF STROKE RATE. The reactions that occur in the SA node are TEMPERATURE dependent. Rise of blood temperature entering the right atrium, such as occurs during exercise, may slightly accelerate the rate of depolarization. This factor is of little significance in the human. CHEMICALS provide a second means of altering stroke rate by changing the rate of the reaction occurring in the SA node. *Acetylcholine* slows heart rate;

epinephrine speeds it; *thyroxin,* a hormone, speeds the rate. NERVOUS REFLEXES provide the greatest and finest degree of control over rate.

The cardiac nerves carry sympathetic impulses that ultimately result in cardioacceleration (rate increase); the vagus nerves carry parasympathetic impulses that cause cardioinhibition (rate decrease). These fibers originate in cardioaccelerator centers (CAC) and cardioinhibitor centers (CIC)

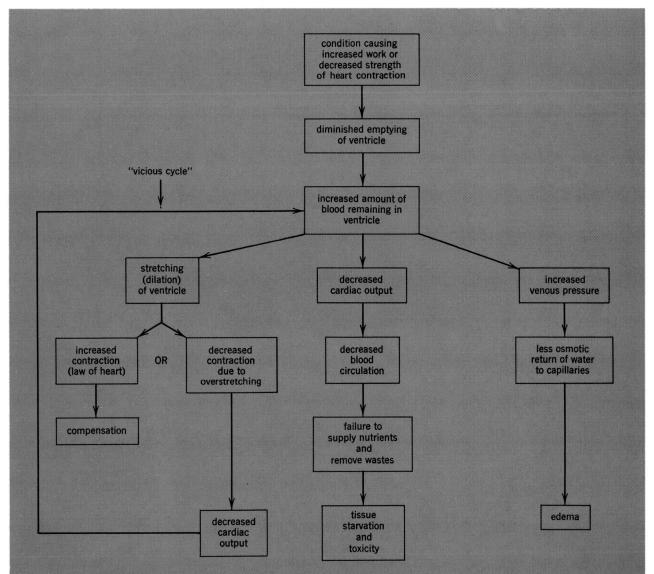

FIGURE 19.17.   The sequence of events in heart failure. If the right ventricle fails, the symptoms generally appear in the systemic circulation; if the left ventricle fails, the symptoms generally appear in the pulmonary circulation.

in the brain stem. In turn, the centers are influenced by impulses arriving at them over afferent nerves from receptors in the carotid sinus, aorta, lungs, coronary arteries, and other body areas. Impulses from the brain also influence the centers, and their effects are seen in the heart rate increase during emotional excitement and stress. The general plan of the cardiac reflexes is shown in Figure 19.15, and the effects of the reflexes are presented in Table 19.2.

CONTROL OF STROKE VOLUME. Stroke volume is controlled by chemical and mechanical factors. Nerves appear to have minimal effects on this component of the cardiac output. The chemical and mechanical factors are integrated as follows:

Stroke volume is determined by the degree of filling of the ventricles during ventricular relaxation and rest, and by the degree of emptying during contraction.

Relaxation of the ventricular muscle, and thus size of the ventricular chamber, is determined by CO_2 levels of the blood, and by pH levels ($\uparrow$ CO_2, or $\downarrow$ pH = more relaxation).

Increased muscular activity causes more blood to be returned to the right atrium by the massaging action of skeletal muscles on body veins, and also releases epinephrine into the circulatory system.

THUS, more blood enters the more relaxed ventricles, more stretch is applied, the muscle contracts more strongly (law of the heart and chemical effect), and more blood is ejected. The blood pressure rises and more blood is delivered to the body tissues. Figure 19.16 relates some of these and other factors to control of cardiac output and blood pressure.

CLINICAL CONSIDERATIONS. If cardiac output falls below 2–2.5 liters per minute, insufficient blood will be circulated to meet the body needs, and HEART FAILURE will be said to have occurred. In some instances, drugs may be employed to stimulate heart action and, thus, raise the output to where it succeeds in meeting the minimum needs of the body for circulation. In this state, the heart failure has been *compensated*. The sequence of events that occurs in heart failure is presented in Figure 19.17. The symptoms of failure, such as, fatigue, difficult breathing, cyanosis, and edema, are attributed to low blood pressure and inadequate circulation.

Summary

1. The heart provides a pressure to circulate blood to the body to provide nutrients and remove wastes.

2. The heart develops early in embryonic life to assure substances necessary for cellular activity.

    a. At 2 weeks, a cardiogenic plate is present.

    b. At 3 weeks, paired heart tubes are formed.

    c. At 4 weeks, fusion of the tubes forms the heart.

    d. Between 5 and 7 weeks, the four chambers are established, the wall is developed, and a circulation of blood is established.

3. The heart has certain structural and locational characteristics.
    a. It is about the size of one's clenched fist (age dependent).

    b. It is located in the chest, about one third to the right of the midline, two thirds to the left.

    c. It is enclosed in a pericardial sac having fibrous and serous portions.

    d. The wall is composed of an inner endocardium, a middle myocardium, and an outer epicardium.

e. It is four chambered, with two atria and two ventricles.

f. The right side of the heart pumps into the pulmonary circulation; the left to the systemic circulation.

g. Valves are present between atria and ventricles, and in the great arteries, and control direction of blood flow.

4. Congenital anomalies may cause defects in formation of the heart; infections may damage it.

5. The heart is supplied with blood through the coronary arteries.

a. Flow through the vessels occurs only when the heart muscle is not contracted.

b. The coronary circulation provides nutrients to sustain heart activity, and ions to govern rate and strength of beat.

c. Blockage of the arteries is associated with "heart attack," tissue death, and possible stoppage of life.

6. Nerves to the heart include:

a. Parasympathetic nerve (vagus) to slow heart rate.

b. Sympathetic nerves (cardiac nerves) to speed heart rate and strength of contraction.

7. Two types of tissue are most important in heart activity.

a. Nodal tissue (SA node, AV node, AV bundle, bundle branches, Purkinje fibers) originates and distributes the stimuli necessary for heart muscle contraction. The SA node is the pacemaker.

b. Cardiac muscle contracts and creates pressure to circulate the blood. It is extensible, involuntary, inherently rhythmic, cannot be thrown into a sustained contraction if ionic concentrations are normal, follows the law of the heart, and the all-or-none law.

8. A cardiac cycle is one complete series of events during cardiac activity.

a. It normally lasts about 0.8 seconds.

b. It creates electrical changes (ECG) that reflect the status of the nodal and cardiac tissue. The ECG is used to determine injury to the heart and to recovery from injury.

c. Pressure changes operate valves and move blood.

d. Sounds are heard as the heart works (see Table 19.1).

e. Murmurs during the cycle usually indicate improper valve functioning.

9. The heart, by its output of blood, establishes a basic pressure that circulates blood. Two factors, stroke rate, and stroke volume, determine output.

a. Stroke rate (beats/min) is controlled primarily by chemicals and nervous reflexes (see Table 19.2).

b. Stroke volume (ml/beat) is determined primarily by chemical and mechanical factors.

c. Heart failure occurs when output does not meet body needs for blood.

Questions

1. What devices are available to the heart to enable increase in volume of blood pumped when demand for increased circulation occurs?

2. Explain the role of the nodal tissue in establishing the rhythmical nature of the heartbeat.

3. In what ways is cardiac muscle especially suited to its tasks of circulation of the blood, and adjustment of amount of blood pumped?

4. Describe the role of nerves in alteration of heart rate.

5. Discuss the role of cardiac reflexes in control of blood pressure.

6. Suppose that a kidney malfunction results in potassium retention and calcium loss. What would be expected in terms of derangement of heart action?

7. What defect is implied in "heart block?" Speculate as to the effect of the various degrees of block on maintenance of adequate cardiac output.

8. What is the value of an electrocardiogram? How does the recording indicate changes in conduction times? In strength of depolarization (and, therefore, condition of the cardiac muscle)?

9. Discuss the pressure changes occurring within the ventricles that lead to: ejection of blood; filling of the ventricles.

10. What are the causes of the first and second heart sounds? How are modifications of these sounds related to valvular defects?

11. How does an increased return of blood to the right atrium stimulate cardiac output? How is the increased venous return brought about?

Readings

Adolph, E. F. "The Heart's Pacemakers." *Sci. Amer. 213*:32 (Mar.) 1967.

Bates, Barbara, and Robert A. Hoekelman. *A Guide to Physical Examination.* University of Rochester, School of Medicine and Dentistry. Rochester, N.Y., 1974.

Berne, R. M., and N. L. Matthew. *Cardiovascular Physiology.* C. V. Mosby Co., St. Louis, Mo., 1972.

Lown, Bernard. "Intensive Heart Care." *Sci. Amer. 219*:19 (July) 1968.

Moore, Keith L. "The Developing Human." *Clinically Oriented Embryology.* Saunders. Philadelphia, 1973.

Ross, J. Jr., and B. E. Sobel. "Regulation of Cardiac Contraction." *Ann. Rev. Physiol. 34*:47, 1972.

Ross Laboratories. Clinical Education Aid, No. 7. Congenital Heart Abnormalities. Ross Laboratories. Columbus, O., 1972.

chapter 20
The Blood and Lymphatic Vessels. Regulation of Blood Pressure

chapter 20

Development of blood vessels

Blood vessels develop (Fig. 20.1) from the meso-dermal blood islands that give rise to the embryonic blood cells. At about three weeks, cavities appear in the islands and the peripheral cells flatten into an endothelial lining. Growth and joining of these cavities produces networks of blood channels, or vessels. "Budding" from these original vessels extends the area of the networks.

Mesodermal cells give rise to additional coats or layers around the endothelial tubes to form arteries and veins. It is beyond the scope of this book to detail the development of specific arteries and veins. The reader is referred to Figure 20.2 and the chapter readings for references on this subject.

Types of vessels

The blood vessels carry the blood pumped by the heart, delivering it to the tissues and back to the heart. There are three main types of vessels:

Arteries carry blood *away from the heart*. Most have smooth muscle in their walls and can actively change their size, thus influencing blood flow and pressure.

FIGURE 20.1.   Early development of blood vessels.

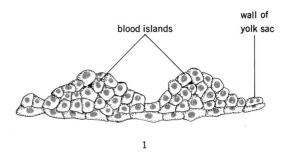

blood islands   wall of yolk sac

1

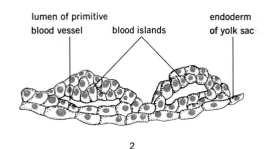

lumen of primitive blood vessel   blood islands   endoderm of yolk sac

2

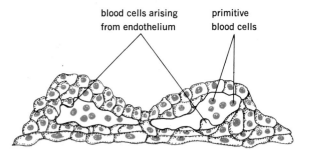

blood cells arising from endothelium   primitive blood cells

3

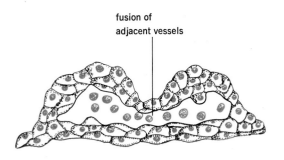

fusion of adjacent vessels

4

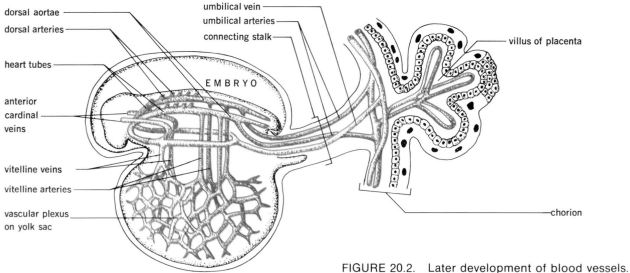

FIGURE 20.2.   Later development of blood vessels.

FIGURE 20.3.   A schematic representation of the circulatory system.

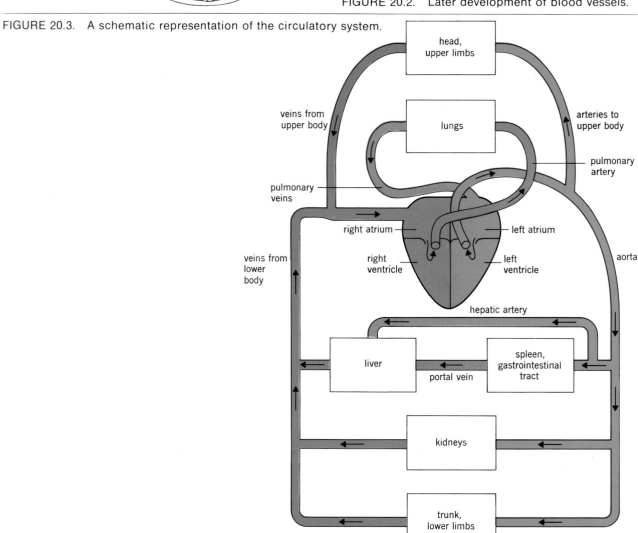

Capillaries are microscopic vessels found *in the body tissues*, and allow *exchange* of materials between the bloodstream and the body cells.

Veins carry blood *toward the heart*, and can also change size because of smooth muscle in their walls. The size change is usually less than that occurring in the arteries, so these vessels are less important than arteries in controlling blood pressure.

The general plan of these vessels is shown in Figure 20.3.

LYMPHATIC VESSELS may be compared to the capillaries and veins of the blood vessel system. Small lymph capillaries originating in the tissues form collecting vessels and lymph ducts. The ducts empty into the veins in the upper chest region.

Smooth (visceral) muscle

We meet smooth muscle for the first time, as a major component of an organ, in the walls of blood vessels. Some of its features and properties are considered to better understand the functions of the vessels.

Structure (Fig. 20.4)

Smooth muscle cells are long (40–100 microns), narrow (3–8 microns), tapered cells. Each cell has one nucleus, and does not carry cross striations (stripes), hence, the name *smooth* muscle. Each cell contains only a few myofibrils, the contractile units of muscle. Individual cells are held together into sheets by reticular connective tissue. In most blood vessels, smooth muscle is arranged circularly around the vessel, so that contraction narrows the lumen (cavity) of the vessel and relaxation of the muscle enlarges the lumen.

Properties

Smooth muscle is generally considered to be INVOLUNTARY, that is, it does not usually depend on outside nerves to cause it to contract. Nerves may, however, modify its activity. The cells can maintain a more or less CONSTANT TENSION, regardless of length, a fact of importance in keeping the blood vessels resistant to expansion by the pressure of the blood inside them. The cells CONTRACT AND RELAX SLOWLY, requiring 3–20 seconds to complete a full series of events. If the muscle did *not* react slowly, the volume and pressure changes in the circulation might occur so fast as

to cause problems in heart action. Those stimuli that are most effective in increasing activity of smooth muscle include CHEMICALS and STRETCHING. The cells are quite sensitive to *epinephrine,* which causes contraction in all vessels except those of heart, lungs, and skeletal muscles, to *acetylcholine,* which dilates vessels except those of heart, lungs, and skeletal muscles, and to CO_2, which causes local dilation, as does *fall of pH.*

FIGURE 20.4.   A diagrammatic representation of the structure of smooth muscle.

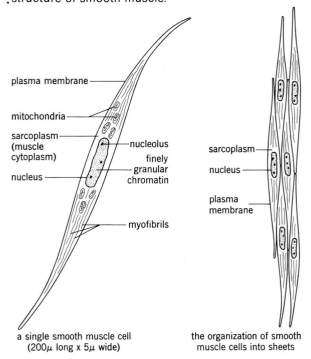

a single smooth muscle cell
(200μ long x 5μ wide)

the organization of smooth muscle cells into sheets

Stretching causes reflex contraction of smooth muscle cells. Autonomic sympathetic and para-sympathetic nerves supply smooth muscle in body organs. Their effects are the same as those presented in Table 14.1, in the chapter on the autonomic nervous system.

These properties suit the smooth muscle's function in blood vessels. For example, changes in size are brought about smoothly, and slow enough so as to not deprive or flood the heart with large amounts of blood; dilation in one area, and constriction in another redistributes blood to organs requiring greater blood flow; a given resistance to blood flow may be maintained in spite of small changes in size of vessels, thus, flow and pressure are maintained.

The structure of blood vessels

Arteries (Fig. 20.5)

Arteries are of three different sizes and structure. They usually have three tunics (coats or layers) of tissue in their walls. An inner TUNICA INTIMA is a thin layer composed of endothelium and connective tissue. A middle TUNICA MEDIA contains many layers of smooth muscle or elastic connective tissue (depending on the specific vessel), and is the major layer in arteries. The TUNICA ADVENTITIA, usually one third to one half as thick as the media, forms the outer wall of the vessel. It consists of loose connective tissue, and contains networks of small blood vessels that nourish the tissues of the arterial wall. These small vessels constitute the VASA VASORUM (literally, blood vessels to blood vessels). The type of tissue predominating in the vessel wall, depends on vessel size and function as summarized in Table 20.1.

CLINICAL CONSIDERATIONS. ARTERIOSCLEROSIS is a generalized term referring to any vascular condition that results in thickening and loss of elasticity and resiliency of the artery wall.

ATHEROSCLEROSIS refers specifically to the changes that occur as fats accumulate in the intima and subintimal layers of the vessel wall. Accumulation of fats in the walls of arteries leads to *nar-*

FIGURE 20.5.   The structure of arteries and veins compared.

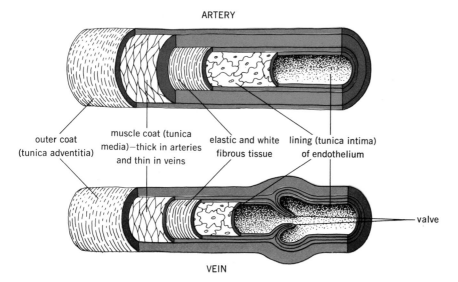

Type of vessel	Average size (mm)	Example(s) or location	Tissue in:			Comments
			Tunica intima	Tunica media	Tunica adventitia	
Large artery	20–25	Aorta, pulmonary artery	Endothelium and c.t. (connective tissue)	Elastic c.t.	Loose c.t.	Elastic vessels expand to contain cardiac output; recoil to push blood onward.
Medium artery	2–10	Named vessels of arms, legs, viscera	As above	25–40 layers of smooth muscle	As above	Muscle makes them important in terms of changing size; not too numerous
Small artery (arteriole)	2 mm to about 20 μ	Unnamed; close to or in tissues and organs	As above	5–10 layers of smooth muscle	As above	Large numbers make these the primary controllers of blood flow and pressure.

TABLE 20.1 A Summary of Artery Structure

rowing of their size, and to *decrease of blood flow* to the tissues supplied by the affected vessel.

Atherosclerosis is present in almost all animal species, and its changes can be detected at any age. It is considered to be an inevitable result of aging. Although the exact cause of the disorder is not known, circumstantial evidence suggests that the intake of a high-fat (cholesterol) diet, cigarette smoking, and lack of exercise accelerate the changes associated with the atherosclerotic process.

The changes that occur have been shown to occur in the following order.

Fatty streaks appear in the arteries in the first days, months, or years of life, and deposition continues into the second decade of life.

Fibrous and fatty plaques appear beginning with the second decade of life and increase until some individuals show the signs of vessel narrowing and diminished blood flow to body organs (cardiac infarction, stroke, gangrene).

Once a lesion is established, it acts as an irritant to the artery wall, causing an inflammatory reaction that further thickens and stiffens the vessel wall. Ulceration may then cause rupture of the wall, or ischemia may cause tissue death.

Treatment of the disorder is centered about reducing dietary intake of lipids, which is thought to reduce the rate of accumulation of fats in the vessel walls. If an artery is of such a size and is in a favorable location, *thromboendarterectomy* may be performed. This procedure in effect "reams out" the vessel, removing its lipid deposits and increasing blood flow. Bypass grafting of blood vessels to renew the supply to an organ or tissue that has been deprived of blood is sometimes successful.

In some individuals, the tunics of the vessel walls do not develop properly, and the wall thins out. Under the pressure of the blood, the artery develops a thin blisterlike aneurysm, which may rupture, leading to hemorrhage, shock, and death.

Veins (Fig. 20.5)

Veins, like arteries, vary in size and structure. They, too, usually have three layers of tissue in their walls, and have a larger size and lumen than corresponding arteries. They have thinner walls than arteries, and the predominant layer is the tunica adventitia that is two to five times thicker than the media. The larger veins have a vasa vasorum, and contain valves that normally allow blood flow only toward the heart. A summary of the structure of veins is presented in Table 20.2.

CLINICAL CONSIDERATIONS. Superficial veins, having thinner walls and less muscle than arteries, and less supporting tissue around them, are more easily enlarged or stretched than deep veins.

VARICOSITIES or VARICOSE VEINS are superficial veins that have become abnormally elongated or dilated. The human's upright posture, combined with the effect of gravity tends to cause pooling of blood in the superficial veins of the legs. Valves in these vessels and the massaging action of the skeletal muscles as they contract and relax tend to move the blood toward the heart. Accumulation of blood in these veins, accompanied by lack of muscular activity, may cause enlargement of the veins to a point where the valves become incom-

petent. Once the cycle of enlargement and valvular incompetence is established, it tends to cause the development of larger and larger varicosities. Support hose, and elevation of the feet may aid the condition, or the veins may be removed surgically.

Capillaries (Fig. 20.6)

Capillaries, as the vessels where exchange of materials occurs, must have thin walls. Thus, all layers but the endothelium are missing. Typical capillaries are 7–9 microns in diameter, and permit the diffusion and filtration of all blood substances except cells and most of the plasma proteins. There are literally miles (one estimate says over 60,000 miles) of capillaries in the body, and their large surface area (800 × that of the aorta) provides not only a great surface for exchange, but also slows the speed of blood flow to allow time for exchange of materials to occur. Sinusoids are large irregular vessels similar to capillaries, but they have a lining of phagocytic cells. They are found in the liver and spleen.

Type of vessel	Average size (mm)	Example(s) or location	Tissue in: Tunica intima	Tunica media	Tunica adventitia	Comments
Large vein	20–30	Vena cava	Endothelium and c.t.	C.t. and scattered smooth muscle cells	Loose c.t.	Few in number, little change in size
Medium vein	2–20	Named veins of appendages and viscera	As above	As above	As above	As above
Small vein (venule)	<2 mm	In tissues and organs	As above	Collagenous c.t. and a few smooth muscle cells	Lacking	Most numerous; can significantly alter venous volume by change in size

TABLE 20.2 A Summary of Vein Structure

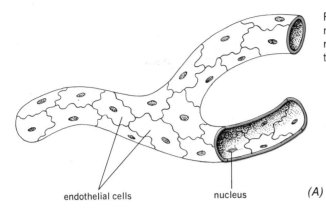

endothelial cells   nucleus   (A)

FIGURE 20.6. (A) The microscopic and (B) ultra-microscopic structure of a capillary (col, collagen; rbc, red blood cell; ec, endothelial cell; ecn, endothelial cell nucleus; pv, pinocytic vesicle).

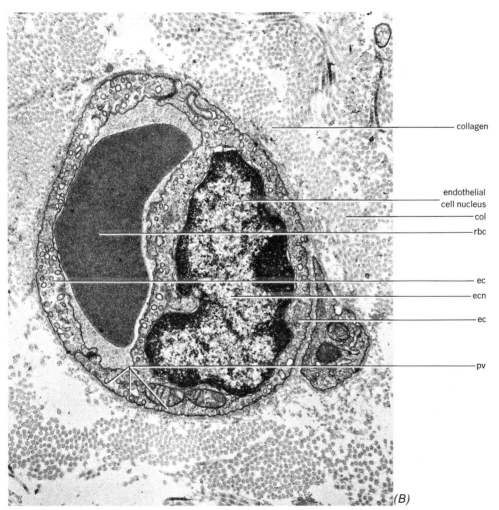

collagen

endothelial
cell nucleus
col
rbc

ec
ecn

ec

pv

(B)

Names of the systemic vessels of the body

Arteries

The major systemic arteries are shown in Figures 20.7 to 20.10, and are summarized in Tables 20.3 to 20.6.

The aorta is the vessel of the systemic circulation from which all other arteries arise.

FIGURE 20.7.   The major arteries originating from the aorta.

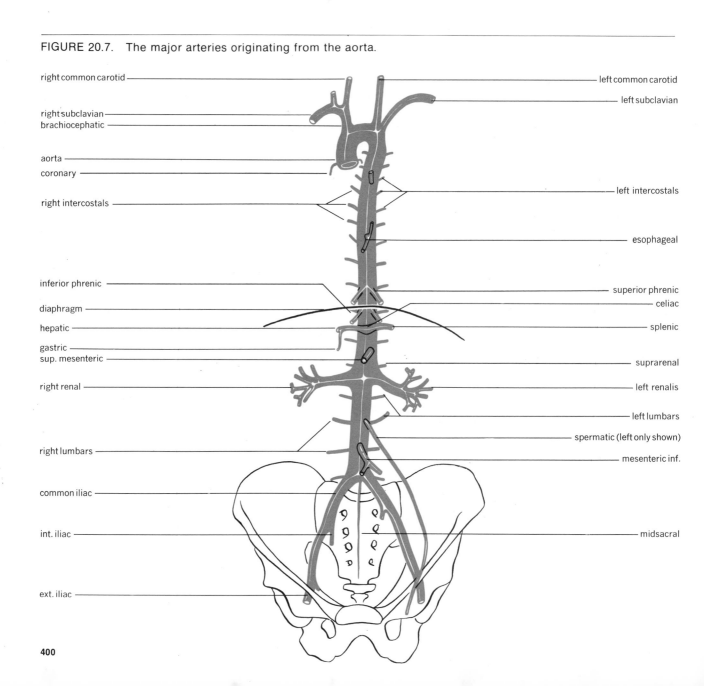

TABLE 20.3   Branches of the Aorta and Area(s) Supplied (Fig. 20.7)

Arising from	Branch	Area supplied by branch
Ascending aorta	Coronary	Heart
Arch of aorta	Brachiocephalic; gives rise to: Right common carotid Right subclavian	 Right side head and neck Right arm
	Left common carotid	Left side head and neck
	Left subclavian	Left arm
Descending aorta: Thoracic portion	Intercostals	Intercostal muscles, chest muscles, pleurae
	Superior phrenics	Posterior and superior surfaces of the diaphragm
	Bronchials	Bronchi of lungs
	Esophageals	Esophagus
	Inferior phrenics	Inferior surface of diaphragm
Abdominal portion	Celiac; gives rise to: Hepatic Left gastric Splenic	 Liver Stomach and esophagus Spleen, part of pancreas and stomach
	Superior mesenteric	Small intestine, cecum, ascending and part of transverse colon
	Suprarenals	Adrenal glands
	Renals	Kidneys
	Spermatics (male) or ovarians (female)	Testes Ovaries
	Inferior mesenteric	Part of transverse colon, descending and sigmoid colon, most of rectum
	Common iliacs, which give rise to: External iliacs Internal iliacs Midsacral	Terminal branches of aorta Lower limbs Uterus, prostate gland, buttock muscles Coccyx

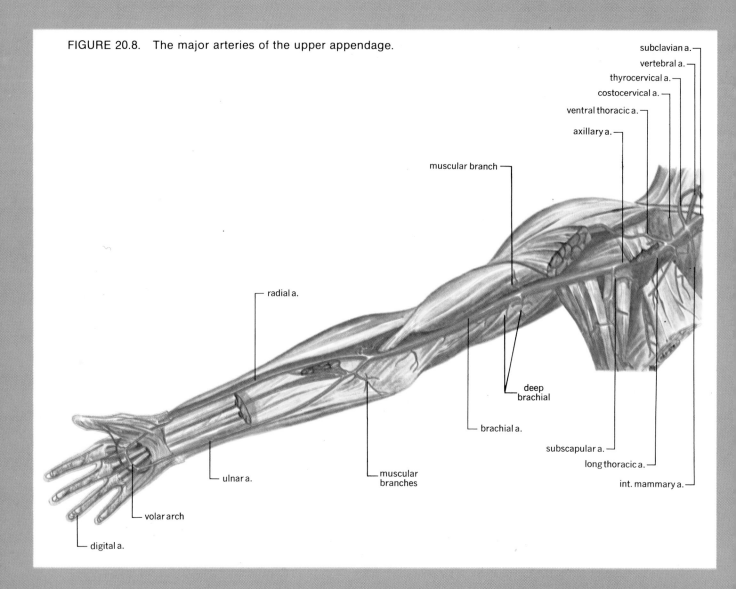

FIGURE 20.8. The major arteries of the upper appendage.

subclavian a.
vertebral a.
thyrocervical a.
costocervical a.
ventral thoracic a.
axillary a.
muscular branch
radial a.
deep brachial
brachial a.
ulnar a.
muscular branches
subscapular a.
long thoracic a.
int. mammary a.
volar arch
digital a.

TABLE 20.4   Arteries of the Upper Appendage and Area(s) Supplied (Figure 20.8)

Artery	Major side branches	Area(s) supplied by branches
Subclavian (beneath clavicle)	Vertebral	Brain and spinal cord
	Internal thoracic (mammary)	Mammary glands, diaphragm, pericardium
	Thyrocervical Costocervical	Muscles, organs, and skin of neck and upper chest
Axillary (armpit)	Long thoracic Ventral thoracic Subscapular	Muscles and skin of shoulder, chest and scapula
Brachial (upper arm)	Muscular	Biceps muscle
	Deep brachial	Triceps muscle. Both also supply skin and other tissues of upper arm
Radial, Ulnar (forearm)	Muscular Muscular	Muscles and skin of the forearm
Palmar (volar) arch	Metacarpals	Muscles and skin of the hand
Digitals	–	Muscles and skin of fingers

TABLE 20.5   Arteries of the Lower Appendage (Fig. 20.9)

Artery	Major side branches	Area(s) supplied by branches
Femoral (thigh)	Deep femoral, pudendals	Muscles and skin of thigh, lower abdomen and external genitalia
Popliteal (knee)	Cutaneous, muscular, genicular	Skin of back of leg, hamstrings, and structures of knee joint
Anterior tibial (anterior side of leg)	Muscular	Anterior crural muscles and skin of leg
Posterior tibial (posterior side of leg)	Muscular, fibular, cutaneous	Posterior crural muscles and skin of leg
Peroneal (lateral side of leg)	Muscular, cutaneous	Lateral crural muscles and skin of leg
Plantar and dorsis pedis Arcuate	Form arcuate arch Metatarsal	Muscles and skin of foot
Digitals		Muscles and skin of toes

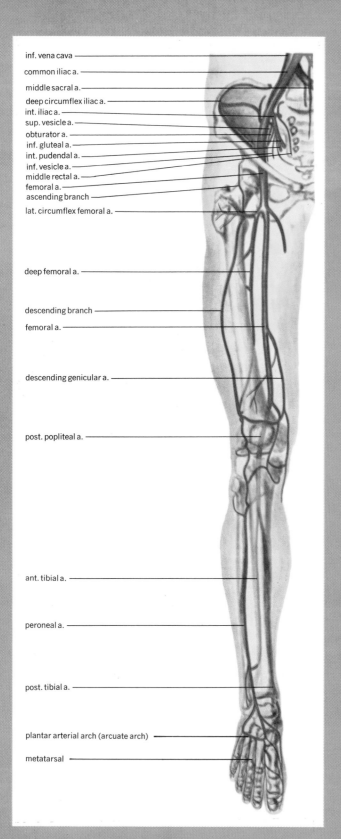

inf. vena cava
common iliac a.
middle sacral a.
deep circumflex iliac a.
int. iliac a.
sup. vesicle a.
obturator a.
inf. gluteal a.
int. pudendal a.
inf. vesicle a.
middle rectal a.
femoral a.
ascending branch
lat. circumflex femoral a.

deep femoral a.

descending branch
femoral a.

descending genicular a.

post. popliteal a.

ant. tibial a.

peroneal a.

post. tibial a.

plantar arterial arch (arcuate arch)
metatarsal

FIGURE 20.9. The major arteries of the lower appendage.

TABLE 20.6   Arteries of the Head and Neck (Fig. 20.10; see Fig. 14.1)

Artery	Major side branches	Area(s) supplied by branches
Common carotid	Internal carotid	Brain and spinal cord
	External carotid: (superior thyroidal, lingual, facial, occipital, superficial temporal, maxillary	Muscles and skin of cranium, face, teeth, and eyes
Basilar (formed from vertebrals)	Cerebellar arteries	Cerebellum
	Pontine	Pons and brain stem
	Auditory	Inner ear structures

Circle of Willis—not a vessel but a circular pathway on the base of the brain. It is composed of the anterior cerebral arteries (from internal carotid), the anterior communicating arteries, the internal carotid arteries, and the posterior communicating arteries.

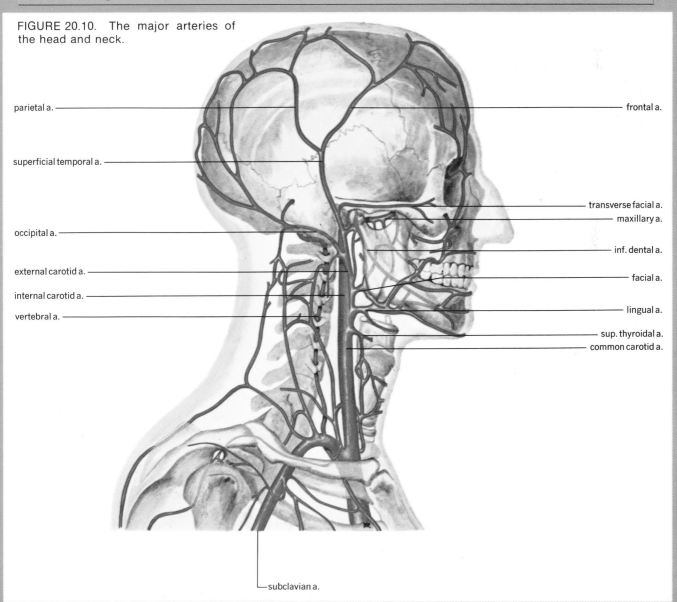

FIGURE 20.10.   The major arteries of the head and neck.

parietal a.

superficial temporal a.

occipital a.

external carotid a.

internal carotid a.

vertebral a.

frontal a.

transverse facial a.

maxillary a.

inf. dental a.

facial a.

lingual a.

sup. thyroidal a.

common carotid a.

subclavian a.

TABLE 20.7   Veins Forming the Superior Vena Cava (Fig. 20.11)

Vein	Formed from	Area(s) drained
Internal Jugular	Dural sinuses	Inside of skull and brain
External Jugular	Veins of face	Muscles and skin of scalp and face
Subclavian	Axillary, cephalic*, basilic*, and their tributaries, scapular, and thoracic veins	Upper appendage, chest, mammary glands
Innominate (brachiocephalic)	Internal jugular, external jugular, and subclavian	

*These are the superficial veins of the extremities.

FIGURE 20.11   The veins forming the superior vena cava, and some tributaries of these veins.

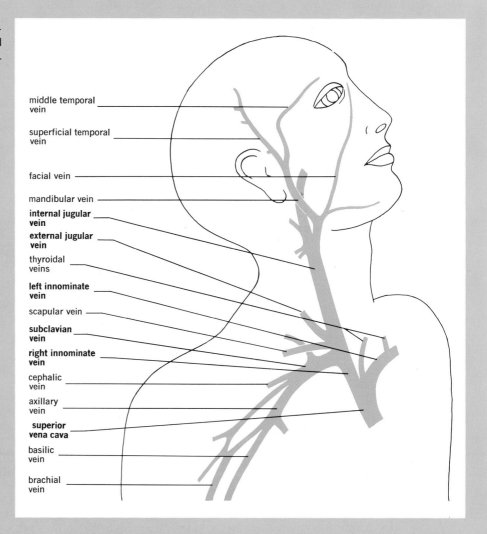

middle temporal vein

superficial temporal vein

facial vein

mandibular vein

internal jugular vein

external jugular vein

thyroidal veins

left innominate vein

scapular vein

subclavian vein

right innominate vein

cephalic vein

axillary vein

superior vena cava

basilic vein

brachial vein

Veins

The major systemic veins are presented in Figures 20.11 and 20.12, and are summarized in Tables 20.7 and 20.8. It may be noted that veins usually accompany their corresponding arteries and are usually named in a similar fashion. In the appendages, a DEEP SET of veins follows close to the arteries, and a SUPERFICIAL SET is found just beneath the skin. The latter set is named differently than the deep set. The SUPERIOR and INFERIOR VENAE CAVAE are the largest systemic veins.

TABLE 20.8   Veins Forming the Inferior Vena Cava (Fig. 20.12)

Vein	Formed from	Area(s) drained
Hepatics	Sinusoids of liver	Liver
Renals	Veins of kidney	Kidney
Gonadals	Veins of gonads	Gonads; testes, and ovaries
Common iliac	External iliac	Lower appendage
	Internal iliac	Organs of lower abdomen
	Saphenous*	Superficial structures of lower appendages

*These are the superficial veins of the extremities.

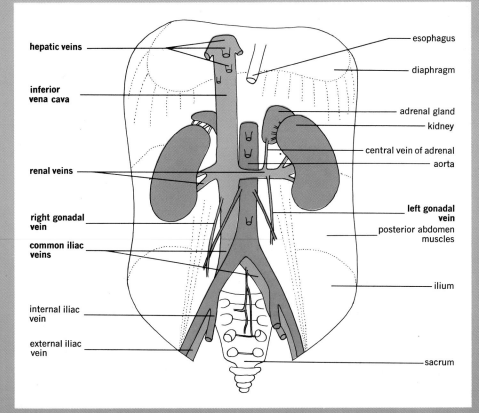

FIGURE 20.12. The veins forming the inferior vena cava with organs shown for reference purposes.

hepatic veins

inferior vena cava

renal veins

right gonadal vein

common iliac veins

internal iliac vein

external iliac vein

esophagus

diaphragm

adrenal gland

kidney

central vein of adrenal

aorta

left gonadal vein

posterior abdomen muscles

ilium

sacrum

Vessels of special body regions

Pulmonary vessels (Fig. 20.13)

The PULMONARY TRUNK arises from the right ventricle, and branches to form the RIGHT and LEFT PULMONARY ARTERIES. These then divide to form the arterioles and capillary networks of the lungs. Four PULMONARY VEINS are formed by the vessels leaving the lungs, and return blood to the left atrium. This system operates at a pressure about one fourth that of the systemic system, and the vessels are fewer and thinner walled.

Portal circulation (Fig. 20.14)

The veins draining the abdominal digestive organs do not empty separately into the inferior vena cava. Instead, they come together to form the hepatic PORTAL VEIN to the liver. The liver thus gets "first choice" of the nutrients absorbed from the intestines. The portal vein empties into the liver sinusoids, and its contents are there mixed with the oxygen-rich blood brought to the liver by the hepatic artery. The liver thus has two afferent blood supplies, both of which empty into the sinusoids. The HEPATIC VEINS are formed by the vessels draining the liver sinusoids, and carry blood to the inferior vena cava.

The role of blood vessels in control of blood pressure and blood flow

Principles governing pressure and flow in tubes

Several basic principles determine the pressure, flow, and speed of flow of liquids in tubes, as in the flow of blood through vessels.

The heart (cardiac) output determines the basic pressure in the system. If the volume of output

FIGURE 20.13.   The major vessels of the pulmonary circulation. (Colors indicate degree of oxygenation: red, high; blue, low.)

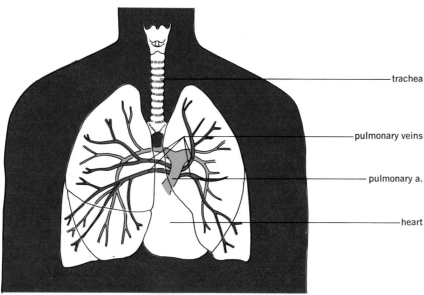

trachea

pulmonary veins

pulmonary a.

heart

Causes

Shock may result from blood loss (hemorrhage), heart failure, trauma (mechanical damage) to the body, or during fright, anesthesia, infections, and drug toxicity.

Compensation

The body makes a number of automatic responses to shock which may, on their own, restore blood

pressure and flow. Where blood is being lost, CLOTTING may prevent further loss. FLUID SHIFTS from interstitial space to plasma, maintain blood volume and pressure. There is VASOCONSTRICTION and epinephrine is released that STIMULATES HEART ACTION. BLOOD IS SHUNTED from nonvital areas (skin, digestive tract) to vital organs (brain, heart). If these measures are not sufficient to restore pressure, a sequence of events known as the "shock cycle" may result. The cycle is shown in Figure 20.17.

Hypertension

The term HYPERTENSION refers to blood pressure that is higher than normal for a given age. It is usually said to occur in the adult, when resting systolic pressures are higher than 140–150 mm

Hg., and resting diastolic pressures are higher than 90–100 mm Hg. Some of the causes and effects of several of the common types of hypertension are presented in Table 20.10.

TABLE 20.10   Some Types, Causes, and Effects of Hypertension

Type	Definition	Cause(s)	Effect(s)
Benign (essential)	Hypertension of slow onset	No apparent cause	Usually without symptoms
Malignant	Severe hypertension with degenerative changes in vessel wall	The common denominator in all these types is an increase in peripheral resistance due to loss of vessel elasticity and/or narrowing of size. Constriction may be due to vascular spasm, or emotional disturbance, or release of vasoconstrictive chemicals. Retention of Na^+ and water also increases pressures.	Reduced blood flow and tissue death
Systolic	Elevation of systolic pressure		Vascular congestion and danger of rupture. High cardiac output; vessels often sclerose
Diastolic	Elevation of diastolic pressure		Reduce blood flow and tissue death
Renal	Hypertension from kidney disease or ischemia		Results in Na^+ and water retention. Ischemia causes renin release*

*An ischemic kidney releases the enzyme renin. Renin converts plasma materials ultimately to angiotensin II, the most powerful vasoconstrictive agent known.

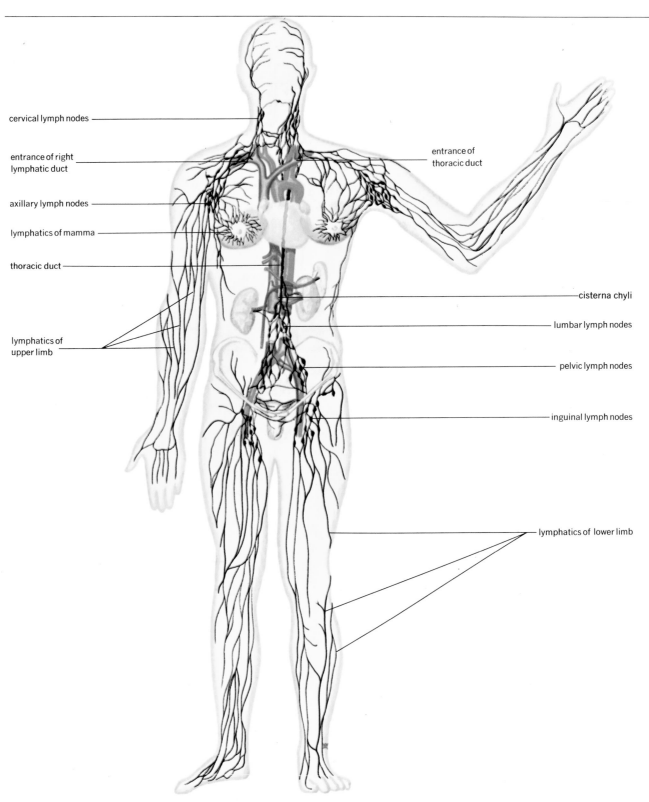

cervical lymph nodes

entrance of right
lymphatic duct

axillary lymph nodes

lymphatics of mamma

thoracic duct

lymphatics of
upper limb

entrance of
thoracic duct

cisterna chyli

lumbar lymph nodes

pelvic lymph nodes

inguinal lymph nodes

lymphatics of lower limb

FIGURE 20.18. The major lymph vessels of the body (left) and their areas of drainage (right).

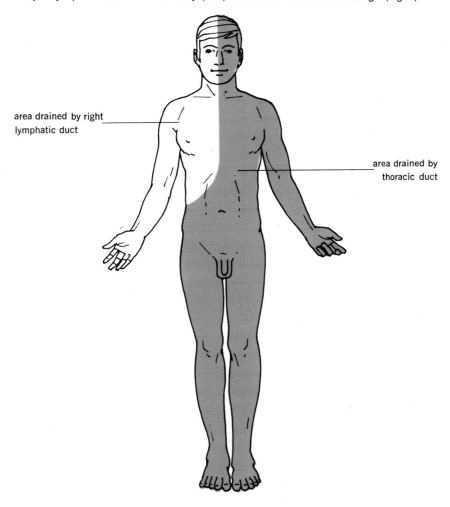

area drained by right
lymphatic duct

area drained by
thoracic duct

Lymphatic vessels

Plan, structure, and functions

The lymphatic vessels of the body (Fig. 20.18) carry lymph or tissue fluid from the body tissues and organs to the blood vascular system. Filtered water, protein, and white blood cells are therefore returned or added to the bloodstream, maintaining water and protein homeostasis.

The LYMPH CAPILLARIES are tiny vessels that are found in all body areas except the brain, spinal cord, and eyes. They have the same basic structure as blood capillaries but are larger and irreg-

ular in size. Dense networks of capillaries form *lymphatic plexuses* in most organs, and collect tissue fluid from the interstitial fluid compartment. Lymph capillaries are much more permeable to large solutes than are blood capillaries, and thus filtered proteins, or large molecules such as enzymes which are produced by cells, easily enter the small lymph vessels. Capillaries form larger lymphatics known as COLLECTING VESSELS along which the lymph nodes are placed. The collecting vessels resemble small or medium-sized veins in structure, having three thin tunics or

415

coats named as in blood vessels. Valves are numerous in the collecting vessels and give a "beaded" appearance to the vessel as they control movement only toward the neck.

The LYMPH DUCTS are formed from the collecting vessels, and resemble large thin-walled veins in structure. They, too, have three tunics and valves. The *thoracic duct* collects lymph from all but the upper chest, right arm, the right side of the neck and head, and empties into the left subclavian vein at its junction with the left internal jugular vein. Its lower end is called the *cisterna chyli,* although its does not form an enlarged bulb as it does in other animal species. The *right lymphatic duct* collects lymph from the areas listed above, and empties into the right innominate (brachiocephalic) vein.

Flow through the vessels

The contraction and relaxation of skeletal muscles, breathing, and the expansion of lymph vessels by radially arranged elastic fibers around them contribute to the entry and movement of lymph in the vessels. Flow averages about 1.5 milliliters per minute from the cut end of a medium-sized lymph vessel.

Clinical considerations

Since the lymphatics return about one tenth of the filtered plasma water to the blood vessels, blockage of a major lymph vessel is associated with the development of EDEMA in the area served by the vessel.

ELEPHANTIASIS is the result of blockage of lymph vessels by a parasite, and used to be common in Africa. Blockage of lymphatics by cancer cells that have metastasized and been trapped may cause mild edema in the body part served by the plugged vessel.

Summary

1. Blood vessels develop as cavities in mesoderm. The cavities grow and join to form networks of vessels. The muscle and connective tissue around the vessels are also of mesodermal origin.

2. There are three major types of vessels in the body.
   a. Arteries carry blood away from the heart.
   b. Capillaries allow for exchange of materials.
   c. Veins carry blood to the heart.

3. Smooth muscle is an important tissue in the walls of larger blood vessels.
   a. It is composed of single, uninucleate, nonstriped, spindle-shaped cells.
   b. It is involuntary, can maintain tone, contracts and relaxes slowly, and is influenced by chemicals, stretching, and nerves.

4. Blood vessels have a typical structure.
   a. Arteries have three layers in their wall (intima, media, adventitia); the media is thickest and contains much smooth muscle or elastic tissue or both.
   b. Arteriosclerosis and atherosclerosis are changes occurring in arteries that narrow their lumina and reduce blood flow. High levels of lipid in the blood accelerates atherosclerosis.

  c. Veins usually have three layers in their walls (intima, media, adventitia). The adventitia is the thickest.

  d. Varicosities develop if veins are stretched and blood collects in them.

  e. Capillaries are very thin-walled tubes through which nutrients and wastes are exchanged. They permeate all tissues, are very small, and give a great surface area for exchange.

5. The major named vessels of the body are presented.

6. Blood vessels act in controlling blood pressure and flow.

  a. Heart action and vessel surface govern pressure and flow in the system.

  b. Speed of flow, amount of flow, direction of flow, and pressure is adjusted by the size of the muscular vessels of the system.

7. Blood moves through the system because of pressure differences, skeletal muscle activity, breathing, gravity, and heart relaxation.

8. Control of blood vessel size depends on nervous and nonnervous factors.

  a. Vasomotor nerves control dilation and constriction of vessels. The nerves come from centers that are influenced by receptors.

  b. Nonnervous control is by chemicals and stretching.

9. Shock results from failure of the circulatory system to maintain adequate blood flow and pressure.

  a. There are several types.

  b. Causes include injury, infection, and blood loss.

  c. Automatic responses may maintain circulation. These include clotting, fluid shifts, vasoconstriction, stimulation of heart action, and shunting of blood to vital organs.

10. Hypertension refers to high blood pressure.

  a. There are several types.

  b. A common denominator for most types is increased peripheral resistance and vasoconstriction due to nervous or chemical factors. Weak heart action may also contribute, by causing water and Na retention.

11. Lymphatic vessels return tissue fluid and protein to the blood vessels.

  a. Lymph capillaries collect fluid from the tissues. They are very permeable to large molecules.

  b. Collecting vessels are larger and have nodes along their course.

  c. Lymph ducts empty the lymph into the subclavian veins. The two ducts are the thoracic duct and the right lymphatic duct.

  d. Blockage of a major lymphatic is associated with development of edema in the body area served by the vessel.

Questions

1. Construct a table comparing velocity, pressure, and surface area in the three main subdivisions of the vascular system (arteries, capillaries, veins).

2. What are the physical principles related to the changes described in question one?

3. Compare and contrast the factors causing blood flow through arteries, veins, and lymphatics.

4. What are the common denominators in the development of shock, regardless of cause?

5. What are some of the causes of hypertension? What are the common denominators of the disorder?

6. What is the relationship between pressure within a blood vessel and the type and thickness of tissues in its wall?

7. Compare arterial and venous systems in terms of total surface area, number of pathways to and from the tissues or organs, and pressure within the systems.

8. How are arteriolar diameter and cardiac output integrated to achieve alterations in blood pressure?

9. It has been said that the whole circulatory system exists to serve the capillaries. Discuss this statement.

10. Trace a drop of lymph from the left leg to the blood vascular system.

Readings

Berne, R. M., and N. L. Matthew. *Cardiovascular Physiology.* C. V. Mosby Co. St. Louis, Mo., 1972.

Johnson, J. A., and J. O. Davis. "Angiotensin II: Important Role in Maintenance of Arterial Blood Pressure." *Science 179:*906, 1973.

Korner, P. I. "Integrative Neural Cardiovascular Control." *Physiol. Rev. 51:*312, 1971.

Mellander, S. "Systemic Circulation." *Ann. Rev. Physiol. 32:*313, 1970.

Moore, Keith L. *The Developing Human. Clinically Oriented Embryology.* Saunders. Philadelphia, 1973.

Science News. "Salt, A Cause of Hypertension." *105,* No. 15, p. 239. April 13, 1974.

Stainsby, W. N. "Local Control of Regional Blood Flow." *Ann. Rev. Physiol. 35:*151, 1973.

Wood, J. Edwin. "The Venous System." *Sci. Amer. 218:*86 (Jan) 1968.

chapter 21
The Respiratory System

chapter 21

Development of the system

The nose and nasal cavities are formed at 3½ to 4 weeks of embryonic life, as the face develops (Fig. 21.1). An OLFACTORY PLACODE, or area of thickened ectoderm appears on the front and lower part of the head. An OLFACTORY PIT develops in the placode, and grows posteriorly to fuse with the front part of the primitive gut (foregut). This latter structure forms the pharynx. The pit extends above the mouth cavity, and is separated from it by a thin membrane. This membrane ruptures, and creates a single large nasal-oral cavity. At about 7--8 weeks of age, division of the cavity into an upper nasal cavity and a lower oral cavity is begun by plates that grow horizontally across the cavity. At about the same time, a vertical plate grows downward from the roof of the nasal cavity. These plates meet and fuse by 3 months of development. The horizontal plate becomes the HARD PALATE, and the vertical plate the NASAL SEPTUM.

If the horizontal processes fail to meet in the midline, a CLEFT PALATE will result. Cleft palate

FIGURE 21.1. The development of the nasal region from fourth to twelfth week. Numbers 1,2,3,4,5, all occur in the *fourth week*.

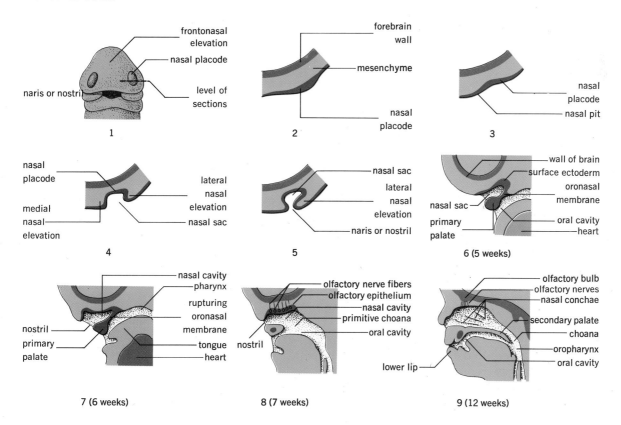

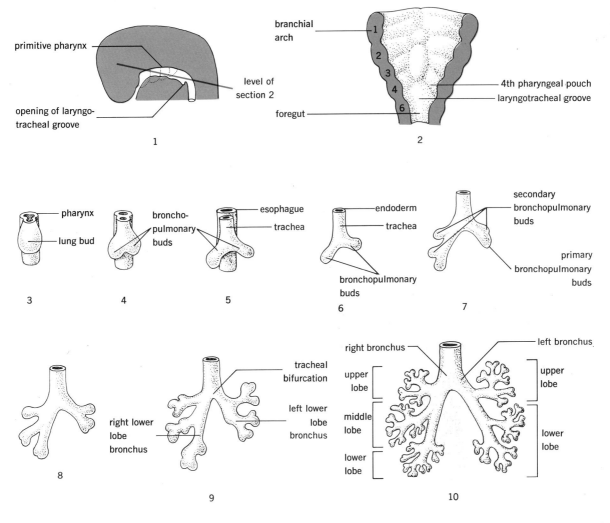

FIGURE 21.2. Development of the lower respiratory system. Numbers 1 and 2, *3 1/2 weeks;* 3,4,5,6, *4 weeks;* 7 and 8, *5 weeks;* 9, *6 weeks;* 10, *8 weeks.*

makes it very difficult for the infant to swallow and create suction to nurse.

The development of the rest of the system (Fig. 21.2) begins as a LARYNGOTRACHEAL GROOVE in the floor of the foregut. The groove deepens, and the walls come together to form a tube. The lower end of the tube forms a lung bud which undergoes growth and repeated branching. The various tubes of the "respiratory tree" and the air sacs are derived from these branchings. In all, some 24 divisions of the original tube occur, to create the vast number and great surface of the respiratory organs.

Until about 26 weeks of age the lungs do not contain alveoli. After this time, there are usually enough alveoli developed to sustain life if the child is born prematurely.

The organs of the system

The organs of the respiratory system (Fig. 21.3) provide a means of bringing air into contact with a surface through which oxygen may diffuse *to* the bloodstream; it also provides a diffusing surface for carbon dioxide to pass *from* the bloodstream and be eliminated.

Accordingly, the organs of the respiratory system may be conveniently divided into those whose function it is to merely transport air, and those that permit gas exchange with the bloodstream. The name CONDUCTING DIVISION is given to the tubes that transport air; the name RESPIRATORY DIVISION is given to the organs that permit gas diffusion.

The conducting division

This division is too thick walled to permit gas exchange with the bloodstream, and is composed of the NASAL CAVITIES and their associated PARANASAL SINUSES, the PHARYNX, LARYNX, TRACHEA,

FIGURE 21.3. The organs of the respiratory system.

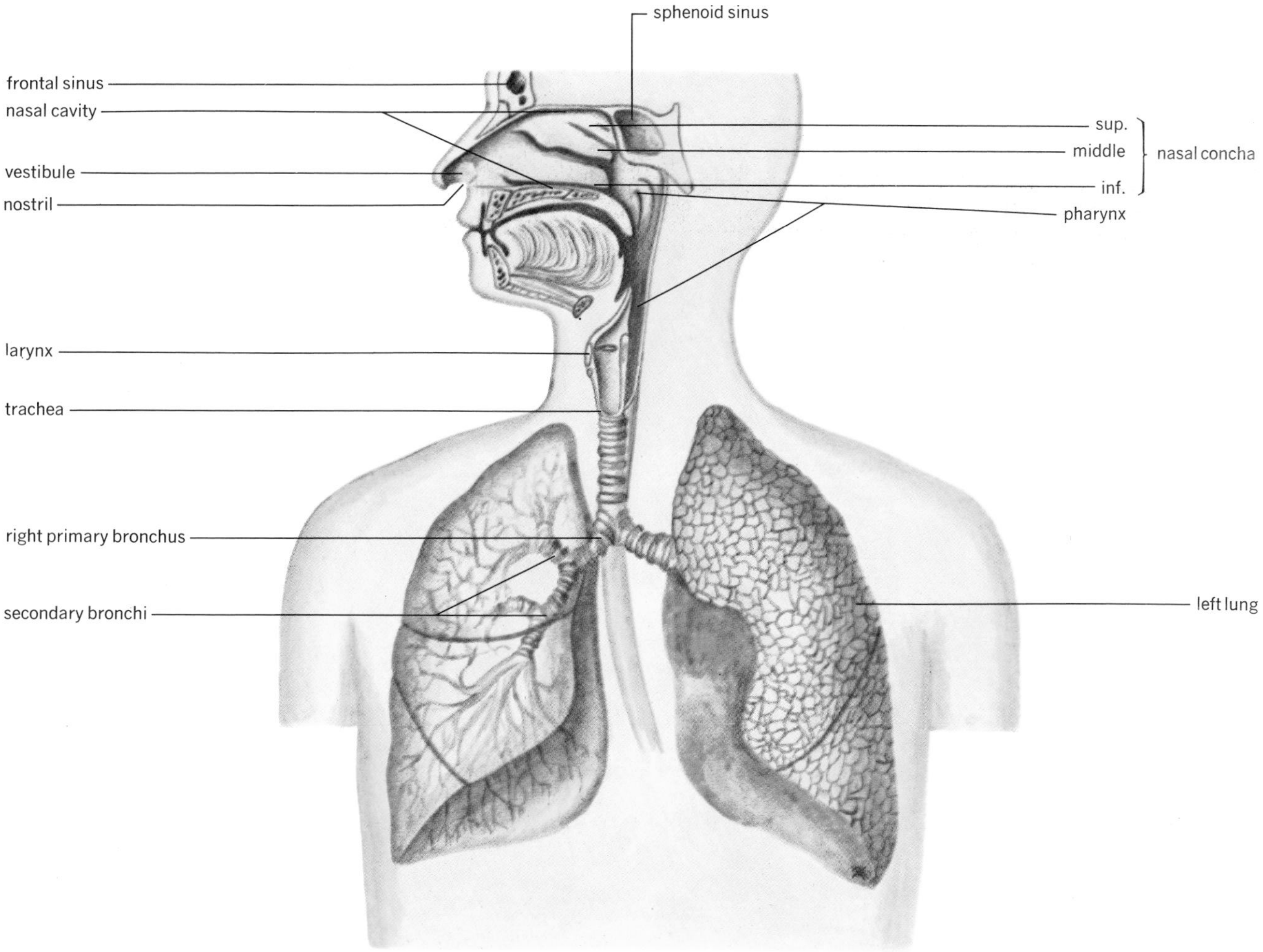

BRONCHI, BRONCHIOLES, and TERMINAL BRONCHI-
OLES. All parts beyond the bronchi are contained
within the lungs.

THE NASAL CAVITIES (Fig. 21.4)

Structure. The anterior openings into the nasal
cavities are the NOSTRILS. The posterior openings
are the CHOANAE. The floor is formed by the hard
palate, and the side walls by the maxillary, inferior concha, ethmoid, and vomer bones. The
lateral walls are irregular due to the CONCHAE or
TURBINATES, which project into the cavities "like
sagging shelves." These "shelves" increase the
surface area of the cavities. The epithelium is
ciliated, and covered with mucus secreted by
mucous cells and glands in the lining tissue.
Many large thin-walled veins are found in the
connective tissue beneath the epithelium. The
paranasal sinuses (frontal, maxillary, ethmoid,
sphenoid) open into spaces or MEATUSES beneath
the conchae. The tear or nasolacrimal duct from
the eye also opens into the nasal cavities. The
olfactory epithelium is found in the roof of each
nasal cavity (see Chapter 15).

Function. The nasal cavities warm, moisten,
and partially cleanse the incoming air. Heat radiates from the blood vessels, moisture is supplied
from the secretions, and the sticky mucus traps
dust, pollens, and other solids. Also, immune
globulin (IgA) is secreted in the nasal cavities.
Ciliary action moves the mucus to the pharynx or
throat where it is swallowed.

Clinical considerations. The most common condition affecting the nose is, of course, the "common cold." It is a viral infection of the linings of
the nose, and produces *rhinitis* (G. *rhin,* nose, +
itis, inflammation). Rhinitis is associated with
swelling of the linings ("plugged nose"), over-
secretion by the glands ("runny nose"), fever,
and general tiredness. It is hoped that, within the
next few years, vaccines may be made against
the viruses causing colds, thus giving protection
against the organisms.

SINUSITIS refers to the inflammation of the paranasal sinuses. It may be caused by microorganisms or allergic reactions. Swelling of the membranes that line the openings of the sinuses into
the nasal cavities block drainage of secretions

FIGURE 21.4.   Medial view of the left nasal cavity.

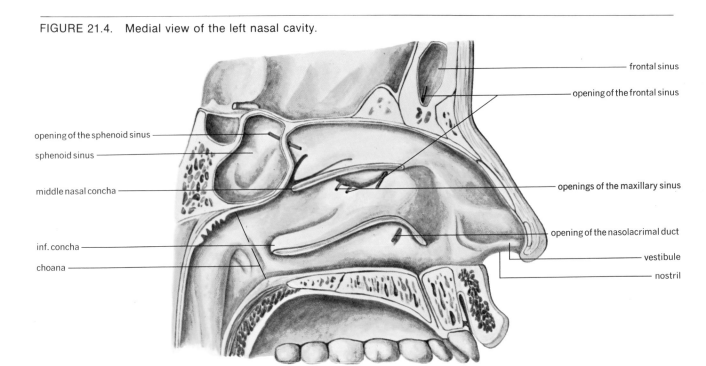

from the sinuses and fever, headache, dizziness, and tenderness when pressing on the affected sinus may develop. Sinusitis may be acute or chronic, and treatment is directed towards opening the exits from the sinuses to promote drainage and relieve pressure.

EPISTAXIS refers to hemorrhage or bleeding from the nose. Trauma, nose picking, fractures of the skull bones, and foreign bodies in the nose are the most common causes of nasal bleeding. Systemic disorders such as leukemia, low platelet levels, and hemophilia are also associated with epistaxis.

THE PHARYNX

Structure. The pharynx (see Fig. 21.3) is a tube common to both the respiratory and digestive systems. It extends from behind the nasal cavities to the level of the larynx ("voice box"). It has three parts.

The *nasal pharynx* lies behind the nasal cavities. The Eustachian tubes from the middle ear cavities open into its lateral walls. Posteriorly, it contains the pharyngeal tonsil or adenoid, a mass of lymphoid tissue. Ciliated and mucus-covered epithelium lines this part.

The *oral pharynx* lies behind the oral cavity. It carries both food and air. It is lined with a tough stratified squamous epithelium that resists the wear and tear of foods hitting its walls.

The *laryngeal pharynx* lies behind the larynx. In its lower end, the tubes for food (esophagus) and air (larynx) separate. It too, has a stratified squamous epithelial lining.

Function. The pharynx conducts either air and/or foods.

Clinical considerations. Enlargement of the adenoids due to inflammation, may interfere with breathing, and may convert an individual to "mouth breathing" with great loss of the warming, moistening, and cleansing functions the nasal cavities serve. Surgical removal of the enlarged organ is usually indicated. Since the body has many other lymphoid organs, loss of the adenoids is not severely missed, although a portion of Waldeyer's ring has been removed.

THE LARYNX (Fig. 21.5)

Structure. The larynx is a cartilagenous box forming the opening into the lower respiratory tract. It is composed of three major and three pairs of minor cartilages. The cartilages are connected by ligaments.

The *thyroid cartilage* is the largest cartilage of the larynx, and forms the "Adam's apple."

The *cricoid cartilage* is "signet-ring" shaped with the widest part of the ring placed posteriorly. It is located beneath the thyroid cartilage.

The *epiglottis* is a leaf-shaped cartilage located above the thyroid cartilage. When one swallows, the epiglottis seals the opening (glottis) into the respiratory system, and prevents entry of food and secretions into the larynx.

The *arytenoid, cuneiform,* and *corniculate cartilages* are the paired cartilages that support the vocal cords.

Function. Besides conducting air, the larynx creates the sounds that tongue and mouth shape into speech. The vocal cords, fibrous bands suspended across the larynx, are caused to vibrate by air forced across them. According to the tension on and length of the cords, the pitch of the voice is changed. Tension on the cords is controlled by skeletal muscles, which attach to the cartilages supporting the cords.

During swallowing, the larynx rises, due to muscular action, the epiglottis is depressed, and the opening into the lower respiratory tract is sealed. Food and liquids are thus not aspirated nor drawn into the respiratory system.

Clinical considerations. Laryngitis, or inflammation of the larynx, may result in swelling of the vocal folds and cords to where the air passageway is obstructed, and breathing becomes difficult or impossible. A hole may be made in the trachea (tracheotomy) below the larynx, to permit air movement to occur. LARYNGECTOMY, or removal of the larynx, is sometimes necessary in cancer of the larynx. Speech is still possible, even though the larynx is removed, by belching air through the esophagus and shaping it into speech (esophageal speech), or by applying a special vibrator to the neck, and shaping the sound into speech.

FIGURE 21.5. Larynx. *(A)* Anterior view. *(B)* Posterior view. *(C)* The vocal folds.

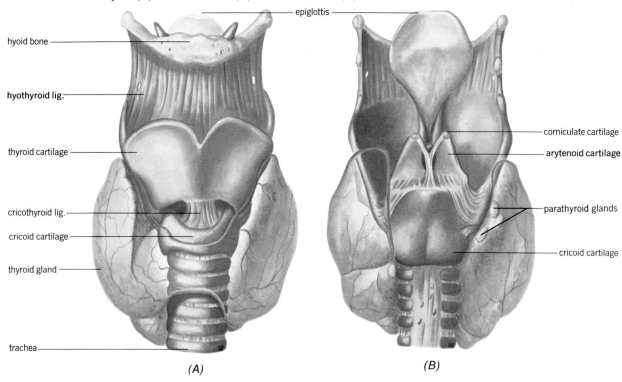

epiglottis

hyoid bone

hyothyroid lig.

thyroid cartilage

cricothyroid lig.

cricoid cartilage

thyroid gland

trachea

(A)

corniculate cartilage

arytenoid cartilage

parathyroid glands

cricoid cartilage

(B)

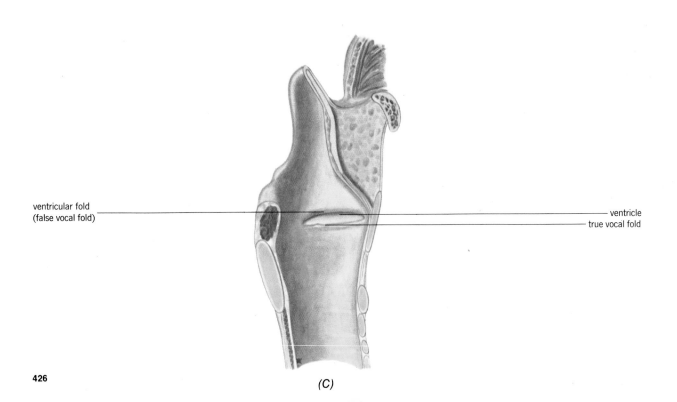

ventricular fold
(false vocal fold)

ventricle

true vocal fold

(C)

LARYNGOSCOPY is a term referring to visual examination of the larynx through an instrument called a LARYNGOSCOPE.

THE TRACHEA, BRONCHI, AND BRONCHIOLES (Fig. 21.6)

Structure. The TRACHEA extends from the larynx about 11 centimeters (4½ inches) into the chest. It divides into the RIGHT and LEFT BRONCHI or primary bronchi which, in turn, branch to form SECONDARY BRONCHI and BRONCHIOLES in some 14–15 generations of branchings. The most characteristic feature of the trachea and bronchi is the presence in their walls of C- or Y-shaped rings of hyaline cartilage. These rings support the tubes, and keep them open as breathing causes pressure changes in the system. In the bronchioles, cartilage rings become scattered plates of cartilage, and finally disappear altogether in the last portion of the conducting division, the terminal bronchioles. As cartilage disappears, smooth muscle increases, to give support to the tube wall. Diameter decreases as branching occurs, so that an increasing number of smaller tubes are produced.

Function. These tubes conduct and cleanse the incoming air.

Clinical considerations. Since the terminal bronchioles have a wall that has no cartilage, and mostly smooth muscle, contraction of that muscle may severely obstruct air flow through the tubes. IN ASTHMA, spasm of the muscle due to irritants may cause extreme difficulty in filling and emptying the alveoli. Any foreign body or tumor that interferes with air flow through these tubes also leads to labored breathing. EMPHYSEMA causes, as one of its many effects, a collapse and kinking of small and terminal bronchioles, thus interfering with air flow through the tubes.

The respiratory division

In the respiratory division, epithelial linings become thinner, until a very delicate diffusion membrane is formed (Fig. 21.7). Also, supporting or surrounding tissue such as connective tissue and muscle virtually disappears. The object here is to get as thin a membrane as possible to allow maximum diffusion of respiratory gases. The organs composing this division are the RESPIRATORY BRONCHIOLES, ALVEOLAR DUCTS, and ALVEOLAR SACS (Fig. 21.8). About nine generations of branchings occur in this division.

STRUCTURE. It is best to consider the changes that occur in the respiratory division as one pro-

---

FIGURE 21.6.   The trachea, bronchi, and bronchioles. (Sternum removed.)

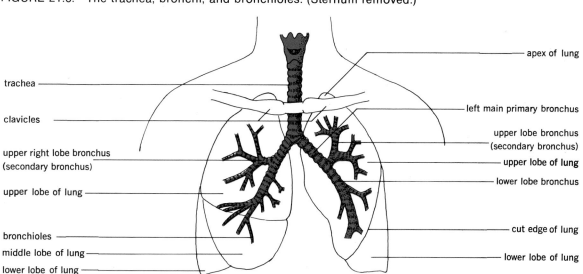

trachea

clavicles

upper right lobe bronchus (secondary bronchus)

upper lobe of lung

bronchioles

middle lobe of lung

lower lobe of lung

apex of lung

left main primary bronchus

upper lobe bronchus (secondary bronchus)

upper lobe of lung

lower lobe bronchus

cut edge of lung

lower lobe of lung

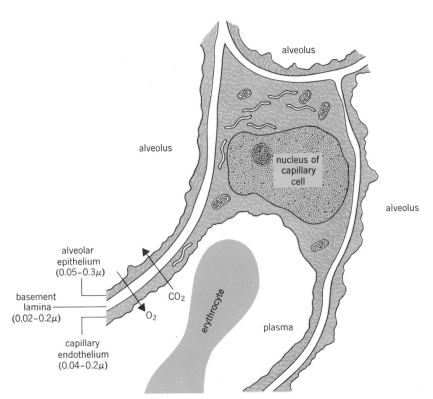

FIGURE 21.7.   A diagrammatic representation of the alveolar-capillary membrane as drawn from an electron micrograph at 20,000X.

FIGURE 21.8.   The terminal portions of the respiratory system.

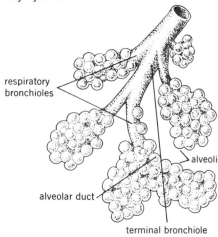

ceeds to the end of the system, rather than to state the structure of each specific organ. Among the changes occurring are the following.

The epithelium changes from a tall to a flat single layer of cells. This allows for increasing diffusion of O_2 and CO_2.

There appear in the walls of the tube, thin-walled ALVEOLI (air sacs) that have many capillaries in their walls. Thus the alveoli are the areas where gas exchange occurs.

FUNCTION. The respiratory division, because it contains alveoli and dense networks of capillaries derived from the pulmonary arteries, serves as the area for O_2 and CO_2 exchange between the lungs and the bloodstream.

Some interesting facts about the two divisions are presented in Table 21.1.

CLINICAL CONSIDERATIONS. The respiratory division, because of its alveoli, is subject to a number of conditions that interfere with gas exchange. Some of the more important are summarized below.

TABLE 21.1 Some characteristics of the respiratory system

Organ	Generation of branching	Number of organs	Diameter (mm)	Total cross-sectional area (cm²)
Trachea	0	1	18–25	2.5
Bronchus	1	2	12	2.3
Lobe bronchi	2	4–5	8	2.1
Small bronchi	5–10	1024	1.3	13.4
Terminal bronchioles	14–15	32,768	0.7	113.0
Respiratory bronchioles	16–18	262,000	0.5	534
Alveolar ducts	19-22	4.2 million	0.4	5880
Alveoli	23–24	300 million	0.2	(50–70m²)

In ATELECTASIS, there is *collapse* of alveoli, or *incomplete expansion* in the newborn. Because alveolar collapse reduces the surface through which gas exchange may occur, symptoms of O_2 lack (cyanosis, labored breathing), or CO_2 retention (hyperventilation, acidosis) may occur.

In PNEUMONIA, production of fluid fills alveoli with secretions that reduce diffusion. Basically the same symptoms occur as in atelectasis.

In EMPHYSEMA, the walls between alveoli disappear, creating large cavities that reduce diffusing surface.

One may rightly conclude, then, that the most important conditions affecting the respiratory division are those that reduce gas diffusion between alveoli and blood.

The lungs (Fig. 21.9)

STRUCTURE. All parts of the respiratory system beyond the bronchi are contained within the lungs. The paired lungs lie within the two lateral PLEURAL CAVITIES of the thorax or chest. The cavities are lined by PARIETAL PLEURA, which is reflected onto the lung surface at the root of the lung as the VISCERAL PLEURA. In life, the lungs fill the cavities, and thus the pleural cavities are potential, not actual, spaces. Other features of the lungs are shown in Figure 21.9. Notice that the shape of the lungs corresponds to the shape of the cavities in which they lie, and that the right lung has three subdivisions or LOBES while the left has only two. The anatomical unit of the lung is the SECONDARY LOBULE, consisting of several terminal bronchioles and all of their branches. The lungs are composed of many such lobules separated by thin connective tissue partitions. The functional unit is the PRIMARY LOBULE, consisting of one respiratory bronchiole and its branches. The parts making up these lobules are shown in Figure 21.8.

Blood supply. The branches of the conducting division from the trachea onward are nourished by blood from the BRONCHIAL ARTERIES. The latter arise from the aorta and the upper intercostal arteries. The respiratory division is nourished by PULMONARY ARTERY blood, which is being oxygenated during its passage through the lungs. The branches of the pulmonary artery form rich

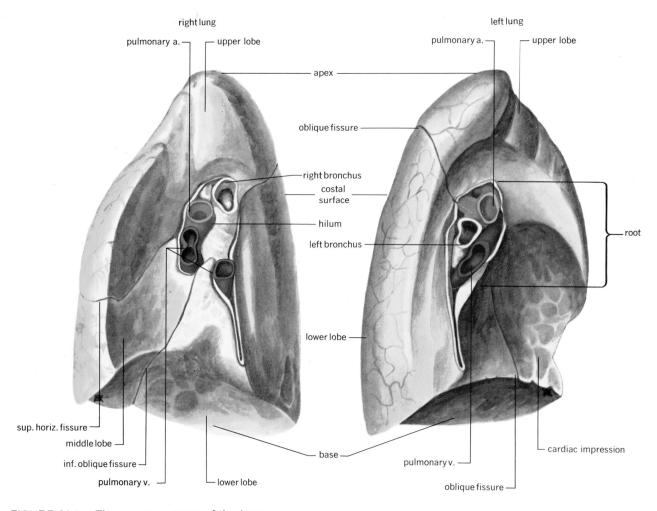

FIGURE 21.9.   The gross anatomy of the lung.

FIGURE 21.10.   The capillary beds around lung alveoli.

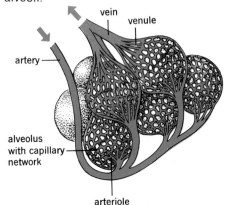

CAPILLARY BEDS around the alveoli (Fig. 21.10) for gas exchange. The PULMONARY VEINS collect blood from these beds and return it to the left atrium.

Nerves. The VAGUS NERVE is the most important nerve serving the lungs. It provides afferent or sensory nerves that serve a number of respiratory reflexes. Vasomotor nerves supply the arteries and veins of the lung but have little effect on blood flow through the lungs.

CLINICAL CONSIDERATIONS. Total or partial LUNG COLLAPSE may follow wounds or disease that cause communication of a pleural cavity with the outside (pneumothorax) or with the lumina of the system's tubes. CANCERS usually require that a part of a lung be removed; if so, a lobe is usually removed because it is more or less a separate unit, and chances of bleeding are reduced. If a whole lung requires removal, the procedure is called a *pneumonectomy*.

The effects of smoking on the respiratory system have been described in detail in the Surgeon General's report on smoking and health. In general, it may be stated that there are changes in the epithelium and loss of cilia that reduce the cleansing and protective functions of the system; there is loss of elastic tissue in the lungs and the respiratory tree that causes collapse of tubes and inability to adequately ventilate the lungs; there is loss of walls between alveoli that results in decrease of diffusing surface. All in all, smoking appears to alter the protective, ventilatory, and diffusing capabilities of the system.

Physiology of respiration

Acquisition of gases by the respiratory system, and delivery to cells may be said to occur in stages.

PULMONARY AND ALVEOLAR VENTILATION. Getting air into (inspiration) and out of (expiration) the lungs is a prerequisite to gas exchange. Pulmonary ventilation implies filling the system with air, but not all of it may reach the alveoli. Thus, alveolar ventilation means air that actually reaches the exchange surfaces of the alveoli.

EXTERNAL RESPIRATION. This refers to exchange of gases between lung and bloodstream.

INTERNAL RESPIRATION. This refers to exchange of gases between bloodstream and cells.

TRANSPORT. This refers to the methods by which the gases are carried by the blood.

All of these activities require control, and may need to be altered according to body activity levels. These activities depend primarily on chemical and nervous factors.

Pulmonary and alveolar ventilation

THE MECHANICS OF BREATHING. INSPIRATION, or intake of air, is brought about by contraction of the diaphragm and external intercostal muscles. Contraction of the diaphragm enlarges the chest cavity in a vertical direction. Contraction of the intercostal muscles causes the ribs to swing up and outward, enlarging the front-to-back and side-to-side dimensions of the chest. Since the chest cavity is a closed cavity with no outside opening, any enlargement of it will cause a pressure drop inside it. The pressure in the chest (intrathoracic pressure) is normally slightly less than atmospheric pressure, and falls even more on inspiration. Because of a thin film of fluid (pleural fluid) lying between the visceral and parietal layers of the pleurae, the lung coheres ("sticks to") the chest wall, and follows it as the chest enlarges. The pressure within the lungs (intrapulmonic pressure) is equal to atmospheric pressure when no breathing is occurring, since the lungs communicate with the atmosphere by way of the bronchial tree. Thus, as the lung expands during inspiration, the pressure falls to less than atmospheric pressure. Air rushes into the lowered pressure area in the lungs. Flow is rapid at first, then slows as the pressures are equalized. Inspiration also stretches the elastic tissue of the lungs as they enlarge.

Normal EXPIRATION is a process not requiring muscular contraction. Relaxation of the diaphragm and external intercostal muscles returns the chest to its original size. Elastic recoil of the lungs creates a higher than atmospheric intrapulmonic pressure that forces air out of the lungs. Again, pressure in the lungs is equalized as the air goes out. Expiration may be made active by contraction of the internal intercostal muscles that depress the ribs, and by contraction of abdominal muscles. The latter press the viscera against the diaphragm, speeding its return to normal position.

Figures 21.11 and 21.12 show the changes occurring in chest size and pressures as breathing occurs.

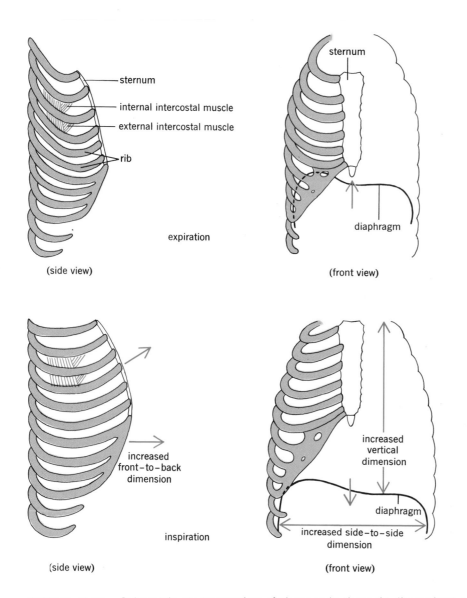

FIGURE 21.11. Schematic representation of changes in thoracic dimensions on expiration and inspiration.

FIGURE 21.12. Changes in intrathoracic and intra-pulmonic pressure during breathing.

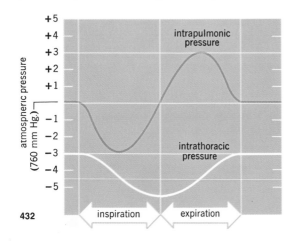

CLINICAL CONSIDERATIONS. Not all the air entering the system reaches the alveoli; some remains in the conducting division. The volume of air in the conducting division is known as DEAD AIR, or air not available for gas exchange with the blood. In certain conditions, such as asthma and emphysema, air cannot enter the alveoli because of blockage of the small bronchioles, and the total surface for diffusion is greatly reduced. In this case, the nonventilated alveoli constitute a PHYS-IOLOGICAL DEAD SPACE, and their volume is added to the previous one. In the normal adult individual, dead air in the conducting division

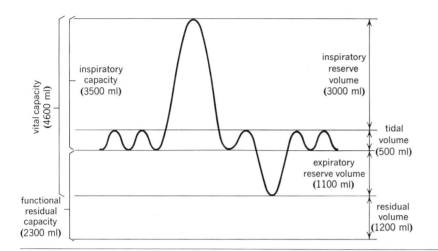

FIGURE 21.13. Respiratory volumes and capacities.

amounts to about 150 milliliters, and the alveoli are all ventilated to some degree. With disease, the nonventilated alveoli and tubes may combine to produce a volume of 1–2 liters.

Surface tension of the alveoli; surfactant

As the lung recoils during expiration, the alveoli, like bubbles, tend to collapse as they get smaller. A force at the surface of an air-water junction, known as SURFACE TENSION, tends to cause a bubble (alveolus) to become as small as it can for its volume. In the lung, alveoli must remain large and open, if gas exchange is to occur through their walls. Thus, surface tension must be lowered to prevent the tendency for collapse. A phospholipid called SURFACTANT (surface active agent) lines the alveoli and reduces surface tension. Hence, when alveoli decrease in size during expiration, surface tension is less, and collapse is prevented.

CLINICAL CONSIDERATIONS. In infants, insufficient amounts of surfactant may be present, particularly if the infant is born prematurely, and the lungs are not fully developed or expanded. A RESPIRATORY DISTRESS SYNDROME (RDS) known as HYALINE MEMBRANE DISEASE (HMD) develops, and alveoli collapse since surface tension of the alveoli is very high due to lack of surfactant.

Exchange of air

VOLUMES. The amount of air exchanged during normal breathing depends on the size, age, and sex of the breather. Obviously, a small person or child will exchange less air than an adult or large person. In the "standard" 70-kilogram male, who is breathing normally at rest, about 500 milliliters of air is moved with each respiration. This volume is called TIDAL AIR. Additional air may be inspired above tidal volume, if forced or during exercise. This volume is called RESERVE INSPIRATORY AIR, and amounts to about 3000 milliliters. The RESERVE EXPIRATORY AIR may be forcibly exhaled after a normal expiration. It amounts to about 1100 milliliters. Even after the most forcible expiration, the lungs do not collapse, and air remains within the alveoli. This air is called RESIDUAL AIR, and amounts to about 1200 milliliters. Even if the lungs are caused to collapse, air remains in small alveoli and in the tissues of the lung. This air is MINIMAL AIR and amounts to about 600 milliliters.

CAPACITIES. Adding the volumes in various ways creates capacities that have physiological significance. Among the important capacities are the following:

Vital capacity. The sum of tidal, reserve inspiratory, and reserve expiratory air, it represents the maximum amount of air that can be moved in the system.

TABLE 21.2 Partial Pressures of Gases in Respired Air						
	Inspired air		Expired air		Alveolar air	
	Percent	Partial Pressure (mm Hg)	Percent	Partial Pressure (mm Hg)	Percent	Partial Pressure (mm Hg)
Nitrogen	78.30	594.70	75.0	569	75.0	570
Oxygen	21.00	160.00	15.2	116	13.6	103
Carbon dioxide	0.05	0.30	3.6	28	5.2	40
Water vapor	0.65	5.00 av	6.2	47	6.2	47
Totals	100.00	760.00	100.0	760	100.0	760

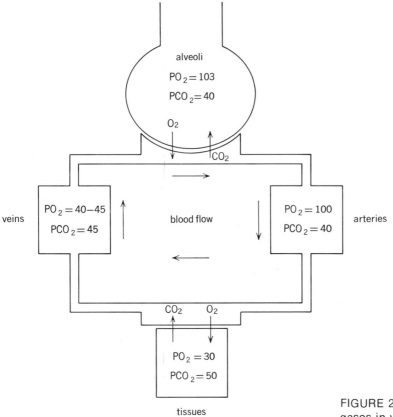

FIGURE 21.14.   Partial pressures of the respiratory gases in various body regions.

Functional residual capacity. The sum of reserve expiratory air and residual volume, this capacity assures continual gas exchange with the blood during expiration.

Total lung capacity. The total of all volumes, this capacity gives an idea of the lung size.

The relationships of lung volumes and capacities are shown in Figure 21.13.

Composition of respired air; internal and external respiration

PARTIAL PRESSURE is commonly used in respiratory physiology to refer to that part of the total pressure contributed by one gas in a mixture of gases. It may be calculated by multiplying the total pressure by the percentage of a gas in the mixture, and is designated by a capital P, followed by the symbol for the gas (e.g., PO_2, PCO_2, PN_2). Table 21.2 shows representative partial pressures for the four main gases in respired air.

It is also known that, under at-rest conditions, the partial pressures of O_2 and CO_2 in the tissues are about 30 and 50 mm Hg, respectively. Thus we can see that as we go from lung to blood to tissues, there is a decreasing pressure of O_2. There is a decreasing pressure for CO_2 in the opposite direction, that is, from tissues to blood to lungs. The principle causing gases to move in a given direction is that of DIFFUSION. We should recall that diffusion is a physical process which causes movement of a substance from higher to lower concentration regions. Thus the direction of *O_2 movement will be toward the tissues,* and that of *CO_2 will be toward the lungs.* Figure 21.14 relates these facts.

Transport of gases

OXYGEN. Ninety-five percent of the oxygen carried by the bloodstream is transported on the iron of the HEMOGLOBIN (Hgb) molecule according to the general equation:

$$Hgb + O_2 \rightleftharpoons HgbO_2$$

reduced
hemoglobin + oxygen $\rightleftharpoons$ oxyhemoglobin

The remaining 5 percent of the O_2 is carried in solution in the plasma.

The hemoglobin (Hgb) carries oxygen in combination on the iron in the molecule. Each molecule of Hgb actually contains four heme groups, and four iron atoms and thus can combine with four molecules of O_2 in steps:

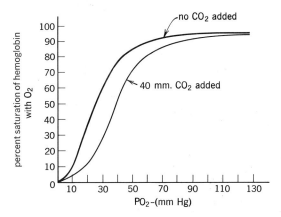

FIGURE 21.15.   Oxygen dissociation curve.

$$Hgb_4 + O_2 \longrightarrow Hgb_4O_2$$
$$Hgb_4O_2 + O_2 \longrightarrow Hgb_4O_4$$
$$Hgb_4O_4 + O_2 \longrightarrow Hgb_4O_6$$
$$Hgb_4O_6 + O_2 \longrightarrow Hgb_4O_8$$

If a curve plotting amount of hemoglobin containing oxygen against oxygen tension (in partial pressure) is constructed, an OXYGEN DISSOCIATION CURVE results (Fig. 21.15). The curve shows several interesting things.

It shows that oxygen tension can fall greatly without large changes in O_2 on Hgb, if in the upper range of available oxygen.

It shows that the presence of CO_2 reduces Hgb carrying ability for O_2.

It shows that, at normal O_2 levels in the alveoli, the blood does not achieve a 100 percent loading with O_2. Thus a few percent reserve of Hgb exists that can take on small additional amounts of O_2.

CARBON DIOXIDE. CO_2 is carried in the bloodstream in three different ways:

About 64 percent is carried in the form of BICARBONATE ION, which is formed from the dissociation of carbonic acid, according to the equation:

$$CO_2 + H_2O \rightleftharpoons H_2CO_3 \rightleftharpoons H^+ + HCO_3^-$$

carbonic   hydrogen   bicarbonate
acid   ion   ion

A maximum of about 27 percent is carried on

HEMOGLOBIN, with the CO_2 attaching to amine groups ($-NH_2$) on the globin protein chains to form carbaminohemoglobin:

$$Hgb + CO_2 \longrightarrow HgbCO_2$$

Binding sites for O_2 and CO_2 on hemoglobin molecules are different as is shown in Figure 21.16.

About 9 percent is carried IN SOLUTION in the plasma. Reactions occurring as O_2 and CO_2 are loaded and unloaded at tissues and lungs are shown in Figures 21.17 and 21.18. The following is an explanation of what is occurring.

At the tissues, carbon dioxide dissolves in plasma water or enters the erythrocytes to react with cellular water. In either case, carbonic acid is formed, which dissociates into hydrogen ion and bicarbonate ion. The hydrogen ions produced in the plasma are buffered by reacting with plasma proteins; the hydrogen ions produced within the erythrocytes do not pass through the cell membrane and are buffered by hemoglobin. Bicarbonate ion accumulates within the erythrocytes and soon begins to diffuse out of the cell into the plasma. A loss of negative charge results, and to return to electrical neutrality, chloride ion moves from plasma into the erythrocytes (chloride shift). Some carbon dioxide displaces O_2 from the hemoglobin, and the excess H^+ in the erythrocyte drops pH and also drives O_2 from the hemoglobin.

At the lungs, the action of carbonic anhydrase liberates CO_2 from carbonic acid within the erythrocytes and plasma. Bicarbonate ion moves into the erythrocyte to form carbonic acid with the H^+ inside the cell and an excess of negative charges accumulates; chloride ion moves back into the plasma to maintain electrical neutrality. The high oxygen levels in the lung region also cause displacement of CO_2 from the hemoglobin as the formation of oxyhemoglobin occurs. Hydrogen ion is released from its combination with protein as bicarbonate enters the eryth-

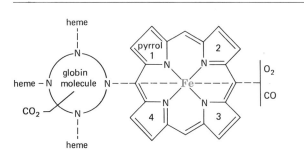

FIGURE 21.16. The chemical skeleton of the hemoglobin molecule showing binding sites for gases.

FIGURE 21.17. Gas exchange between the blood and tissues.

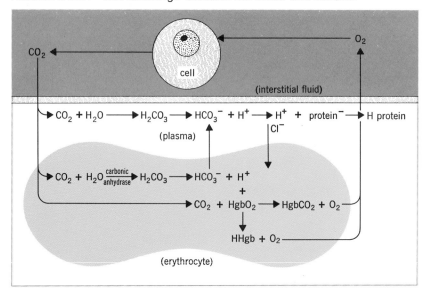

rocyte, and the protein is made available to buffer more H[+].

Acid-base regulation by the lungs

The lung is the most important body organ involved in the regulation of acid-base balance. The production of carbon dioxide by body cells results in the addition of 13,000 to 20,000 meq. per day of H[+] to the body fluids. The reaction involved in the formation of H[+] from carbon dioxide is

$$CO_2 + H_2O \longrightarrow H_2CO_3 \longrightarrow H^+ + HCO_3$$

This reaction is reversed in the lungs by the elimination of carbon dioxide, which shifts the chemical equilibria of the reaction toward the formation of carbonic acid:

$$H^+ + HCO_3^- \longrightarrow H_2CO_3 \longrightarrow H_2O + CO_2$$
$$\text{(eliminated)}$$

The rate of elimination of carbon dioxide is dependent on the process of ventilation, and, in turn, the arterial PCO_2 is a function of carbon dioxide production and ventilation according to the equation:

$$\text{arterial } PCO_2 = \frac{PCO_2 \text{ production}}{\text{alveolar ventilation}}$$

We thus have a system for SELF-REGULATION of arterial PCO_2 and, therefore, pH. The respiratory centers controlling respiration are stimulated by increase in PCO_2 and/or a decrease of pH, and elimination of carbon dioxide increases. Conversely, decreases of PCO_2 and/or rise of pH result in decreased stimulation of the centers, and carbon dioxide elimination is decreased. The entire system is thus controlled within very narrow limits by a feedback mechanism that monitors the PCO_2 as determined by the balance between carbon dioxide production and elimination. Adjustments are made rapidly and accurately.

CLINICAL CONSIDERATIONS. The reader should review the sections on respiratory acidosis and alkalosis in Chapter 16.

Control of respiration

The basic desire to breathe, regardless of activity levels, is a function of the action of blood CO_2 and/or H[+] levels acting on neurons located

---

FIGURE 21.18.   Gas exchange between the blood and the lungs.

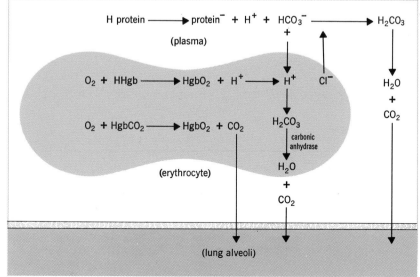

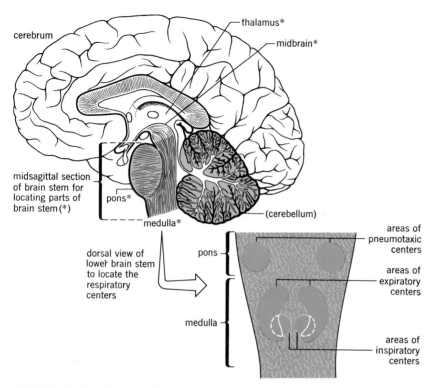

FIGURE 21.19. The location of the respiratory centers.

within the pons and medulla of the brain stem. From these neurons impulses pass over efferent nerve pathways to the muscles of respiration. These neurons form the CENTRAL CONTROL OF RESPIRATION. Additionally, reflex controls, originating in peripherally located receptors, provide a means of changing respiratory activity according to body needs. These reflexes form the PERIPHERAL CONTROL OF RESPIRATION.

CENTRAL INFLUENCES. Located in the medulla and pons of the brain stem are several groups of neurons (Fig. 21.19) that are called the RESPIRATORY CENTERS. The paired INSPIRATORY CENTERS show a continuous cyclical activity, which is caused by blood CO_2 levels, or by the H^+ resulting from dissociation of carbonic acid. Nerves leave these centers and go to the diaphragm (phrenic nerves) and to the external intercostal muscles (intercostal nerves) to cause the inspiratory phase of breathing. Nervous connec-

tions also pass to paired EXPIRATORY CENTERS and PNEUMOTAXIC CENTERS. As the inspiratory centers generate nerve impulses, these impulses are passed to the other centers and partially activate these centers; in turn, these centers send connections to the inspiratory centers and exert an inhibitory effect on the inspiratory centers' activity. While some inhibition to the inspiratory center is provided, the impulses from these centers cannot, on their own, stop inspiratory center activity and allow expiration to occur.

Respiration may also be stimulated by emotions which are controlled by the cerebrum. For example, excitement, fear, and hysteria are usually associated with acceleration of breathing.

PERIPHERAL INFLUENCES. BARORECEPTORS (stretch sensitive receptors) located in the lungs, aorta, and carotid sinuses, provide the final inhibition to inspiratory activity. As the lungs inflate during inspiration, baroreceptors in the

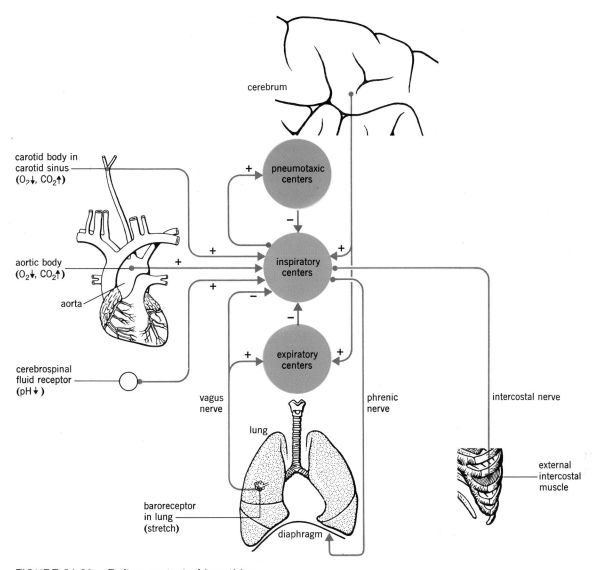

FIGURE 21.20.   Reflex control of breathing.

lungs are stimulated to discharge impulses over branches of the vagus nerve. These impulses are directed primarily to the inspiratory centers, where they provide the final influence needed to stop inspiratory center activity. The muscles of inspiration (diaphragm, external intercostals) are thus allowed to relax, and expiration results. This reflex is the HERING-BREUER REFLEX.

In the carotid sinuses and aorta are CHEMORE-CEPTORS which monitor blood CO_2 and O_2 levels,

and which can alter breathing patterns. If blood CO_2 levels rise, or O_2 levels fall, a stimulation to breathing is provided, which eliminates the excess CO_2 and speeds intake of O_2. There also appear to be chemoreceptors located in the brain, which monitor the pH [H^+] of the cerebrospinal fluid, and which stimulate breathing if pH falls. These reflex controls are shown in Figure 21.20.

It may be noted that these lung and brain re-

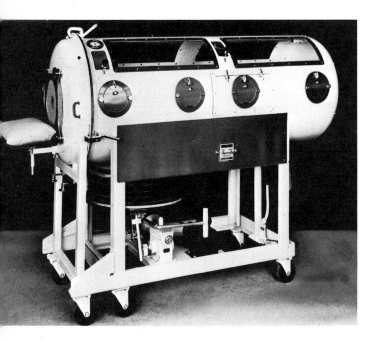

FIGURE 21.21.   An "iron lung."

ceptors also have connections with the cardiac and vasomotor centers, so that increases in breathing are associated with stimulation of heart action, and vasoconstriction, to raise blood pressure.

CLINICAL CONSIDERATIONS.   Damage to the respiratory centers may result in insufficient or no breathing and the failure to eliminate CO_2 and to supply O_2 to the body. Before the advent of vaccination for poliomyelitis ("infantile paralysis"), the virus commonly attacked the inspiratory center neurons in the brain stem. Failure of inspiration occurred, and patients had to be placed in "iron lungs" (Fig. 21.21). The "lung" alternately squeezes and releases the chest (by pressure changes) and causes air to enter and leave the lungs. Obviously, any condition that results in destruction of the inspiratory centers or the efferent pathways to the muscles of respiration, will paralyze breathing. In interruption of the peripheral afferent pathways, the rhythmical nature of breathing continues, but becomes slower and deeper.

Summary

1. The respiratory system develops between 3½ and 12 weeks of age.

   a. Nasal cavities and bronchial tubes develop prior to 12 weeks.

   b. Alveoli develop after 26 weeks of age.

2. The organs of the system provide a means of conducting air to the alveoli, and the alveoli provide a means of allowing gas exchange between lungs and bloodstream.

   a. The conducting division includes nasal cavities, pharynx, larynx, trachea, bronchi, and bronchioles. It is too thick walled to allow gas exchange between lungs and blood.

   b. The respiratory division includes the respiratory bronchioles, alveolar ducts and sacs, and has alveoli that allow gas exchange between lungs and blood.

   c. Each part of the system has special functions including sound production, warming, moistening and cleansing of inhaled air, and secretion of immune globulins.

3. The nasal cavities warm, moisten, and cleanse the inhaled air.

4. The pharynx is a tube common to both respiratory and digestive systems and conducts air. It has nasal, oral, and laryngeal subdivisions.

5. The larynx is a cartilagenous structure which conducts air, and creates sound that is shaped into speech by the tongue and mouth.

6. The trachea, bronchi, and bronchioles conduct air and form the branching system of tubes commonly referred to as the "respiratory tree."

7. The respiratory division allows gas exchange between lung and blood because of alveoli.

8. The respiratory system below the throat is nourished by blood from bronchial and pulmonary arteries, and receives nerve fibers primarily from the vagus nerve.

9. The physiology of respiration involves intake and output of air, diffusion between lungs and blood and tissues, transport of gases, and control of breathing.

10. Pulmonary and alveolar ventilation refer to the processes that bring air into and out of the system.

    a. Air is exchanged by pressure differences created by muscular activity, lung expansion, and elastic recoil of the lungs.

    b. Inspiration (intake of air) is active, requiring muscular activity; normal expiration is passive, and depends on elastic recoil of the lung tissue.

11. Collapse of alveoli on expiration is prevented by a phospholipid called surfactant. It acts to reduce surface tension of the alveoli.

12. Several subdivisions of the respired air, in the adult, are recognized.

    a. Tidal air (500 ml) is exchanged during normal breathing.

    b. Reserve inspiratory air (3000 ml) is air inhaled above tidal air.

    c. Reserve expiratory air (1100 ml) is air exhaled beyond tidal air.

    d. Residual air (1200 ml) remains in the lungs even after forced expiration.

    e. Minimal air (600 ml) is air in alveoli that cannot be removed even by lung collapse.

    f. These volumes are combined to create lung capacities.

13. The composition of the gases in lung, blood, and tissues is such that, by the process of diffusion, oxygen moves from lung to blood to tissue, and carbon dioxide moves from tissue to blood to lungs.

14. Transport of gases in the blood occurs by several methods.

    a. Oxygen is transported primarily on hemoglobin.

    b. Carbon dioxide is carried as bicarbonate (64%), on hemoglobin (as carbaminohemoglobin, to 27%), and in solution in the plasma (9%).

15. The lung, by eliminating CO_2, is an important regulator of the body acid-base balance.

16. Control of respiration depends on central and peripheral influences.

    a. Central influences include the effect of CO_2 on the respiratory centers to establish inspiratory ability.

b. Central and peripheral influences determine expiration and modification of breathing patterns. These influences originate within baroreceptors and chemoreceptors in various body areas.

Questions

1. What is the functional significance of the changes in epithelial type that occur as we pass from the nasal cavities to the alveoli?

2. Compare the blood supplies of conducting and respiratory divisions, and comment on the effect on the tissues that a decreased blood flow in each system might have.

3. Compare lobes, and primary and secondary lobules of the lungs in terms of anatomical extent and functional significance.

4. Commencing with the effect of carbon dioxide on the inspiratory center, catalogue all events that occur to achieve inspiration and expiration.

5. Explain how the lung is involved in acid-base regulation.

6. What factors govern air flow through the respiratory system? Which are changed, and how, in the disease emphysema?

7. Describe the protective mechanisms employed by the respiratory system. Against what type of threat is each designed to protect?

Readings

Avery, Mary E. et al. "The Lung of the Newborn Infant." *Sci. Amer. 228:* (Apr) 1973.

Comroe, Julius H. *Physiology of Respiration.* Year Book Medical Pubs. Chicago, 1965.

Comroe, Julius H. "The Lung." *Sci. Amer. 214:*57 (Feb) 1966.

Hock, Raymond J. "The Physiology of High Altitude." *Sci. Amer. 222:*52 (Feb) 1970.

Fraser, R. G., and J. A. P. Pare. *Structure and Function of the Lung.* Saunders. Philadelphia, 1971.

Heinemann, H. O., and A. P. Fishman. "Nonrespiratory Functions of the Mammalian Lung." *Physiol. Rev. 49:*1, 1969.

Smith, Kline, and French Laboratories. *Emphysema and Related Diseases.* Cliggott Publishing Co. Hackensack, N. J., 1967.

chapter 22
The Digestive System

chapter 22

The digestive system is composed of the tube-like ALIMENTARY TRACT and the ACCESSORY ORGANS that lie inside and outside of the tract. It performs the essential tasks of RECEIVING FOODS and nutrients (ingestion), PREPARATION of food-stuffs for absorption and utilization by cells (digestion), ABSORPTION of end products of digestion, and ELIMINATION (egestion) of unusable residues of the digestive process.

Development of the system

Formation of the gut

Development of the embryo in the first 3 weeks produces a cavity lined with endoderm that is the primitive gut (see Fig. 3.4). Growth of the embryo results in the formation of a gut tube which, by 3.5 weeks, may be subdivided into an anterior FOREGUT, a middle MIDGUT, and a posterior HINDGUT (Fig. 22.1). Each section is supplied by a major branch of the aorta that will form its blood supply throughout life. The celiac artery supplies the foregut, the superior mesenteric artery supplies the midgut, and the inferior mesenteric artery supplies the hindgut. Derivatives of each portion are given in Table 22.1, with indication of when each feature becomes obvious. Formation of the mouth and pharynx is outlined in Chapter 21.

FIGURE 22.1. Development of the digestive system at 3½ weeks.

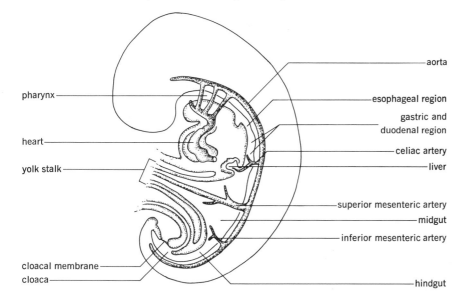

pharynx

heart

yolk stalk

cloacal membrane
cloaca

aorta

esophageal region

gastric and
duodenal region

celiac artery

liver

superior mesenteric artery

midgut

inferior mesenteric artery

hindgut

TABLE 22.1   Derivatives of the Gut Tube

Part of tube	Derivatives	Comments
Foregut	Esophagus	Recognizable at 4 weeks.
	Stomach	Recognizable at 4 weeks, becomes baglike at about 8–10 weeks, and assumes typical form at about 12 weeks.
	Duodenum to entrance of bile duct	Recognizable at 4 weeks.
	Liver } Pancreas }	Develop as outgrowths of gut at about 4 weeks; liver lobes form by 6 weeks; pancreas complete by 10 weeks.
	Bile ducts and gall bladder	Connection retained to duodenum forms ducts. Bladder is an outgrowth of duct.
Midgut	Rest of small intestine	Recognizable at 4 weeks; elongates and coils at 5 weeks; villi at 8 weeks.
	Cecum, appendix, one half of large intestine	Separated at 5 weeks; completed by 8 weeks.
Hindgut	Remainder of large intestine, rectum, anal canal	Formed by 4 weeks; completed by 7 weeks.

TABLE 22.2   Some Common Congenital Disorders of the Digestive System

Disorder	Frequency (no. per births)	Cause of disorder	Comments
Esophageal atresia	1/2500–3000	No recanalization of esophagus, improper separation from respiratory system. No passage to stomach.	Infant shows excess of saliva and poor nutrition since foods cannot reach stomach.
Pyloric stenosis	1/200 male 1/1000 female	Excessive development of muscle fibers of distal end of stomach	Blockage of stomach exit causes vomiting of feedings, weight loss, and dehydration.
Imperforate anus	1/5000	Failure of anal membrane to rupture. No anus formed.	Surgery necessary.
Megacolon	1/25,000	Failure of nerve cells to innervate a section of the colon.	Part without nerves does not move contents onward, accumulates feces and dilates.

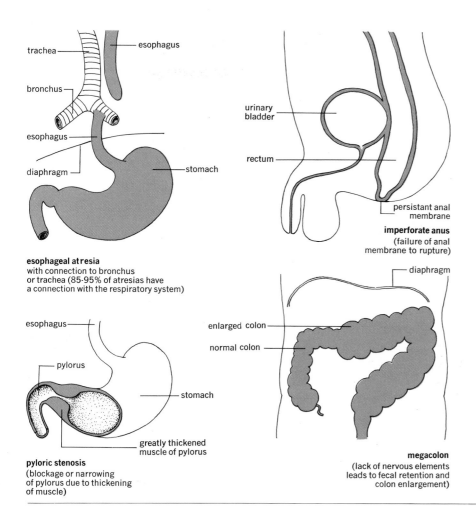

esophageal atresia
with connection to bronchus
or trachea (85-95% of atresias have
a connection with the respiratory system)

pyloric stenosis
(blockage or narrowing
of pylorus due to thickening
of muscle)

imperforate anus
(failure of anal
membrane to rupture)

megacolon
(lack of nervous elements
leads to fecal retention and
colon enlargement)

FIGURE 22.2. Some common congenital disorders of the digestive system.

Clinical considerations

As the major changes mentioned above take place, several other processes are occurring. For example, the open cavities of the gut tube are obliterated by proliferation of the lining epithelium; later, recanalization occurs. The membranes separating the mouth and anal cavities from the outside normally rupture to provide a tube open at both ends. Nerve cells, muscular and connective tissue coats are forming around the tube (see also Table 2.4). Alterations in such processes as these create a wide variety of congenital disorders. Several of the more common disorders are presented in Table 22.2 and are shown in Figure 22.2.

The organs of the digestive system

The term ALIMENTARY TRACT is used to refer to the tubular organs of the system, including the mouth, pharynx, esophagus, stomach, small and large intestines, rectum, and anal canal. The ACCESSORY ORGANS of the system include those that develop from the tube, but which (usually)

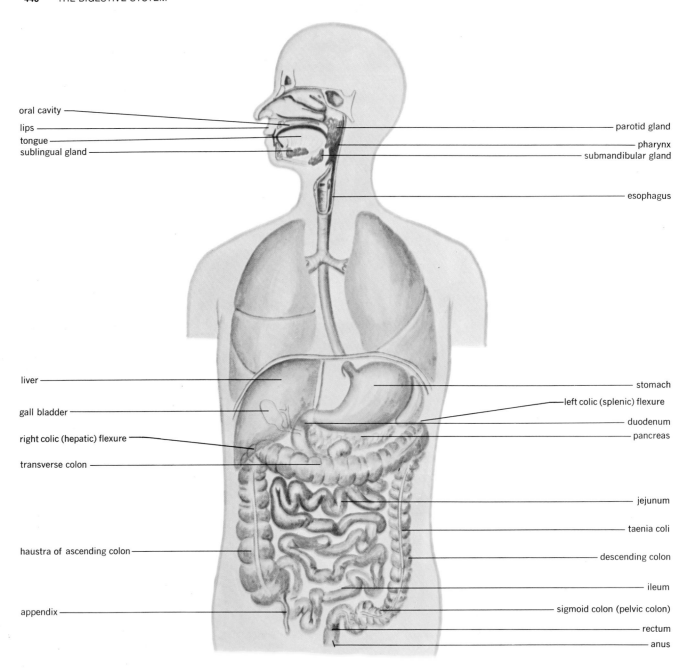

oral cavity

lips

tongue

sublingual gland

parotid gland

pharynx

submandibular gland

esophagus

liver

gall bladder

right colic (hepatic) flexure

transverse colon

haustra of ascending colon

appendix

stomach

left colic (splenic) flexure

duodenum

pancreas

jejunum

taenia coli

descending colon

ileum

sigmoid colon (pelvic colon)

rectum

anus

FIGURE 22.3. The organs of the digestive system.

lie outside the tube and communicate with it by ducts. The salivary glands, liver, and pancreas are the major organs in this category. Also included as accessory organs are the tongue and teeth. The major organs of the system are shown in Figure 22.3.

The structure and functions of the organs of the system

Structural layers of the alimentary tract

The organs of the alimentary tract, particularly those from esophagus to anus, have a basic organization of tissues in their walls (Fig. 22.4). Four main layers are described from inside outward.

1. The *mucosa* (mucous membrane) consists of an epithelial lining of a type that varies according to the jobs each organ is carrying out. The epithelium is underlain by connective tissue that is rich in glands. These glands produce most of the substances necessary for digestion of foods. Absorption also occurs through this layer.

2. The *submucosa* is a layer of vascular connective tissue serving to nourish the tissues of the tube wall, and it also holds the mucosa to the muscular layers. Nerve fibers and cells form the *submucosal plexus* (of Meissner) in the connective tissue of this layer. The nerve cells of this plexus supply the smooth muscle of the villi with impulses.

3. The *muscularis externa* typically consists of two layers of smooth muscle that are responsible for the movements that occur in the tract. The inner layer is circularly arranged (actually a tight spiral), and the outer layer is longitudinally arranged (actually a very loose spiral). The *myenteric plexus* (of Auerbach) is formed by nerve fibers and cells lying between the two layers of muscle, and controls the activity of the muscularis externa.

4. An outer *serosa* or *fibrosa* completes the tube wall. The term serosa is used to describe the layer of connective tissue and epithelium that forms the outer covering of most of the organs located in the abdominopelvic cavity; a fibrosa is a layer of connective tissue that is not separated from the surrounding tissues. Fibrosas occur on the organs lying in the thoracic cavity, such as the trachea and the esophagus.

FIGURE 22.4. A diagram of a cross section of the intestinal tract to show its tissue layers.

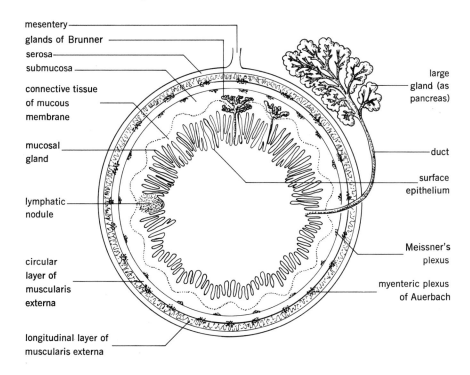

mesentery
glands of Brunner
serosa
submucosa
connective tissue of mucous membrane
mucosal gland
lymphatic nodule
circular layer of muscularis externa
longitudinal layer of muscularis externa

large gland (as pancreas)
duct
surface epithelium
Meissner's plexus
myenteric plexus of Auerbach

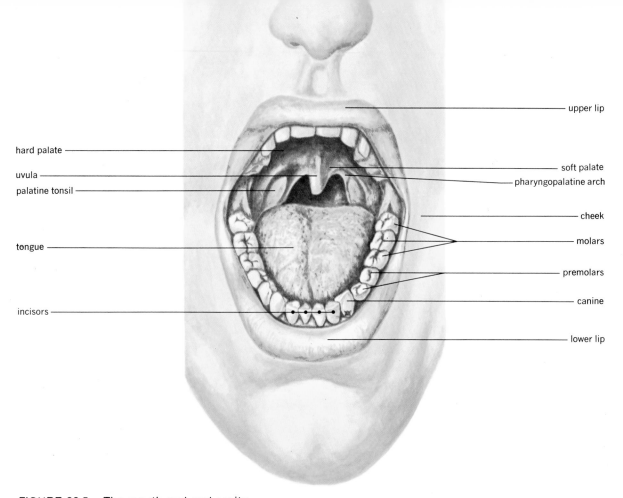

FIGURE 22.5.   The mouth and oral cavity.

FIGURE 22.6.   The superior surface of the tongue.

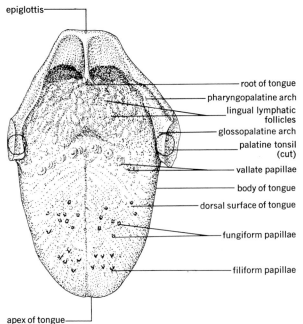

The mouth and oral cavity (Fig. 22.5)

STRUCTURE. The MOUTH is that part of the digestive system bounded by the lips, cheeks, hard palate, tongue, and soft palate. The ORAL CAVITY is that part of the mouth enclosed by the teeth and gums. The VESTIBULE lies between the teeth and the cheeks and lips. The lips and cheeks are important in guiding food between the teeth during chewing, and in speech. The epithelium in this area is stratified squamous to withstand the wear-and-tear the mouth must take, and is translucent, allowing the blood vessels beneath to show their red color.

FUNCTION. The mouth receives food, and mechanical and chemical digestion is begun here.

CLINICAL CONSIDERATIONS. Many types of lesions may occur in the mouth linings. *Koplik's* *spots* are reddish spots with blue-white centers that appear on the side walls of the mouth before the rash of measles appears. *Canker sores* are lesions of the mouth that resemble tiny volcanoes; that is, they are elevated with an ulcer-like depression in their center. *Epstein's pearls* are whitish plugs of keratin in the roof of the mouth of newborns. They have no clinical significance.

The tongue (Fig. 22.6)

STRUCTURE. The tongue is primarily a muscular organ, composed of skeletal muscle covered with mucous membrane. The mucosa on the upper surface is formed into a variety of folds or PA-PILLAE. FILIFORM PAPILLAE are more or less pointed projections, and roughen the surface of the tongue to enable it to more efficiently guide foods during chewing and swallowing. They also contain nerves for touch sensations. FUNGIFORM PAPILLAE are rounded projections and also serve to roughen the tongue surface. Most also contain taste buds, serving the sense of taste. Vallate (circumvallate) papillae are large papillae forming a V-shaped line on the posterior one third of the tongue. They are sunken into the tongue and contain taste buds.

FUNCTION. As indicated above, the tongue serves in swallowing, manipulating food for chewing, and in taste. It is also an important organ in the articulation of speech. It is anchored anteriorly to the floor of the oral cavity by the FRENULUM. The sense of taste is discussed in Chapter 15.

CLINICAL CONSIDERATIONS.   An abnormally short frenulum restricts tongue movements and may interfere with speech (*tongue-tie*). Inflammation of the tongue is GLOSSITIS. "Coatings" on the tongue are most commonly associated with digestive upsets and smoking. The appearance of the tongue, its size, and mobility are often used to aid the diagnosis of pernicious anemia, endocrine disorders, and nerve defects.

In PERNICIOUS ANEMIA, the tongue is often sore, appears beefy red in color, and may develop patchy white vesicles on its surface. Loss of pa-

pillae and the development of a slick smooth surface is a later development in this disease.

In ACROMEGALY, characterized by excessive growth hormone production in an adult, the tongue becomes enlarged. In HYPOTHYROIDISM, the tongue also becomes enlarged.

DAMAGE TO THE XIITH CRANIAL NERVE (hypoglossal) may be reflected by the tongue deviating to one side or the other when its owner is told to "stick out your tongue."

The teeth

Each person receives two sets of teeth during his lifetime. The DECIDUOUS or milk teeth consist of 20 teeth that usually begin to erupt about 6 months of age, with one appearing about each month thereafter until the full set has erupted. These teeth include incisors, canines and premolars. The permanent teeth are 32 in number, include

8 incisors

4 canines

8 premolars

12 molars

and appear between 6 and 17 years of age. The last 3 teeth in each half of each jaw come in only once; all other teeth are replaced.

There are several types of teeth (see Fig. 22.5).

Incisors are chisel-shaped and exert a scissorslike action useful in biting.

Canines are conical, and are most useful in tearing or shredding food.

Premolars (bicuspids), and *molars* (tricuspids) are specialized for grinding foods and are the most important teeth in subdividing food (mechanical digestion).

STRUCTURE. Each tooth has a CROWN, a NECK, and a ROOT. A section of a tooth shows these and other features (Fig. 22.7). The CLINICAL CROWN is the visible part, the ANATOMICAL CROWN extends to the neck and is covered with ENAMEL, the hardest substance in the body. The root is covered with CEMENTUM, a bonelike material. Where

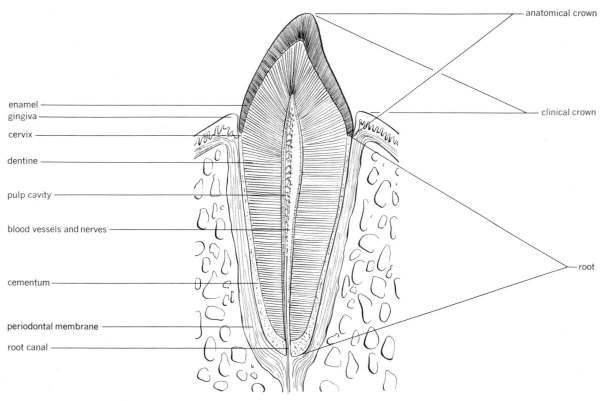

enamel
gingiva
cervix
dentine
pulp cavity
blood vessels and nerves
cementum
periodontal membrane
root canal

anatomical crown
clinical crown
root

FIGURE 22.7.   Longitudinal section of a tooth to show its structure.

enamel and cementum meet is the NECK of the tooth. The main mass of the tooth is composed of DENTINE, a soft, yellowish, bonelike substance. The roots of the teeth fit in the sockets (alveoli) in the jaw bones. A PERIODONTAL MEMBRANE holds the tooth, but not rigidly, in its socket.

FUNCTION. The teeth mechanically subdivide the food for easier digestion.

CLINICAL CONSIDERATIONS. Cavity formation (caries) in the teeth appears to be one of the chief penalties of civilization and the consumption of soft, sweet foods. Acid production from bacterial action on foods in the mouth is regarded as the agent that dissolves tooth substance and leads to cavity formation. Adequate brushing within five minutes after the intake of food or drink is a major aid in the prevention of cavities. Brushing should begin when the first tooth erupts. Coating the teeth with fluoride containing compounds and various types of plastic coatings appears to increase the resistance of the tooth to acid action.

The term PERIODONTAL DISEASE is used to refer to inflammation or degeneration or both of any of the tissues surrounding the teeth. These tissues include the gums (gingivae), bone, periodontal membrane, and the cementum of the teeth themselves. The diseases are characterized by loosening of the teeth, resorption of bone and shrinking of the gums.

GINGIVITIS refers to inflammation, swelling, bleeding, and redness of the gums. It may be caused by local factors such as infection, foods jammed between the teeth and gums, and malocclusion, or by generalized factors such as vitamin C deficiency, allergy, diabetes, and leukemia.

PERIODONTITIS develops when gingivitis progresses to the point where it results in bone destruction. Most periodontal disease can be reduced by attention to oral hygiene, and regular professional care.

The salivary glands   (Figs. 22.8 and 22.9)

STRUCTURE. Three pairs of salivary glands lie outside the mouth and empty their secretions into the mouth. Two types of cells are found in vari-

ous combinations in the glands: *serous cells* are small and granular, and produce a watery secretion rich in digestive enzyme; *mucous cells* are large and pale staining and produce a slimy mucus that is used mainly for purposes of lubrica-

FIGURE 22.8.   The locations of the salivary glands.

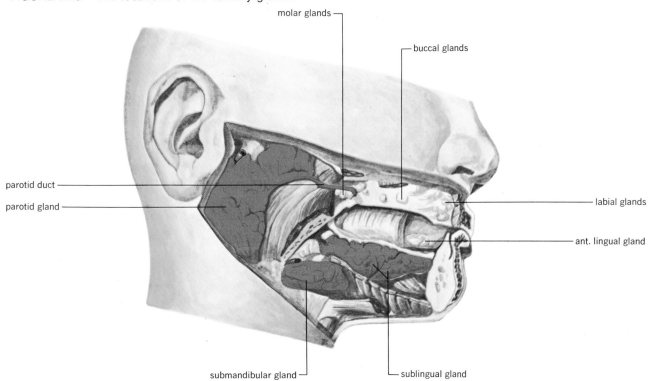

molar glands

buccal glands

parotid duct

parotid gland

labial glands

ant. lingual gland

submandibular gland

sublingual gland

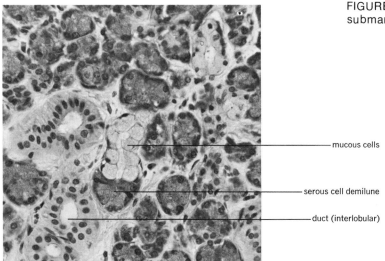

FIGURE 22.9.   Photomicrograph of a section of a submandibular salivary gland.

mucous cells

serous cell demilune

duct (interlobular)

TABLE 22.3	The Salivary Glands				
Name of gland	Location	Cellular composition	Name of duct	Entry of duct into mouth	Secretion contains
Parotid	Side of mandible in front of ear	All serous	Stensen's	Lateral to upper second molar	Water, salts, enzyme
Submandibular	Beneath the base of the tongue	Mostly serous, some mucous	Wharton's	Papillum lateral to frenulum	Water, salts, enzyme, some mucus
Sublingual	Anterior to submandibular under tongue	Mostly mucous, some serous	Rivinus'	With duct of submandibular	Mostly mucus, a little water, salt, and enzyme

TABLE 22.4 Some Characteristics and Constituents of Human Saliva	
Characteristic or constituent (examples)	Value or amount
Average daily volume	1000–1500 ml
Water	99.5 %

Solids
  Inorganic salts (NaCl, KCl, NaHCO₃, KHCO₃, Na₃PO₄, K₃PO₄)
  Organic substances (urea, uric acid, proteins, salivary amylase, mucus)
	0.5%

pH	5.8–7.1

tion. Table 22.3 summarizes the important facts concerning the structure of the salivary glands.

FUNCTION. The composite fluid of all three pairs of salivary glands is the SALIVA. Some of its characteristics and constituents are presented in Table 22.4

Saliva functions to MOISTEN and SOFTEN ingested foods, to LUBRICATE them for swallowing, to CLEANSE the teeth and mouth, and acts as a route of EXCRETION for materials such as urea and uric acid.

The digestive function of the saliva centers around its content of SALIVARY AMYLASE (ptyalin) an enzyme that starts carbohydrate digestion according to the equation:

$$\text{Starch} \xrightarrow{\text{amylase}} \underset{(95-97\%)}{\text{dextrins}} \xrightarrow{\text{amylase}} \underset{(3-5\%)}{\text{disaccharides}}$$

Dextrins are polysaccharides containing fewer simple sugar units than starches. Foods normally remain in the mouth for only a short time, so that breakdown to disaccharides (double sugars) occurs only to a slight degree.

CONTROL OF SALIVARY SECRETION (Fig. 22.10). The secretion of saliva is controlled by nervous reflexes, with the taste buds as receptors, cranial nerves VII and IX as afferent and efferent pathways, the salivary nuclei as centers, and the glands themselves as the effectors. The nature as well as quantity of the saliva may be altered by this mechanism. For example, dry foods stimulate both serous and mucus secretion; soft moist foods stimulate less secretion of mucus. Cerebral influences also cause salivary secretion as evidenced by "watering of the mouth" on sight, smell, or sound of food in preparation.

CLINICAL CONSIDERATIONS. PAROTITIS refers to inflammation of the parotid gland, as by mumps virus.

The pharynx (see Fig. 21.3)

The basic structure of the pharynx has been considered in Chapter 21.

The esophagus

STRUCTURE. The esophagus is a muscular tube about 25 centimeters (10 inches) in length that connects the pharynx to the stomach. Its wall structure follows the plan described previously. The epithelium is stratified squamous to resist abrasion created by swallowing food. A thickened circular layer of muscle forms a functional sphincter at the junction of esophagus and stomach, and tends to prevent regurgitation of stomach contents into the esophagus. The thickened muscle is named the *gastroesophageal* (cardiac) *constrictor*.

FUNCTION. The esophagus conducts food to the stomach; it has no digestive function.

CLINICAL CONSIDERATIONS. ACHALASIA refers to failure of the gastroesophageal constrictor to relax and permit substances to enter the stomach. Enlargement of the esophagus (MEGAESOPHAGUS) may occur as materials accumulate in the lower portion. If gastric contents do regurgitate into the lower portion of the esophagus, the acid within it may irritate the esophagus; HEARTBURN results.

Mesenteries and omenta (Fig. 22.11)

The abdominopelvic (peritoneal) cavity, in which most of the organs of the alimentary tract lie, has a serous membrane lining known as the PERITONEUM. Organs developing in tissues of the wall of the cavity may grow into the cavity, pushing the lining ahead of them. They may then lose their connections with the wall and become suspended by a double-layered fold of the peritoneum. The term PARIETAL PERITONEUM is given to the lining membrane of the cavity; VISCERAL PERITONEUM is the name given to the peritoneum on the organ surface; the double-layered suspending structure is known as a MESENTERY. Specific mesenteries are often named according to the organs they suspend, for example: mesogastrium (stomach), mesocolon (large intestine).

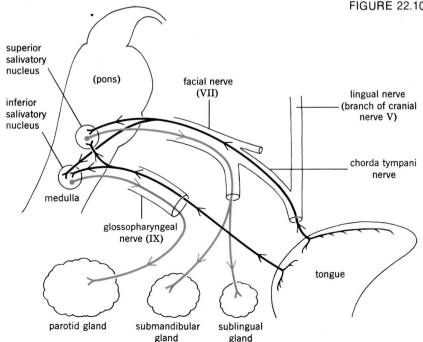

FIGURE 22.10. The control of salivary secretion.

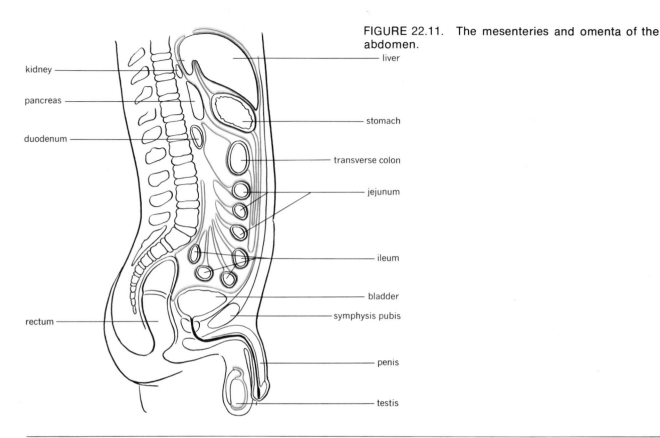

kidney

pancreas

duodenum

rectum

FIGURE 22.11. The mesenteries and omenta of the abdomen.

liver

stomach

transverse colon

jejunum

ileum

bladder

symphysis pubis

penis

testis

OMENTA (sing.: omentum) are mesenteries that lie between two organs, such as the liver and stomach, and stomach and duodenum. They arise when two organs develop one behind another and move into the cavity. They do not suspend organs. Adipose tissue is commonly stored in the omenta.

Some organs, such as pancreas and kidney, remain against the body wall, and are not suspended by mesenteries in the cavity. Such organs lie behind (retro) the peritoneum and are said to be RETROPERITONEAL in position.

CLINICAL CONSIDERATIONS. PERITONITIS is inflammation of the peritoneum due to bacterial invasion of the cavity. It may follow rupture of a hollow organ, or organisms may enter through the female genital tract, bloodstream, lymph stream, or after surgery within the cavity. The continuous nature of the lining and covering membranes allows rapid spread of the inflamma-

tion and may result in involvement of most of the organs in the cavity.

The stomach

STRUCTURE. The stomach is a J-shaped, baglike organ lying under the left side of the diaphragm. Its gross anatomy is illustrated in Figure 22.12. Wall structure follows the typical plan, with three layers of smooth muscle in the externa. The epithelium is simple columnar, with the outer portions of the cells filled with mucus. The mucus normally prevents the destruction of the stomach by the enzymes and acid within it. Some 35 million gastric glands lie in the mucosa, and these produce the gastric juice.

FUNCTION. The stomach RECEIVES FOOD, STORES IT while the initial stages of digestion are occurring, and COMMENCES THE DIGESTION OF PROTEINS.

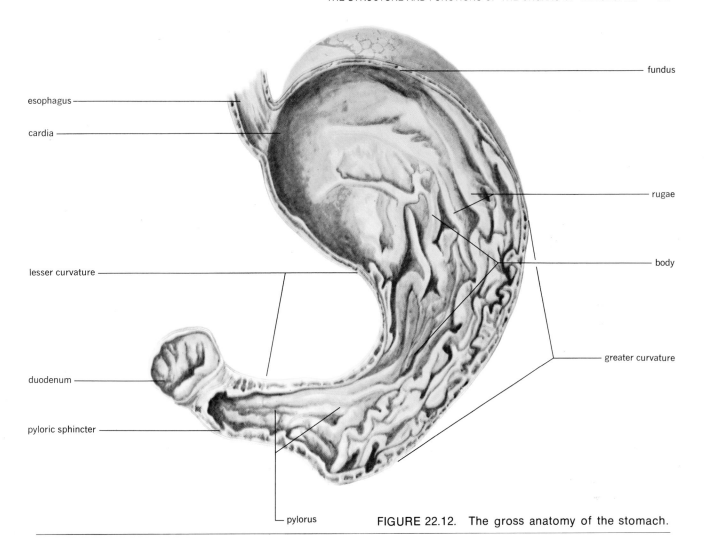

esophagus

cardia

lesser curvature

duodenum

pyloric sphincter

pylorus

fundus

rugae

body

greater curvature

FIGURE 22.12. The gross anatomy of the stomach.

Gastric juice is a watery solution of HYDROCHLORIC ACID (HCl), containing PEPSIN, and small quantities of lipase. Its pH is between 0.9 and 2.0. Steps in the formation of active pepsin are shown in the following equations:

In cells of the gastric glands

$$CO_2 + H_2O + NaCl \longrightarrow HCl + NaHCO_3$$

Secretion of pepsinogen (an inactive precursor of pepsin)

Then

$$HCl + pepsinogen \longrightarrow pepsin$$

Pepsin attacks large protein molecules and breaks them into units known as *proteoses* and *peptones*. These units contain 4–12 amino acids. Gastric lipase is of little value in the human.

CONTROL OF GASTRIC SECRETION. Gastric secretion occurs in three phases.

The CEPHALIC PHASE occurs when food is seen, smelled, or tasted. Impulses pass from brain to stomach over the vagus nerves, and about 50–150 milliliters of juice is produced by this phase.

The GASTRIC PHASE is controlled by two mechanisms. When food enters and stretches the stomach, and when the products of protein digestion act on the pyloric portion of the stomach, a hor-

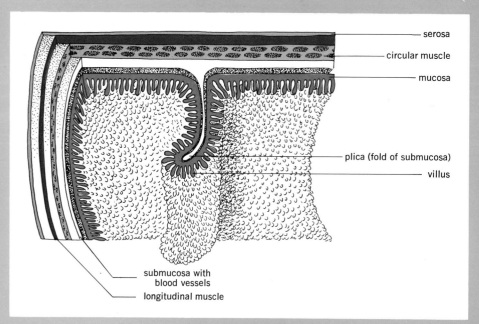

FIGURE 22.13. Villi and plicae of the small intestine. These devices increase surface area.

FIGURE 22.14.
The structure of a villus.

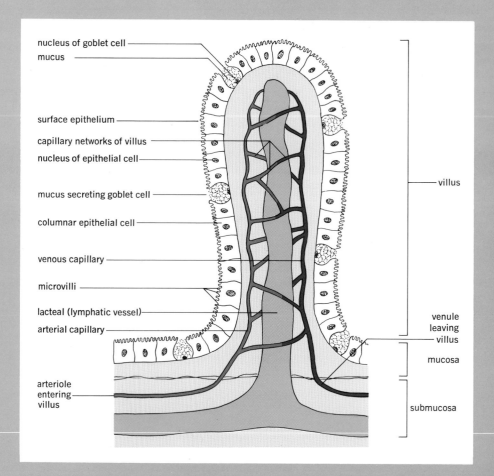

monelike substance called GASTRIN is produced. Gastrin is distributed through the bloodstream to the entire stomach, and causes production of 600–750 milliliters of gastric juice.

The INTESTINAL PHASE adds additional small quantities of juice to the previous phases. It occurs when the digesting food mass (chyme) enters the duodenum of the small intestine. This produces a hormone, unnamed at present, that stimulates gastric secretion.

The intestinal mucosa also produces a hormone called ENTEROGASTRONE that reaches the stomach via the bloodstream. This material slows the muscular activity of the stomach and allows a longer time for gastric digestion of proteins to occur.

CLINICAL CONSIDERATIONS. PEPTIC ULCERS may occur in the stomach if the mucus lining is destroyed or is not secreted normally. The acid and pepsin quite literally digest the stomach wall.

The cause of peptic ulcer of the stomach is not definitely known, but excessive secretion of gastric juice when there is no food in the stomach is an important factor in the production of the ulcer. Such production of acid commonly occurs during stress, and has been related to the emotional pressures generated by a competitive modern society. The "executive ulcer" is a classical example of this type of disorder. The ulceration usually extends only into the mucosa, but occasionally penetrates the muscularis externa and perforates into the abdominal cavity.

GASTRITIS is inflammation and irritation of the stomach lining produced by irritant foods or drink. Avoidance of a food creating the irritation is the only known help. LAVAGE means "to wash," and refers to flushing toxins out of the stomach ("having the stomach pumped").

The small intestine

STRUCTURE. The small intestine extends from the pyloric valve of the stomach some 3.4–4 meters (10–12 feet*) to the large intestine. It is divided into three parts on the basis of microscopic struc-

*Measurement in the living person; postmortem measurements may reach 20–21 feet.

ture: the DUODENUM, the first 25–30 centimeters (10–12 inches); the JEJUNUM, the next 1–1.5 meters (3–4 feet); and the ILEUM, the last 2–2.5 meters (6–7 feet). All parts have a wall structure following the typical plan described earlier. VILLI, and PLICAE (Fig. 22.13) increase surface area of the intestine for production of digestive enzymes and absorption of end products of digestion. The villi contain networks of arteries, capillaries, and veins, and a lymph capillary known as a lacteal (Fig. 22.14).

FUNCTION. The small intestine receives the secretions of the pancreas, liver (bile), and its own glands, and digestion of all three basic foodstuff groups is completed here. Most of the absorption of nutrients occurs from the small intestine. The digestive juices secreted into the small intestine include the following.

Pancreatic juice. Pancreatic juice is an alkaline fluid (pH 7.1–8.2) that stops pepsin action, and creates the proper environment (pH 7–8) for the action of all enzymes operating in the intestine. Enzymes in pancreatic juice act on all three main foodstuff groups and include:

1. *Trypsin*, secreted as trypsinogen, and activated by enterokinase, an intestinal enzyme.
2. *Chymotrypsin*, secreted as chymotrypsinogen and activated by trypsin.
3. *Carboxypeptidase*, secreted as procarboxypeptidase and activated by trypsin.

These three enzymes further the digestion of proteins by converting proteoses and peptones to *dipeptides* units containing two amino acids.

4. *Pancreatic amylase* (amylopsin) converts the dextrins of salivary digestion to *disaccharides*, such as maltose, lactose, and sucrose.
5. *Pancreatic lipase* (steapsin) splits triglycerides into three *fatty acids* and *glycerol*, products that are ready for absorption.

Bile. Bile, secreted by the liver, aids lipase in its action by preventing the coming together of the small fat droplets in the intestine and thus allows more rapid digestion of the fats. This action is called the emulsifying action of bile. The bile salts also combine with the fatty acids and render them more easily absorbed. This is called the hydrotropic action of the bile.

Intestinal juice (*succus entericus*). Intestinal juice is produced by the glands of the intestine itself. It completes the digestion of proteins and carbohydrates. Its enzyme include:

Erepsin. Erepsin is a name for a group of enzymes that split dipeptides into their individual amino acids.

Specific amylases attack a particular disaccharide and reduce it to simple sugars.

   Maltase acts on maltose and releases two glucose units.

   Sucrase acts on sucrose and releases glucose and fructose.

   Lactase acts on lactose and releases glucose and galactose.

Foodstuff digestion has been completed, and the end products, amino acids, simple sugars, glycerol and fatty acids, are ready for absorption. The digestion of the three main foodstuff groups is summarized below.

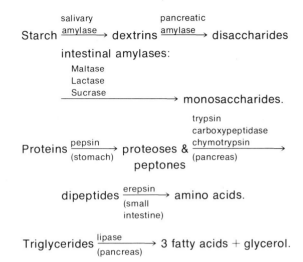

$$\text{Starch} \xrightarrow[\text{amylase}]{\text{salivary}} \text{dextrins} \xrightarrow[\text{amylase}]{\text{pancreatic}} \text{disaccharides}$$

intestinal amylases:

$$\left.\begin{array}{l}\text{Maltase}\\\text{Lactase}\\\text{Sucrase}\end{array}\right\} \longrightarrow \text{monosaccharides.}$$

$$\text{Proteins} \xrightarrow[\text{(stomach)}]{\text{pepsin}} \begin{array}{c}\text{proteoses \&}\\\text{peptones}\end{array} \xrightarrow[\text{(pancreas)}]{\substack{\text{trypsin}\\\text{carboxypeptidase}\\\text{chymotrypsin}}}$$

$$\text{dipeptides} \xrightarrow[\substack{\text{(small}\\\text{intestine)}}]{\text{erepsin}} \text{amino acids.}$$

$$\text{Triglycerides} \xrightarrow[\text{(pancreas)}]{\text{lipase}} \text{3 fatty acids + glycerol.}$$

Control of secretion of pancreas, liver, intestinal wall. Control of secretion of the three organs is carried out primarily by hormonal materials that are produced by the intestine and are carried by the blood to the organ controled. The control mechanisms are summarized in Table 22.5.

TABLE 22.5 Control mechanisms for secretion of pancreas, liver, and intestine			
Organ	Stimulus	Hormone(s) produced	Effect of hormone
Pancreas	HCl in intestine (also water, meat juices, fats, alcohol)	Secretin	Stimulates secretion of watery, buffered fluid
	HCl alone	Pancreozymin*	Stimulates secretion of enzymes
Liver and gall bladder	Fats in intestine	Secretin	Stimulates liver secretion of bile
	as above	Cholecystokinin*	Stimulates contraction of gall bladder and emptying of stored bile
Intestine	Stroking, stretching, flooding of intestine	Enterocrinin	Stimulates secretion of intestinal glands
*Pancreozymin and cholecystokinin are apparently the same chemical substance.			

CLINICAL CONSIDERATIONS. DUODENAL ULCER may result from the action of stomach acid and pepsin on the wall of the duodenum. A variety of irritants may cause inflammation of various parts of the intestine (duodenitis, jejunitis, ileitis). Severe inflammation may require removal of the affected portion and joining of the cut ends.

The large intestine and associated structures (Fig. 22.15)

STRUCTURE. The LARGE INTESTINE is about 1½ meters (5 feet) long, and consists of the CECUM, APPENDIX, COLON, RECTUM, and ANAL CANAL. It is larger in diameter than the small intestine and its greater length is easily recognized by its HAUSTRA or sacculations. The APPENDIX arises from the blind pouch of the cecum. The COLON consists of *ascending, transverse, descending,* and *sigmoid* portions, and has two *colic flexures* (see Fig. 22.3). The rectum (Fig. 22.16) lacks haustra, and has longitudinal RECTAL COLUMNS containing hemorrhoidal blood vessels. The anus, or opening to the exterior, is surrounded by an INTERNAL SPHINCTER of smooth muscle, and an EXTERNAL SPHINCTER of voluntarily controlled skeletal muscle.

Microscopically, the large intestine and its parts are distinguished by lack of villi, and by the great numbers of goblet (mucus secreting) cells in the epithelium. The mucus lubricates the fecal material as water is absorbed from it. Additionally, the outer longitudinal layer of the muscularis externa

FIGURE 22.15. The ileocecal region.

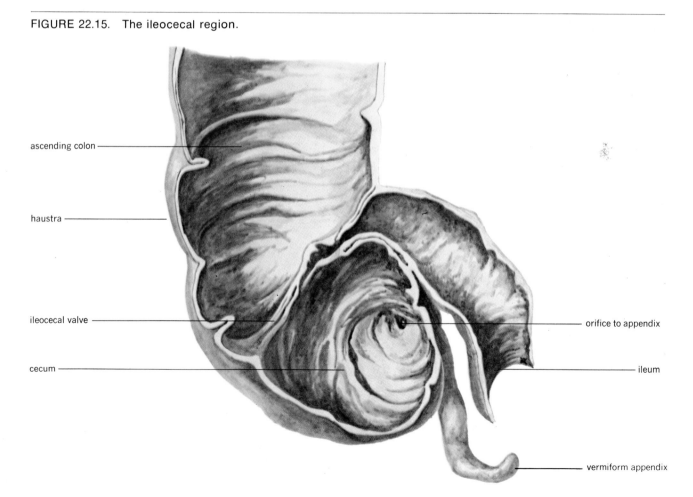

ascending colon

haustra

ileocecal valve

cecum

orifice to appendix

ileum

vermiform appendix

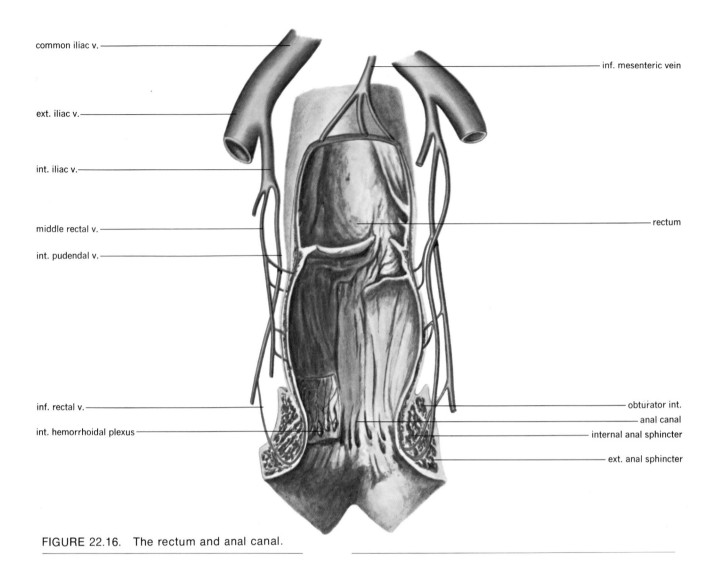

common iliac v.

ext. iliac v.

int. iliac v.

middle rectal v.

int. pudendal v.

inf. rectal v.

int. hemorrhoidal plexus

inf. mesenteric vein

rectum

obturator int.

anal canal

internal anal sphincter

ext. anal sphincter

FIGURE 22.16. The rectum and anal canal.

is in the form of three bands known as the TAENIA COLI.

FUNCTION. The large intestine has no digestive function. It ABSORBS some 400 milliliters of water and much inorganic salt per day. It contains a large flora of microorganisms that produce vitamins K and B, and amino acids. These products are absorbed by the host. FECES ARE FORMED in the colon, and consist of 60 percent solid matter (bacteria, food residues, digestive secretions), and 40 percent water.

CLINICAL CONSIDERATIONS. COLITIS refers to inflammation of the colon. APPENDICITIS occurs when infectious agents enter an appendix that does not or cannot empty itself. The organ may require surgical removal before it ruptures and causes peritonitis. DIVERTICULOSIS describes a sac-like protrusion of the colon wall where arteries enter the organ and weaken its wall. They may appear on X-ray photographs, but usually do not cause symptoms. If the outpocketings accumulate fecal material and become infected, DIVERTICULITIS occurs. Diverticulitis produces pain in the lower left part of the abdomen, a "left-sided appendicitis."

HEMORRHOIDS are enlarged hemorrhoidal veins. They may bleed profusely and often require surgery.

The pancreas (Fig. 22.17)

STRUCTURE. The pancreas is a carrot-shaped gland located along the greater curvature of the stomach. It has a HEAD, BODY, and tapering TAIL, is about 20 centimeters (8 inches) long, and weighs from 65–160 grams. It is a double gland (Fig. 22.18) having an EXOCRINE PORTION producing digestive enzymes, and an ENDOCRINE PORTION (islets of Langerhans) producing hormones.

FUNCTION. The digestive function of the pancreas has already been considered. The endocrine function includes the production of INSULIN, a hormone increasing the formation of glycogen and increasing cellular uptake of glucose, and of GLUCAGON, a hormone causing glycogen breakdown to glucose and rise of blood sugar levels.

CLINICAL CONSIDERATIONS. CYSTIC FIBROSIS is a genetically transmitted abnormality that affects many exocrine glands, including the pancreas. Fibrous tissue invades the glands, reducing secretory capacity. Pancreatic insufficiency and failure to properly digest foods may result, with failure to thrive. Disorders of the endocrine portion will be discussed in Chapter 26.

The liver (Figs. 22.19 and 22.20)

STRUCTURE. The liver is the largest gland in the body, weighing about 1500 grams (about 3 pounds) in the adult. It has four major LOBES, and

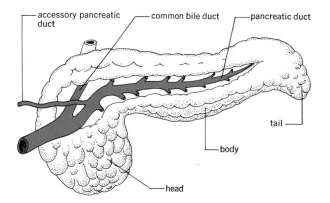

FIGURE 22.17.   The gross anatomy of the pancreas.

FIGURE 22.18.   Photomicrograph of the exocrine and endocrine portions of the pancreas.

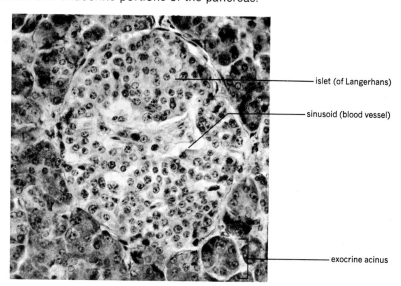

FIGURE 22.19. The gross anatomy of the liver.

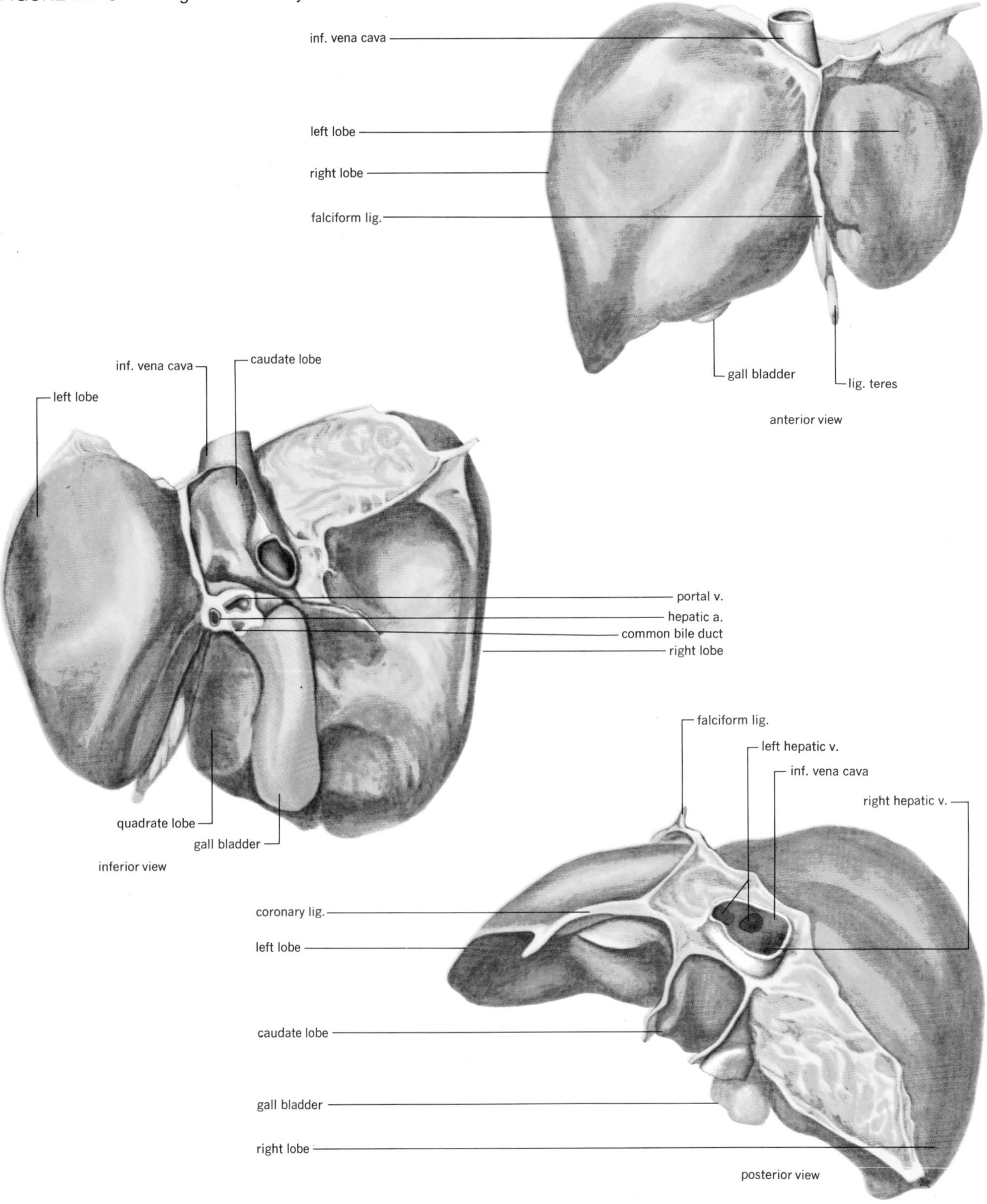

inf. vena cava

left lobe

right lobe

falciform lig.

gall bladder

lig. teres

anterior view

inf. vena cava

caudate lobe

left lobe

portal v.

hepatic a.

common bile duct

right lobe

quadrate lobe

gall bladder

inferior view

falciform lig.

left hepatic v.

inf. vena cava

right hepatic v.

coronary lig.

left lobe

caudate lobe

gall bladder

right lobe

posterior view

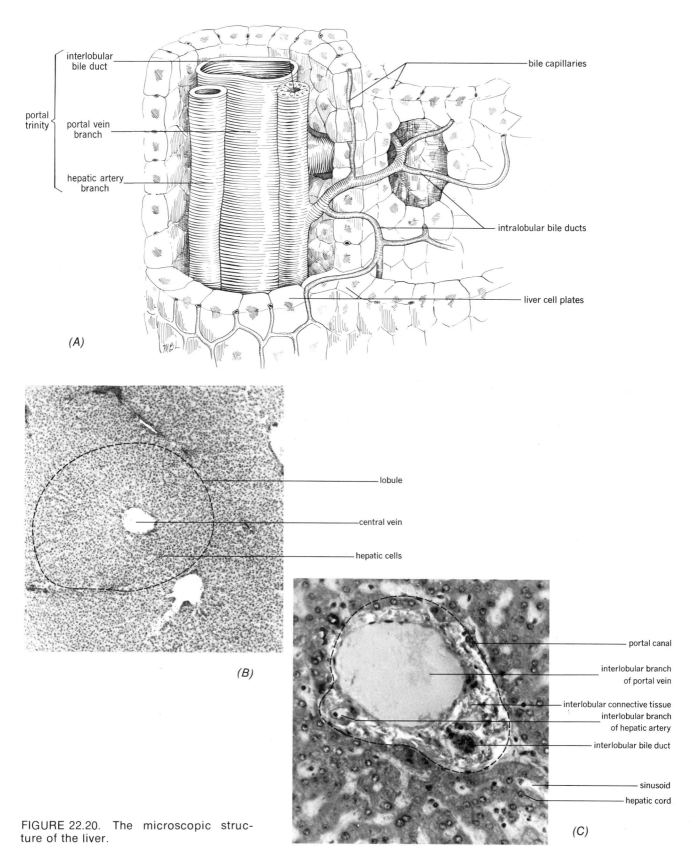

interlobular
bile duct

portal
trinity

portal vein
branch

hepatic artery
branch

bile capillaries

intralobular bile ducts

liver cell plates

(A)

lobule

central vein

hepatic cells

(B)

portal canal

interlobular branch
of portal vein

interlobular connective tissue

interlobular branch
of hepatic artery

interlobular bile duct

sinusoid

hepatic cord

(C)

FIGURE 22.20. The microscopic structure of the liver.

is fixed in position by several ligaments. The unit of structure is the LOBULE, which consists of plates or cords of HEPATIC CELLS radiating from a CENTRAL VEIN, with SINUSOIDS lying between the plates of cells. The blood supply to the liver is double; the HEPATIC ARTERIES bring O_2 rich blood to the sinusoids, and the PORTAL VEIN brings nutrient-laden blood from the intestines to the sinusoids. Outgoing blood empties via the HEPATIC VEINS into the inferior vena cava.

A system of bile ducts begins in the lobules and ultimately carries bile to the gall bladder and small intestine.

FUNCTIONS. The liver produces an exocrine secretion (bile) whose functions have already been considered. Bile is stored in the gall bladder until required for digestion. Other functions include:

1. *Synthetic reactions.*

a. Synthesis of certain amino acids, plasma proteins, prothrombin, antibodies, urea, creatine, cholesterol.

b. Glycogen and glucose (gluconeogenesis).

c. Phospholipids (for cell membranes, etc.).

d. Fatty acids.

e. Ketone bodies.

2. *Metabolic reactions.*

a. Breakdown of glucose, glycogen, fatty acids, glycerol, amino acids, metabolism of hormones (insulin, aldosterone, testosterone, estrogens, and thyroid hormone).

b. Interconversion of certain amino acids to simple sugars and vice versa.

c. ATP formation.

3. *Embryonic formation of blood cells.*

4. *Destruction of aged red cells.*

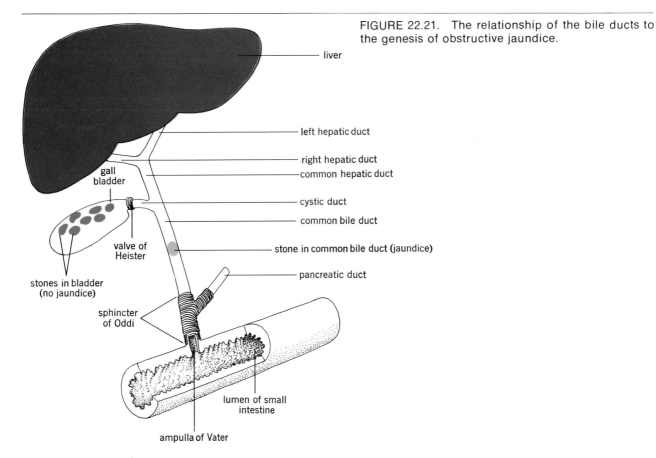

FIGURE 22.21. The relationship of the bile ducts to the genesis of obstructive jaundice.

liver

left hepatic duct

right hepatic duct

common hepatic duct

gall bladder

cystic duct

common bile duct

valve of Heister

stone in common bile duct (jaundice)

pancreatic duct

stones in bladder (no jaundice)

sphincter of Oddi

lumen of small intestine

ampulla of Vater

5. *Storage.*
   a. Glycogen.
   b. Amino acids.
   c. Fats.
   d. Vitamins (A, B complex, D).
   e. Iron and copper.
6. *Detoxification* (conversion of toxic substances to harmless compounds).

Specific metabolic functions of the liver are considered in Chapter 23.

CLINICAL CONSIDERATIONS. Because of the many and essential functions carried out by the liver, it can be considered to be one of the vital body organs. Interestingly, there is such a great reserve of liver tissue that a person can get along with only about one fifth of his tissue functional. CIRRHOSIS of the liver involves replacement of cells by fibrous tissue when cells are destroyed by infection, parasites, alcohol, or other causes. HEPATITIS refers to inflammation of the liver. It is most commonly viral in origin, with the virus transmitted from one person to another by transfusion, needles, or other instruments, or by an oral route. The gall bladder not only stores bile, but concentrates it by absorption of water. If the solution in the organ becomes saturated, precipitation may occur with formation of GALLSTONES. The stones may block the bile ducts (Fig. 22.21) and cause JAUNDICE.

Motility in the tract

Movements in the digestive system, caused by muscular tissue, are essential to mechanical subdivision of foods and to transfer of foods along the tube. The main types of movement are briefly described below, and are summarized in Table 22.6.

CHEWING is started voluntarily by opening the mouth to receive food and closing it to start chewing. Rhythmical continuation of the activity is controlled by the brain.

SWALLOWING starts with collection of a mass of food (bolus) on the tongue, and movement of the bolus into the throat by the tongue. A positive pressure must be created by sealing the nasal and oral cavities to swallow. (A cleft palate prevents creation of a positive pressure.)

PERISTALSIS is the most characteristic type of movement occurring in the tract. It is a wavelike constriction of the gut that pushes ahead of it a mass of chyme. It is used mainly for transport of chyme along the length of the tube, and normally occurs only in one direction, from mouth to anus.

TONIC CONTRACTIONS and SEGMENTING CONTRACTIONS are local contractions of gut muscle that serve to mix and churn the contents of the organ.

The stomach shows a RECEPTIVE RELAXATION as food enters it, and thus it expands to contain a meal. The stomach also exhibits "HUNGER CONTRACTIONS," very strong peristaltic waves, which usually signal a need for food intake.

The small intestine also shows PENDULAR MOVEMENTS, a to-and-fro type of movement up and down short segments of the tube. They mix and churn intestinal contents. MOVEMENTS OF THE VILLI aid in absorption of substances by keeping "fresh" material next to the villus.

In the colon, DEFEACTION is a strong peristaltic type movement caused by stretching of the rectum by fecal material. Material is moved into the rectum by MASS MOVEMENT. Expulsion of material is aided by "straining," or using abdominal muscles to "push."

Control of motility

A variety of nervous, chemical, and physical factors control the strength and duration of motility in various parts of the tract.

In the esophagus, movement is controlled primarily by the VAGUS NERVE, which supplies the esophageal muscle.

Motility of the stomach depends on: the amount

TABLE 22.6 A Summary of Motility in the Alimentary Tract

Area	Type of motility	Frequency	Control mechanism	Result
Mouth	Chewing	Variable	Initiated voluntarily, proceeds reflexly	Subdivision, mixing with saliva
Pharynx	Swallowing	Maximum 20 per min	Initiated voluntarily, reflexly controlled by swallowing center	Clears mouth of food
Esophagus	Peristalsis	Depends on frequency of swallowing	Initiated by swallowing	Transport through esophagus
Stomach	Receptive relaxation	Matches frequency of swallowing	Unknown	Allows filling of stomach
	Tonic contraction	15–20 per min	Inherent by plexuses	Mix and churn
	Peristalsis	1–2 per min	Inherent	Evacuation of stomach
	"Hunger contractions"	3 per min	Low blood sugar level	"Feeding"
Small intestine	Peristalsis	17–18 per min	Inherent	Transfer through intestine
	Segmenting	12–16 per min	Inherent	Mixing
	Pendular	Variable	Inherent	Mixing
	Villus movements shortening and waving	Variable	Villikinin	Facilitates absorption
Colon	Peristalsis	3–12 per min	Inherent	Transport
	Mass movement	3–4 per day	Stretch	Fills pelvic colon
	Tonic	3–12 per min	Inherent	Mixing
	Segmenting	3–12 per min	Inherent	Mixing
	Defecation	Variable 1 per day-3 per week	Reflex triggered by rectal distension	Evacuation of rectum

of fat it contains (more fat = slower activity); products of protein digestion (a nervous reflex called the ENTEROGASTRIC REFLEX is triggered by proteoses and peptones); osmotic pressure of stomach contents (hyper- and hypotonic solutions decrease motility).

In the intestine, the MYENTERIC REFLEX is a contraction caused by stretching the gut.

In the colon, the GASTROCOLIC REFLEX stimulates colon evacuation when the stomach is stretched.

The MUCOSAL REFLEX results in an increase of secretion and stimulation of muscular activity anywhere in the tract when it is irritated. The reactions tend to dilute and remove the irritating agent.

The PERITONEAL REFLEX causes an inhibition of motility in the tract when the peritoneum is traumatized, as after an abdominal operation.

Summary

1. The digestive system takes in food, digests it to absorbable end products, absorbs those products, and eliminates unusable materials.

2. The system begins its development during the third week of embryonic life with the formation of a fore-, mid-, and hindgut. Each part develops certain portions of the gut and its accessory structures (see Table 22.1).

   a. Abnormal development may cause relatively common congenital disorders of the system, including atresia, stenosis, and imperforate anus.

3. The organs of the system are divided into two groups.
   a. The tubelike alimentary tract includes the mouth, pharynx, esophagus, stomach, small and large intestines, appendix, rectum, and anal canal.
   b. The accessory organs include tongue, teeth, salivary glands, liver, pancreas, and gall bladder.

4. The alimentary tract has a typical layering of tissues in its wall that includes four layers.
   a. The mucosa is internal, glandular, and absorbs or protects.
   b. The submucosa is vascular connective tissue, and nourishes.
   c. The muscularis externa is muscular and creates the motility in the tract.
   d. An outer serous or fibrous layer surrounds and protects the tube.

5. The mouth receives food, and shows many lesions helpful in diagnosing disease.

6. The tongue is a muscular organ surfaced with a variety of papillae for roughening and taste. The organ manipulates food in the mouth and articulates speech.

7. The teeth subdivide food, occur in two sets (deciduous and permanent), are of four types (incisors, canines, premolars, and molars) and have a typical structure.

8. There are three pairs of salivary glands which secrete saliva.
   a. Saliva moistens and softens foods, cleanses the mouth, and excretes substances.

b. Salivary amylase (ptyalin) splits starches to dextrins.

c. Control of salivary secretion is by nerves.

9. The esophagus is a muscular tube conducting food to the stomach.

10. The peritoneum lines the abdominopelvic cavity, and covers organs. Mesenteries suspend organs within the cavity; omenta lie between organs and may store fat.

11. The stomach stores food, and secretes gastric juice.

a. Gastric juice contains HCl, and pepsin.

b. Pepsin breaks proteins to proteoses and peptones.

c. Gastric secretion is controlled in three phases, cephalic, gastric, and intestinal. The first is under the control of the vagus nerve, the last two are hormonally controlled by gastrin and an unnamed hormone.

12. The small intestine completes the digestion of foods and absorbs the end products of digestion.

a. It has a duodenum, jejunum, and ileum.

b. Villi and plicae increase surface area for secretion and absorption.

c. Pancreatic juice, emptied into the small intestine, contains three protein digesting enzymes called trypsin, chymotrypsin, and carboxypeptidase, that create dipeptides; amylase to create disaccharides; lipase to break down fats to glycerol and fatty acids.

d. Bile, secreted by the liver and released from the gall bladder, emulsifies fats and renders fatty acids water soluble.

e. Intestinal juice, secreted by the intestine wall, contains amylases which liberate simple sugars from double sugars; and erepsin for liberating amino acids from dipeptides.

13. Control of pancreatic, liver, and intestinal secretion is by hormones.

a. Secretin and pancreozymin control pancreatic secretion.

b. Secretin stimulates liver production of bile, and cholecystokinin releases bile from the gall bladder.

c. Enterocrinin stimulates secretion by intestinal glands.

14. The large intestine has haustra, taenia coli, lacks villi, and consists of appendix, cecum, colon, rectum, and anal canal. The colon consists of ascending, transverse, descending, and sigmoid portions. It absorbs water, salts, products of bacterial activity, and forms feces.

15. The pancreas is carrot shaped, and produces digestive enzymes and true hormones.

16. The liver is the largest gland of the body, has lobes and lobules, and carries on a variety of functions, including synthesis, metabolism, interconversion of substances, formation and destruction of blood cells, and storage of nutrients.

17. Movements (motility) of the tract subdivide foods, mix and churn food with enzymes, and move materials through and out of the tract (see Table 22.6 for types and details).

    a. Control of motility is by nerves, chemicals, and physical factors (osmotic pressure, stretch).

Questions

1. Now that you have studied the entire digestive system, explain the concept that the inside of the digestive organs is a part of the outside environment.

2. What is indicated by the phrase that in its lumen each digestive organ had its "own peculiar ecology"?

3. Name the types of teeth in the human mouth and describe the anatomy of a tooth.

4. Name the tunics in the walls of a typical hollow, tubular, organ of the digestive system, and their tissue components.

5. List some of the ways by which the absorptive surface of the small intestine is increased.

6. What are some of the movements involved in the small intestine during digestion and absorption? How is each controlled?

7. What is the function of bile in the digestive system?

8. Describe the functions of the small intestine.

9. Describe the functions of the large intestine.

10. Name and locate the lobes of the liver.

11. Describe a liver lobule and its relationship to bile and blood channels.

12. Explain the statement: "The liver has a double blood supply."

13. The pancreas is a double gland with a twofold function. Explain this statement.

14. Trace, in proper order, the enzymes acting upon starch; list the products formed at each step. Do the same for protein.

Readings

Davenport, H. W. *Physiology of the Digestive Tract.* 3rd ed. Yearbook Medical Publishers. Chicago, 1971.

Davenport, H. W. "Why the Stomach Does Not Digest Itself." *Sci. Amer. 226*:86 (Jan) 1972.

Rabinowitz, Melba. "Why Didn't Anyone Tell Me About Bottle Mouth Cavities?" *Children Today,* vol 3, no 2, pp 18–20. Children's Bureau, Office of Human Development. Washington, D.C., Mar–Apr 1974.

Science News. "Smoking and Gum Disease." 105, no 15, p 240. April 13, 1974.

chapter 23
Absorption of Foodstuffs, Metabolism, and Nutrition

chapter 23

Absorption of foodstuffs, metabolism, and nutrition

As is indicated in the previous chapter, the process of digestion has as its major aim the production of absorbable end products from more complex foodstuffs. Until these end products pass through the wall of the alimentary tract, they are not truly in the body, and are unavailable for cellular use. ABSORPTION involves the TRANSFER OF MATERIALS through the mucosa of the alimentary tract into blood and lymph vessels. Transfer may be active, as by active transport and pinocytosis, or passive, as by diffusion and osmosis.

METABOLISM is defined as the sum total of the chemical reactions occurring in the body. It may be recalled that metabolism may be constructive or result in synthesis of new substances (anabolism), or may be energy releasing and result in the destruction of a compound (catabolism). The absorbed end products of digestion form the basis for anabolic and catabolic reactions. INTERMEDIARY METABOLISM refers to the specific steps a cell employs in metabolizing a given compound.

NUTRITION emphasizes the role of foodstuffs in meeting the energy and specific requirements of the body for continuation of cellular activity. It also studies the mechanisms that determine the rate at which foodstuffs are metabolized, and the use to which the products are put.

Absorption of foodstuffs

From the mouth

There is no absorption of foods from the mouth, because the molecules are too large to pass through the epithelium and the stratified squamous epithelium is rather impervious to most molecules. Certain drugs, in the form of lozenges, may be absorbed through the mouth lining particularly when held under the tongue.

From the stomach

Absorption from the stomach is limited to small molecules and inorganic salts. Water, alcohol, and compounds of Na and K apparently can pass through the stomach wall. Absorption of salts seems to occur by active transport, that of water and alcohol by physical processes.

From the small intestine

The bulk of nutrient absorption occurs from the small intestine. In this organ, foodstuffs have been reduced to a size where they can pass through the epithelium, the epithelial cells possess transport systems to move materials, the villi and plicae (Fig. 23.1) provide a large surface through which substances may pass, and the intestine is relatively long, providing time for absorption to occur.

ABSORPTION OF SIMPLE SUGARS. Absorption of glucose and galactose occurs by active transport, and places the sugars into the blood vessels of the villi for transport to the liver. Other sugars, including pentoses (e.g., ribose) are absorbed by carrier-facilitated diffusion or diffusion alone. Glucose is absorbed at a maximum rate of about 120 gms/hour; galactose passes about 10 percent faster than glucose.

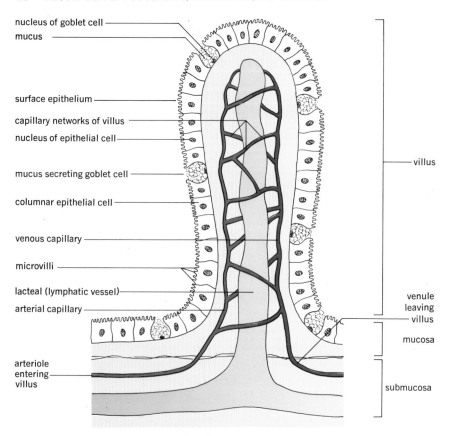

FIGURE 23.1.   The structure of a villus.

ABSORPTION OF AMINO ACIDS. Most amino acids are actively transported from the intestine to the mucosal cells, and then diffuse into the blood vessels of the villi. Three separate transport systems appear to exist: one for amino acids that contain one carboxyl (—COOH) and one amine (—NH$_2$) group; one for basic amino acids containing two amine groups; one for acids containing ring structures. In infants, permeability of the intestine appears to be greater than in later life, and some large protein molecules such as antibodies in the mother's milk and protein allergens pass by diffusion through the wall. Pinocytosis appears to account for the absorption of these large molecules in children and adults.

ABSORPTION OF LIPIDS. Lipids are absorbed by diffusion into the epithelial cells of the mucosa. Fatty acids containing less than 10 or 12 carbons enter the blood vessels of the villus and are carried to the liver as free fatty acids (FFA). Those containing more than 10–12 carbons are recombined with glycerol in the mucosal cells, and are coated with protein, cholesterol, or phospholipid to form tiny globules called *chylomicrons*. These chylomicrons then leave the cells and enter the lymphatics of the villi (lacteal). Most of the absorption of fatty acids and glycerol occurs in the upper portion of the intestine; that of cholesterol occurs mainly in the lower intestine.

ABSORPTION OF VITAMINS, WATER AND MINERALS. Pinocytosis appears to account for the absorption of vitamins from the intestine. Fat soluble vitamins (A, D, E, K) follow the large lipids into the lacteal, while water soluble vitamins (C, B complex) go into blood vessels. Many vitamins then form parts of enzymes necessary for metabolic processes.

Water moves by osmosis secondarily to removal of solutes from the intestine.

Sodium, potassium, and calcium are transported actively from the intestine. Since they are cations, anions (e.g., Cl^-, HCO_3^-) follow by electrostatic attraction (+ draws −).

From the large intestine

The large intestine absorbs salts, mostly sodium (60 meq/day), and water, some 400 milliliters per day, follows osmotically. Absorption of some of the products of bacterial activity in the colon such as vitamins B and K, and amino acids occurs from this organ by the same processes that absorb them elsewhere.

Clinical considerations

Defects in absorption of foods result mainly from genetic abnormalities that cause malformation of the villi of the intestine (celiac disease), or from failure to synthesize digestive enzymes and transport systems because of disease processes or genetic abnormalities. In any event, two primary things occur: foods are not brought to an absorbable state; surface area of the gut is reduced so that absorption, even if normal, is very slow, and malnutrition results. Interestingly, in gross obesity, where an individual is fat because of absorption of vast amounts of consumed food, removal of a large part of the intestine may be carried out as a device to reduce absorption when the subject cannot or will not control his food intake.

Intermediary metabolism of foods (see also Chapter 4)

The main reason that foods are consumed is to degrade them to supply energy for the synthesis of adenosine triphosphate (ATP). ATP may be characterized as the universal biological energy carrier that forms the immediate source of energy for cellular activity such as contraction, secretion, and active transport. Thus intermediary metabolism of foodstuffs becomes a series of catabolic reactions that provide energy for the anabolic reactions leading to formation of ATP.

Carbohydrates

GLUCOSE forms the primary source of energy for synthesis of ATP. Depending on the economy of the body at any given moment, the fate of glucose entering the body follows one or the other of two main pathways.

It may be *catabolized* for energy and the synthesis of ATP.

It may be *converted* into a storage form from which it may be recovered as needed.

The basic schemes involved in each choice are summarized below. The catabolic scheme shows the fate of a single glucose molecule.

The presence of glucose within the cells in quantities greater than needed to run cellular machinery results in storage of the glucose as the polysaccharide glycogen in the process called GLYCOGENESIS. It occurs mostly in liver and muscle cells. *Insulin,* a hormone from the pancreas, aids entry of glucose into most body cells and speeds its conversion into glycogen. When blood levels of glucose are diminished, glucose may be recovered from glycogen in the process known as GLYCOGENOLYSIS. This process is speeded by a pancreatic hormone *glucagon,* and by *epinephrine* from the adrenal medulla.

Initial degradation of glucose is handled by GLYCOLYSIS, an anaerobic metabolic scheme that produces 2 three-carbon units called pyruvic acid from the six-carbon glucose molecule. The scheme also releases nearly 60,000 kilocalories* of energy and results in the production of 2 new ATP molecules for cellular energy stores. Further combustion of pyruvic acid requires the presence of oxygen. The pyruvic acid has a CO_2 removed from it to form acetic acid, and then enters the KREBS CYCLE that produces 2 more CO_2, and 1

*A calorie is the amount of heat required to raise the temperature of 1 gram of water 1°C.
A kilocalorie is 1000 times larger than a calorie, and raises the temperature of 1 kilogram of water 1°C.

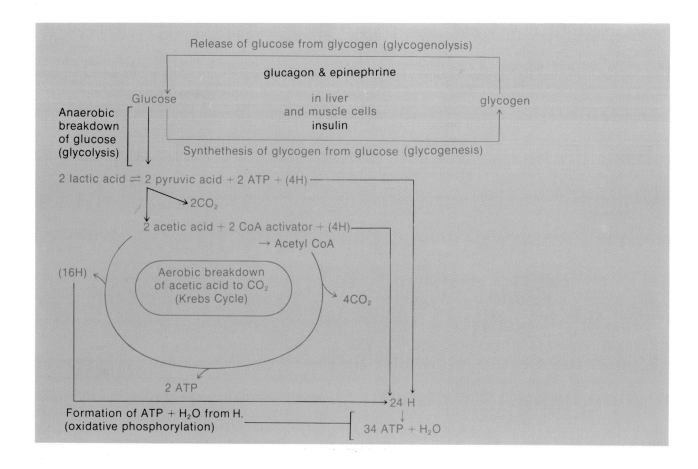

new ATP for each acetic acid molecule combusted. Per mol of acetic acid combusted, nearly 630,000 kilocalories of heat are also produced. Hydrogens are released by glycolysis and the Krebs cycle. They are collected by *hydrogen acceptors* and are carried to OXIDATIVE PHOSPHORYLATION. This scheme converts the hydrogen into water in the presence of oxygen, and releases enough energy to produce 34 new ATP molecules for body use. Therefore, 1 mol of glucose may be degraded to produce a total of 38 new ATP molecules for body use.

We emphasize several things about these interrelationships.

1. Glycogen synthesis will occur only if there is more glucose available than is required to run the body machinery.

2. Initial stages of glycogen breakdown do not require oxygen (anaerobic); later stages do (aerobic). If oxygen is not available to assure aerobic metabolism, lactic acid is formed as a means of temporary storage to conserve the energy in pyruvic acid until O_2 *is* available.

3. The *complete* combustion of glucose produces CO_2 and H_2O, with ATP formation. The majority of the ATP formation comes from oxidative phosphorylation that converts H to ATP.

Lipids

FATTY ACIDS AND GLYCEROL. Fatty acids are metabolized mainly in liver and muscle by removal of two carbon units in the process known as BETA-OXIDATION. The two carbon units are acetic acid which, when combined with Coenzyme A, enter the Krebs cycle for further oxidation. The Krebs cycle is, therefore, a sort of final common path for biological oxidation. Glycerol is metabolized by conversion to a compound found in glycolysis. It may then follow glycolysis and the Krebs cycle as does glucose. If body metabolism does not require breakdown of these lipids, they may be recombined into triglycerides and stored in the adipose cells of the body. The adipose tissue thus formed is like a savings account; it may be withdrawn when needed, but can accumulate (with interest!) if food intake exceeds utilization.

Unrestricted or excessive catabolism of fatty acids may produce more Acetyl CoA than can be combusted in the Krebs cycle. Excess AcCoA forms *acetone* and a compound called *beta-hydroxybutyric acid*. These are the KETONE BODIES, and their accumulation causes KETOSIS. Some of the interrelationships of fatty acid and glycerol metabolism are shown below.

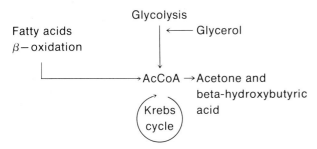

CHOLESTEROL. Cholesterol is a complex lipid containing the "sterol nucleus"

It is the basic material for the synthesis of bile acids, many steroid hormones, and has been widely discussed as to its involvement in the production of atherosclerosis in the blood vessels. Cholesterol is synthesized from AcCoA, mainly by the liver. Restriction of dietary intake of cholesterol has been suggested to reduce the development of sclerosis of blood vessels, but no definitive evidence is available to prove the suggestion.

ESSENTIAL FATTY ACIDS. Several fatty acids that contain unsaturated carbon bonds ($-C=C-$) are necessary for normal lipid metabolism and to maintain growth, and are not synthesized in the body. Among these "essential fatty acids" are linoleic, linolenic, and arachidonic acids.

Proteins

Proteins, by their content of amino acids, provide the basis for synthesis of enzymes, many hormones, and the structural material(e.g., collagen, elastin) of the body. In general, proteins and amino acids are not oxidized for their energy content to any great degree, unless glucose and fatty acids have virtually disappeared from the body.

The amino acids composing proteins are of two general types: ESSENTIAL AMINO ACIDS* must be taken in the diet either because they are not synthesized in the body, or are synthesized in amounts too small to be significant; NONESSENTIAL AMINO ACIDS† may be synthesized by the liver usually from nonprotein precursors.

Essential to the formation of amino acids is the process of TRANSAMINATION, by which amine ($-NH_2$) groups are transferred from an amine donor to an organic acid (a keto acid) to form an amino acid. Conversely, DEAMINATION removes amine groups from amino acids and creates compounds (keto acids) that may be utilized in the Krebs cycle or glycolysis. Ammonia may be one product of deamination. It is very toxic and is converted in the liver to urea by the ornithine cycle, and is excreted in that form. Some of these interrelationships are shown below.

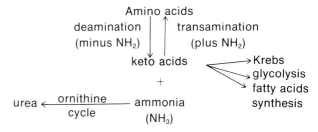

Interconversion of compounds

Synthesis of nonessential amino acids is an example of "creating" one substance from another.

---

*Threonine, Methionine, Valine, Leucine, Isoleucine, Lysine, Arginine, Phenylalanine, Tryptophane, and Histidine.
†Glycine, Alanine, Serine, Cysteine, Aspartic Acid, Glutamic Acid, Hydroxylysine, Cystine, Tyrosine, Diiodotyrosine Thyroxine, Proline, and Hydroxyproline.

In general, it may be stated that lipids, carbohydrates, and amino acids may be converted into one another *if* a compound common to the metabolism of two or more materials is present and, *if* the metabolic cycles involved are reversible. Most cycles are reversible, and "crossover compounds" do exist. The body cells (again mainly the liver) therefore can assure themselves of their required building blocks even though the diet may not contain enough to supply their needs. Some of the interrelationships between the basic foodstuffs are shown below, with the cycles involved named and crossover compounds indicated in italic type. This scheme of interrelationships is sometimes called the "metabolic mill." These schemes explain how one may become fat by eating starches, or how the nonessential amino acids may be produced from carbohydrates.

Control of metabolism

The term metabolic rate is used to describe how fast the body is catabolizing foodstuffs.

The term basal metabolic rate (BMR) describes how many calories the body liberates when carrying on only those activities essential to life, with no extra activity involved. BMR is influenced by several factors, including:

Size, or surface area. The skin serves as a surface through which heat is lost. In general, the larger the skin surface area, the higher the BMR, since more heat is lost through the greater surface and must be replaced by foodstuff catabolism.

Sex. Males oxidize foods about 7 percent faster than females, presumably due to their great amount of muscular tissue that calls for glucose to fuel its activities.

Age. In general, BMR declines as one ages because of loss of muscular tissue and general overall slowdown of activity.

Hormones. Thyroid hormone appears to "set the thermostat" of combustion in the body. Excessive hormone raises the BMR greatly.

Drugs. Benzedrine is a drug raising BMR by stimulating CNS activity and increasing alertness and feelings of exhilaration.

Other factors. Pregnancy raises the BMR of the

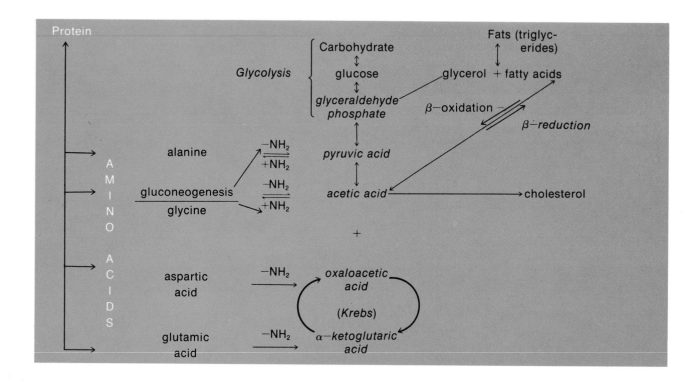

pregnant female; emotions, such as anger, also raises BMR.

We thus conclude that genetic, chemical, and environmental factors combine to govern the "intensity with which the fires of life burn."

The body temperature

Combustion of foodstuffs by glycolysis, the Krebs cycle, oxidative phosphorylation, and B-oxidation, releases energy. Only about 30 percent of the energy released from foods is channeled into the synthesis of ATP. The rest is released as heat that primarily enables the human to maintain body temperature with the narrow limits necessary for continuation of *normal* cellular function. These limits are usually given as $98.6 \pm 1.8°F$, or $37 \pm 1°C$, when the temperature is taken orally. Rectal temperatures are $0.5-1°F$ higher, and internal organ temperatures (e.g., liver) may be as high as $105°F$. Since heat is continually being produced by metabolic activity, there must be heat loss to maintain a near constant body temperature.

Heat production

Basal metabolism produces, in the average adult, about 1700 calories of heat per day. Muscular activity, ingestion of foods, emotions, and other factors raise this output to higher values. Ingestion of foods is associated with secretion of materials, digestion, and absorption. These reactions are energy releasing and raise metabolic rate. This effect on metabolic rate is called the *specific dynamic action* (SDA) of foods. Moderate work, may cause an output of 3000 calories; heavy work, 7000 calories. Involuntary muscular activity, such as shivering, occurs when body temperature is lowered.

Heat loss

Heat loss occurs mainly through the skin. Small amounts of heat are lost by breathing warm air out through the lungs and by loss in urine and feces.

RADIATION is transfer of heat from a warmer to a cooler medium. It is effective only when there is a considerable difference (about $10°F$) between the body and its environment. As environmental temperature approaches that of the body, radiation diminishes.

CONDUCTION is transfer of heat from a warmer to a cooler area when they are in contact. Thus, in a swimming pool, heat is transferred to the water from the body.

CONVECTION is heat transfer to a moving medium, such as moving air.

VAPORIZATION (evaporation) of water from the skin surface carries away heat. This method assumes increasing importance as a cooling device when environmental temperature approaches and exceeds that of the body.

The small "scale" shown below summarizes the methods of heat production and loss.

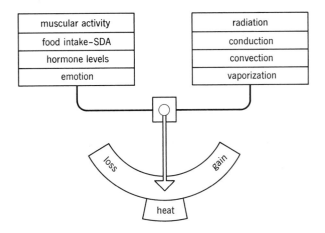

Control of body temperature

The hypothalamus of the brain controls body temperature. It contains heat gain and heat loss centers. Impulses reach these centers over nerves from the heat and cold receptors of the body, and the centers also monitor the temperature of the blood reaching them. Efferent connections of the centers are made to skeletal muscles, sweat glands, and skin blood vessels. If body temperature needs to be raised, muscles tense and may shiver, (increase production of heat), and skin blood vessels constrict (heat conservation). If temperature must be lowered, muscles relax, skin vessels dilate, and sweating occurs. These processes decrease heat production and increase heat loss.

Alterations in body temperature

FEVER is a term used to describe a body temperature that is above the usual limits of normal. It may result from damage to the brain, from chemicals that affect the hypothalamic thermostat, from failure of the normal cooling mechanisms of the body (heat stroke), or from dehydration. In many diseases caused by bacteria or viruses, the toxic products produced by the microorganism, or the products of cell destruction, "reset" the thermostat to a higher value. Chemicals that have this effect are called PYROGENS (*pyro* = fire, *gen* = to produce). Brain tumors may affect the thermo-

stat. Dehydration results in a smaller volume of body water to get rid of metabolic heat, and its temperature rises. Elevated body temperature stimulates chemical reactions, more heat is produced, and a "vicious circle" is established. Heat stroke most commonly occurs when heavy work is performed in a hot humid environment. To prevent this condition, one must cool off by whatever means are available. The upper limits of temperature tolerance appear to be about 43°C (109.4°F) rectally. Death is common from heat stroke at this temperature. A 41°C (106°F) may cause convulsions and brain damage.

HYPOTHERMIA (low body temperature) is often created artificially during surgical operations (e.g., heart surgery). The combination of an anesthetic and cooling of the body can reduce core temperature to 70–75°F without permanent ill effects. Such low core temperatures result in less demand for nutrients by the brain, and promote the ability of the patient to tolerate the surgery. Core temperatures cannot be reduced below about 70°F without running the danger of damaging the hypothalamic centers.

Nutrition

To supply basic caloric and nutritional requirements per day, one should select a diet that contains all three basic foodstuffs. Recommendations suggest that 10–15 percent of the daily adult caloric intake be in the form of protein, 55 to 70 percent as carbohydrate, and 20 to 30 percent as fat. These percentages are roughly equal to 33 to 42 grams of protein, 250 to 500 grams of CH_2O, and 66 to 83 grams of fat. This type of intake may be achieved by insuring that the diet includes the foods listed below. Such a diet will also insure adequate intake of essential minerals and vitamins.

Milk and milk products. Three or more glasses per day for children; four or more for teen-agers; two or more for adults. Cheese, ice cream, yogurt, cottage cheese, and other milk products may substitute for part of the milk requirement.

Meat group. Two or more servings daily of meat,

fish, eggs, or cheese. Nuts, dry beans, and peas may substitute for one of the meats.

Vegetables and fruits. Four or more servings per day to include green and yellow vegetables, and citrus fruit or tomatoes. One serving per day should be raw.

Bread and cereals. Four or more servings per day. Servings should be composed of whole grain or enriched products.

Intake of specific foods

CARBOHYDRATES. Fruits, vegetables, and whole grains (in breakfast foods, bread) provide the major sources of carbohydrate after weaning. Sufficient carbohydrate should be provided in the diet to assure normal weight gain in terms of the activity level of the individual. It may be noted that until about 6 months to 1 year of age, the child has little salivary amylase and has low levels of pancreatic amylase. Thus he should not be fed a diet containing a large percentage of polysaccharides. He does possess the ability to efficiently utilize lactose, and his diet should emphasize intake of this disaccharide.

PROTEIN. Protein is regarded as the mainstay of life, because of its incorporation into enzymes, hormones, contractile tissues, and other body components. Egg and milk proteins appear to be the best sources of essential and nonessential amino acids. These sources are closely followed by animal proteins like those of meat, fish, and poultry. Plant proteins are deficient in certain amino acids, or contain too low an amount to sustain optimum nutrition. Inadequate protein intake is usually reflected in muscular wastage, and in edema due to decreased plasma protein levels. Children need more protein to sustain their growth and maturation than do adults. A good measure of whether protein intake is adequate is to determine the nitrogen balance of the individual; that is, is nitrogen intake greater than, equal to, or less than nitrogen excretion? In general, greater nitrogen intake than output is associated with states of growth, while the reverse situation generally is associated with malnutrition and starvation.

FATS. Fats are important as energy sources and as the vehicle for absorption of fat soluble vitamins (A, D, E, K). Intake must include the essential fatty acids, which are necessary to combust other lipids. Fats are synthesized or are converted into a wide variety of compounds essential to the structure of cell membranes and myelin sheaths around nerves. Evidence is accumulating that the feeding of large amounts of fats early in life results in the development of excess numbers of adipose cells. These cells "like to be filled" and tend to accumulate fat, and this may lead to fat babies, which are not necessarily healthy babies. Adults who are obese tend to have been obese as children, possibly from having more adipose cells "crying for fat."

VITAMINS. Vitamins are classed as trace sub-

stances or accessory food factors that are required in very small amounts for normal cellular metabolism. None of these substances can be synthesized by the animal body, and it must therefore depend mainly on dietary intake. Supplementation may occur through synthesis by intestinal flora and subsequent absorption through the gut (colon) wall. Some 16 organic substances are recognized as essential to normal body functions; 13 are designated as vitamins.

Two groups of vitamins are recognized: the fat soluble vitamins include A, D, E, and K; the water soluble vitamins are those of the B complex (thiamine or B_1, riboflavin or B_2, niacin, pyridoxine or B_6), vitamin C, vitamin B_{12}, folic acid, pantothenic acid, and biotin. Choline, inositol, and para-aminobenzoic acid may be essential to body function, but are not usually designated as vitamins.

FAT SOLUBLE VITAMINS are ingested and absorbed along with dietary fats. Thus any condition affecting fat digestion and absorption affects the intake of fat soluble vitamins. The use of oils as laxatives may cause dissolving of fat soluble vitamins in the oil and their excretion in the feces. Fat soluble vitamins are not present in all dietary fat sources. Liver tends to concentrate fat soluble vitamins and is thus an excellent source of these vitamins. Cod-liver oil has for years been a standard source of vitamins A and D. Vitamin K is widely distributed in all foods and is essential for prothrombin formation; vitamin E is present in eggs and grain oils. Vitamin E is an antioxidant, and may be necessary to maintain reduced states of hydrogen acceptors and electron transport systems.

WATER SOLUBLE VITAMINS are widely distributed in all foodstuffs, but many are destroyed by heat (cooking), sunlight, and are easily oxidized, as by standing in a vitamin pill bottle. Many water soluble vitamins are lost from foods if they are soaked in water or are cooked in large volumes of water. The water soluble vitamins are, to a large extent, incorporated into enzymes or cofactors for enzymes, and are utilized in the basic metabolic schemes of cells. Table 23.1 lists the vitamins, some of their sources, and minimum daily requirements (where known).

TABLE 23.1   The vitamins

Designation letter and name	Major properties	Requirement per day	Major sources	Metabolism	Function	Deficiency symptoms
A—Carotene	Fat soluble yellow crystals, easily oxidized	5000 I.U.	Egg yolk, green or yellow vegetables and fruits	Absorbed from gut; bile aids, in liver	Formation of visual pigments; maintenance of normal epithelial structure	Night blindness, skin lesions
D$_3$—Calciferol	Fat soluble needlelike crystals, very stable	400 I.u. much made through irradiation of precursors in skin	Fish oils, liver	Absorbed from gut; little storage	Increase Ca absorption from gut; important in bone and tooth formation	Rickets (defective bone formation)
E—Tocopherol	Fat soluble yellow oil, easily oxidized	Not known for humans	Green leafy vegetables	Absorbed from gut; stored in adipose and muscle tissue	Humans—maintain resistance of red cells to hemolysis	Increased RBC fragility
					Animals— maintain normal course of pregnancy	Abortion, muscular wastage
K—Naphthoquinone	Fat soluble yellow oil, stable	Unknown	Synthesis by intestinal flora; liver	Absorbed from gut; little storage; excreted in feces	Enables pro-thrombin synthesis by liver	Failure of coagulation
B$_1$—Thiamine	Water soluble white powder, not oxidized	1.5 mg	Brain, liver, kidney, heart; whole grains	Absorbed from gut; stored in liver, brain, kidney, heart; excreted in urine	Formation of cocarboxylase enzyme involved in decarboxylation (Krebs cycle)	Stoppage of CH$_2$O metabolism at pyruvate, beri-beri, neuritis, heart failure, mental disturbance
B$_2$—Riboflavin	Water soluble; orange-yellow powder; stable except to light and alkalies	1.5–2.0 mg	Milk, eggs, liver, whole cereals	Absorbed from gut; stored in kidney, liver, heart; excreted in urine	Flavoproteins in oxidative phosphorylation (hydrogen transport)	Photophobia; fissuring of skin
Niacin	Water soluble; colorless needles; very stable	17–20 mg	Whole grains	Absorbed from gut; distributed to all tissues; 40% excreted in urine	Coenzyme in H transport, (NAD, NADP)	Pellagra; skin lesions; digestive disturbances, dementia

INORGANIC SUBSTANCES. Fourteen of the elements appearing in the periodic table have been shown to be essential for health. They are: calcium, phosphorus, magnesium, sodium, potassium, sulfur, chlorine, iron, copper, cobalt, iodine, manganese, zinc, and fluorine. Table 23.2 presents facts relative to acquisition, requirements, and uses of these inorganic materials.

HYPERALIMENTATION AND OBESITY. Malnutrition is usually associated with undernutrition. A greater problem in many parts of the world is overnutrition, or the intake of calories in excess of body needs (hyperalimentation) leading to the development of obesity. Obesity is a more threatening form of mal (poor) nutrition than deficiency of caloric intake. A definition of obesity is that the person is 20 percent over his "ideal

Designation letter and name	Major properties	Requirement per day	Major sources	Metabolism	Function	Deficiency symptoms
B_{12}—Cyanocobal-amin	Water soluble; red crystals; stable except in acids and alkalies	2–5 mg	Liver, kidney, brain. Bacterial synthesis in gut	Absorbed from gut; stored in liver, kidney, brain; excreted in feces and urine	Nucleoprotein synthesis (RNA); prevents pernicious anemia	Pernicious anemia; malformed erythrocytes
Folic acid (Vitamin B_c, M, pteroyl glutamate)	Slightly soluble in water; yellow crystals; deteriorates easily	500 micrograms or less	Meats	Absorbed from gut; utilized as taken in	Nucleoprotein synthesis; formation of erythrocytes	Failure of erythrocytes to mature; anemia
Pyridoxine (B_6)	Soluble in water; white crystals; stable except to light	1–2 mg	Whole grains	Absorbed from gut; one-half appears in urine	Coenzyme for amino acid metabolism and fatty acid metabolism	Dermatitis; nervous disorders
Pantothenic acid	Water soluble; yellow oil; stable in neutral solutions	8.5–10 mg	?	Absorbed from gut; stored in all tissues; urine	Forms part of coenzyme A (CoA)	Neuromotor disorders, cardiovascular disorders; GI distress
Biotin	Water soluble; colorless needles; stable except to oxidation	150–300 mg	Egg white; synthesis by flora of GI tract	Absorbed from gut; excreted in urine and feces	Concerned with protein synthesis, CO_2 fixation and transamination	Scaly dermatitis; muscle pains, weakness
Choline (maybe not a vitamin)	Soluble in water; colorless liquid; unstable to alkalies	500 mg	?	Absorbed from gut; not stored	Concerned with fat transport; aids in fat oxidation	Fatty liver; inadequate fat absorption
Inositol	Water soluble; white crystals	No recommended allowance	?	Absorbed from gut; metabolized	Aids in fat metabolism; prevents fatty liver	Fatty liver
Para-amino benzoic acid (PABA)	Slightly water soluble; white crystals	No evidence for requirement	?	Absorbed from gut; little storage; excreted in urine	Essential nutrient for bacteria; aids in folic acid synthesis	No symptoms established for humans
Ascorbic acid (Vitamin C)	Water soluble; white crystals; oxidizable	75 mg/day	Citrus	Absorbed from gut; stored; excreted in urine	Vital to collagen and ground substance	Scurvy—failure to form c.t. fibers

weight." In the USA, it has been estimated that 15 million persons are 20 percent over their ideal weight, and are thus obese. What causes obesity? Overeating, because of boredom, unhappiness, or other emotional problems accounts for 95 percent of obesity. Genetic causes account for less than 5 percent. The penalties one pays for obesity include a higher mortality (death rate) from heart disease, the development of diabetes mellitus, digestive disorders, cerebral hemorrhage, and a higher morbidity (sickness) rate.

Basically, the control of obesity involves motivation to reduce, and a caloric intake less than that required for activity and general living. One must have an appreciation of the caloric costs of activity and must adjust his intake accordingly, if he is serious about losing weight. A loss of a pound per week requires about a 500 calorie

Substance	Requirements per day	High level sources	Where absorbed	Where found in body	Functions	Effects of	
						Excess	Deficiency
Calcium	About 1 gm	Dairy products, eggs, fish, soybeans	Small intestine	Bones, teeth, nerve, bloodstream, muscle	Bone structure, blood clotting, muscle contraction, excitability, synapses	None	Tetany of muscles, loss of bone minerals
Phosphorus	About 55mg/kg	Dairy products, meat, beans, grains	Small intestine	Bones, teeth, nerve, bloodstream, muscle, ATP	Bone structure, intermediary metabolism, buffers, membranes	None	Unknown; related to rickets, loss of bone mineral
Magnesium	Estimated 13 mg	Green vegetables, milk, meat	Small intestine	Bone, enzymes, nerve, muscle	Bone structure, factors with enzymes, regulation of nerve and muscle action	None	Tetany
Sodium	Newborn 0.25 gm Infant 1 gm Child 3 gm Others 6 gm	All foods, table salt	Stomach, small and large intestine	Extracellular fluids	Ionic equilibrium, osmotic gradients, excitability in all cells	Edema hypertension	Dehydration, muscle cramps, renal shutdown
Potassium	1–2 gms	All foods, meats, vegetables, milk	Stomach, small and large intestine	Intracellular fluids	Buffering, muscle and nerve function	Heart block (>10 meq/L)	Changes in ECG, alteration in muscle contraction
Sulfur	0.5–1 gm ?	All protein containing foods	Small intestine, as amino acids primarily	Amino acids, bile acids, hormones, nerve	Structural as amino acids are made into proteins	Unknown	Unknown
Chlorine	2–3 gm	All foods, table salt	Stomach, small and large intestine	Extracellular fluids	Acid-base balance, osmotic equilibria	Edema	Alkalosis, muscle cramps
Iron	Infant 0.4–1 mg/kg Child 0.4 mg/kg Adult 16 mg	Liver, eggs, red meat, beans, nuts, raisins	Small intestine	Respiratory proteins (hemoglobin, myoglobin, cytochromes)	O_2 and electron transport	May be toxic	Anemia (insufficient hemoglobin in red cells)
Copper	Infant and Child 0.1 mg/kg Adult 2 mg	Liver, meats	Small intestine	Bone marrow	Necessary for hemoglobin formation	None	Anemia
Cobalt	Unknown	Meats	Small intestine	Liver (Vitamin B_{12})	Essential to hemoglobin formation	None	Pernicious anemia
Iodine	Children 40–100 micrograms Adult 100–200 micrograms	Iodized table salt, fish	Small intestine	Thyroid hormone	Synthesis of thyroid hormone	None	Goiter, cretinism
Manganese	Unknown	Bananas, bran, beans, leafy vegetables, whole grains	Small intestine	Bone marrow, enzymes	Formation of hemoglobin, activation of enzymes	Muscular weakness, nervous system disturbance	Subnormal tissue respiration
Zinc	Unknown	Meat, eggs, legumes, milk, green vegetables	Small intestine	Enzymes, insulin	Part of carbonic anhydrase, insulin, enzymes	Unknown	Unknown
Fluorine	0.7 part/million in water is optimum	Fluoridated water, dentifrices, milk	Small intestine	Bones, teeth	Hardens bones and teeth, suppresses bacterial action in mouth	Mottling of teeth	Tendency to dental caries

TABLE 23.2 Inorganic Substances

reduction in intake, or an increase in activity sufficient to increase caloric consumption by 500 calories per day. All in all, it seems easier to reduce the caloric intake than to increase the exercise level. For example, dietary omission of an ice cream sundae, or a piece of pie à la mode, eliminates the necessary calories; to burn up an equivalent amount of calories would involve running for an hour, walking (at 4 miles per hour)

for nearly 2 hours, or washing dishes for more than 7 hours.

The use of diet pills, usually amphetamines, is effective for about 2 weeks. Such drugs diminish appetite and elevate the spirits. Beyond 2 weeks time, the body adjusts to the drug, and appetite tends to return to previous levels. To get the effect again, a user may take more drug, or may take it more often, leading to a dependency on it.

Summary

1. Absorption involves transfer of foodstuffs across the linings of the alimentary tract.

   a. The mouth, pharynx, and esophagus absorb no food.

   b. Salts, water, and alcohol are absorbed from the stomach.

   c. All basic foodstuffs are absorbed by the small intestine.

   d. Salts, water, and bacterial products are absorbed by the colon.

2. Absorption of lipids, water, and certain sugars is by diffusion. Salts, amino acids, glucose, galactose, vitamins, and certain proteins is active, involving active transport and pinocytosis. Absorbed substances pass into blood vessels and lymphatics of the tract.

3. Intermediary metabolism of foods describes the changes cells make in absorbed products of digestion.

   a. Carbohydrates (glucose) are catabolized by glycolysis, the Krebs cycle, and oxidative phosphorylation. They are synthesized into glycogen (glycogenesis) from which they may be recovered (glycolysis).

   b. Fatty acids are oxidized by beta-oxidation which produces acetic acid that goes through the Krebs cycle. Glycerol is converted into a product found in glycolysis. Cholesterol is synthesized by the liver, and forms bile acids and steroid hormones. Essential fatty acids must be included in the diet.

   c. Proteins are composed of essential and nonessential amino acids, undergo conversion to compounds of glycolysis and Krebs cycle, and are oxidized for energy or incorporated into body structure.

4. The basic building blocks of the three major foodstuffs are, to a great degree, interconvertible.

5. Control of metabolism is by genetic, chemical, and environmental factors.

6. Body temperature is maintained nearly constant in humans.

   a. Heat production occurs through activity, food intake, and emotions.

   b. Heat loss is by radiation, conduction, convection, and vaporization.

   c. Control of body temperature is exerted by the hypothalamus and its connections to muscles, sweat glands, and blood vessels.

7. Nutrition is concerned with intake of foods necessary to maintain body function.

    a. Protein should make 10–15 percent of the diet, carbohydrate 55–70 percent, fats 20–30 percent.

    b. The diet should include milk and milk products, meats, vegetables and fruits, and breads and cereals.

    c. Tables summarizing vitamin and inorganic salt needs and uses are included.

8. Malnutrition includes excessive food intake, as well as undernutrition.

    a. Obesity exists when weight is 20 percent over ideal weight.

    b. Obesity is caused primarily by overeating.

    c. Obesity causes an increase in morbidity and mortality.

    d. Diet control, involving reduction of caloric intake, is the only effective means of reducing weight.

Questions

1. Define absorption, metabolism, intermediary metabolism, and nutrition.

2. List the several parts of the alimentary tract and describe what is absorbed, by what method, and where the absorbed substance goes.

3. How is (are) glucose metabolized? Fatty acids? Amino acids?

4. How can one get fat by eating potatoes that are composed of carbohydrate?

5. What factors are involved in control of metabolism?

6. What are the normal limits of body temperature and how is it maintained?

7. What is fever? Hypothermia? What may cause each?

8. What constitutes an adequate diet?

9. To what uses are vitamins and minerals put in the body?

10. Why is obesity a threat to health, and how does it develop?

Readings

Armstrong, W. and A. S. Nunn. *Intestinal Transport of Electrolytes, Amino Acids, and Sugars.* Thomas Pubs. Springfield, Ill., 1971.

Harper, Harold A. *Review of Physiological Chemistry.* 13th ed. Lange Medical Publications. Los Altos, Cal. 1971.

Folk, G. E. *Introduction to Environmental Physiology. Extremes and Mammalian Survival.* Lea & Febiger. Philadelphia, 1966.

Gale, C. C. "Neuroendocrine Aspects of Thermoregulation." *Ann. Rev. Physiol.* 35:391, 1973.

Brady, Roscoe O. "Hereditary Fat Metabolism Diseases." *Sci. Amer. 229*:88 (Aug) 1973.

Bruch, Hilde. *Eating Disorders: Obesity, Anorexia Nervosa, and the Person Within.* Basic Books. Reading, Mass., 1972.

Maugh II, Thos. H. Trace Elements: "A Growing Appreciation of Their Effects on Man." *Science. 181*:253, 1973.

Science News. "High Protein Diet and Cholesterol." 105, no 15, p 240. April 13, 1974.

Screening Children for Nutritional Status: suggestions for child health programs. Public Health Service, HEW, Rockville, Md., 1971.

Underwood, E. J. *Trace Elements in Human and Animal Nutrition.* Academic Press. New York, 1971.

Young, V. R. and N. S. Scrimshaw. "The Physiology of Starvation." *Sci. Amer. 225*:14 (Oct) 1971.

chapter 24
The Urinary System

chapter 24

Pathways of excretion

Metabolism of substances by body cells produces a variety of waste products that must be removed from the body. Carbon dioxide, nitrogen containing substances such as urea, ammonia, and uric acid, heat, and excess water are the chief metabolic wastes. The lungs account for the elimination of the greater part of the carbon dioxide, and small quantities of water and heat. The alimentary tract eliminates some carbon dioxide, water, and heat, and rids the body of unused ingested material and digestive secretions. The skin plays the major role in elimination of heat, but eliminates only small quantities of solids in the sweat.

Also treated as wastes, in that they will ultimately be eliminated from the body, are substances present in greater amounts than are needed for normal cellular function. Water, inorganic salts, H^+, sulfates, and phosphates are examples of such materials. Their levels in the extracellular fluid must be carefully controlled to assure the presence of only those amounts necessary for maintenance of homeostasis. Excretion of detoxified materials, antibiotics, and drugs also occurs. The organs of the urinary system, especially the kidney, are of primary importance in the regulation of composition of the extracellular fluid, and in the elimination of excess, detoxified and waste materials. The organs serving excretory functions in the body, and some of the substances they deal with are summarized in Table 24.1.

TABLE 24.1	Excretory Organs of the Body and Substances Dealt with	
Organ	Substance(s)	Comments
Kidney	Water	Regulates body hydration
	Nitrogen containing wastes of metabolism (NH_3, urea, uric acid)	Originate from protein metabolism. Toxic if retained.
	Inorganic salts (Na^+, Cl^-, K^+, $PO_4^{\equiv}$, $SO_4^=$), H^+, HCO_3^-	Regulates osmotic pressure, pH of ECF; maintains excitability of cells.
	Drugs	
	Detoxified substances	Detoxification occurs in the liver; kidney rids body of end product.
Lungs	Carbon dioxide	Regulation of acid-base balance
	Water	Fixed loss
	Heat	Fixed loss
Skin	Heat	Regulates temperature
	Water	Cooling
Alimentary tract	Digestive wastes	—
	Salts (Ca^{++})	—

Development of the system (Fig. 24.1)

The first signs of development of the urinary system are seen at about 3½ weeks of embryonic growth. A ridge of mesoderm, the UROGENITAL RIDGE, located in the dorsal part of the celom (primitive peritoneal cavity) gives rise to a series of tubular structures. This series of tubular struc-

FIGURE 24.1. The development of the urinary system. (A) Lateral view, (B) ventral view.

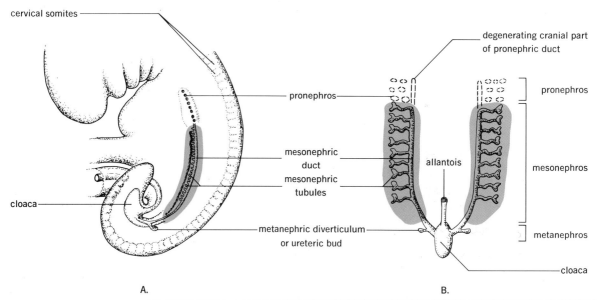

A.

B.

TABLE 24.2 Some Congenital Anomalies of the Kidney			
Disorder	Frequency	Cause of disorder	Comments
Agenesis Bilateral	1/6000 autopsies	Failure of both kidneys to develop	Fetus makes it to term, is using placenta to excrete. Dies several days after birth of uremia.
Unilateral	1/1200 autopsies	Failure of one kidney to develop	One kidney suffices to maintain life.
Aberrant blood vessels	1/105 autopsies	Extra blood vessels develop to supply kidney	Most common disorder. Not serious.
Horseshoe kidney	1/1200 individuals	Lower poles of kidney fused across midline	No symptoms.
Bifid (two) pelves	10% of individuals	2 pelves develop on a single kidney	No symptoms.

tures constitutes the first of three kidneys that will form during development, and is called the PRONEPHROS. At 4 weeks, the pronephros degenerates and is replaced by a MESONEPHROS. This in turn degenerates at about 7 weeks and is replaced by the METANEPHROS or permanent kidney, which contains the functional units of that organ.

As the pronephros develops, it sends a tubule, the PRONEPHRIC DUCT, to join the cloaca (the primitive common terminal cavity for digestive, urinary, and reproductive organs). The distal parts of the pronephric duct also serve the mesonephros as the mesonephric duct, and eventually connect to the gonads to form part of their duct system. The portion of the duct connecting with the cloaca will form an outgrowth that will form the bladder, ureters, pelvis, and the collecting tubules of the kidney.

CLINICAL CONSIDERATIONS. Anomalies of the kidney are relatively common, but most are not severe enough to cause symptoms. Table 24.2 presents several congenital anomalies of the kidney

The organs of the urinary system

The organs of the system include paired KIDNEYS, paired URETERS, and a single URINARY BLADDER and URETHRA (Fig. 24.2).

FIGURE 24.2.   The organs of the urinary system and associated structures.

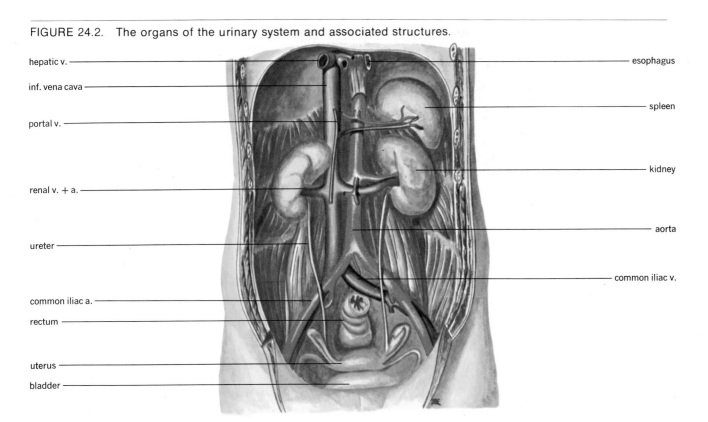

The kidneys

Size, location, attachments

Living kidneys are reddish, bean-shaped organs, located behind the peritoneum (retroperitoneal) at the level of the twelfth thoracic to third lumbar vertebral bodies. They are about one half covered by the eleventh and twelfth ribs. They measure about 11 centimeters long, 6 centimeters wide, and 2.5 centimeters thick (about $4 \times 2 \times 1$ inches). The organs are attached to the body wall by adipose tissue and fibrous tissue that form the ADIPOSE CAPSULE and FIBROUS CAPSULE respectively.

Gross anatomy (Fig. 24.3)

The kidneys have a medially directed indentation, the HILUS (hilum) that marks the entry and exit of all blood vessels, nerves, and the excretory tube of the organ. A longitudinal (frontal) section of the kidney shows the RENAL PELVIS, its branches, the MAJOR and MINOR CALYCES (sing.: calyx), the RENAL SINUS, and the CORTEX and MEDULLA of the kidney. The medulla is composed of 8–18 MEDULLARY PYRAMIDS, whose tips contain duct openings that carry urine into the calyces for eventual elimination. RENAL COLUMNS are cortical substance lying between the pyramids.

Microscopic anatomy

The cortical and medullary portions of each kidney contain an estimated 1–1½ million microscopic units known as NEPHRONS (Fig. 24.4). Each nephron contains a CAPSULE, PROXIMAL CONVOLUTED

FIGURE 24.3.   Frontal section of the kidney.

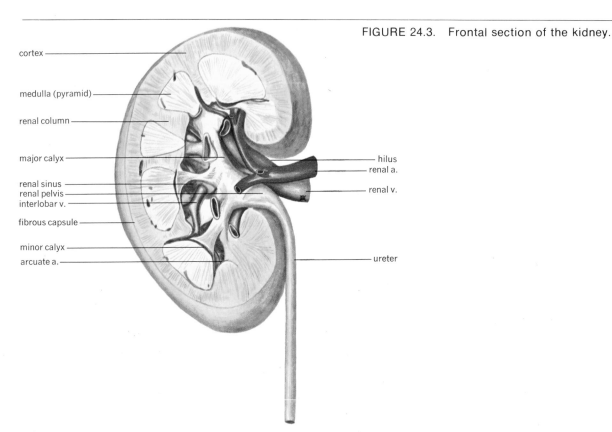

cortex

medulla (pyramid)

renal column

major calyx

renal sinus
renal pelvis
interlobar v.

fibrous capsule

minor calyx
arcuate a.

hilus
renal a.

renal v.

ureter

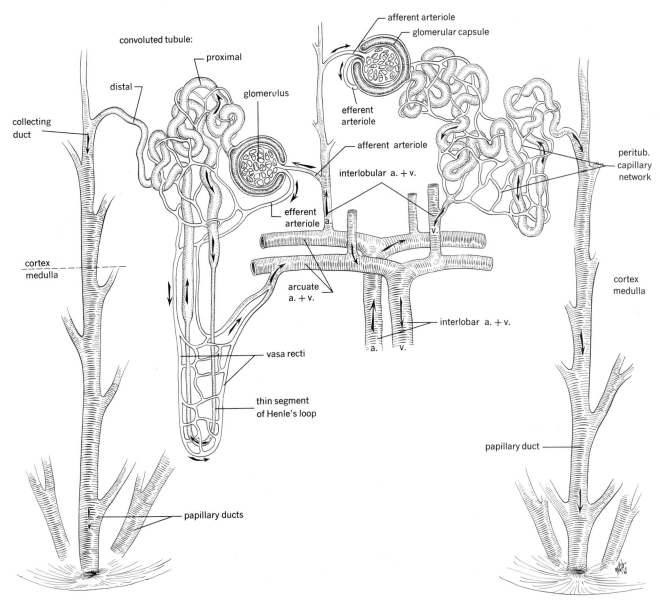

FIGURE 24.4. Two nephrons of the kidney and their associated blood vessels.

TUBULE, LOOP OF HENLE, and DISTAL CONVOLUTED TUBULE, all of which are derivatives of the mesoderm, and a COLLECTING TUBULE and PAPILLARY DUCT, derivatives of the cloacal outgrowth. The nephrons are the units that control composition of the ECF and excrete solid wastes.

Blood supply (Fig. 24.5)

The aorta normally gives rise to a single RENAL ARTERY supplying each kidney. They carry about 1300 milliliters of blood per minute. Each renal artery branches to form, in order, INTERLOBAR ARTERIES, ARCUATE (arched) ARTERIES, and INTER-

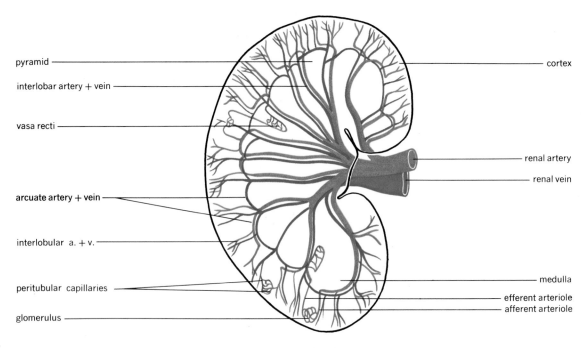

FIGURE 24.5.   The blood vessels of the kidney.

LOBULAR ARTERIES. AFFERENT ARTERIOLES (see Fig. 24.4) lead to the GLOMERULI which lie in the nephron capsules. EFFERENT ARTERIOLES leaves the glomeruli and form PERITUBULAR CAPILLARY NETWORKS or looped vessels (vasa recti) around the kidney tubules. INTERLOBULAR VEINS drain these capillary beds, and form ARCUATE VEINS, INTERLOBAR VEINS and usually, coalesce into a single RENAL VEIN draining each kidney. The

attachment between vessels and nephron at the glomerulus is the critical one for urine formation. Also, the afferent arteriole, just before connecting with the glomerulus, and the adjacent part of the distal convoluted tubule, bear a curious mass of modified smooth muscle and tubule cells in their walls called the JUXTA-GLOMERULAR APPARATUS or JG. These cells produce a substance that can ultimately result in hypertension.

The formation of urine

Urine formation, and regulation of ECF composition involves several processes that are carried on by the nephron in a more or less stepwise fashion. These processes are:

Glomerular filtration

Tubular reabsorption and secretion

Acidification

Concentration

Glomerular filtration

Blood reaches the glomerulus and its capsule under a pressure of about 75 mm Hg. The membranes act like filters or sieves, and permit the passage through their walls, of any substance small enough to go through the "sieve pores." A fluid called the filtrate is formed, and it resembles plasma, except for a lack of blood cells and low protein content.

The filtrate contains not only the nitrogenous wastes of metabolism, formed mainly in the liver and excess substances not required for body use, but also substances (salts, water, glucose, amino acids) that are still useful for body function. Therefore, these useful materials must not be lost from the body.

Tubular reabsorption

By utilizing active and passive processes, the tubules remove substances from the filtrate and place them in the outgoing blood flow for return to the body generally.

Active removal of Na^+, K^+, amino acids, glucose, vitamins, and the small amounts of filtered protein, returns these substances to the body. Negatively charged ions such as Cl^-, HCO_3^- follow the positive ions out of the tubules by electrostatic attraction. Some 80–90 percent removal or reabsorption of such materials occurs in the proximal convoluted tubule. Glucose is reabsorbed completely if present in the filtrate in normal blood concentration.

As active and passive removal of solutes occurs, water concentration in the tubules tends to rise. Thus osmosis of water occurs from tubule to the surrounding ECF in proportion to solute removal, and the filtrate remains isotonic to the plasma. From the ECF, water enters blood vessels and returns to the circulation. Of about 190 liters of water filtered per day in the glomeruli, only about 1½ liters is excreted; the rest is reabsorbed. As water leaves the tubules, nitrogenous wastes accumulate in the tubule. Enough accumulates so that there is diffusion of these materials back into the circulation. Thus both filtration and reabsorption of wastes are going on at the same time; fortunately, reabsorption of these wastes is less than filtration, and so the blood concentrations of these substances are kept below toxic levels. It should *not* be assumed that all of a waste is removed during a trip through the kidney. It is not; rather it is removed to the extent that we are not poisoned by it, but it is still there.

Tubular secretion

Active transport can move materials from the bloodstream into the tubules, as well as passing them out of the tubules. Secretion describes the

FIGURE 24.6.   Two mechanisms of acidification of the urine.

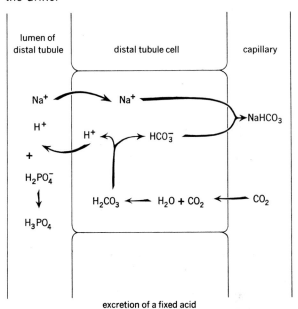

excretion of a fixed acid

secretion of ammonia

movement of substances into the tubules. Both proximal and distal convoluted tubules secrete materials such as H^+, K^+, organic acids, and bases. One of the results of secretion is the acidification of the filtrate.

Acidification

Acid-producing foods predominate over alkali- producing foods in a normal diet. The body thus faces a real threat to its acid-base balance. Secretion of H^+ by the kidney provides a means of completely compensating a metabolic acidosis, given several days for the compensation to occur. The proximal and distal tubules are the primary sites of H^+ secretion. The ions are secreted in exchange for Na^+ or K^+ in the filtrate to maintain electrical neutrality of the cells. Some of the mechanisms of acidification are shown in Figure 24.6.

FIGURE 24.7.   The result of the operation of the countercurrent multiplier. The heavy lines indicate the portion of the loop of Henle that is not permeable to water.

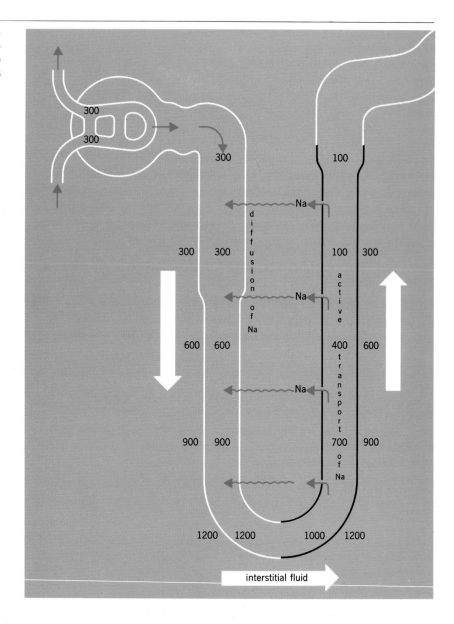

Concentration of the filtrate

As the filtrate moves through the proximal tubule, it loses solute and water in equal ratios, so that its volume is reduced but the osmotic pressure is not changed. As the fluid enters the loop of Henle, it is subjected, in the ascending limb of the loop, to a selective active transport of sodium out of the fluid, *without* water following. The transported sodium is retained in the ECF around the loop, and tends to increase in concentration the deeper one goes toward the interior of the kidney. The mechanism responsible for this sodium concentrating effect is termed the COUNTERCURRENT MULTIPLIER, because sodium concentration is increased in the ECF by the mechanism. The multiplier ultimately results in the sodium concentrations shown in Figure 24.7. The fluid passing from the loop into the distal convoluted tubule is thus poor in solute (sodium) and rich in water, and is hypotonic to the plasma. The fluid now passes into the collecting tubules and papillary ducts that run through the extracellular fluid that has the increasing concentration of sodium ion.

An osmotic gradient for water is thus created between the tubule and the surrounding fluids. The height of the gradient is 3–4 times the osmotic pressure of the blood.

Therefore, water has a tendency to leave the tubule, *but* cannot do so unless a particular hormone is present.

ANTIDIURETIC HORMONE (ADH) is released from the posterior pituitary gland and allows water to pass through the membranes of the cells of the collecting tubules and papillary ducts. This hormone has the effect of matching need for reabsorption of water to the body requirements for water, and is the single most important device for controlling blood-water concentration. The center controlling how much ADH is released is in the hypothalamus, and monitors the osmotic pressure of the blood reaching it. If osmotic pressure of the blood increases, more ADH is released, more water is reabsorbed, and the blood is diluted. The reverse situation also occurs to concentrate the blood when it becomes too dilute.

The material leaving the papillary ducts has no further changes made in its composition, and is now properly called URINE. A summary of the processes occurring in urine formation is provided in Table 24.3

TABLE 24.3 Summary of the Processes Occurring in Urine Formation			
Process	Where occurring	Force responsible	Result
Filtration	Glomerulus	Blood pressure, opposed by osmotic, interstitial, and intratubular pressures	Formation of fluid having no formed elements and low protein concentration
Reabsorption	Proximal tubule Distal tubule Loop of Henle	Active transport	Return to bloodstream of physiologically important solutes
Secretion	Proximal tubule Distal tubule	Active transport	Excretion of materials Acidification of urine
Acidification (acid and base regulation)	Distal tubule	Active transport and exchange of alkali for acid	Excretion of excess H^+ Conservation of base (Na^+ and HCO_3^-)
Concentration of filtrate	Collecting tubule and papillary ducts	Osmosis of water under permissive action of ADH	Formation of hypertonic urine

Urine

Characteristics and composition

Normal urine differs from plasma, as is shown in Table 24.4, and contains a variety of substances, as is shown in Table 24.5. It is an amber or yellow, transparent, usually acidic fluid, and has a solute concentration about three times that of the plasma. Its volume is determined by several factors including the following.

The height of the blood pressure. Filtration is proportional to blood pressure, so that a higher blood pressure causes the formation of a larger volume of filtrate, from which relatively less water is reabsorbed; more water is therefore eliminated.

A high level of filtered solutes. High solute concentrations in the filtrate tend to cause water to remain in the tubules by "osmotic attraction." Thus a greater volume is eliminated.

Amount of ADH present. This has been explained above.

Loss by other routes. Kidney loss of water is really the only route of loss that can be altered. If one sweats heavily, or loses water by vomiting or diarrhea, less is excreted by the kidney to offset losses by other means.

Presence of diuretic substances. A diuretic is a substance promoting water loss by the kidney. It may act by slowing solute removal from the filtrate or by increasing solute filtration. Thus the volume of water in the filtrate is increased, and more water is eliminated.

Control of urine formation

At least two hormones and one enzyme are involved in regulating both the volume and concentration of urine formation.

ADH has been discussed above.

ALDOSTERONE, a hormone of the adrenal cortex, controls the rate of Na^+ reabsorption and thus influences not only Na^+ reabsorption but also that of Cl^-, HCO_3^-, and water. Anions follow cations by electrostatic attraction; water follows solute removal.

RENIN, an enzyme, is secreted by the JG apparatus when the kidney receives an inadequate flow of blood. It acts as follows:

Renin + angiotensinogen → angiotensin I
(enzyme) (plasma substrate) (in plasma)

+

plasma enzyme

↓

angiotensin II
(potent vasocon-
strictor)

Angiotensin II causes a generalized vasoconstriction that raises blood pressure to the kidneys and assures adequate filtration.

Property or Substance	Plasma	Urine
Osmolarity (mOs)	300	600–1200
pH	7.4 ± 0.02	4.6–8.2
Specific gravity	1.008–1.010	1.015–1.025
Sodium	0.3	0.35
Potassium	0.02	0.15
Chloride	0.37	0.60
Ammonia	0.001	0.04
Urea	0.03	2.0
Sulfate	0.002	0.18
Creatinine	0.001	0.075
Glucose	0.1	0
Protein	7–9	0

TABLE 24.4 Comparison of Plasma and Urine for Certain Constituents*

*Values, unless otherwise indicated, are gm %

TABLE 24.5 Some Normal Constituents, Origins, and Amounts Excreted per Day in Normal Urine

Constituent	Origin	Amount per day
Water	Diet and metabolism	1200–1500 ml
Urea	Ornithine cycle	30 g
Uric acid (purine)	Catabolism of nucleic acids	0.7 g
Hippuric acid	Liver detoxification of benzoic acid	Trace
Creatinine	Destruction of intracellular creatine phosphate of muscle	1–2 g
Ammonia	Deamination of amino acids	0.45 g
Chloride (as NaCl)	Diet	12.5 g
Phosphate	Diet and metabolism of phosphate containing compounds	3 g
Sulfate	Diet, metabolism of sulfate containing compounds, formation of H_2SO_4 in kidney tubules	2.5 g
Calcium	Diet	200 mg

The ureters

Gross anatomy and relationships

The ureters extend from the hilus of the kidney to the urinary bladder, a distance of 28–35 centimeters (about 11–14 inches). They are retroperitoneal in placement and have an increasing diameter as they course toward the bladder.

Microscopic anatomy

Three coats of tissue form the wall of the ureter. A MUCOSA, composed of transitional epithelium and connective tissue, forms the inner layer. The central MUSCULARIS is composed of inner longitudinal and outer circular layers of smooth muscle throughout most of the length of the ureter. On the lower one third of the organ, a third layer of muscle (outer longitudinal) is added. The outer FIBROUS COAT (adventitia) blends without demarcation into the surrounding fascia.

Innervation

The ureters in their upper part receive sympathetic fibers from the renal plexus; in the middle part, they receive fibers from the ovarian or spermatic plexus; and near the bladder, they receive fibers from the hypogastric nerves (Fig. 24.8). These fibers exert primarily a motor effect. They cause the ureters to exhibit rhythmical peristaltic contractions traveling at a speed of 20–25 millimeters per second and at a frequency of 1–5 per minute. The urine, therefore, enters the bladder, not in a continuous stream, but in separate squirts synchronous with the arrival of the peristaltic wave.

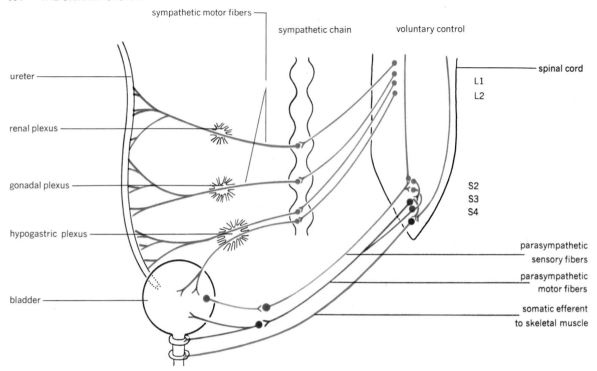

sympathetic motor fibers

sympathetic chain

voluntary control

spinal cord

L1
L2

ureter

renal plexus

gonadal plexus

hypogastric plexus

S2
S3
S4

bladder

parasympathetic
sensory fibers

parasympathetic
motor fibers

somatic efferent
to skeletal muscle

FIGURE 24.8.   The innervation of the ureter, bladder, and urethra.

The urinary bladder

Gross anatomy and relationships

The urinary bladder serves as a reservoir for the urine until it is voided. The organ is located posterior to the symphysis pubis and is separated from the symphysis by a prevesicular space. The space, filled with loose connective tissue, allows for expansion of the filling bladder. Internally, three openings may be found in the bladder wall: the two ureters, and the urethra. An imaginary line drawn to connect these three openings outlines the *trigone.*

Microscopic anatomy

Although similar to the ureter in structure, the bladder has more cell layers in the transitional epithelial lining, a submucous layer of loose tis-

sue between mucosa and muscularis, and three heavy layers of smooth muscle in the muscularis, disposed longitudinally, circularly, and longitudinally. Around the urethral opening, a dense mass of circularly oriented smooth muscle forms the internal sphincter of the bladder. A serous layer is formed by the peritoneum over the superior surface of the organ.

Innervation

The efferent nerves to the bladder and urethra (Fig. 24.9) are from both sympathetic and parasympathetic divisions. The sympathetic fibers furnish inhibitory fibers to the muscle of the bladder; they furnish motor fibers to the trigone and internal sphincter and the muscle of the upper part of the urethra. These fibers arise in the

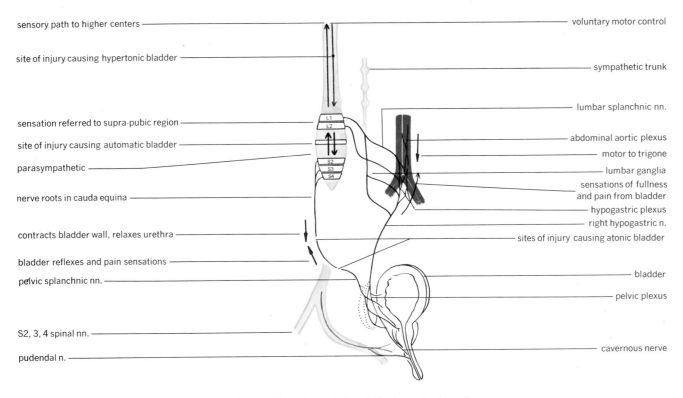

sensory path to higher centers

site of injury causing hypertonic bladder

sensation referred to supra-pubic region

site of injury causing automatic bladder

parasympathetic

nerve roots in cauda equina

contracts bladder wall, relaxes urethra

bladder reflexes and pain sensations

pelvic splanchnic nn.

S2, 3, 4 spinal nn.

pudendal n.

L1
L2
S2
S3
S4

voluntary motor control

sympathetic trunk

lumbar splanchnic nn.

abdominal aortic plexus

motor to trigone

lumbar ganglia

sensations of fullness
and pain from bladder

hypogastric plexus

right hypogastric n.

sites of injury causing atonic bladder

bladder

pelvic plexus

cavernous nerve

FIGURE 24.9. Nerves of the bladder illustrating the origin of bladder dysfunction.

lumbar spinal segments and pass to the bladder, via the inferior hypogastric plexus. The parasympathetic nerves supply motor fibers to the detrusor muscle (the muscle of the bladder) and inhibitory fibers to the internal sphincter. The desire to urinate occurs when a volume of 200–300 milliliters of urine has accumulated in the bladder. Stretch on the muscle of the bladder, brought about by filling, evokes an afferent impulse in the pelvic nerves. This impulse ascends through the spinal cord to a center in the hindbrain. An efferent discharge of motor impulses to the muscle of the bladder is accomplished through descending pathways in the cord and through the pelvic motor nerves. The same efferent or motor fibers bring about a simultaneous relaxation of the internal sphincter so that urine may be emptied from the bladder into the urethra (micturition). Urination may occur by reflex action not involving the hindbrain center. Filling evokes a reflex contraction of the detrusor and relaxation of the sphincter, which is served by lower cord segments only. The reflex is seen in infants, where the voluntary control over sphincters has not yet been achieved.

The urethra

The female urethra is about 4 centimeters (1¾ inches) in length, is completely separate from the reproductive system, and bears no regional differences. A mucosa is present, consisting of transitional epithelium near the bladder and stratified squamous elsewhere, and is underlain by connective tissue. A muscularis is present, consisting of circularly arranged fibers of smooth muscle. The female urethra opens just anterior to the vaginal orifice.

The male urethra is about 20 centimeters (8 inches) in length, serves as a common tube for the terminal portions of both urinary and reproductive systems, and does show regional varia-

tions. The PROSTATIC URETHRA is the first 3 centimeters of the organ; it is surrounded by the prostate gland and is lined with transitional epithelium. The MEMBRANOUS URETHRA is 1–2 centimeters long; it penetrates the pelvic floor, is very thin, and has a pseudostratified epithelium lining. The CAVERNOUS URETHRA is about 15 centimeters in length and lies within the penis.

Clinical considerations involving the urinary system

Kidney

Glomerulonephritis, nephrotic syndrome, and pyelonephritis are probably the most frequently encountered kidney disorders.

GLOMERULONEPHRITIS is almost always associated with a previous infection by a beta-hemolytic streptococcus bacterium elsewhere in the body (usually the pharynx) and is generally regarded as an "autoimmune disease." One of the components of the basement membrane of the glomerular capsule, sialic acid, is thought to be hydrolyzed by the bacterial toxin. The body then recognizes the hydrolyzed acid as a "new" or "foreign" substance and produces antibodies to neutralize it. In the process, the basement membrane is injured and allows red cells and more protein to leak into the filtrate. Edema is also a common occurrence because the osmotically active protein molecules are diminished in the blood, resulting in failure to return water to the circulation.

The NEPHROTIC SYNDROME also involves an increased glomerular permeability, but appears to involve the tubules to a greater degree than the glomeruli. The chief difference from glomerulonephritis lies in the fact that a massive loss of protein from the blood occurs, with cast formation in the tubules and a near shutdown of tubular function.

PYELONEPHRITIS is a bacterial infection of the kidney that usually begins in the lower tract and spreads upward to involve the kidney. Females are more susceptible to this disease as a consequence of the short urethra and the greater possibility of bladder infection occurring. It is the most

common disease affecting the urinary system. Chills, fever, flank pain, nausea, and vomiting characterize the infection. Treatment is with antibiotics to which the organism cultured from the urine is sensitive.

Bladder and urethra

Nerve injuries produce three types of bladder dysfunction (see Fig. 24.9).

ATONIC BLADDER. Interruption of sensory supply results in loss of bladder tone, and the organ may become extremely distended with no development of an urge to urinate.

HYPERTONIC BLADDER. Interruption of the voluntary pathways results in excessive tone, and very small distentions create an uncontrolled desire to urinate.

AUTOMATIC BLADDER. Complete section of the cord above the first sacral nerve exit (S1) produces automatic emptying in response to filling, by the described cord reflex.

Other disorders associated with the lower urinary tract include cystitis, urethritis, and obstruction of the urethra.

CYSTITIS, or inflammation of the bladder, is usually secondary to infection of the prostate, kidney, or urethra.

INFECTION of the male urethra is most commonly associated with gonorrheal infections; in the female, almost any infection of the perineum may invade the urethra.

Blockage of the urethra is most commonly associated with calculi (stones) in the bladder. The stone may be voided in the urine, if small enough, or may be crushed by use of a cystoscope inserted through the urethra.

Stones may be found anywhere in the tract and are caused by infections, or metabolic disorders that cause excretion of large amounts of organic and inorganic substances. As the kidney concentrates the filtrate, the substances in the fluid may reach high enough levels to cause precipitation of the substance in the form of a tiny granule or stone. This granule then serves as a nucleus for further precipitation of materials to form the large stones. Stones are often described according to shape (e.g., staghorn stone in renal pelvis), or by what they contain (cystine, uric acid, calcium stones).

Individuals that form only an occasional stone may require only a large water intake (to 3 liters per day) to reduce stone formation; chronic stone formers usually are treated with substances to render the urine alkaline and to reduce the tendency for precipitation to occur.

Summary

1. Several pathways are utilized to rid the body of wastes of metabolism, excess materials, and undigestible substances.
   a. The lungs eliminate CO_2.
   b. The alimentary tract eliminates CO_2, water, undigested materials, and secretions.
   c. The skin eliminates heat.
   d. The kidney eliminates water, salts, nitrogenous wastes, and controls ECF fluid composition.

2. The system develops from mesoderm and endoderm, beginning at about 3½ weeks.
   a. The kidney is mesodermal.
   b. The ducts are endodermal.
   c. The kidney is a common organ for developmental anomalies. Most produce no symptoms.

3. The kidneys
   a. Are reddish, bean shaped, and are high against the back wall of peritoneal cavity.
   b. Measure $4 \times 2 \times 1$ inches.
   c. Have cortex, medulla, and several branches of the ureter in them.
   d. Contain microscopic nephrons as the functional units.
   e. Have a "separate circulation" designed to supply large volumes of blood for cleansing by the kidney.

4. Urine is formed by several processes.
   a. Glomerular filtration removes all but the largest substances or cells from the blood.
   b. Tubular reabsorption actively removes needed solutes from the filtrate, and passively removes water and some wastes.

c. Tubular secretion places into the tubules certain organic acids, bases, and ions.

d. Acidification, by secretion of H^+, changes the filtrate to an acidic state.

e. Concentration of the urine occurs by water reabsorption in the presence of antidiuretic hormone (ADH).

5. Urine differs from plasma and filtrate in both substances present, and amounts present.

a. Its volume is determined by blood pressure, solute load, ADH, loss of water by other routes, and diuretics.

b. ADH, aldosterone, and renin control composition.

6. The ureters are 11–14 inches long and convey urine to the bladder.

a. They have a three-layered wall.

b. Their muscularis moves urine to the bladder by peristalsis.

c. They receive nerves from lumbar and sacral spinal cord segments.

7. The bladder stores urine for voiding.

a. Structure is similar to that of the ureters, but thicker.

b. Nerves supply the bladder from lumbar and sacral cord, and contain both sensory and motor fibers.

c. Micturition (urination) occurs reflexly, but may be voluntarily inhibited.

8. The urethra carries urine to the exterior. It differs in structure in male and female.

9. A variety of diseases and disorders involving the organs of the system are discussed.

Questions

1. Describe the location of the kidneys and their supporting structures.

2. What structures enter or leave the hilus of the kidney?

3. Describe the distribution of blood vessels within the kidney.

4. What is a nephron and what are its functions?

5. What are the processes the nephron uses in forming urine? Describe the main result of the operation of each process.

6. List five factors controlling urine volume, and explain how each works.

7. Compare plasma, filtrate, and urine for the following properties: specific gravity, water content, glucose content, protein content, pH. What accounts for any differences?

8. Compare the structure of ureters and bladder.

9. Describe the mechanism involved in emptying of the bladder.

10. Describe differences between the male and female urethra.

11. Describe two disorders of the kidney, their cause, and their effect on normal nephron functioning.

12. What contribution do the tubules make to the formation of urine?

Readings

DeWardener, H. E. *The Kidney,* 3rd ed. Little, Brown. Boston, 1968.

Fisher, J. E. (ed). *Kidney Hormones.* Academic Press. New York, 1971.

Frazier, H. S. "Renal Regulation of Sodium Balance." *New Eng. J. Med. 279:*868, 1968.

Hamburger, Jean et al. *Structure and Function of the Kidney.* Vol. I. Saunders. Philadelphia, 1971.

Keitzer, W. A., and G. C. Huffman. *Urodynamics.* Thomas Pubs. Springfield, Ill., 1971.

Orloff, J., and M. Burg. "Kidney." *Ann. Rev. Physiol. 33:*83, 1971.

Pitts, Robert F. *Physiology of the Kidney and Body Fluids.* 2nd ed. Year Book Medical Publishers. Chicago, 1968.

chapter 25
The Reproductive Systems

chapter 25

The male and female reproductive systems provide a means of creation of a new human individual. Certain of the organs are responsible for the production of sex cells, others transport and nourish these cells, or provide a haven for development of a new individual. Some of the organs produce hormones that are essential for the process of reproduction of the species, and which may also affect the functions of the body generally.

Development of the systems

Development of the reproductive systems is evident after about 5 weeks of embryonic life. They originate from the urogenital ridge that gave rise to the kidney. At about 5 weeks, the structures shown in Figure 25.1 are present. The tubules shown in this figure were originally those developed in association with the kidney and become portions of the duct system of the reproductive organs. At about 8 weeks, the testes and ovaries assume an internal structure characteristic for each organ, with primitive sex cells visible. The testes also secrete small amounts of male sex hormone (testosterone) about this time.

Until about 8 weeks of development, the external genitalia are in an undifferentiated state, and are not characteristic of a particular sex. Genetic influences assure development of female external genitalia in the absence of hormones. Male external genitalia development requires testosterone production by the primitive testis to assure normal male differentiation. The changes occurring in the external genitalia from the undifferentiated state are shown in Figure 25.2. Organs in the two sexes having the same origin (homologous organs) are shown in the same color.

The male reproductive system

Organs

The organs of the male reproductive system (Fig. 25.3) may be grouped under three headings:

The *essential organs*, or *testes*, produce male sex cells and secrete the hormone *testosterone.*

The *excretory ducts* store or convey sperm, and include the straight tubules, rete tubules, efferent ductules, epididymis, ductus (vas) deferens, ejaculatory duct, and urethra.

The *accessory glands* include the seminal vesicles, prostate gland, and bulbourethral (Cowper's) glands.

The testes

LOCATION. After about 8 months of embryonic life, the testes descend from their region of development in the abdomen into the SCROTUM. The latter structure is an outpocketing of skin and connective tissue of the lower anterior abdominal wall. It is divided internally into two lateral compartments, each of which houses a testis and its associated structures. Each scrotal compartment communicates with the abdominopelvic cavity through the inguinal canal. Descent of the testes into the scrotum insures a temperature low enough to allow sperm cell development at puberty. Failure of descent results in *cryptorchidism* and sterility.

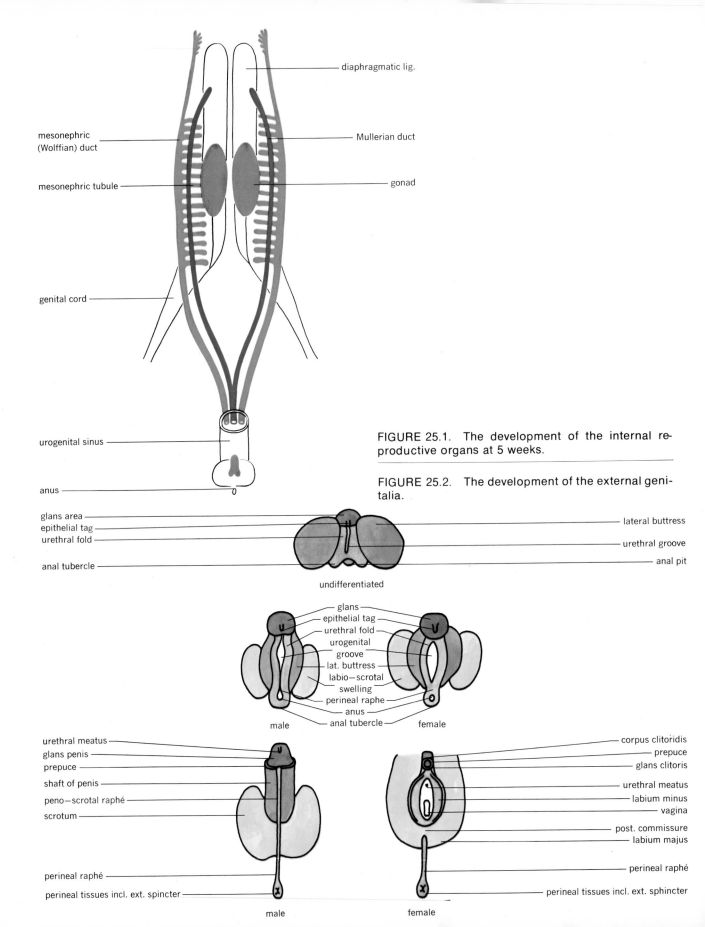

FIGURE 25.1. The development of the internal reproductive organs at 5 weeks.

FIGURE 25.2. The development of the external genitalia.

diaphragmatic lig.

mesonephric (Wolffian) duct

Mullerian duct

mesonephric tubule

gonad

genital cord

urogenital sinus

anus

glans area
epithelial tag
urethral fold

lateral buttress

urethral groove

anal tubercle

anal pit

undifferentiated

glans
epithelial tag
urethral fold
urogenital groove
lat. buttress
labio—scrotal swelling
perineal raphe
anus
anal tubercle

male   female

urethral meatus
glans penis
prepuce
shaft of penis
peno—scrotal raphé
scrotum

perineal raphé
perineal tissues incl. ext. spincter

male

corpus clitoridis
prepuce
glans clitoris
urethral meatus
labium minus
vagina
post. commissure
labium majus
perineal raphé
perineal tissues incl. ext. sphincter

female

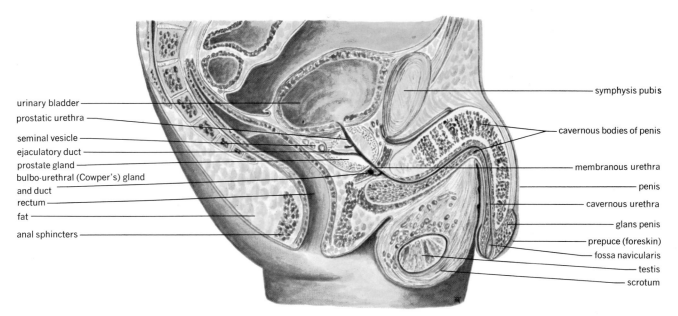

urinary bladder
prostatic urethra
seminal vesicle
ejaculatory duct
prostate gland
bulbo-urethral (Cowper's) gland
and duct
rectum
fat
anal sphincters

symphysis pubis
cavernous bodies of penis
membranous urethra
penis
cavernous urethra
glans penis
prepuce (foreskin)
fossa navicularis
testis
scrotum

FIGURE 25.3.   The male reproductive organs and associated structures.

STRUCTURE (Fig. 25.4). Each adult testis is ovoid in shape, measuring 4–5 centimeters (1¾–2 inches) long, and 2.5 centimeters (1 inch) in diameter. Several tunics or coats of tissue form coverings for the testis. The most obvious tunic is one made of fibrous tissue, and is called the TUNICA ALBUGINEA. It acts as a capsule for the testis, sends partitions or SEPTAE into the organ, dividing it into about 250 LOBULES, and is thickened on the medial side of the testis to form the MEDIASTINUM (TESTES).

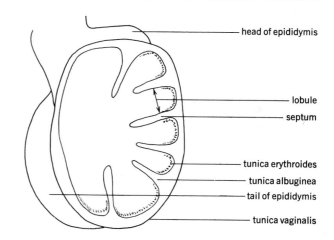

head of epididymis

lobule
septum

tunica erythroides
tunica albuginea
tail of epididymis

tunica vaginalis

FIGURE 25.4.   The internal structure of the testis.

FIGURE 25.5.   The organization of the seminiferous tubules and ducts of the testes.

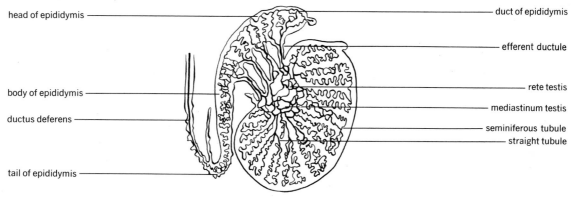

head of epididymis

body of epididymis

ductus deferens

tail of epididymis

duct of epididymis
efferent ductule
rete testis
mediastinum testis
seminiferous tubule
straight tubule

FUNCTION. Each lobule contains 1–4 highly FOLDED SEMINIFEROUS TUBULES (Fig. 25.5), each of which is lined with a germinal epithelium (Fig. 25.6). At puberty, the germinal epithelium begins production of sperm cells, and shows several stages in the development of these sperm cells (see Fig. 25.6). The more primitive cells are located at the periphery of the tubules, and

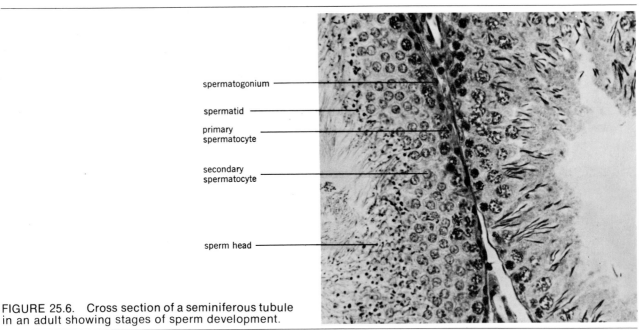

FIGURE 25.6. Cross section of a seminiferous tubule in an adult showing stages of sperm development.

FIGURE 25.7. A spermatozoan.

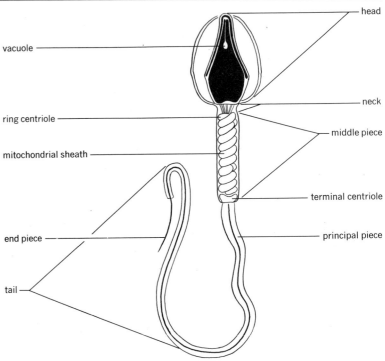

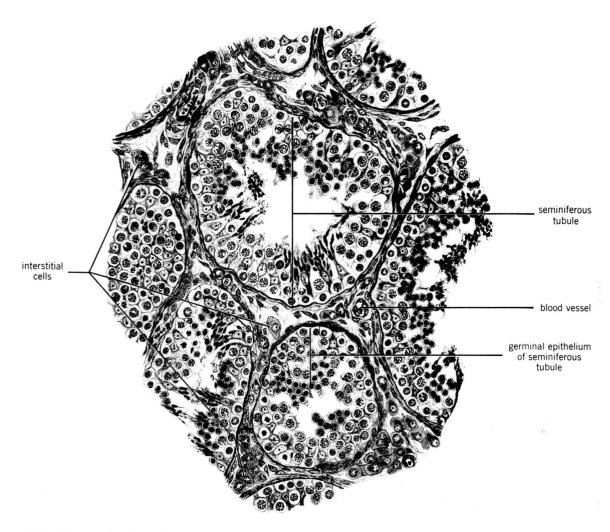

interstitial
cells

seminiferous
tubule

blood vessel

germinal epithelium
of seminiferous
tubule

FIGURE 25.8.   Section of the testes to show interstitial cells.

maturation is more advanced toward the center of the tubules. Sperm are highly differentiated cells (Fig. 25.7) that carry a haploid chromosome number.

The hormone-producing cells of the testis are located in the connective tissue between seminiferous tubules, and are called INTERSTITIAL CELLS (Fig. 25.8). Their product, testosterone, is necessary for development and growth of the external genitalia, and development of male secondary sex characteristics.

At puberty, pituitary secretion of FSH (follicle stimulating hormone) initiates maturation of spermatozoa, and ICSH (interstitial cell stimu-

lating hormone) stimulates secretion of the interstitial cells. Testosterone secretion then causes development of the genital ducts, external genitalia and secondary sex characteristics.

The duct system
(Fig. 25.9; see also Fig. 25.4)

STRUCTURE. The STRAIGHT TUBULES connect the seminiferous tubules to the RETE TUBULES in the mediastinum. Only these two parts of the duct system lie within the testis. The EFFERENT DUCTULES are 12–16 in number and carry sperm to

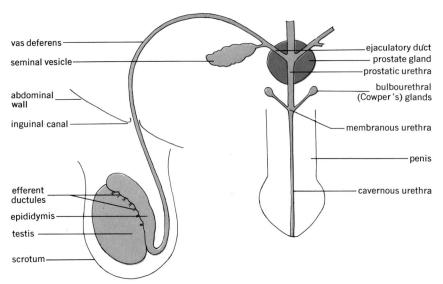

FIGURE 25.9.   A diagrammatic representation of the extratesticular ducts of the male reproductive system and associated organs.

FIGURE 25.10.   The spermatic cord is shown passing from the testis through the inguinal ring.

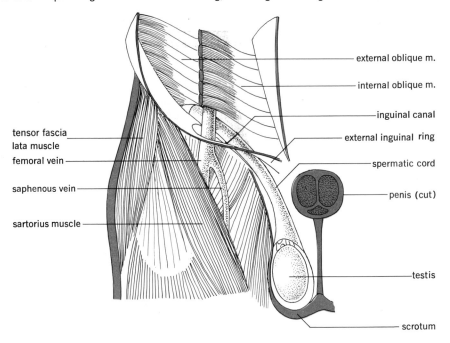

the EPIDIDYMIS where they are stored until released. The epididymis is a single coiled tube about 20 feet in length. It is held by connective tissue to the testis. The DUCTUS (*vas*) DEFERENS is a muscular tube arising from the lower end of the epididymis, and passes out of the scrotum, through the inguinal canal, to behind the bladder. It conveys sperm from the scrotum with

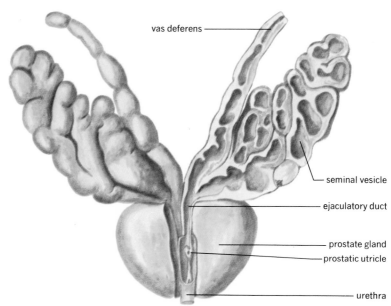

FIGURE 25.11.   The prostate, seminal vesicles, and associated structures.

much force during ejaculation. As the deferens passes out of the scrotum, it is bound by connective tissue along with nerves, arteries, veins, and lymphatics into a structure known as the SPERMATIC CORD (Fig. 25.10). The components of the cord separate within the abdominopelvic cavity. Behind the bladder, the deferens is joined by the duct of the seminal vesicle, and the common tube, or EJACULATORY DUCT, joins the urethra. The structure of the urethra has been described in Chapter 24.

FUNCTION. The ducts have smooth muscle in their walls, and serve primarily to move sperm eventually to the exterior. The epididymis is so long that the sperm within it, though moving, are considered to be "in storage" as they pass slowly through. The deferens is a propulsive tube; the urethra is a conduction tube that directs sperm (via the penis) to the female reproductive system.

The accessory glands (Fig. 25.11)

LOCATION AND STRUCTURE. The SEMINAL VESI-CLES are paired tortuous sacs lying on the posterior aspect of the urinary bladder. The PROSTATE GLAND surrounds the upper urethra, just below the neck of the bladder. It is about the size of a walnut, and consists of 30–50 separate glands all bound together by connective tissue. The paired BULBOURETHRAL (*Cowper's*) GLANDS are pea-sized, and empty into the membranous urethra.

FUNCTION. The secretion of the accessory glands, together with the sperm, constitutes the SEMEN. The ejaculate is the total volume of semen expelled during *ejaculation*. Vesicle secretion is alkaline, contains much fructose and vitamin C, and constitutes about 60 percent of the average 3 milliliters of the ejaculate. This secretion suspends and nourishes the sperm, and aids in neutralizing acidity in the vagina. Prostatic secretion is rich in cholesterol, buffering salts, and phospholipids, makes up 40 percent of the ejaculate, and is thought to activate the sperm to make them motile. It also aids in neutralizing vaginal acidity. Some 300–400 million sperm are found in the semen, and are very sensitive to acid pH's. The secretion of the bulbourethral glands is an alkaline mucus that neutralizes any urine remaining in the male urethra. The bulbouretheral glands secrete their material before ejaculation occurs.

The penis (Figs. 25.12 and 25.13)

The penis serves as a copulatory organ to introduce sperm into the vagina of the female. It is formed by three cylindrical masses of spongy

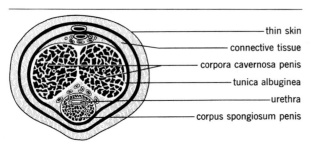

thin skin
connective tissue
corpora cavernosa penis
tunica albuginea
urethra
corpus spongiosum penis

FIGURE 25.12.   Cross section of the penis.

tissue known as the cavernous bodies. Two of the bodies are dorsally placed in the penis and are known as the CORPORA CAVERNOSA PENIS; a smaller, single, ventrally placed CORPUS SPONGIOSUM (*cavernosum*) URETHRAE carries the urethra. The cavernous bodies contain large venous sinuses which, when filled with blood, erect or stiffen the penis and make it an effective copulatory organ.

Clinical considerations

MALE INFERTILITY is defined as inability to produce children. Congenital causes, such as mal-

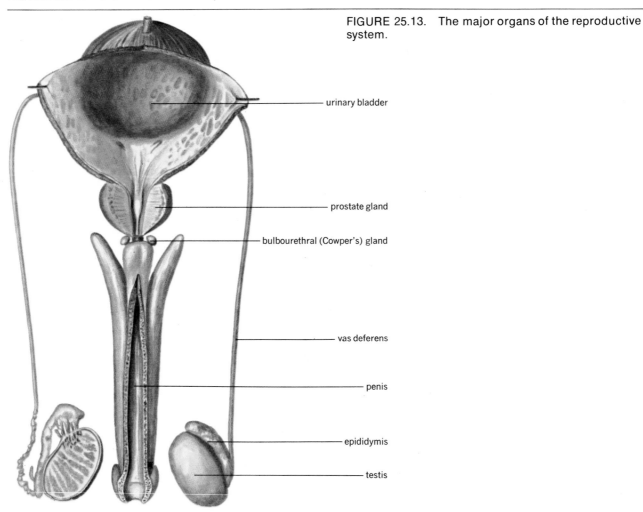

FIGURE 25.13.   The major organs of the reproductive system.

urinary bladder

prostate gland

bulbourethral (Cowper's) gland

vas deferens

penis

epididymis

testis

development of the reproductive organs may result in infertility, as may diseases (e.g., venereal diseases, mumps) that attack the testes and destroy or reduce their ability to produce sperm.

Among the most common INFECTIONS affecting the male reproductive system are gonorrhea and syphilis. The microorganisms causing these diseases may be transmitted by sexual intercourse, body-to-body, or oral contact with an infected person. Recent evidence indicates that live gonorrhea organisms may be isolated from contaminated toilets, silverware, or doorknobs up to 8 hours after deposition of the bacteria.

IMPOTENCE, or an inability to create and sustain an erection may result in inability of the male to copulate successfully with the female. Impotence appears to be an increasingly common affliction of the male in "civilized" societies.

Sperm counts less than about 50 million per milliliter of semen make it difficult for the male to fertilize a normal female because not enough sperm are provided to:

Resist the loss of numbers due to acids in the vagina.

Survive the 6 inch trip through the uterus and uterine tubes to fertilize the ovum.

Provide sufficient hyaluronidase, an enzyme that liquefies the intercellular cement holding several layers of follicular cells to the released ovum.

Sperm are capable of surviving up to 3 days within the female reproductive tract, so that pregnancy can occur even though copulation has not occurred for several days. If ovulation occurs and sperm are present, pregnancy may result.

Enlargement of the prostate gland is a common occurrence with aging. If the gland enlarges greatly, it may compress the urethra to where both reproductive function and urination become difficult or impossible. Surgery is usually indicated to remove the enlarged organ. Cancer of the prostate is the second most common neoplasm in the male, and is the sixth leading cause of cancer death in the male.

The female reproductive system

Organs

The organs of the female reproductive system (Figs. 25.14; 25.15) may be grouped under two headings:

The *internal organs* are located within the abdominopelvic cavity and include the ovaries, uterine tubes, uterus, and vagina.

The *external organs* include the external genitalia (clitoris, labia majora, and minora) and the mammary glands. The latter are actually highly modified sweat glands, but are considered with this system because of their close functional relationship to the reproductive system.

The ovaries (Fig. 25.16)

LOCATION AND STRUCTURE. The ovaries are paired, ovoid, organs lying within the pelvic portion of the abdominopelvic cavity, and measure about 3 centimeters (1½ inches) long, 2 centimeters (¾ inches) wide, and 1 centimeter (½ inch) deep. Each ovary has an indented HILUS, that marks the entry and exit of ovarian blood vessels, nerves, and lymphatics. The ovaries are supported by a mesentery, the MESOVARIUM, and OVARIAN and SUSPENSORY LIGAMENTS.

Microscopically (Fig. 25.17), each ovary is seen to be covered with a layer of cuboidal cells called the GERMINAL EPITHELIUM. This is a misnomer, inasmuch as the cells do not give rise to sex cells, but are merely a lining or covering tissue. A capsulelike TUNICA ALBUGINEA of fibrous tissue, forms the major tunic of the ovary, and surrounds a connective tissue STROMA. The stroma is further subdivided into an outer CORTEX, containing OVARIAN FOLLICLES, and an inner vascular MEDULLA, which provides for the nutrition of the organ.

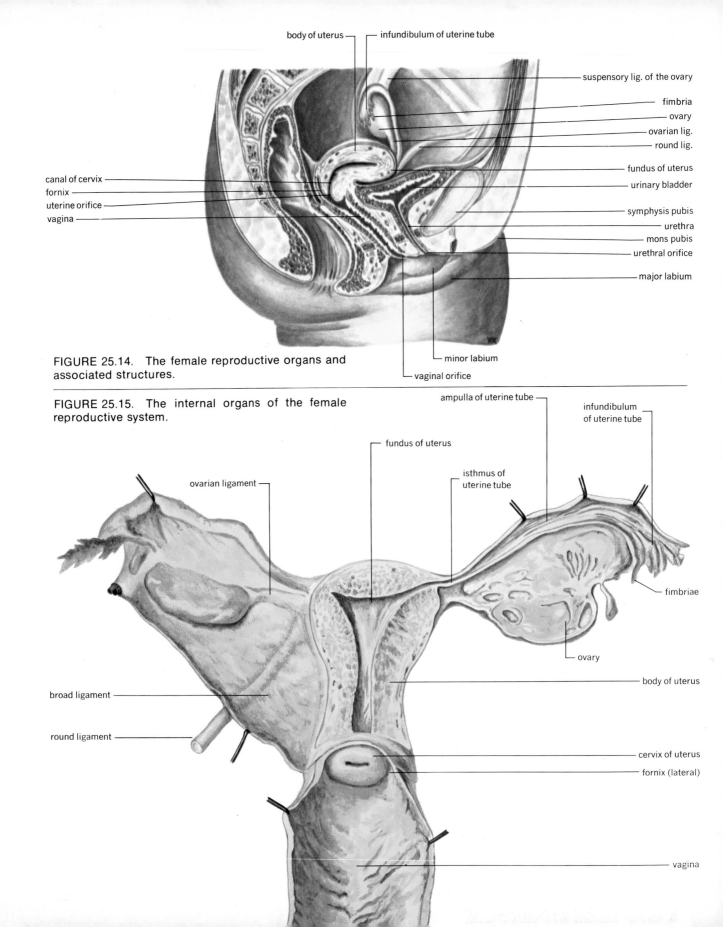

body of uterus
infundibulum of uterine tube
suspensory lig. of the ovary
fimbria
ovary
ovarian lig.
round lig.
fundus of uterus
urinary bladder
symphysis pubis
urethra
mons pubis
urethral orifice
major labium

canal of cervix
fornix
uterine orifice
vagina

minor labium
vaginal orifice

FIGURE 25.14. The female reproductive organs and associated structures.

FIGURE 25.15. The internal organs of the female reproductive system.

ampulla of uterine tube
infundibulum of uterine tube
fundus of uterus
isthmus of uterine tube
ovarian ligament
fimbriae
ovary
body of uterus
broad ligament
round ligament
cervix of uterus
fornix (lateral)
vagina

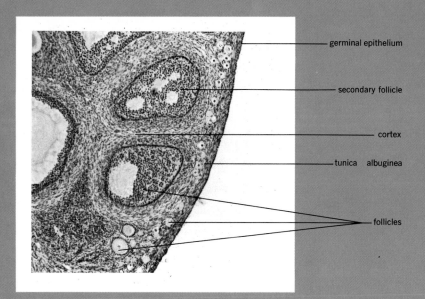

germinal epithelium

secondary follicle

cortex

tunica albuginea

follicles

FIGURE 25.16. Microscopic structure of the ovary, general view.

FIGURE 25.17. Photomicrographs of the ovaries showing the development of follicles.

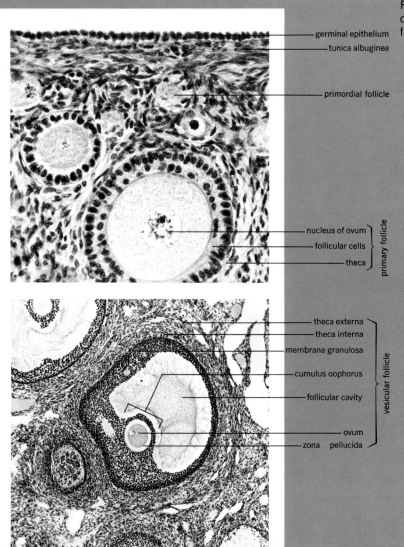

germinal epithelium
tunica albuginea

primordial follicle

nucleus of ovum ⎱
follicular cells ⎰ primary follicle
theca ⎰

theca externa ⎱
theca interna ⎰
membrana granulosa ⎱ vesicular follicle
cumulus oophorus ⎰
follicular cavity ⎰
ovum ⎰
zona pellucida ⎰

FUNCTION. The ovary, after puberty is the site of maturation of sex cells or OVA, and secretes several HORMONES necessary for development of the other internal organs, and of the female secondary sex characteristics.

The ovarian follicles (Fig. 25.18)

At birth, approximately 400,000 PRIMORDIAL FOLLICLES are present in the ovaries. These are composed of a small primitive ovum, surrounded by a single layer of stromal cells. They will remain in this form until puberty, when pituitary hormones cause their further development. Of these 400,000 original ova, about 400 will mature during the reproductive life of the female. The rest undergo a degenerative process known as ATRESIA.

At puberty, follicular growth begins, caused by a pituitary hormone, FSH (follicle stimulating hormone). PRIMARY FOLLICLES are larger structures surrounded by several layers of cells that develop from the primitive follicles. Cavities known as ANTRA (sing.: antrum) next develop within the layers of cells, and the structure is known as a SECONDARY FOLLICLE. The separate cavities fuse to form a single large cavity and, at this stage, the structure is called a VESICULAR OR MATURE FOLLICLE. It may ultimately reach a size of ½ inch. Maturation requires 8 to 15 days, and is associated with production of a hormone known as ESTROGEN. This hormone affects the lining of the uterus (causes proliferation of cells), and develops the ducts of the mammary glands.

OVULATION, or release of the mature ovum and its surrounding coat of several layers of follicle cells, occurs next under the influence of pituitary luteinizing hormone. The tissues remaining in the ovary turn into a CORPUS LUTEUM (yellow body) also under the influence of LH or luteinizing hormone. The corpus luteum produces PROGESTERONE, a hormone that prepares the uterus to receive an egg if it is fertilized, and that develops secretory tissue in the breasts. The luteum will last about 2 weeks if fertilization does not occur; it will last about 7 months of a pregnancy if fertilization does occur. In either case, it ultimately degenerates, is replaced by scar tissue, and forms a corpus albicans (white body).

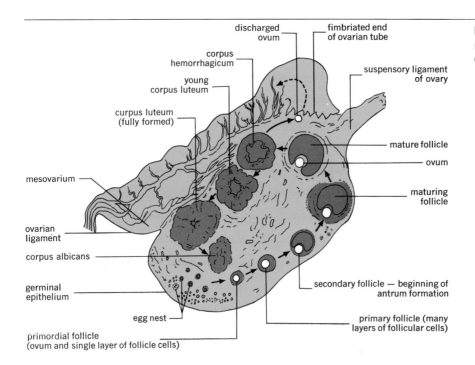

FIGURE 25.18. Mammalian ovaries showing stages in the development of an ovarian follicle.

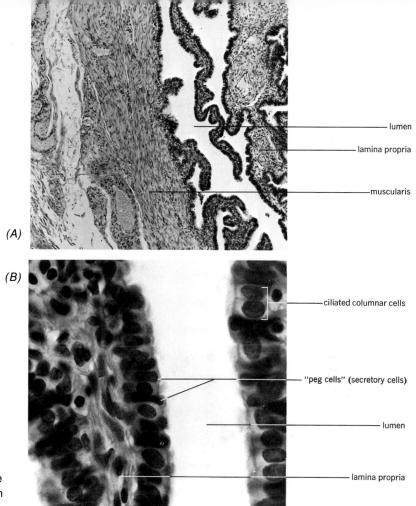

lumen

lamina propria

muscularis

(A)

(B)

ciliated columnar cells

"peg cells" (secretory cells)

lumen

lamina propria

FIGURE 25.19. Photomicrographs of the uterine tube. *(A)* Low power. *(B)* high power of epithelium.

FIGURE 25.20. Diagram illustrating the three tissue layers of the uterus.

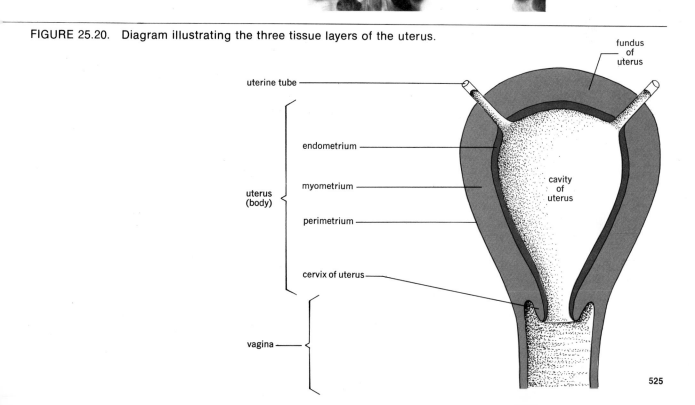

fundus of uterus

uterine tube

endometrium

uterus (body)

myometrium

perimetrium

cavity of uterus

cervix of uterus

vagina

The uterine tubes
(oviducts, Fallopian tubes;
see Figs. 25.14, 25.15)

STRUCTURE. The paired uterine tubes are about 4 inches in length, and connect to the uterus at their proximal ends. An outer funnel-shaped INFUNDIBULUM bears many fingerlike fimbrae. An intermediate AMPULLA lies between the infundibulum and the ISTHMUS that passes through the uterine wall. The tubes are lined (Fig. 25.19) by a ciliated and secretory epithelium, and have considerable smooth muscle in their walls.

FUNCTION. The uterine tubes, by ciliary action and peristaltic movements of the smooth muscle, propel the egg to the uterus in 3–4 days. A released egg will degenerate if it is not fertilized within 18–24 hours after its release. Therefore, if fertilization is to occur, it *must* occur in the distal third of the uterine tubes.

The uterus
(see Figs. 25.14 and 25.15)

LOCATION AND STRUCTURE. The single uterus is, in the female who has never been pregnant, a pear-shaped organ about 7.5 centimeters in length, 5 centimeters wide, and 2½ centimeters thick (2½ × 2 × 1 inches). It lies behind and above the urinary bladder, at an angle of about 90 degrees to the vagina. It is supported by eight ligaments.

The *broad ligaments* run laterally from the sides of the uterus.
The *round ligaments* run anteriorly from the uterus to anterior abdominal wall.
Four *uterosacral ligaments* run from the uterus to the posterior body wall.

The uterus itself has an upper FUNDUS, a tapering BODY, and a NECK or CERVIX that projects into the upper portion of the vagina.
The wall of the uterus is composed of three main layers of tissue (Fig. 25.20). An outer PERIMETRIUM consists of a thin mesothelial and connective tissue layer. A middle MYOMETRIUM contains smooth muscle and forms most of the thickness of the wall. An inner ENDOMETRIUM undergoes cyclical hormone controlled changes after puberty (Fig. 25.21). The cyclical changes occurring in the endometrium constitute the menstrual cycles.

FUNCTIONS. The uterus is an organ designed to retain a fertilized egg, and to sustain and nourish it during the approximate 280 days of a pregnancy. It expels the placenta after birth of the new individual.

The menstrual cycle

The menstrual cycle is hormonally controlled, and depends on the ovarian production of estrogen and progesterone. The cycle is divided into four phases:

The *menstrual phase* is characterized by a bloody discharge from the uterus. It occupies the first 3–5 days of a cycle, and is due to a lack of hormonal effect on the uterus. A small band of endometrial tissue remains in the uterus after menstruation has occurred.
The *proliferative (follicular or preovulatory) phase* is characterized by regrowth of endometrial tissue from the narrow band remaining in the uterus. It is under the control of estrogen from the maturing follicle, and causes about a 2-millimeter thick endometrial lining consisting of connective tissue, blood vessels, and glands. It occupies the next 7–15 days of the cycle.
The *secretory (luteal or postovulatory) phase* is the next 14–15 days of the cycle, and is characterized by an increase in glands and blood supply to the endometrium, in preparation for nourishing a fertilized ovum. If no fertilization occurs, the corpus luteum, which produces the progesterone responsible for this phase, degenerates.
The *premenstrual phase* is characterized by regression of blood vessels and glands, and by degeneration of tissue, as progesterone levels decrease. This phase terminates with the first external show of blood.

The length of a cycle varies from 24–35 days, de-

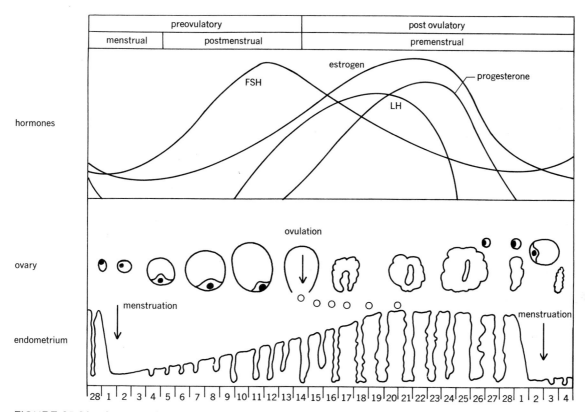

FIGURE 25.21.   Interrelationships of hormones and ovarian and uterine changes.

pending on individual differences. It should not be assumed that all cycles are 28 days, or that ovulation occurs 14 days after the external show of blood. The most variable period is the length of the proliferative phase. Misinterpretation of these times can lead to accidental or unwanted pregnancy.

Some relationships between pituitary hormones, and ovarian and uterine changes are shown in Figure 25.21.

The vagina (see Fig. 25.14)

STRUCTURE. The vagina is a tubular organ, about 8 centimeters in length, which is normally in a collapsed state; that is, its cavity (lumen) is usually not open. The cervix of the uterus projects into the upper portion of the vagina, and the moatlike FORNICES surround the cervix. A mem-

branous fold of tissue, known as the HYMEN, may partially or completely seal the distal end of the vagina. Three layers of tissue compose the wall of the organ: an inner MUCOSA is lined with stratified squamous epithelium containing much glycogen in the cells; a middle MUSCULARIS contains smooth muscle; an outer CONNECTIVE TISSUE LAYER fixes the organ in position. The glycogen is broken down into organic acids that create an acid environment to retard microorganism growth.

FUNCTION. The vagina receives the penis during sexual intercourse, and serves as the birth canal during childbirth.

The external genitalia (Fig. 25.22)

Collectively, the external genitalia are known as the VULVA or PUDENDUM. The MONS PUBIS is a

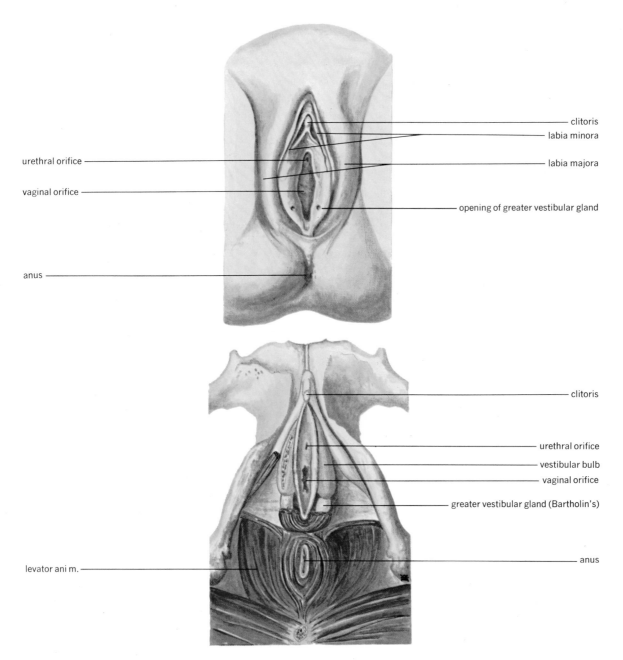

FIGURE 25.22. The external organs of the female reproductive system and associated structures.

rounded pad of fat covering the pubic symphysis. The LABIA MAJORA are two fat-filled and (after puberty) hair-covered folds of skin extending from the mons pubis towards the anus. The LABIA MINORA are smaller folds of skin, lacking hair and fat, that closely surround the clitoris and vaginal opening. The CLITORIS is a single midline organ lying anterior to the urethral opening. The term OBSTETRIC PERINEUM refers to the area of the pelvic floor (perineum) between the vagina and anus. It may be incised (cut) to permit easier passage of an infant during birth, and to avoid tearing the perineum, in what is termed an EPISIOTOMY.

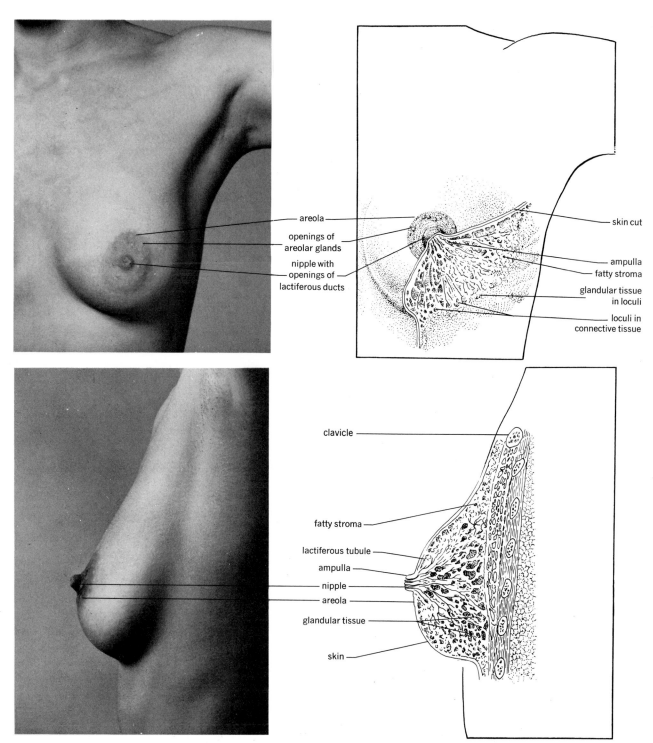

areola
openings of
areolar glands
nipple with
openings of
lactiferous ducts

skin cut

ampulla
fatty stroma

glandular tissue
in loculi

loculi in
connective tissue

clavicle

fatty stroma

lactiferous tubule
ampulla
nipple
areola
glandular tissue

skin

FIGURE 25.23.   The mammary glands.

The mammary glands (Fig. 25.23)

The mammary glands are modified sweat glands, functionally related to the reproductive system. They produce milk after childbirth for nourishment of the newborn.

Each gland consists of a comma-shaped mass of connective and glandular tissue, located mainly between the second and sixth ribs on the anterior chest wall. Within each gland, there are 15–20 individual lobes of glandular tissue, each served by a milk (lactiferous) duct. These ducts form 3–5 larger ducts that empty to the exterior through the apex of the NIPPLE. Each larger duct contains an enlargement known as the lactiferous sinus that can store secreted milk until the infant removes it by suckling. The areola is the circular, pigmented region surrounding the nipple.

Clinical considerations

CARCINOMA (cancer) of the breast is the most common form of cancer in the female, and is the leading cause of death from cancer in the female. Cancer of the uterine cervix is the third most common type of cancer in the female. Breast cancer may be easily discovered by routine self-examination of the breasts, and the Papanicolaou (Pap) test can discover many cervical cancers before they become life threatening. Chronic irritation, trauma, and viral infections appear to be the most common causes of cervical cancer.

AMENORRHEA refers to failure of a female to menstruate by 18 years of age or cessation of menstruation in a menstruating female, and calls for a complete investigation of the functional status of the hypothalamus, pituitary, ovaries, and uterus.

DYSMENORRHEA refers to painful menstruation. Its cause is not known, and is associated with cramps and pain 24 to 48 hours before menstrual flow.

Venereal diseases

The VENEREAL DISEASES (VD) are the number one communicable diseases in the United States today. Explanations for the continuing increase in incidence of the diseases may be due to several basic causes.

1. There have been and will continue to be changes in attitudes toward sexual intercourse that separate it from its reproductive function. Increased sexual contact has therefore increased the possibility of spread of the diseases among the population.

2. The increasing number of younger people in the population has increased the incidence of sexual contact with greater possibility of spread of the diseases. (The highest incidence is presently in the 15–30 age group.)

3. A large part of the social stigma associated with the diseases has disappeared and reporting of cases has increased. It may be hoped that increased reporting of cases has been an incentive for the establish-

ment or increase in funding of programs designed for case finding and treatment of venereal diseases.

4. The discontinuance of indiscriminate use of antibiotics for "colds" and other infections that do not warrant their use has resulted in the removal of a "brake" on the spread of VD, in that many previously undiagnosed cases are no longer being treated inadvertently. Use of antibiotics can and has resulted in the development of antibiotic strains of the VD organisms.

5. Better epidemiological studies have resulted in the discovery of reservoirs of infection that may not have been brought to light previously.

6. The belief of many segments of the public that a "shot" of antibiotic will cure the disease(s) has perhaps led to the development of complacency, as well as a false sense of security. All patients must be rechecked at a certain interval following treatment to insure that the tests for organisms are negative. Also, after an initial set of symptoms, several of the

venereal diseases may enter an asymptomatic period that may persist for years, leading to the belief that the disease has disappeared. During this time, damage may continue to be wrought in various body systems although no overt changes become apparent.

7. Reinfection is common since no immunity is developed to the organisms. Contacts must be examined and adequately treated if they are infected to prevent further spread. Sexual partners, diagnosed as infected, must be treated at the same time or they will be playing "ping-pong" with the organism. Persons having a gonococcal infection must also be checked for syphilis as both infections may be present at the same time but the amount and type of treatment may differ.

Gonorrhea ("clap") and syphilis ("bad blood") are presently in epidemic proportions throughout the United States, and some other countries. These diseases have no respect for race, creed, or color, and one cannot tell an infected person by just looking at him or her.

All personal information regarding cases, contacts, and suspects of VD is legally CONFIDENTIAL, which should encourage infected persons to seek early treatment of the disease(s).

Free treatment is available through your local health department. Other clinic facilities or private physicians or both are available in most communities. Information on your state's law regarding the treatment of minors may be obtained through your health department. Some states are now treating minors without the parents' consent, since the child's future welfare is at stake, and as an effort to cut down the VD incidence in the sexually active teenagers.

Some characteristics and effects of gonorrhea and syphilis are presented in Table 25.1.

TABLE 25.1 Some Characteristics and Effects of Gonorrhea and Syphilis		
	Gonorrhea	**Syphilis**
Causative organism	Neisseria gonorrhoeae	Treponema pallidum
Incubation period	2–14 days (usually 3 days)	7–90 days (usually 3 weeks)
Method of transmission	Sexual, oral, or physical contact with an infected person. Contaminated object up to 8 hours after organisms deposited. *Infants*—during birth through vagina.	Sexual, oral, or physical contact with an infected person. Blood transfusion. On contaminated objects, dies quickly by drying. *Infants*—may acquire during birth, or through the placenta if mother not treated before third trimester and adequately.
Contact examination *(all sex contacts exposed within the following time periods)*	2 weeks (male) 1 month (female)	*Primary.* 3 months (+ duration of symptoms). *Secondary.* 6 months (+ duration of symptoms). *Early latent.* 1 year. *All syphilis.* "Family" contacts as indicated.

TABLE 25.1   Some Characteristics and Effects of Gonorrhea and Syphilis

	Gonorrhea	Syphilis
Clinical characteristics	Discharge; burning, pain, swelling of genitals and glands.	**Primary.** Chancre present, solitary, nonpainful ulcer on genital or mucous membranes.
	Male. Purulent urethral discharge, hematuria, chordee. Urethritis, prostatitis, seminal vesiculitis, epididymitis, occasional involvement of testes.	**Secondary.** Rashes or mucous patches. Macules or papules on hands, feet, oral cavity, genitoanal area, trunk, extremities.

Tertiary:
Latency. No symptoms, positive serology. Profound changes are produced in the skin, mucous membranes, skeleton, GI tract, kidney, brain; heart and blood vessels show destructive changes (abscesses, scarring, tissue destruction). Tabes dorsalis in spinal cord destroys dorsal columns. Paresis and psychosis may result. |
	If untreated, symptoms disappear in about 6 weeks, and organism persists in the prostate gland (Gc carrier).	*Relapse.* Recurrence of infectious lesions after disappearance of secondary lesions.
	Female. *Child,* vaginitis. Leukorrhea, tubal abscess, urethritis, cervicitis, pelvic inflammation (Peritonitis).	*Late.* Cardiovascular, central nervous system, gummata. Obvious systemic damage appears.
	Tubal stricture, possible sterility due to closure of tubes.	
	In both male and female, healing is by scar tissue formation. Strictures and closing of tubular structures may result. Arthritis, endocarditis, meningitis may occur.	

	Gonorrhea	Syphilis
Diagnostic procedures	Culture of discharge. Smear. Fluorescent antibody test (FAT). Currettage to get tissue containing cocci. Several cultures may be necessary as not all tests consistently show cocci. History, clinical and contacts. (Serologic test for syphilis).	Darkfield examination—microscopic for spirochete. Serologic (blood) test for antibody (reagin) to organism. [(Wasserman, VDRL—Venereal Disease Research Lab) False positive te ts may be reported by smallpox antibodies, hepatitis, mononucleosis, and high fevers.] Spinal fluid test. X-rays of long bones of infants. History, clinical and contacts.
Treatment	Penicillin Broad spectrum antibiotics (e.g. sulfonamides, streptomycin). Organism must be sensitive to the drug of choice.	Penicillin Broad spectrum antibiotics (e.g., Erythromycin, Tetracycline). Organism must be sensitive to drug of choice. Reexamination at 6 months and 1 year to evaluate treatment results.

Summary

1. The male and female reproductive organs provide for creation, development, and nourishment of new individuals. Several hormones are produced by certain of the organs and affect the whole body as well as the organs of reproduction.

2. The system develops from mesoderm at about 5 weeks. The organs go through an undifferentiated state and, at about 8 weeks, assume the appearance of male or female.

3. The male reproductive organs include the testes, a system of ducts, and accessory glands.

   a. The testes are found in the scrotum, are ovoid, are surrounded by tunics, and produce sperm cells in their seminiferous tubules, and a hormone (testosterone) in the interstitial cells.

   b. The ducts (straight and rete tubules, efferent ductules, epididymis, and vas deferens) transport and store sperm.

   c. The accessory glands (seminal vesicles, prostate, bulbourethral glands) produce the semen in which sperm are suspended. Semen nourishes, and activates the sperm, and neutralizes acid in male urethra and vagina.

d. The penis serves as a copulatory organ, and is composed of three cavernous bodies.

4. The female reproductive organs include internal organs (ovaries, uterine tubes, uterus, and vagina), the external genitalia (labia majora and minora, and the clitoris), and the mammary glands.

   a. The ovaries are found in the pelvic cavity, are ovoid, are covered with germinal epithelium, a fibrous tunic, and contain a stroma (cortex and medulla) and follicles. They produce ova and two main hormones.

   b. The follicles are present in a primordial state at birth. At puberty they develop through primary, secondary, and vesicular follicles. The vesicular follicle produces estrogen, primarily responsible for female sexual maturity. Ovulation is followed by corpus luteum formation and progesterone production; progesterone insures uterine wall development and mammary development for milk secretion. A scar is finally formed by the follicle.

   c. The uterine tubes are about 4 inches long and pass eggs to the uterus. Fertilization occurs in the tube.

   d. The uterus is an organ in which new individuals develop. It is supported by eight ligaments, has three parts (fundus, body, and cervix), and three layers in its wall (perimetrium, myometrium, and endometrium). The inner endometrium undergoes cyclical changes after puberty in the menstrual cycles.

   (1) Each cycle has four stages: menstrual (bleeding), proliferative (regrowth), secretory (development of many glands), and premenstrual (degeneration of blood vessels).

   (2) The proliferative and secretory stages are controlled by estrogen and progesterone respectively.

   e. The vagina is a tubular organ serving as the receptacle for the penis, and as the birth canal.

   f. The external genitalia consist of labia majora and minora, and the clitoris.

   g. The mammary glands are lobed structures that produce milk for nourishment of offspring.

5. The characteristics and effects of venereal disease are summarized in Table 25.1.

Questions

1. Compare ovaries and testes as to structure, function, and hormone production.

2. Trace an ovum's development in the ovary.

3. Describe the changes occurring in the uterus during a menstrual cycle.

4. How is maturation of eggs and sperm controlled in the two sexes?

5. With the continuing increase in venereal diseases, what do you think could be done to reduce the incidence of the diseases, or to discover undiagnosed cases?

Readings

Benson, Ralph C. *Handbook of Obstetrics and Gynecology.* 4th ed. Lange Medical Pubs. Los Altos, Cal., 1971.

Johnson, A. D., W. R. Gomez, and N. L. Vandemark (eds). *The Testis.* 3 vols. Academic Press. New York, 1971.

Masters, Wm. H., and V. E. Johnson. *Human Sexual Response.* Little, Brown. Boston, 1966.

Rugh, Roberts, and L. B. Shettles. *From Conception to Birth.* Harper & Row. New York, 1971.

Trussel, James, and Steve Chandler. *The Loving Book.* Red Clay Pubs. Charlotte, N. C., 1971.

chapter 26
The Endocrines

chapter 26

Working with the nervous system, the endocrine system provides a means of controlling many body functions. Endocrine control is exerted by chemical substances that are secreted into and carried by the bloodstream, and usually requires a somewhat longer time to exert an effect than does the nervous control. Additionally, endocrines control more widespread processes in the body, rather than controlling organs, as is more true of nervous control. For example, endocrines control processes such as metabolism, growth, and development.

Development of the system

Many of the endocrine glands are derivatives of the mouth, pharynx, and gut. Others are derivatives of the ectoderm that forms the nervous system, or of the mesoderm. Table 26.1 indicates when the listed endocrines appear, and their area or germ layer of origin.

TABLE 26.1	Development of Some of the Endocrines			
Endocrine	Age of first appearance (weeks)	Age of typical structure (weeks)	Area(s) formed in or from	Germ layer of origin
Thyroid gland	3½	12	Pharynx floor	Endoderm
Pituitary	4	16	Roof of mouth and brain	Ectoderm
Adrenal cortex	5	7	Dorsal celom	Mesoderm
Adrenal medulla	5	7	Neural crest by spinal cord	Ectoderm
Pancreas islets	12	14	Foregut	Endoderm
Parathyroids	7	9	Pharynx	Endoderm
Testes and ovaries	6	8	Pelvic cavity	Mesoderm
Pineal gland	7	11	Brain stem	Ectoderm

Organs of the system and general principles of endocrinology

Organs: Criteria of endocrine status

Although not a connected series of organs forming a system such as we have studied in previous chapters, all endocrines have several things in common: they have no ducts; they secrete specific chemicals; removal of an endocrine causes clearly defined alterations in function that may be restored to normal by administration of an extract of the suspected endocrine; they are extremely vascular and consist of well-defined groups of cells. The recognized endocrines are shown in Figure 26.1.

Hormones

The specific chemical(s) secreted by an endocrine gland is a hormone. Strictly speaking, chemicals such as gastrin, secretin, cholecystokinin, and other substances are not true hormones, since their source is not definitely known; nevertheless, such substances are often called hormones.

Chemically, hormones fall into three classes: proteins or polypeptides, small amines, or steroids. In general, protein hormones are secreted by endodermal endocrines, amines by ectodermal, and steroids by mesodermal endocrines.

Control of secretion

Hormones are secreted according to body need for them. This obviously implies the presence of control mechanisms to govern production and release. Control is supplied by several methods:

FIGURE 26.1.   Locations and names of endocrine glands.

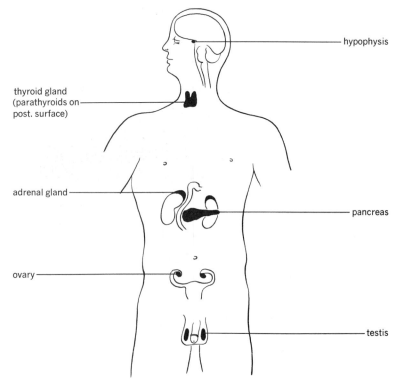

NEGATIVE FEEDBACK INVOLVING TWO HORMONES. One endocrine secretes a hormone, it acts on another endocrine, and the hormone of the second endocrine inhibits secretion by the first endocrine.

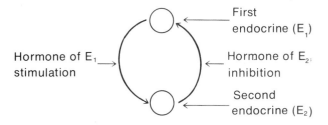

The thyroid, adrenals, and gonads are controlled in this manner.

NERVES. The posterior part of the pituitary, and the adrenal medulla release hormones upon nervous stimulation. Response is thus very rapid.

PRODUCTION OF NEUROHUMORS. Chemicals produced by nerve cells are called neurohumors. They may pass over nerve fibers or be placed into blood vessels to control an endocrine. The hy-

TABLE 26.2 General Summary of Endocrine Glands	
Item	Comments
Functions of endocrines	To integrate, correlate, and control body processes by chemical means.
Criteria for establishing function as endocrine	Cells are morphologically distinct.
	Cells produce specific chemicals not produced elsewhere.
	Chemicals exert specific effects; effect is lost if gland removed, restored if chemical administered.
Hormones	Hormones are produced by cells qualifying as endocrines.
Chemical nature of hormones	Steroid—mesodermal origin (e.g.: adrenal cortex, gonads).
	Polypeptides—endodermal origin (e.g.: pancreas).
	Small MW amines—ectodermal origin (e.g.: posterior pituitary, adrenal medulla).
Methods of control available	
1. Neurohumor	Nerve cells produce a chemical, it goes to endocrine and controls secretion. Stimulus is chemical.
2. Nerves	Nerve fibers pass to endocrine. Stimulus is electrical (nerve impulse).
3. Feedback	Target organ hormone influences secretion of another endocrine which stimulated target organ.
4. Nonhormonal organic substances in blood	Glucose acting on pancreatic islets. Rise causes increased secretion (usually).
5. Total osmolarity of blood	Requires nervous system to detect it. Nervous system then signals endocrine.
6. Inorganic substances in blood	Ca^{++} on parathyroid. Effect usually direct, not inverse.

pothalamus controls the anterior part of the pituitary in this manner.

BLOOD LEVELS of substances other than hormones. Blood sugar (glucose) levels control pancreas hormone secretion; blood calcium levels control parathyroid secretion; blood levels of total solutes (osmotic pressure) control release of certain posterior pituitary hormones.

A summary of these generalities is presented in Table 26.2.

The pituitary (hypophysis)

Location and description

The pituitary gland is about the size of a large pea (about 10 millimeters in diameter), weighs ½ gram, and lies in the sella turcica of the sphenoid bone just beneath the hypothalamus of the brain. Part of it, the ADENOHYPOPHYSIS, is derived from the roof of the mouth, while the remainder, the NEUROHYPOPHYSIS, is a downgrowth of the hypothalamus. The adenohypophysis is, in turn, composed of a large ANTERIOR LOBE, and a narrow INTERMEDIATE LOBE. The neurohypophysis includes the STALK that attaches the pituitary to the brain, and the POSTERIOR LOBE. These divisions are shown in Figure 26.2.

The anterior lobe is composed of acidophil (red staining), basophil (blue staining), and chromophobe cells (little or no staining). Of these, the first two are regarded as being the sources of the anterior lobe hormones. The anterior lobe is connected to the hypothalamus by a system of capillaries called the PITUITARY PORTAL SYSTEM. The intermediate lobe is a narrow band of red- and blue-staining cells.

FIGURE 26.2.   The divisions of the pituitary gland.

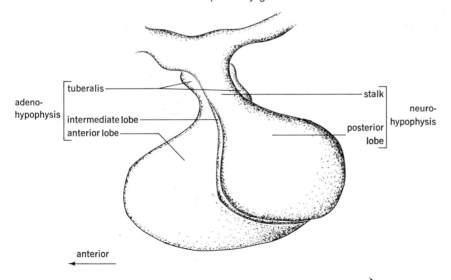

actual size
of gland

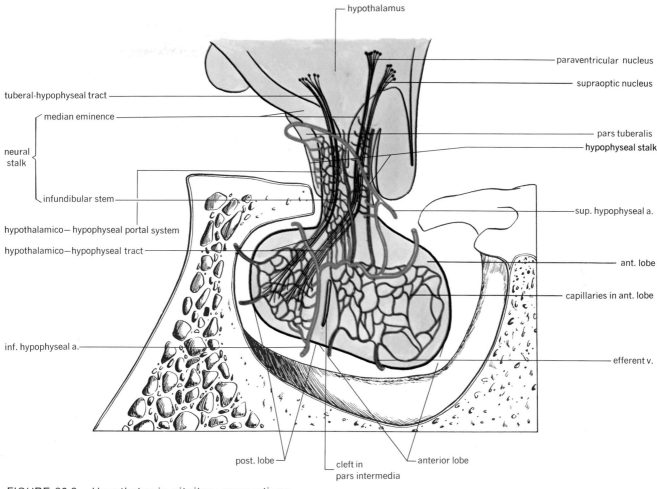

FIGURE 26.3. Hypothalamic-pituitary connections.

The posterior lobe is composed of modified nerve cells called PITUICYTES, and many nerve fibers. The posterior lobe is connected to the hypothalamus by these fibers, whose cell bodies are in the hypothalamus itself. These relationships are shown in Figure 26.3.

Hormones of the anterior lobe and their effects

At least six hormones are known to be secreted by the acidophils and basophils of the anterior lobe.

Growth hormone (somatotropin, Fig. 26.4) is a giant molecule composed of 188 amino acids. It increases protein synthesis in all body organs and thus aids in determining their rate of enlargement; it controls cartilage production in the ends of the long bones of the body and thus controls bone growth in length; it promotes fat metabolism and decreases use of carbohydrates.

Thyroid stimulating hormone (TSH, thyrotropin) is concerned with controlling all phases of thyroid gland activity, including cell division, synthesis, and release of thyroid hormones.

Adrenocorticotropic hormone (ACTH, corticotropin) controls the growth and secretory activity of the

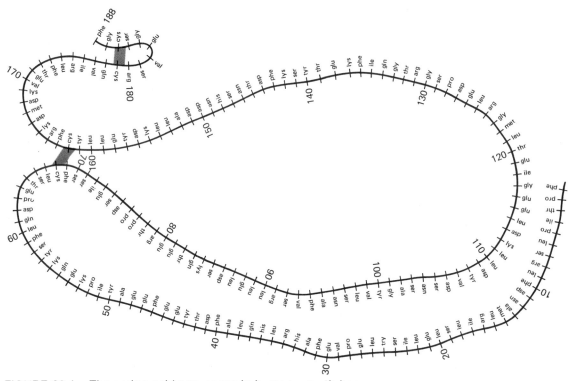

FIGURE 26.4.   The amino acid sequences in human growth hormone.

FIGURE 26.5.   Hormones involved in lactation.

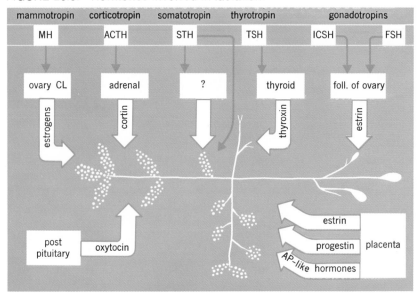

adrenal cortex. It also provides resistance to stress imposed on the body by stimulating adrenal cortical secretion of steroids.

Two *gonadotropic hormones,* known as *follicle stimulating hormone* (FSH) and *luteinizing hormone* (LH, luteotropin) also known in the male as *interstitial cell stimulating hormone* (ICSH), are produced by the anterior lobe. FSH stimulates maturation of ovarian follicles, and sperm in the testes. LH is reponsible for ovulation, corpus luteum formation and, as ICSH, secretion of testicular interstitial cells.

Prolactin (lactogenic hormone) aids in maintenance of the corpus luteum of pregnancy, and in maintaining milk secretion, if the mammary glands have been brought to a "ready state" by several other hormones (Fig. 26.5).

TABLE 26.3   Hypothalamic Neurohumors* Controlling the Pituitary Anterior Lobe

Factor	Controls secretion of (hormone)	Effect on secretion	Comments
Growth hormone releasing factor (GHRF)	GH	Stimulates	GHRF is released by low blood sugar, increased amino acid intake, stress, sleep.
Growth hormone inhibiting factor (GIF)	GH	Inhibits	
Prolactin releasing factor (PRF)	Prolactin	Stimulates	Produced after childbirth; allows lactation to occur.
Prolactin inhibiting factor (PIF)	Prolactin	Inhibits	Present continually in nonpregnant and antenatal pregnant female.
Corticotropin releasing factor (CRF)	ACTH	Stimulates	Factor released by direct effect of hormones of target glands or via hypothalamus.
Thyrotropin releasing factor (TRF)	TSH	Stimulates	
Luteinizing hormone releasing factor (LRF)	LH; ICSH	Stimulates	
Follicle stimulating hormone releasing factor (FSHRF)	FSH	Stimulates	

*These factors were formerly called *releasing factors* (RF). Since it is now known that these factors may both stimulate or inhibit secretion, it is appropriate to call them *regulatory factors* when speaking collectively of the chemicals.

Control of anterior lobe secretion

Primary control of anterior lobe secretion is afforded by, at least, eight neurohumors that are produced in the hypothalamus in response to nervous, hormonal, and other influences. These neurohumors are secreted into the pituitary portal system and are delivered to the anterior lobe to exert their effects. The pituitary secretion of TSH, ACTH, FSH, and LH (ICSH) may also be directly influenced by blood levels of thyroid, adrenal, ovarian, and testicular hormones in a negative feedback mechanism. Table 26.3 summarizes the known hypothalamic neurohumors.

The intermediate and "posterior lobe hormones" and their effects

MELANOCYTE STIMULATING HORMONE (MSH) is produced by the intermediate lobe. Its significance in the human has not been established. In lower animals, the hormone is secreted in response to changes in light intensity and causes contraction of certain pigment cells of the skin, enabling camouflaging of the organism for survival purposes.

The posterior lobe produces no hormones. OXYTOCIN and ANTIDIURETIC HORMONE (ADH) are secreted by nerve cells in the hypothalamus, and

TABLE 26.4 A Summary of Pituitary Hormones and Their Effects		
Hormone	Site of production	Effects
Growth hormone	Anterior lobe	Controls growth of body; influences fat and sugar metabolism
Thyroid stimulating hormone	Anterior lobe	Controls thyroid activity
Adrenocorticotropic hormone	Anterior lobe	Controls adrenal cortex activity
Follicle stimulating hormone	Anterior lobe	Causes ovarian follicle and sperm maturation
Luteinizing hormone	Anterior lobe	Causes ovulation and corpus luteum formation
Interstitial cell stimulating hormone	Anterior lobe	Causes secretion of interstitial cells of testes
Prolactin	Anterior lobe	Sustains lactation
Melanocyte stimulating hormone	Intermediate lobe	Contracts melanocytes
Oxytocin	Posterior lobe	Stimulates uterine contraction
Antidiuretic hormone	Posterior lobe	Permits kidney to reabsorb water

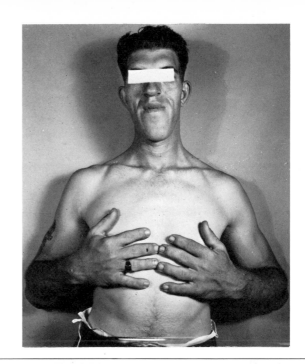

FIGURE 26.6. Acromegaly in an adult. Note the coarseness of the facial features, the enlarged mandible, and the large hands with thick, blunt fingers. (Armed Forces Institute of Pathology.)

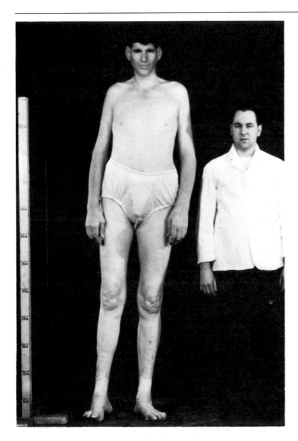

FIGURE 26.7. Giantism in a 42-year-old man. He is 7 feet, 6 inches tall. His companion is normal. The stick is 6 feet tall. (From the teaching collection of the late Dr. Fuller Albright. Courtesy Endocrine Unit and Department of Medicine, Massachusetts General Hospital.)

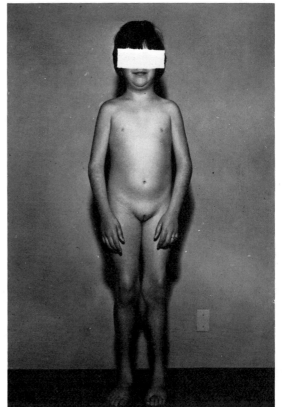

FIGURE 26.8. Hypophyseal infantilism in a 15-year-old female. Notice shortness of stature (4 feet, 3¾ inches tall), and failure to develop sexually. (Armed Forces Institute of Pathology.)

pass over nerve fibers to the posterior lobe where they are stored in the ends of the axons of the nerve cells that produced them. Release is due to nerve impulses arriving over these same axons. Oxytocin is a stimulant to uterine contraction during childbirth, and is important as an agent in causing uterine contraction after childbirth. This latter effect prevents hemorrhage. ADH is released according to the osmotic pressure of the hypothalamic blood, and promotes water reabsorption by the kidney.

In cases where a pregnant female is long overdue in delivery of her child, or in one who has been in prolonged labor without delivering, her uterus is sometimes caused to start contractions (induction) or to contract more strongly by injection of *oxytocics*. These are substances that act in the same manner on the uterus as does oxytocin. *Posterior pituitary injection* is an extract of the posterior lobe of domestic animals and contains both hormones. *Oxytocin* itself, or *synthetic oxytocin* may also be employed.

Table 26.4 summarizes the pituitary hormones and their effects.

Clinical considerations

In theory at least, one or several pituitary hormones may be affected by pathological conditions that cause the cells to either oversecrete or undersecrete hormones. Table 26.5 and Figures 26.6 to 26.8, present the essential features of several pituitary disorders.

TABLE 26.5 Some Disorders of the Pituitary

Condition	Cause	Hormone(s) involved	Secretion is Excess	Secretion is Deficient	Characteristics	Comments
Giantism	Anterior lobe tumor before maturity	Growth hormone (GH)	X		Large body size	Proportioned but large
Acromegaly	Anterior lobe tumor after maturity	Growth hormone (GH)	X		Misshapened bones, hands, face, feet	Disproportional growth of face, and extremities
Hypophyseal infantilism	Destruction of anterior lobe cells by disease, accident, etc.	Growth hormone (GH)		X	Juvenile appearance, sexually and in height	Body properly proportioned
Water retention + demonstration of high ADH blood levels	Hypothalamic tumor	Antidiuretic hormone	X		Dilution of body fluids, weight gain	Must differentiate from H_2O retention due to other causes
Diabetes insipidus	Hypothalamic damage	Antidiuretic hormone		X	Excessive urine excretion	No threat to life

The thyroid gland

Location and description

The thyroid gland (Fig. 26.9) is a two-lobed organ lying on the anterior upper trachea and lower larynx. The two LATERAL LOBES are connected across the midline by the ISTHMUS, which often shows a PYRAMIDAL LOBE. It weighs about 20 grams in the adult, and receives a great blood supply (80–120 milliliters per minute).

The structural and functional units of the gland are spherical hollow units known as THYROID FOLLICLES. They are lined with an epithelium which synthesizes thyroid hormone. The follicles contain a pink-staining material known as the THYROID COLLOID that is composed of a substance called THYROGLOBULIN. Thyroglobin is a storage form for thyroid hormone. Between the follicles, and scattered in between the follicular epithelial cells are PARAFOLLICULAR CELLS. These cells also produce a hormone.

Hormones and effect

The thyroid follicular cells synthesize THYROXIN (T_4) and TRIIODOTHYRONINE (T_3) (Fig. 26.10) by iodination (adding iodine) of the amino acid tyrosine in the thyroglobulin of the colloid. The hormones, when needed, are released from the colloid by an enzyme and enter the bloodstream. T_4 represents about 95 percent of the hormone secretion, T_3 represents about 5 percent. Both hormones govern the overall rate of biological oxidation in the body (metabolic rate), are essential for normal growth and development, especially of the brain, and exert effects on the metabolism of foods, such as increasing cell uptake of glucose, increasing glycogenolysis, and promoting protein synthesis.

CALCITONIN (thyrocalcitonin) is believed to be produced in the human, by the parafollicular cells. It is involved in the metabolism of calcium

FIGURE 26.9. Thyroid gland. *(A)* Gross anatomy and location. *(B)* Microscopic anatomy.

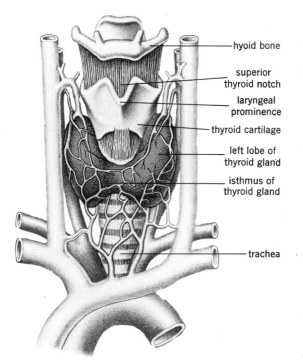

- hyoid bone
- superior thyroid notch
- laryngeal prominence
- thyroid cartilage
- left lobe of thyroid gland
- isthmus of thyroid gland
- trachea

(A)

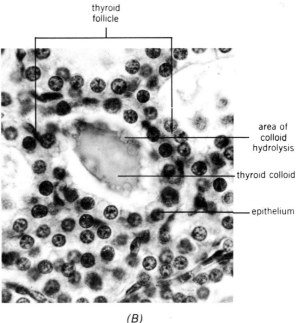

thyroid follicle

- area of colloid hydrolysis
- thyroid colloid
- epithelium

(B)

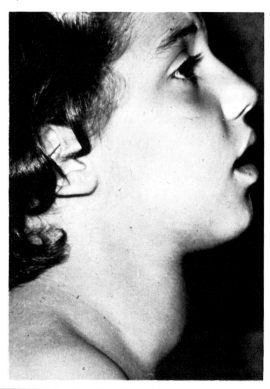

FIGURE 26.10. The formulae of Thyroxin (T_4) and Triiodothyronine (T_3).

HO—⬡—O—⬡—CH_2—$CHNH_2$—COOH

T_4 − thyroxin

HO—⬡—O—⬡—CH_2—$CHNH_2$—COOH

T_3 − triiodothyronine

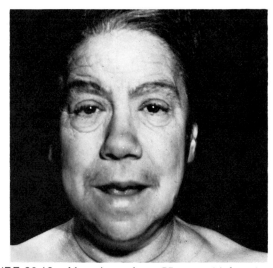

FIGURE 26.12. Myxedema in a 55-year-old female. Her PBI was less than 1 microgram per 100 milliliter (normal 4-8), and metabolic rate was 51 percent below normal. Notice puffy appearance of face and generalized fatigued appearance. (Armed Forces Institute of Pathology.)

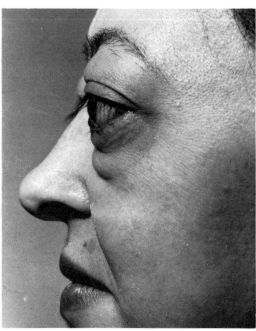

FIGURE 26.13. Hyperthyroidism. Note the protrusion of the eyeball (exophthalmos) that may accompany the disease.

and phosphate in the body. It causes a fall of blood Ca^{++} and $PO_4^{\equiv}$ levels, probably by increasing their incorporation into bony tissue.

Control of secretion

All phases of thyroid activity, including the production and release of T_4 and T_3 are controlled by pituitary TSH. T_4 and T_3, in turn, exert a negative feedback effect on TSH secretion. Calcitonin secretion is triggered by a rise of blood Ca^{++} levels.

Clinical considerations

If dietary iodine intake is insufficient, the thyroid may enlarge and forms a SIMPLE GOITER (Fig. 26.11). Using iodized salt will prevent this disorder.

HYPOTHYROIDISM (Fig. 26.12) or failure of the gland to produce enough T_3 and T_4 to meet body needs, causes a lowering of metabolic rate, the development of dry flaky skin, and lack of resistance to cold temperatures. Myxedema is a term sometimes used to describe hypothyroidism occurring after birth. If hypothyroidism occurs in a child before birth, CRETINISM results, with dwarfing and mental retardation common.

HYPERTHYROIDISM (Fig. 26.13) results from overstimulation by TSH, or may be due to a tumor in the gland. In this case, metabolic rate is elevated, heart action is very rapid, blood pressure increases, there is weight loss, and the gland may enlarge (goiter). The eyes may bulge (exophthalmus) due to increase in mass and water content of the tissues behind the eyes.

STATUS OF THYROID FUNCTION is commonly determined by analyzing a blood sample for its content of *protein bound iodine* (PBI). T_4 and T_3, when secreted into the bloodstream, bind to plasma proteins; thus a PBI determination is an indirect measure of iodine containing hormones. A normal PBI is 4–8 micrograms per 100 milliliters of plasma. To make hormones, the gland

TABLE 26.6 Summary of the Thyroid	
Item	Comments
Location and parts; weight	Two-lobed, plus connecting isthmus; located on lower larynx and upper trachea; 20 grams in weight
Requirements for hormone synthesis	The amino acid tyrosine; iodide in diet; TSH to control all steps in synthesis

Hormones produced and effects:
   Thyroxin
   Thyrocalcitonin
	
	Main controller of catabolic metabolism
	Lowers blood Ca^{++} and $PO_4^{\equiv}$ levels

Control of hormone secretion:
   Thyroxin
   Thyrocalcitonin
	
	By thyroid stimulating hormone
	By blood Ca^{++} level

Disorders:
   Hypersecretion
 	
   Hyposecretion
	
	Creates hyperthyroidism:
	   Elevated BMR, elevated heart action,
	   exophthalmus, goiter.
	Creates:
	   Goiter-enlarged gland,
	   Hypothyroidism (cretinism, myxedema)
	   Low BMR, low heart rate, blood pressure,
	   body temperature

must accumulate iodine as one of its building blocks. Radioactive iodine may be given, and its rate of uptake by the gland determined by a counting device. Normally, 25--50 percent of a given dose will be picked up by the gland in 24 hours.

Table 26.6 summarizes facts about thyroid gland.

The parathyroid glands

Location and description (Fig. 26.14)

There are usually four parathyroid glands, located on the posterior aspect of the thyroid lobes. Each measures about 5 millimeters in diameter, and all four weigh about 120 milligrams.

Microscopically, the gland shows tightly packed cells (principal cells) that produce a hormone.

Hormone and effects

The parathyroid glands produce PARATHYROID HORMONE (PTH), also called parathormone. The hormone controls, with calcitonin, the metabolism of calcium and phosphate in the body. Specifically, PTH increases absorption of these minerals from the gut, increases the reabsorption of calcium by the kidney, and governs blood calcium and phosphate levels by controlling the destruction of bony tissue.

Control of secretion

A fall of blood calcium level increases PTH secretion. More minerals are released from bone, and the blood calcium level rises.

Clinical considerations

Insufficient production of PTH produces HYPOPARATHYROIDISM or TETANY. Because of the low blood calcium levels, muscle excitability is disturbed, and the skeletal muscles undergo painful cramps (tetany). Convulsions may also occur. Treatment requires raising the blood calcium level by intravenous injection of calcium salts.

HYPERPARATHYROIDISM is rare, and is usually due to a tumor of the glands. Excessive PTH secretion causes bone destruction, formation of kidney stones from the high levels of calcium filtered from the bloodstream, and may lead to calcium rigor (paralysis) in the heart and skeletal muscles.

FIGURE 26.14. The location of the parathyroid glands on the posterior aspect of the thyroid gland.

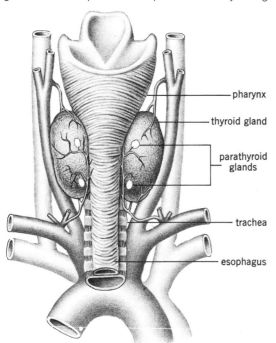

pharynx

thyroid gland

parathyroid glands

trachea

esophagus

The adrenal glands

Location and description

The adrenal glands are paired, hat-shaped organs, located above the kidneys. Each gland weighs 7–8 grams, and is composed of an outer CORTEX, and an inner MEDULLA. The cortex is further subdivided into three zones, each of which produces hormones with different physiological effects. The cortex is essential for life, the medulla is not.

Hormones and effects

MEDULLA. Two hormones, EPINEPHRINE (*adrenalin*) and NOREPINEPHRINE are produced by the adrenal medulla. Both are sympathomimetic, that is, they have the same effects on the body as does stimulation of the sympathetic nervous system. For example, the hormones increase heart rate and output, stimulate breathing, raise blood pressure by causing vasoconstriction, and epinephrine alone causes acceleration of glycogenolysis.

CORTEX. The cortex produces steroid hormones generally known as CORTICOIDS. These may be grouped into three general types.

MINERALOCORTICOIDS, produced primarily by the outer zone of the cortex, are essential to the fluid and electrolyte balance of the body. ALDOSTERONE is the most potent mineralocorticoid, and deals with the transport of Na^+ in the kidney. Recall that both negatively charged ions and water follow active transport of sodium from the kidney tubules.

GLUCOCORTICOIDS are secreted by the middle zone of the cortex, and deal with metabolism of carbohydrate and proteins. For example, CORTISOL and CORTISONE, the major glucocorticoids, increase synthesis of glucose from amino acids (gluconeogenesis), increase glycogen formation in the liver, and stimulate ATP formation. Glucocorticoids are increased when the body is stressed, and confer resistance to stress of any sort. However, continued high levels of glucocorticoids in the body may cause ulcer formation, loss of resistance to disease, and high blood pressure. Glucocorticoids also have an *anti-inflammatory effect,* and are used in the treatment of diseases such as rheumatoid arthritis.

CORTICAL SEX HORMONES are produced by the inner zone of the cortex. Both male- and female-type hormones are produced. Their role in body function has not been established.

FIGURE 26.15 Cushing's Syndrome in a 12-year-old female. Notice the moon-face, pendulous abdomen, abdominal striae, and concentration of adiposity on the trunk. (From F. Albright and E. C. Reifenstein, "The Parathyroid Glands and Metabolic Bone Diseases," Williams & Wilkins, 1948. Courtesy Mass. General Hospital.)

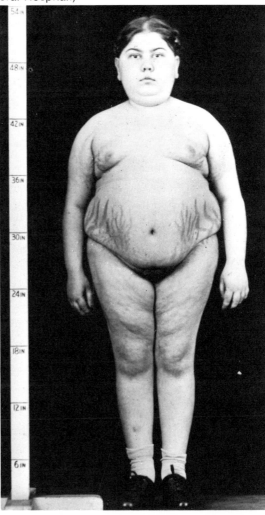

553

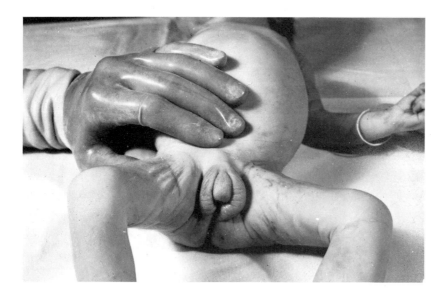

FIGURE 26.16.   Adrenogenital syndrome in a 24-day-old female. Notice masculinization, with enlarged clitoris (center of genitalia) and scrotumlike development of the labia. (Armed Forces Institute of Pathology.)

Control of secretion

MEDULLA. The medulla receives nerve fibers from the sympathetic nervous system that stimulate release of medullary hormones very rapidly. The hormones are rapidly destroyed by the liver so that their effect, though great, is short-lived.

CORTEX. The secretion of aldosterone is increased by ACTH, angiotensin II, fall of blood sodium levels, and hemorrhage. Increased reabsorption of salts and water tends to restore blood volume and pressure.

Glucocorticoid secretion is controlled by ACTH, and secreted cortisol exerts a negative feedback effect on ACTH secretion.

Cortical sex hormone secretion is also controlled by ACTH, with the sex hormones exerting a negative feedback effect.

Clinical considerations

MEDULLA. Since the medulla is not essential for life, there is no disorder associated with hyposecretion. A tumor of the medulla called a *pheochromocytoma,* liberates vast amounts of hormones into the blood and can cause circulatory damage if not removed.

CORTEX. ADDISON'S DISEASE is the main disorder of cortical deficiency, and usually involves all zones of the cortex. Deficiency of glucocorticoids causes anemia, weakness, alimentary tract disturbances, and mineralocorticoid deficiency causes low blood sodium levels. Administration of the proper hormone corrects the deficiency. CUSHING'S SYNDROME (Fig. 26.15) results from excess cortisol secretion and causes a moonlike face with red cheeks, a "buffalo-hump" obesity, and stripes on the abdomen. ADRENOGENITAL SYNDROME (Fig. 26.16) results from overproduction of cortical sex hormones of male type. If the excess secretion occurs in a female, she will be virilized (develop male-type secondary sex characteristics). A male will have his masculine development accelerated.

The islets of the pancreas

Location and description

Small islets (a small mass of one type of tissue in another) of endocrine tissue are found in the body and tail of the pancreas. Their numbers have been estimated to lie between 500,000 and two million. Each islet is well supplied with blood vessels, and consists of two main types of cells called ALPHA CELLS and BETA CELLS.

Hormones and effects

The alpha cells produce a hormone called GLU-CAGON. It is a substance that speeds the conversion of glycogen to glucose, and thus raises the blood glucose level (*hyperglycemia*).

The beta cells produce INSULIN, a hormone that increases cell uptake of glucose, and that speeds synthesis of glycogen in the liver. Since both these actions draw sugar from the blood and lower blood sugar levels, insulin may be said to have a *hypoglycemic* effect.

Control of secretion

The secretion of both hormones is controlled by the blood glucose level. A fall of blood sugar causes glucagon secretion; a rise causes insulin secretion. Thus blood sugar is maintained within narrow limits (80–100 mg/100 ml).

Clinical considerations

No named disease is associated with disorders of glucagon secretion. However, the literature contains reports of patients with chronic low blood sugar levels associated with lack or deficiency of alpha cells in the islets. The disorder appears to be hereditary.

Failure to produce insulin causes DIABETES MELLITUS. While primarily a disorder of carbohydrate uptake and metabolism, it affects the metabolism of all three basic foodstuffs. The following list of symptoms shows how metabolism is affected.

Insulin deficiency causes:

Hyperglycemia, due to failure of liver glycogen synthesis and cellular uptake.

Glucosuria, or sugar in the urine, caused by loss through the kidney. The tubules reabsorb only so much, the excess is lost. Cells, unable to take in glucose for metabolic activity, increase metabolism of fatty acids.

Ketosis and *acidosis* result from the production of ketone bodies derived from the excess acetyl CoA released during accelerated fatty acid oxidation.

Polyuria, or the secretion of large quantities of urine, results from osmotic attraction for water exerted by the large amounts of glucose in the kidney tubules.

Polyphagia, or excessive food intake, results from the caloric loss represented by loss of glucose.

Polydipsia, or excessive thirst, is the consequence of water loss.

Atherosclerosis may be accelerated by the high blood levels of fats, and the accumulation of fats inside blood vessels may cause a decrease of blood flow to organs and tissues. If the tissues die, gangrene may set in.

COMA may result if the hyperglycemia is not corrected. The coma is believed to result from dehydration of brain cells and by the increased H^+ concentration occurring in diabetes (diabetic acidosis). Insulin administration by injection may be required to control the disease. If there is deficiency, but not lack of insulin, diet control and orally administered drugs (e.g., Diabinese, Orinase, Tolinase, Dymelar) may control the disorder. The drugs stimulate secretion of remaining beta cells, increase cellular uptake of glucose, and inhibit liver glycogen breakdown.

Excessive secretion of insulin or insulin overdose produces a hypoglycemia and INSULIN SHOCK. Since the brain depends entirely on glucose for its metabolism, the main symptoms of hypoglycemia are nervous in origin and include: disturbances in walking, mental confusion, respiratory disturbances, depression of body temperature, and unconsciousness which may be fatal. Intake of glucose, either orally or intravenously, is required to control the condition.

The testis and ovary (gonads)

The anatomy of the testes and ovaries is described in Chapter 25. Hormones of both organs influence the same types of processes. For example:

An individual's maleness or femaleness depends on his chromosomal constitution, reinforced by hormones.

The secondary sex characteristics of fat distribution, hair patterns, muscular development, and voice changes, are dependent on gonadal hormones.

The ovarian hormones are responsible for the changes in the uterus during menstrual cycles.

The ovary: hormones and effects

ESTROGEN is a steroid hormone produced by the vesicular follicle. It is responsible for the proliferative phase of the menstrual cycle, increases mammary growth, stimulates contraction of the uterus, causes and maintains the development of the accessory sex organs, and develops the female secondary sex characteristics. PROGESTERONE (progestin) is produced by the corpus luteum after ovulation has occurred. It is responsible for the secretory phase of the menstrual cycle, causes

FIGURE 26.17.   Interrelationships of hormones and ovarian and uterine changes.

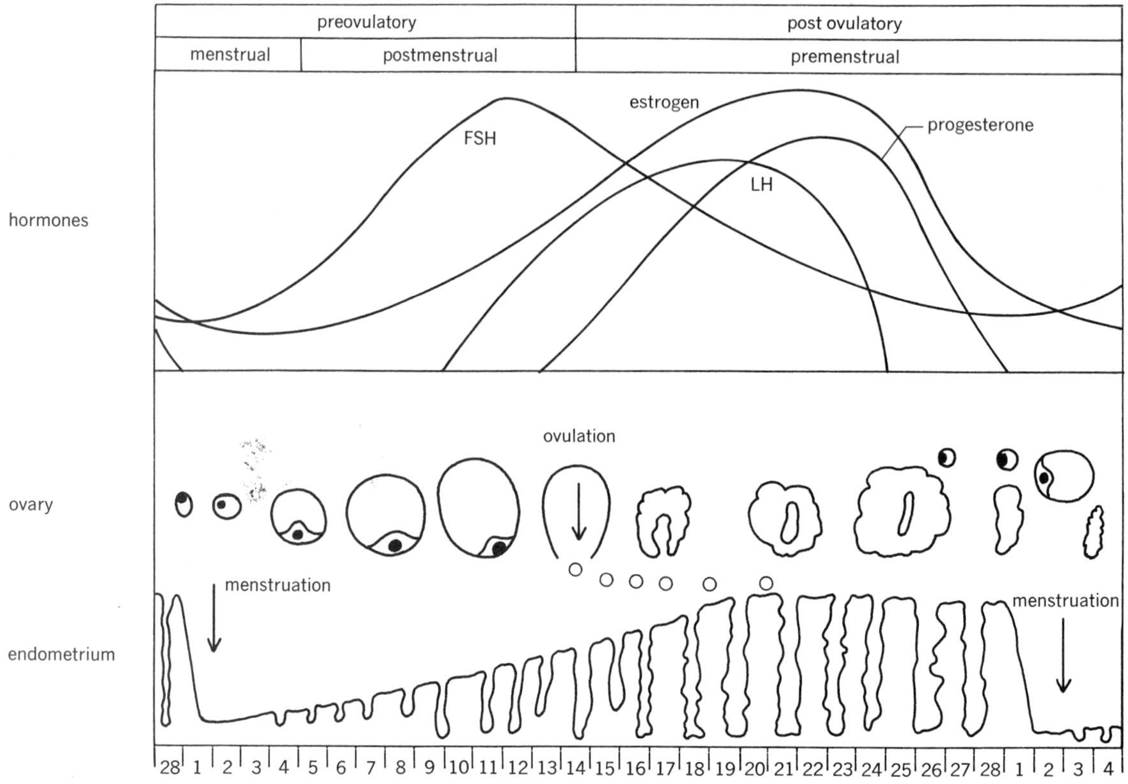

milk production (but not secretion) by the mammary glands, causes ovulation, and is required for placenta formation.

Control of ovarian secretion

Four hormones are involved in a complex relationship in control of ovarian secretion. FSH from the pituitary causes follicular development and estrogen secretion. Estrogen, in turn, inhibits FSH secretion in a negative feedback mechanism. This relationship prevents other follicles from developing until the fate of the first one has been determined. LH from the pituitary causes ovulation, corpus luteum formation, and progesterone secretion. Progesterone, in turn, inhibits both FSH and LH production. The reciprocal relationships between these hormones is shown in Figure 26.17.

Oral contraception

The fact that estrogen and progesterone inhibit FSH secretion, and thus maturation of follicles, is the basis for "the Pill" to achieve contraception. The first substances utilized in this manner were like progesterone, but created many unpleasant side effects including nausea, water retention, and uncomfortable mammary swelling. Newer compounds are estrogenlike or are combinations of estrogens and progesterone in low doses to imitate the natural secretion of hormones while preventing ovulation and ovum development. Tables 26.7 and 26.8 present the Pill and other methods of contraception that are available or are under investigation.

Clinical considerations: ovary (Fig. 26.18)

Failure to produce sufficient estrogen results in female EUNUCHOIDISM, in which sexual development remains infantile. Secondary characteristics develop minimally if at all. Excessive secretion of estrogen, usually due to tumors, may create PRECOCIOUS PUBERTY, and adult sexual function at tender ages. The youngest person ever to become pregnant was a sufferer from precocious puberty, and conceived at $5\frac{1}{5}$ years of age.

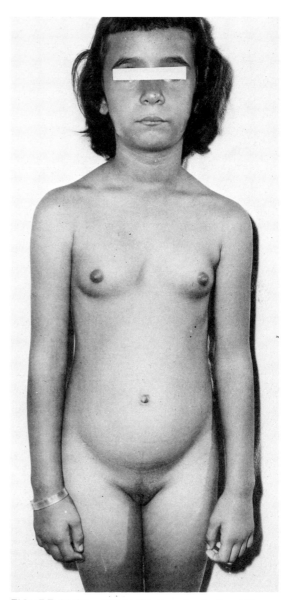

FIGURE 26.18. Precocious puberty in a 7-year-old female. Height was 4 feet, 3¼ inches. Notice development of pubic hair and mammary glands. (Lester V. Bergman & Associates.)

Method	Mode of action	Effectiveness if used correctly	Action needed at time of coitus	Requires resupply of materials used	Requires instruction in use	Requires services of physician	Suitable for menstrually irregular women	Side effects
Oral pill 21 day adminis- tration	Prevents follicle maturation and ovulation	Highest	None	Yes	Yes— timing	Yes—pre- scription	Yes	Early—some water retention, breast tenderness. Late—possible embolism, hypertension
Intrauterine device (coil, loop)	Prevents implantation	High	None	No	No	Yes—to insert	Yes	Some do not retain device. Some have menstrual discomfort
Diaphragm with jelly	Prevents sperm from entering uterus, plus jelly spermicidal	High	Previous to coitus	Yes	Yes—must be inserted correctly each time	Yes—for sizing and instruction on use	Yes	None
Condom (worn by male)	Prevents sperm entry into vagina	High	Yes	Yes	Not usually	No	—	Some deadening of sensation in male
Temperature rhythm	Determines ovulation time by noting body temperature at ovulation. T ↑	Medium	No	No	Definitely. Must learn to interpret chart correctly	No. Phy- sician should advise	Yes, if are skilled in reading graph	None. Requires abstinence during part of cycle
Calendar rhythm	Abstinence during part of cycle	Medium to low	No—no coitus	No	Definitely. Must know when to abstain	No. Physi- cian should advise	No!	None (pregnancy?)
Vaginal foams	Spermicidal	Medium to low	Yes. Requires application before coitus	Yes	No	No	Yes	None usually. May irritate
Withdrawal	Remove penis before ejaculation	Low	Yes. With- drawal	No	No	No	Yes	Frustration in some
Douche	Wash out .sperm	Lowest	Yes. Immedi- ately after	No	No	No	Yes	None

TABLE 26.7 Methods of Contraception Currently Available

TABLE 26.8	Newer Methods of Contraception under Investigation							
Method	Mode of action	Effectiveness if used correctly	Action needed at time of coitus	Requires resupply of materials used	Requires instruction in use	Requires services of physician	Side effects
"Mini-pill"— very low content of progesterone (¼ mg.)	Inhibits follicle development	High	No	Yes	Yes	Yes— prescription	Irregular cycles and bleeding (25%)
"Morning-after pill"	Arrests pregnancy probably by preventing implantation. 50 × Normal dose of estrogen	By currently available data, high	No. For 1–5 days after coitus	Yes	No	Yes	Breast swelling, nausea, water retention
Vaginal ring— inserted in vagina; contains progesteroid in it	"Leaks" progesteroid into bloodstream through vagina at constant rate. Thereby inhibits follicle maturation	Studies are "promising"	No	Yes. Perhaps at yearly intervals	Yes	Yes	Spotting, some discomfort
Once-a-month pill	Injected in oil base into muscle. Slow passage of birth control drug into circulation inhibits follicle maturation	Said to be 100 percent	No	Yes. On monthly basis	No	Yes	Similar to oral pill

The testis: hormone and effects

The interstitial cells of the testes produce the steriod TESTOSTERONE. Testosterone is responsible for the development of the male secondary sex characteristics, the external genitalia, and the accessory organs of the system. Testosterone is also anabolic, in that it stimulates protein synthesis in muscle. Certain athletes (both male and female) have taken such hormones in the belief that they will increase muscular development and strength, and thus will improve athletic performance.

Control of interstitial cell secretion

ICSH (interstitial cell stimulating hormone) is the male equivalent of LH. It stimulates secretion of testosterone, and testosterone inhibits ICSH secretion.

Clinical considerations: testis

Males may suffer EUNUCHOIDISM and PRECOCIOUS PUBERTY as described for females. Symptoms include failure of sexual development, and acceleration of development and appearance of secondary characteristics, respectively. Figures 26.19 and 26.20 show the two conditions.

Climacterics

Males and females, as they age, undergo a cessa-

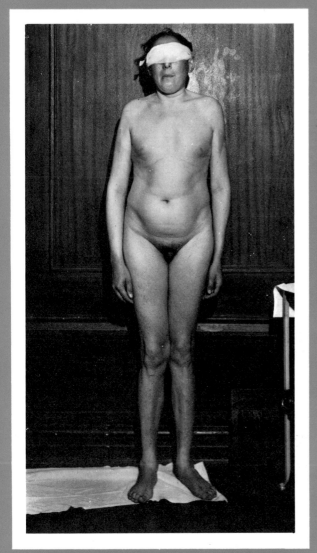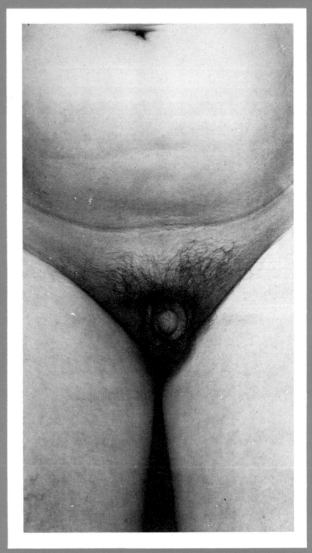

FIGURE 26.19. Eunuchoidism in a 40-year-old male. Notice disproportionately long arms and legs for height (6 feet, 1 inch), and failure of the genitalia and secondary characteristics to develop normally. (Lester V. Bergman & Associates.)

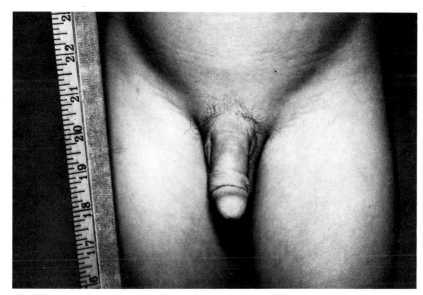

FIGURE 26.20.   Precocious puberty in a 3-year-old male. Notice development of genitalia equivalent to that of an adolescent, and appearance of pubic hair. (Armed Forces Institute of Pathology.)

FIGURE 26.21.   The location of the pineal gland, as shown on a midsagittal section of the brain.

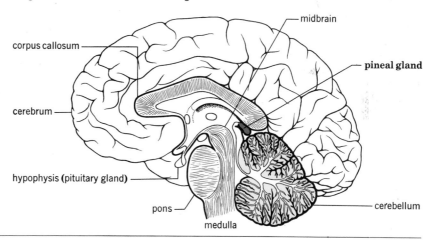

tion or diminution of gonadal secretion of hormones. This period is known as the climacteric or, in the female, as the menopause. Both sexes may show:

Psychic symptoms: nervousness, irritability, spells of emotional upheaval, and some loss of mental "sharpness."

Vasomotor symptoms: sweating, hot and cold "flashes," and headache.

Constitutional symptoms: fatigue, muscular weakness, and "lack of ambition."

Sexual symptoms: decrease of sexual drive, interest, and (in the female) ultimate sterility.

In the female, the symptoms appear to be related to decrease of progesterone secretion. In the male, about 15 percent of men show interstitial cell atrophy beginning about 40 years of age, and are restored by testosterone administration. The remainder seem to retain nearly full interest and potency as they age, with only small gradual decrease of testosterone production.

The placenta

The placenta serves not only as the organ for exchange of nutrients and wastes between the fetal and maternal circulations, but produces several hormones that assure maintenance of pregnancy.

CHORIONIC GONADOTROPIN (CG) resembles LH in structure and effects. It maintains the corpus luteum of early pregnancy, and assures its production of progesterone.

CHORIONIC ESTROGEN and PROGESTERONE are secreted in sufficient quantities, after 3 months of fetal life, to sustain a pregnancy even though the ovaries (and thus the corpus luteum) are removed.

CHORIONIC GROWTH-HORMONE-PROLACTIN (CGP) has lactogenic and growth-stimulating activity.

Finally, a hormone with TSH-like activity has been isolated from the placenta.

The placenta thus acts as insurance that the pregnancy will not suffer deficiencies of hormones that are necessary for continuation of development.

The pineal gland

The pineal gland (Fig. 26.21) has been variously regarded throughout man's recorded history as the residence of the soul, a valve to control movement of "spirits" through the body, and as an endocrine. In recent years, the pineal has acquired new status as an endocrine structure. Several known hormones or chemical substances have been shown to be synthesized in the pineal. Of these, MELATONIN is the only one that is synthesized only in the pineal. Melatonin causes contraction of pigment cells in amphibian skin. The eye apparently controls, by detecting changes in light intensity, the secretion of melatonin, and "matches" the skin color of the animal to light intensity. In this sense, the mechanism may have survival value. In humans, effects of the hormone are uncertain and its exact role in human physiology remains to be determined. It has been suggested that melatonin may be involved in the timing of human adolescence.

Summary

1. The endocrine system provides control over metabolism, growth and development.

2. The organs of the system develop from the alimentary tract, the nervous system or mesoderm of the body.

3. The organs or glands of the system include: the pituitary, thyroid, parathyroids, adrenals, pancreatic islets, ovaries and testes, and pineal gland.

   a. All endocrines are vascular, ductless, hormone-secreting structures whose removal disturbs body function or development.

   b. Hormones are the products of endocrine glands. They are proteins, steroids, or small amines.

4. Control of endocrine secretion is by nerves, neurohumors, negative feedback, or blood levels of specific or all solutes.

5. The pituitary gland lies on the undersurface of the brain. It has anterior, intermediate, and posterior lobes.

   a. The anterior lobe produces six hormones controlling growth (GH), thyroid activity (TSH), adrenal cortex activity (ACTH), gonadal activity (FSH, LH, or ICSH), and milk secretion (prolactin).

   b. Secretion of anterior lobe hormones is controlled by hypothalamic regulating hormones.

   c. The intermediate and posterior lobes release three hormones. Oxytocin and ADH are actually produced in the hypothalamus and are stored in and released from the posterior lobe by nerve impulses. Oxytocin stimulates uterine contraction; ADH insures water reabsorption by the kidney. MSH (intermediate lobe) has no known function in humans.

   d. Disorders of pituitary function include disturbances in growth, and water metabolism (see Table 26.5).

6. The thyroid gland is located in the neck. It has follicles as its functional unit.

   a. The thyroid produces thyroxin (T_4), triiodothyronine (T_3), and calcitonin as its hormones.

   b. T_4 and T_3 control body metabolism and growth.

   c. Calcitonin aids in Ca^{++} and $PO_4^{\equiv}$ deposition in bones.

   d. Goiter is an enlarged thyroid gland; cretinism and myxedema indicate hyposecretion; hyperthyroidism is excess T_4 and T_3 secretion.

   e. Control of T_4 and T_3 secretion is by TSH; calcitonin by blood calcium level.

   f. The PBI and iodine uptake tests are good tests for determining thyroid function.·

7. The parathyroid glands lie on the thyroid. They contain principal cells that secrete a hormone.

   a. The parathyroid hormone (PTH) is concerned with increasing absorption, retention, and metabolism of calcium and phosphate.

   b. Secretion is controlled by blood calcium level.

   c. Hypothyroidism (tetany) is due to disturbed blood calcium levels and muscular paralysis. Excess PTH secretion causes bone destruction.

8. The adrenal glands lie above the kidneys and have an inner medulla and an outer cortex.

   a. The medulla produces epinephrine and norepinephrine which elevate body functions.

   b. The cortex produces cortical steroids. Mineralocorticoids deal with sodium (and thus Cl^-, HCO_3^-, and H_2O) metabolism. Glucocorticoids govern glucose, fat, and protein metabolism, and exert stress resistance and anti-inflammatory effects. Cortical sex hormones have no known effects in normal people.

      c. The medulla is controlled by nerves; the cortex is controlled by several chemicals, including ACTH.

      d. Deficient medulla secretion is no threat to life; deficient cortical secretion is associated with weakness; excess cortical secretion is associated with disturbances of metabolism.

9. The islets of the pancreas contain alpha and beta cells.

      a. Alpha cells produce glucagon, which raises blood sugar levels; beta cells produce insulin which lowers blood sugar levels, and controls carbohydrate, fat, and protein metabolism.

      b. Secretion of both hormones is controlled by blood glucose levels; fall increases glucagon secretion, rise increases insulin secretion.

      c. Diabetes (mellitus) is the result of insufficient insulin secretion. It is associated with disturbances of foodstuff metabolism, appetite, and acid-base balance. Excessive insulin creates disturbances in the brain due to lack of glucose for brain metabolism.

10. The testes and ovaries secrete hormones that control secondary sex characteristics and development of the sex organs.

      a. The ovaries produce estrogen which develops organs and characteristics; and progesterone, which insures ovulation, placenta formation, and milk production.

      b. The ovaries are controlled by pituitary FSH and LH.

      c. Oral contraception is based on the effects ovarian hormones have on the pituitary.

      d. Deficient ovarian hormone secretion results in failure of sexual maturity; excess causes early sexual development and function.

      e. The testes produce testosterone which controls male characteristics and organ development.

      f. The testes are controlled by pituitary ICSH.

      g. Clinical considerations are the same as for the ovary.

      h. Climacterics (menopause in the female) are associated with decrease of gonadal hormone secretion. Disturbing sexual, body, and neural symptoms may develop.

11. The pineal secretes melatonin, which alters skin color in "lower animals."

Questions

1. The hypophysis has sometimes been designated as "the master gland" of the endocrine system. What, in your mind, justifies this exalted position?

2. What hormones are required for normal body growth and development? Give the contribution of each to these processes.

3. What hormones are concerned with carbohydrate metabolism, and how is each involved?

4. What endocrines exhibit a negative feedback mechanism of control? Which are controlled by blood levels of substances other than hormones?

5. What hormones enable the body to meet situations that are stressful or to meet sudden demands for accelerated activity?

6. What hormones are concerned with bone and tooth formation? Where are they produced and how are they involved?

7. Describe hypothalamic control of the hypophysis.

8. What justifies the statement that nervous and endocrine function are related?

9. What hormones insure proper sexual development? Discuss contribution of each.

Readings

Ciba Foundation. *The Pineal Gland.* G. E. W. Wolstenholme (ed.) Churchill, Livingston. London, 1971.

Danowski, T. S. Diabetes Mellitus. Diagnosis and Treatment. I. 21–24, American Diabetic Assn. New York, 1964.

Dillon, Richard S. *Handbook of Endocrinology.* Lea & Febiger. Philadelphia, 1973.

Duffy, Jr., Benedict A., and Sister M. Jean Wallace, CSC. *Biological and Medical Aspects of Contraception.* Univ. of Notre Dame Press. Notre Dame, Ind., 1969.

Frohman, L. A. "Clinical Neuropharmacology of Hypothalamic Releasing Factors." *New Eng. Jour. Med. 286*:1391, 1972.

Gillie, R. Bruce. "Endemic Goiter." *Sci. Amer. 224*:92 (June) 1971.

Jaffe, Frederick S. "Public Policy on Fertility Control." *Sci. Amer. 229*:17 (July) 1973.

Martin, Joseph P. "Neural Regulation of Growth Hormone Secretion." *New Eng. Jour. Med. 288*:1384, 1973.

Rudel, Harry W., M. D. et al. *Birth Control.* Macmillam. New York, 1973.

Schally, A. V., A. Arimura, and A. J. Kastin. "Hypothalamic Regulatory Hormones." *Science 179*:341, 1973.

Tepperman, J. *Metabolic and Endocrine Physiology.* 2nd ed. Year Book Medical Pubs. Chicago, 1968.

Turner, C. D. *General Endocrinology.* 5th ed. Saunders. Philadelphia, 1971.

Epilogue

The text concentrates primarily on the structure and function of the normal adult. Development of organs and systems has been considered. Few introductory physiology books consider that there is a life beyond "maturity," into old age and senescence. Inasmuch as we shall all arrive, sooner or later, at this terminal part of our life span, it is appropriate to conclude this volume with a few remarks about what lies ahead.

Senescence

Physiological processes reach a peak during maturity (about 30 years of age), and thereafter undergo gradual declines. At some point, one or more functions decline below the point necessary to sustain life and death ensues. Heredity, environment, diet, and "speed of living" all influence the rate of decline and time of death.

Three general types of changes contribute to the aging process: SECULAR CHANGES are the result of natural wear and tear; SENESCENT CHANGES are the result of aging of tissues and organs, especially those with low mitotic rate; PATHOLOGIC COMPLICATIONS are those resulting from disease processes developing in the aging organs.

Secular changes

As one grows older, tissues appear to demand more metabolic support than the body can supply. In short, secular changes appear to be the result of an imbalance between vascular supply and tissue demand. Thus the amount of active tissue declines in proportion to its vascularity, and body weight tends to be reduced. Little attempt appears to be made by the body to restore the balance. The changes are usually seen first in endocrine-dependent structures such as breast, prostate, internal reproductive organs, and thyroid.

Senescent changes

All body systems share in these changes. Some of the more noteworthy changes in systems are described below. Changes are gradual and may not be noticed by the individual.

Aging of the integument. The epidermis thins, becomes more translucent and dry, and the dermis becomes dehydrated and suffers loss of elastic fibers. The skin thus tends to "sag" on the body. There is usually loss or thinning of hair as a result of lowered sex hormone levels, and lowered synthesis of proteins. Sweat gland secretion diminishes, making the older individual more susceptible to the effects of high environmental temperature. Skin lesions (cancer, ruptured blood vessels, etc.) are more common in the aged. Diabetic vasculopathy is also more common in the aged.

Aging of the eye is reflected by a high incidence of cataracts, occlusion of retinal vessels with subsequent retinal changes, glaucoma, and changes in the transparency of the cornea. About one person in six in the over-65 age group exhibits some ocular pathology.

Aging of the ear is evidenced by sclerotic alterations in the ear drum and ossicles which may result in loss of hearing. Vascular blockage may result in sudden hearing loss or loss of ability to maintain body equilibrium.

Neurological disorders are most commonly associated with change in vascular supply, either gradually (Parkinson's disease) or suddenly ("stroke"). Peripheral loss of sensation acuity, neuritis, and neuralgia appear to be more common in the aged.

Circulatory disorders may include hardening of arteries (arteriosclerosis) and deposition of lipids in the vessel walls (atherosclerosis). If deposition occurs in coronary vessels, the individual becomes a candidate for "heart attack." Heart action lessens, with fall in cardiac output. Shortness of breath may develop as a result of low cardiac output and low pulmonary perfusion. Hypertension is common.

At the cellular level, there is *diminution of DNA*

and RNA synthesis, with failure to replace worn-out body proteins.

The *lungs* lose elasticity and dyspnea may develop. Diffusing capacity of gases diminishes, due primarily to loss of lung capillaries.

The *alimentary tract* undergoes changes in function to a greater degree than changes in structure. More than one half (56%) of aged persons show functional disorders such as heartburn, belching, nausea, diarrhea, constipation, and flatus. Many of these complaints have an emotional basis, based on fear of death and disease, and loss of contact with offspring. An organic basis may be demonstrated in some cerebral arteriosclerosis patients. Malignancy of the lower tract is more common in elderly persons (11%).

Renal disorders include lowered filtration rates, reduced reabsorption, and renal hypertension. Pyelonephritis (infection) is the most common renal disease. Calculi (stones) in the kidney and ureter are also more common in the aged.

Gonadal function decreases. The ovary atrophies, producing menopause in females; testicular function declines more slowly. Internal organs atrophy and often prolapse ("drop down") as, for example, the uterus entering the upper vagina.

Loss of the inorganic component of the *skeleton* produces osteoporosis and greater liability to fracture. The *muscles* become smaller and weaker and are prone to cramps and effects of altered electrolyte balance (hypokalemia, low blood potassium).

Articulations are affected by arthritis. *Osteoarthritis* is a noninflammatory condition in which joints undergo degenerative changes in cartilage and synovial membranes. *Rheumatoid arthritis* is typically inflammatory in nature. *Gouty arthritis* is a metabolic disorder in which uric acid crystals accumulate in the joints.

Endocrines may increase activity (as in acromegaly) or may decrease activity (diabetes, hypothyroidism). Gonadotropins appear to show the widest secretion ranges.

The *central nervous system* shows the greatest change in the brain, where neuron loss (estimated at 30 cells per minute from 30 years onward) may cause increased reaction times, loss of memory, and personality alterations.

Pathologic changes

The diseases of old age are too numerous to consider here. Suffice it to say that the aged person is more susceptible to the diseases we all are heir to, and that the presence of a disease process tends to accelerate the aging changes described previously.

Theories of aging

There are two general groups of theories as to why the body ages.

External theories

The external theories of aging suggest that factors acting upon the body from outside determine the health of our cells and the length of time that they will continue to function. Environment plays a role in that air, food, or noise pollution may cause cell death. Nutrition exerts an effect in that diets slightly limited (caloric, not in essential foods) create an underweight but longer-living organism. Bacteria and viruses or their products or both may accelerate the aging process. Radiation may contribute to chemical changes within cells and to their inability to function normally.

Internal theories

The internal theories of aging suggest that the forces causing aging originate within the organism itself. Direct genetic programming may cause vital chemical reactions to cease, accelerate, or to escape control at definite times. If this theory is correct, there would have to be an "aging gene" present, which is turned on or off. Such a gene has not been shown to exist to this time. Mutations of existing genes might result in the production of abnormal proteins that would stop some vital chemical reaction(s). The production of specific youth or aging chemicals has also been investigated. No evidence exists to suggest that aging is associated with the disappearance from the body of some "youth potion," or the accumulation in the body of an "aging substance." Defective immune responses, the result of inability of plasma cells to react to antigens has been suggested to contribute to cell death. Establishment of cross-linking in collagen molecules has been shown to occur as aging processes. This implies that links which did not exist before may be detrimental to cell health. Thus far, no relationship between this phenomenon and cell death has been established.

If there is a common denominator in all these theories, it is that they all lead to an inability to adapt or adjust to environmental changes, and perhaps, in the last analysis, this is why we die.

Death

While old age itself does not cause death, organs and systems age at different rates. If function of a vital organ drops below the critical level for maintenance of life, life ceases. The operation of bodily systems is interrelated so that if one vital organ fails, all fail. "The chain is only as strong as its weakest link." Loss of recovery ability is obvious, so that alterations that would be rapidly returned to normal by a young person become life threatening to the aged.

Definitions of death are extremely varied, and a strict definition is almost impossible. Among the definitions considered are:

Total irreversible cessation of cerebral function (as evidenced by a "flat" electroencephalogram) for at least 48 hours.
No spontaneous heart beat or respiration.
Permanent cessation of all vital functions.

At one time, stoppage of kidney function, cardiac standstill, or stoppage of respirations may have been considered signs of death. The use of the artificial kidney, respirators, and heart-lung machines has caused a reevaluation of the definition of death, as has the problem of securing organs from a donor for transplant. At present, spontaneous brain activity appears to be the most valid criterion of death, although no specific legal definition has as yet been advanced.

A philosophical statement

To be sure, there are demonstrable changes that occur in all systems of the body as life progresses. Some of these, such as deposition of fats in blood vessel walls, may be demonstrated in infants. Others may be seen to begin in adolescence, such as decline in auditory function. It thus becomes a difficult and highly individual matter as to when one "begins to age." Certain agencies suggest that 65 years is the time of beginning of "old age." Perhaps the best criterion is the old adage that "you're as young as you feel." Certainly, blanket application of some specific criterion that forces a person suddenly to quit as a productive member of society is not justified.

Research in gerontology may eventually result in great extensions of the human life span. But, without simultaneous efforts to increase the quality, as well as the quantity of life, no effort will be deemed worthwhile. We must, in the last analysis, "add years to life," and "life to years," regardless of our age.

Similarly, it is becoming clear that the care of our bodies must begin with conception and terminate only with the death of a given individual. Thus genetics, nutrition, environment, and the mental and physical health of the individual set the foundations for the next generation. We cannot expect to improve or maintain the quality of human resources without an all-out commitment to prevention of disease and disorders as the primary measure for promotion of optimum health throughout life.

Readings

Chinn, Austin B. "Working With Older People." Vol. IV. *Clinical Aspects of Aging.* U. S. Dept. of Health, Education, and Welfare. Pub. No. 1459. 1971.

Hayflick, L. "Human Cells and Aging." *Sci. Amer. 218*:32 (Mar) 1968.

Masters, Wm. H., and V. E. Johnson. *Human Sexual Response.* Little, Brown. Boston, 1966.

Selye, Hans. The Stress of Life. McGraw-Hill. New York, 1956.

Tanner, J. M. "Earlier Maturation in Man." *Sci. Amer. 218*:21 (Jan) 1968.

Timiras, P. S. *Developmental Physiology and Aging.* Macmillan. New York, 1972.

Weiss, Jay M. "Psychological Factors in Stress and Disease." *Sci. Amer. 226*:104 (June) 1972.

General References

Arey, L. B. *Developmental Anatomy*. 7th ed. Saunders. Philadelphia, 1965.

Baker, A. B. *An Outline of Clinical Neurology*. Wm. C. Brown Book Co. Dubuque, Ia., 1958.

Barr, Murray L. The Human Nervous System. Harper and Row. New York, 1972.

Beland, Irene L. Clinical Nursing. 2nd ed. Macmillan. London, 1970.

Benenson, A. S. (ed) *Control of C.D. in Man*. 11th ed. American Public Health Association. Washington, D.C., 1970.

Bergersen, Betty S., and Elsie E. Krug. *Pharmacology in Nursing*. C. V. Mosby Co. St. Louis, Mo., 1969.

Bloom, Wm., and Don W. Fawcett. *A Textbook of Histology*. 9th ed. Saunders. Philadelphia, 1968.

"Children Are Different—Relation of Age to Physiologic Function." Ross Laboratories. Columbus, O., October 1972.

de la Cruz, Felix F. et al (eds). Minimal Brain Dysfunction. Annals, New York Acad. Sciences, Vol. 205, 1973.

Diem, K., and C. Lentner. Scientific Tables. J. R. Geigy, S. A. Basle, Switzerland, 1970.

Fox, John P. et al. Epidemiology; Man and Disease. Macmillan. London, 1970.

Gairdner, Douglas, and David Hull (eds). *Recent Advances in Pediatrics*. J. and A. Churchill. London, 1971.

Ganong, Wm. F. *Review of Medical Physiology*. 5th ed. Lange Medical Pubs. Los Altos, Cal., 1971.

Gray, Henry. Anatomy of the Human Body. Edited by C. M. Goss. 29th ed. Lea and Febiger. Philadelphia, 1973.

Guyton, Arthur C. *Textbook of Medical Physiology*. 4th ed. Saunders. Philadelphia, 1971.

Harper, Harold A. *Review of Physiological Chemistry*. 13th ed. Lange Medical Pubs. Los Altos, Cal., 1971.

Leach, Gerald. *The Biocrats. Ethics and the New Medicine*. Penguin Books. Baltimore, 1972.

Macbryde, Cyril M. *Signs and Symptoms*. Lippincott. Philadelphia, 1970.

Moidel, Harriett C. et al. *Nursing Care of the Patient with Medical-Surgical Disorders*. McGraw-Hill. New York, 1971.

Netter, Frank E. Ciba Collection of Medical Illustrations. Ciba Pharmaceutical Products. Newark, N. J. Vol. I—Nervous System; Vol. II—Reproductive Systems; Vol. III—Digestive System; Vol. IV—Endocrine System and Selected Metabolic Diseases; Vol. V—Heart.

Rubin, Alan (ed). *Handbook of Congenital Malformations.* Saunders. Philadelphia, 1967.

Thompson, James S., and M. W. Thompson. *Genetics in Medicine.* Saunders. Philadelphia, 1973.

Wallace, Jacques. *Interpretation of Diagnostic Tests.* Little, Brown. Boston, 1970.

Ziai, Mohsen (ed). *Pediatrics.* Little, Brown. Boston, 1969.

Glossary

Abbreviations

abbr.	abbreviation
A.S.	Anglo-Saxon
Colloq.	Colloquial
e.g.	for example
esp.	especially
Fr.	French
G.	Greek
Ger.	German
i.e.	that is
L.	Latin
M.E.	Middle English
pert.	pertaining
pl.	plural
sing.	singular

Pronunciation

Pronunciations may be only approximately indicated unless all the markings in Webster's New International Dictionary are used, which is not practical in a glossary.

Accent. Indicates the stress upon certain syllables.

Diacritics. Marks over the vowels to indicate the pronunciations. Only two diacritics are used: The *macron,* showing the long or name sound of the vowels, and the *breve,* showing the short vowel sounds, as

a in rāte	**a** in căt
e in ēat	**e** in ĕver
ee in sēed	
i in īsle	**i** in ĭt
o in ōver	**o** in nŏt
oo in mōon	**oo** in bŏok
u in ūnit	**u** in cŭt

a-, an- [G. *alpha*]. Prefix: without, away from, not.
ab- [L.]. Prefix: from, away from, negative, absent.
abdomen (ab-dō′men) [L.]. The part of the trunk between the chest and the pelvis.
absorb [L. *absorbere,* to suck in]. To take in.
absorption (ab-sorp′shun) Taking into body fluids or tissues, usually across a membrane.

abducens (ab-dū'senz) [L. drawing away from]. The sixth cranial nerve; innervates the lateral rectus muscle of the eye which moves the eyeball outward or away from the median line of the body.

abduction (ab-dukt'shun) [L. *abducere,* to lead away]. Lateral movement away from the median plane of the body.

acceleration (ak-sel'e-rā-shun) [L. *acceleratus,* to hasten]. Increasing the speed of pulse, respiration, rate or motion.

accept'or [L. *accipere,* to accept]. Compounds taking up or chemically binding other substances, e.g., hydrogen acceptor.

accom'modate [L. *accomodare,* to suit]. To adjust or adapt; a decrease in nerve impulse discharge with continued stimulation; the adjustment of the eye for various distances by changing the curvature of the lens.

acetabulum (as-ĕ-tab'ū-lum) [L. a little saucer for vinegar]. The rounded cavity of the os coxae that receives the head of the femur.

acetic acid (a-sē'tik a'cid) [L. *acetum,* vinegar, + *acidus,* sour]. The acid of vinegar. CH_3COOH.

acetone (as'ĕ-tōn). A chemical [$(CH_3)_2CO$] produced when fats are not properly oxidized in the body. It is one of several substances called *ketone bodies* or *acetone bodies.*

acetylcholine (ă-sĕt-ĭl-kō'lēn). An ester of choline released from certain nerve endings; thought to play a role in transmission of nerve impulses at synapses and myoneural junctions. Easily destroyed by the enzyme, *cholinesterase.*

Achilles tendon (ăh-kil'ēz). Heel tendon of the gastrocnemius and soleus muscles of the leg.

acid [L. *acidus,* sour]. Any substance releasing H ion and neutralizing basic substances; a pH less than 7.00.

acid-base balance. State of equilibrium between acids and alkalies so that the H^+ concentration of the arterial blood is maintained between pH 7.35–7.45.

acid'ification [L. *acidum,* acid, + *factus,* to make]. Process of making acidic; becoming sour.

acidosis (as-ĭ-dō'sĭs) [" + G. *-osis,* condition]. Decrease of alkali in proportion to acid, or lowering of pH of extracellular fluid, caused by excess acid or loss of base.

acinus (as'ĭ-nus) (pl. *acini*) [L. grape]. Smallest division of a gland; a saclike group of secretory cells surrounding a cavity.

acne (ak'nē) [G. *acme,* point]. Chronic inflammatory disease of the sebaceous glands and hair follicles of the skin.

acoustic (a-koos'tik) [G. *akoustikos,* hearing]. Pert. to sound or to the sense of hearing. The cochlear portion of the eighth cranial nerve.

acromegaly (ak-ro-meg'ă-lī) [G. *akros,* extreme, + *megas,* big]. Enlargement of the extremities (hands, feet) and jaw due to hypersecretion of the somatotrophic or growth hormone from the anterior pituitary, after full growth has been achieved.

actin (ak'tīn). A muscle protein responsible for the shortening of the muscle when it contracts. Forms a combination with the protein *myosin* when muscle contracts.

ac′tion potential (pō-tēn′shăl) [L. *actio*, from *agere*, to do + *potentia*, power]. The measurable electrical changes associated with conduction of a nerve impulse or contraction of a muscle.

ac′tivator. Substances converting inactive materials to active ones, or causing activity.

ac′tive transport′ [L. *trans*, across, + *porta*, to carry]. Using carriers, energy (ATP), and enzymes of cell to cause a substance to cross a membrane.

acu′ity [L. *acuere*, to sharpen]. Clearness or sharpness, usually in reference to vision or hearing.

adap′tation [L. *adaptare*, to adjust]. The ability of an organism to adjust to environmental change; the ability of the eye to adjust to various intensities of light by change in size of the pupil.

adduction (ăd-dŭk′shŭn) [L. *ad*, toward, + *ducere*, to bring]. Movement toward the median plane of the body.

adenine (ad′ē-nēn). A nitrogenous base found in DNA and RNA (nucleic acids).

adenohypophysis (ad-ē-no-hi-poph′i-sis) [G. *adēn*, gland + *ypo*, under, + *physis*, growth]. The glandular lobes of the hypophysis or pituitary gland. Derived from the mouth, and it includes the anterior lobe *(pars distalis)* and intermediate lobe *(pars intermedia)*.

adenoid (ad′ē-noid) [G. *adenoeides*, glandular]. Lymphoid tissue; the pharyngeal tonsil.

adenosine triphosphate (ad′ē-nō-sen). The major source of cellular energy; found in all cells. Composed of a nitrogenous base *(adenine)*, ribose sugar, and 3 phosphoric acid radicals. Abbr: **ATP.**

adhe′sive [Fr. *adhesif*]. Sticky; that which causes two materials to adhere.

ad′ipose [L. *adeps*, fat]. Refers to a type of connective tissue composed of fat containing cells; fatty.

adolescence (ad-o-les′ens) [L. *adolescere*, to grow up]. The period from the beginning of puberty until maturity; variable among individuals.

adrenal (ăd-rē′năl) [L. *ad*, to, + *ren*, kidney]. An internally secreting gland located above the kidney. Outer *cortex* produces hormones essential to life; inner *medulla* not essential to life.

adrenalin (ă-dren′ă-lin) [″ + ″ + *in*, within]. Proprietary name for *epinephrine*, the main hormone of the adrenal medulla.

adrenergic (ad-ren-er′jik) [″ + ″ + G. *ergon*, work]. Term used to describe nerve fibers that release *norepinepherine* (noradrenalin or sympathin) at their endings when stimulated.

adsorption [″ + *sorbere*, to suck in]. ″Sticking together″ of a gas, liquid, or dissolved substance on a surface. No bonding.

adventitia (ad′ven-tish′ē-ā) [L. *adventitius*, coming from abroad]. The outermost covering of an organ or structure.

aerobic (ā-er-ō′bik) [G. *aer*, air + *bios*, life]. Requiring the presence of oxygen for life or function.

af′ferent [L. *ad*, to, + *ferre*, to bear]. Carrying toward a center or toward some specific reference point, e.g., afferent glomerular vessels, or afferent nerves to the brain.

agglutinin (ă-glū′tĭ-nĭn) [L. *agglutinare*, to glue a thing]. An antibody in the red blood plasma capable of causing clumping of specific antigens on red blood cells.

agglutinogen (ă-glū-tĭn′ō-gĕn) [″ + G. *gennan,* to produce]. An antigen on red blood cells stimulating the production of a specific agglutinin.

ag′onist [G. *agōn,* a contest]. The contracting muscle; a muscle whose contraction is permitted by relaxation of a muscle (antagonist) having the opposite action to the agonist.

albumin (al-bū′min) [L. *albus,* white; *albumen,* coagulated egg white]. One of a group of simple proteins in both animal and vegetable cells or fluids.

aldehyde (al′de-hīd) [*al,* abbr. alcohol, + *dehyd,* abbr. dehydrogenatum, alcohol deprived of H⁺]. A compound containing the CHO group; derived by oxidation of a primary alcohol.

aldosterone (al-dos′tĕr′ōn). The most active mineralocorticoid secreted by the cortex of the adrenal gland; regulates metabolism of sodium, potassium, and chloride.

-algia (al-je-a) [G. *algeis,* sense of pain]. Suffix: denoting pain.

alimen′tary [L. *alimentum,* nourishment]. Pert. to nutrition.

alkali [Arabic *al-qili,* ashes of salt wort]. A substance that can neutralize acids; combines with acid to form soaps; basic in reaction (pH).

alkali(ne) reserve. The amount of base in the blood, mainly bicarbonates, available to neutralize fixed acids.

alkalo′sis [″ + G. *-osis,* condition of]. Excessive amount of alkali, or rise of pH of extracellular fluid caused by excess base or loss of acid.

allantois (a-lan′tō-ĭs) [G. *allanto,* sausage, + *eidos,* resemblance]. A kind of long bladder located between the chorion and amnion of the fetus; its blood vessels contribute to the formation of the umbilical vessels.

allele (ă-lēl′) [G. *allēlōn,* reciprocally]. One of two or more genes occuring at the same position on a specific pair of chromosomes and controlling the expression of a given characteristic.

al′lergy [G. *allos,* other, + *ergeia,* work]. A condition characterized by reaction of body tissues to specific antigens without production of immunity.

allograft (al′ō-graft) [″ + L. *graphium,* grafting knife]. Grafts or transplants between genetically dissimilar members of the same species.

alveolus (al-vē′ō-lus) (pl. *alve′oli*) [L. small hollow or cavity]. A little hollow; an air sac of the lungs; a socket of a tooth.

ame′boid [G. *amoibē,* change, + *eidos,* resemblance]. Like an ameba; ameboid movements imply cellular motion by use of *pseudopods* (false feet or cytoplasmic outflows).

amenorrhea (a-men-ō-rē-ă) [G. *a-,* without, + *mēn,* month, + *rein,* to flow]. Absence or suppression of menstruation.

amine (ă-mēn′). A nitrogen containing organic compound with one or more hydrogens of ammonia (NH_3) replaced by organic radicals.

amino acid (ă-mē′no). Unit of structure of proteins; a compound containing both an amine group (-NH_2), and a carboxyl group (-COOH). Two groups: *essential,* not synthesized in body or not produced in amounts necessary for normal function; *nonessential,* synthesized in body.

am′nion (am′ni-on) [G. little lamb]. The inner fetal membrane that holds the fetus suspended in amniotic fluid. By 8 weeks it fuses with the *chorion* to form the *bag of waters* or *caul.*

amorphous (a-mor'fus) [G. *a-*, without, + *morphē*, form]. Without definite shape or structure.

amphiarthrosis (am-fi-ar-thrō'sis) [G. *amphi*, on both sides, + *arthrosis*, joint]. A form of articulation in which the mobility in all directions is slight.

ampulla (ăm-pŭl'a) [L. little jar]. A saclike dilation of a duct or canal.

anabolism (an-ab'ō-lizm) [G. *anabolē*, a building up]. Synthetic or constructive chemical reactions.

anaerobic (an-ā-er-ōb'ic) [G. *an-*, without, + *aer*, air, + *bios*, life]. Not requiring oxygen for life or function.

anaphase (an'a-fāz) [G. *ana*, up, + *phainein*, to appear]. A stage in mitosis characterized by separation and movement of chromosomes toward poles of the dividing cell.

anastamosis (a-nas-to-mō'sis) [G. opening]. An opening of one vessel into another, or the joining of parts to form a passage between any two spaces or organs.

androgen (ăn'drō-jĕn) [G. *anēr, andros*, man, + *gen*, to produce]. A substance stimulating or producing male characteristics, as testosterone, the male sex hormone.

anemia (an-e'mĭ-a) [G. *an*, without, + *aima*, blood]. A condition where there is reduction of circulating red blood cells or of hemoglobin.

anesthesia (an-es-the'zĭ-ă) [G. *an-*, without, + *aisthesis*, sensation]. Partial or complete loss of sensation with or without loss of consciousness.

aneurysm (an'ū-rism) [G. *aneurysma*, a widening]. Abnormal dilation of a blood vessel due to weakness of the vessel wall or to a congenital defect.

angina (ănjĭ'na) [L. *angere*, to choke]. Any disease characterized by suffocation or choking. *Angina pectoris* occurs when the heart muscle is deprived of blood flow and includes severe pain referred to the chest and left arm.

angiotensin (an-jĭ-ō-ten'sin) [G. *angeion*, vessel, + L. *tensio*, tension]. A pressor substance that increases arterial muscular tone, producing vasoconstriction and a rise of blood pressure. (Formerly called *angiotonin* or *hypertensin*.)

Angström unit (ong'strum) [Anders J. Angstrom, Swedish physicist, 1814-1874]. An international unit of wave length measurement. $1/10$ millionth of a millimeter, or $1/250$ millionth of an inch. Abbr: **Å** or A.

anion (an'ī-on) [G. *ana*, up, + *ion*, going]. A negatively charged ion or radical. In electrolysis, anions go toward the anode.

anisocytosis (an-ĭ-sō-sī-tō'sis) [G. *anisos*, unequal, + *kytos*, cell, + *-osis*, condition]. An abnormal condition with inequality in size of cells, esp. erythrocytes.

ann'ulus [L]. a fibrous ring surrounding an opening.

anoxemia (an-oks-ē'mĭ-ă) [" + oxygen + "]. Deficiency of oxygen in the arterial blood. Also, *hypoxemia*.

anoxia (an-ox'ĭ-ă) [" + oxia, oxygen]. Deficiency of oxygen in body tissues. Also, *hypoxia*.

antag'onist [G. *antagōnizesthai*, to struggle against]. That which counteracts or has the opposite action of something else *(agonist)*; esp. muscular contraction.

antenatal (ăn-tē-nā'tal) [L. *ante*, before, + *natus*, birth]. Occurring before birth.

ante'rior [L.] Before, or in front of, or the ventral (belly) portion.

antibiotic (ăn-tĭ-bī-ŏt'ik) [G. *anti*, against, + *bios*, life]. A substance produced by bacteria, molds, and other fungi, that has power to inhibit growth of or destroy other organisms, esp. bacteria.

antibody (an'tĭ-bod-ē). A protein (usually a globulin) developed against a specific substance (antigen), whether the antigen is foreign or not.

antigen (an'tĭ-jĕn) [" + *gennan,* to produce]. A substance causing production of antibodies.

antitox'in [" + G. *toxikon,* poison]. An antibody capable of neutralizing a specific chemical (toxin) produced by microorganisms.

an'trum (pl. *antra*) [G. *antrom,* cavity]. Any nearly closed chamber or cavity, e.g., the antrum of the stomach *(pylorus).*

anus (ā'nŭs) [L.]. The outlet of the rectum lying in the fold between the buttocks.

aorta [G. *aorte,* aorta]. The main trunk of the arterial system of the body arising from the left ventricle of the heart. It has a thoracic portion (ascending, arch, and descending subdivisions) and an abdominal portion (part of descending aorta). Branches from these supply the body with arterial (oxygenated) blood.

aperture (ap'er-tūr) [L. *apertura,* opening]. An opening or orifice.

apex (ā'peks) (pl. *apices*) [L. apex, tip]. Summit or peak; the narrow portion or tip of an organ shaped in a pyramidal form.

Apgar score [Virginia Apgar, M. D.]. System of scoring an infant's physiological condition at 1- and 5-minute intervals after birth.

aphasia (ā-fā'zĭ-ā) [G. *a-,* not, + *phasis,* speaking]. Loss of verbal comprehension, or inability to express oneself properly through speech.

apnea (ap'nē-a) [G. " + *pnoē,* breath]. Temporary stoppage or cessation of breathing.

apocrine (ap'ō-krēn, -krĭn) [G. *apo,* from, + *krinein,* to separate]. A method of production of a secretion by an externally secreting gland in which the cells lose part of their cytoplasm in forming the secretion, e.g., mammary gland, apocrine sweat glands.

aponeurosis (ăp-ō-nū-rō'sis) [G. *apo,* from, + *neuron,* sinew]. A white, flat, fibrous sheet of connective tissue that attaches muscle to bone or other tissues at their origin or insertion.

appendage (ă-pĕn'dĭj) (pl. *appendices*) [L. *appendere,* hang to]. Any part attached or appended to a larger part, as a limb or the appendix.

apposition (ap-ō-zĭ-shun) [L. *ad,* to, + *ponere,* to place]. A fitting together or contact of two substances, usually at a surface, e.g., appositional growth of cartilage by addition of new cartilage on the surface.

aqueduct (ak'we-dukt) [L. *aqua,* water, + *ductus,* duct]. A passage or canal to convey fluids; cerebrospinal fluid flows from the third to fourth ventricle through the cerebral aqueduct (of Sylvius).

aqueous (ā'kwē-ŭs) [L. *aqua,* water]. Watery; having the nature of water. Aqueous humor.

a. humor [L. fluid]. Watery fluid in the anterior and posterior chambers of the eye.

arachnoid (ā-rak'noid) [G. *arachne,* spider, + *eidos,* form]. The central membrane covering the brain and spinal cord, having the appearance of a spider web.

arborization (ar-bor-ĭ-zā'shun) [L. *arbor,* tree]. A treelike structure; terminations of nerve fibers and arterioles.

ar'bor vi'tae [" + *vita,* life]. In the cerebellum the treelike outline of the internal white matter; a series of branching ridges on the uterine cervix.

areola (ā-rē'ō-lă) (pl. areolae) [L. a small space]. A small cavity in a tissue; around the nipple, the ringlike pigmented or red area; the part of the iris enclosing the pupil.

arrhythmia (ă-rith′mĭ-ă) [G. *a-*, not, + *rhythmos*, rhythm]. Lack of rhythm; irregularity.

arteriole (ar-tē′rĭ-ōl) [L. *arteriola*, small artery]. A small or tiny artery; its distal end leads into a capillary or capillary network.

arteriosclero′sis [″ + G. *sklērōsis*, hardening]. General term referring to many conditions where there is thickening, hardening, and loss of elasticity of the walls of blood vessels.

artery (ar′ter-ī) (pl. arteries) [G. *arteria*, windpipe]. Any vessel carrying blood away from the heart to the tissues. Usually refers to the larger vessels.

arthritis (ar-thrī′tis) (pl. *arthritides*) [G. *arthron*, joint, + *itis*, inflammation]. Inflammation of a joint, usually accompanied by pain and frequently structural changes.

arthrosis [″ + -*osis*, increased]. A degenerative condition of a joint; a joint.

artic′ulate [L. *articulatus*, jointed]. To join together as in a joint; to speak clearly.

artic′ulation [L.]. A connection between bones; a joint.

-ase. Suffix used in forming the name of an enzyme; added to the part of or name of substance on which it acts.

asphyxia (ăs-fĭk′sĭ-ă) [″ + *sphyxis*, pulse]. Decreased amount of O_2 and an increased amount of CO_2 in the body as a result of some interference with respiration.

associa′tion. Uniting or joining together; coordination with another structure, idea; a neuron lying between and joining an afferent and efferent neuron.

asthenia (as-thē′nĭ-ă) [″ + *sthenos*, strength]. Loss or lack of strength; weakness in muscular or cerebellar disease.

athero- (ath-er-ō) [G. *athērē*, porridge]. Prefix: referring to lipid (fatty) substances.

atherosclerosis. Arteriosclerosis characterized by lipid deposits in the inner layer of blood vessel walls (chiefly arteries).

asthma (az′mă) [G. panting]. Intermittent severe difficulty of breathing accompanied by wheezing and cough; due to spasms or swelling of the bronchioles.

astigmatism (a-stig′mă-tism) [″ + *stigma*, point, + *ismos*, condition of]. Refractive surfaces of the eyeball are different in one or more planes; usually due to unequal curvature of the cornea and lens in vertical and horizontal planes.

astrocyte (as′trō-sīt) [G. *astron*, star, + *kytos*, cell]. A star-shaped neuroglial cell possessing many branching processes.

ataxia (ă-taks′ĭ-ă) [G. lack of order]. Mascular incoordination when voluntary muscular movements are attempted.

atlas. The first cervical vertebra by which the head articulates with the occipital bone. Named for Atlas who was supposed to have supported the world on his shoulders.

atom [G. *atomos*, indivisible]. The smallest particle of an element that can take part in a chemical change, keeping its identity, and which cannot be further divided without changing its structure; composed of protons, neutrons, electrons.

atre′sia [″ + *trēsis*, a perforation]. Congenital absence or pathological closure of a normal opening; degeneration of ovarian follicles.

ATP. Abbr. for *adenosine triphosphate*.

atrium (ā′trĭ-um) (pl. *atria*) [L. corridor]. A cavity; one of the two upper chambers of the heart; tympanic cavity of the ear; a space that opens into the air sacs of the lungs.

at'rophy [G. " + *trophe*, nourishment]. Wasting or decrease in size of a part due to lack of nutrition, loss of nerve supply, or failure to use the part.

aud'itory [L. *auditiō*, hearing]. Pert. to the sense of hearing.

auditory nerve. Eighth cranial nerve, sensory. Also, *vestibulocochlear* n.

auricle (aw'rĭ-kl) [L. *auricula*, the ear]. The outer ear flap on the side of the head; a small conical portion of the upper chambers of the heart, *NOT* synonomous with atrium.

auscultation (aws-kul-tā'shun) [L. *auscultāre*, listen to]. Listening for sounds in body cavities, esp. chest and abdomen, to detect or judge some abnormal condition.

autograft [G. *autos*, self, + L. *graphium*, grafting knife]. A graft or transplant taken from one part of the body to fill in another part on the same person.

autoimmune [" + L. *immunis*, safe]. A condition in which antibodies are produced against a person's own tissue.

autonomic (aw-tō-nom'ik) [" + *nomus*, law]. Self-controlling; refers to the involuntary or self-regulating portion of the peripheral nervous system.

autosome (aw'to-sōm) [" + *soma*, body]. Any of the paired chromosomes other than the sex (X and Y) chromosomes.

axial (aks'ĭ-al) [L. *axis*, lever]. Pert. to or situated in an axis.

axil'la [L. little pivot]. The armpit.

ax'is (pl. *axes*). A line running through the center of a body, or about which a part revolves. The second cervical vertebrae that has the dens (*odontoid process*) about which the atlas rotates.

ax'on [G. axis]. A neuron process that conducts impulses away from (*efferent*) the cell body.

balance. State of equilibrium; state in which the intake and output or concentrations of substances are approximately equal (homeostasis).

baroreceptors (baro-rē-sēp'tor) [G. *baros*, weight + L. a receiver]. Receptor organs sensitive to pressure or stretching.

bar'rier. An obstacle or impediment.

bā'sal [G. *basis*, base]. The base of something; of primary importance; the lowest level to maintain function.

ba'sal metab'olism [G. *basis*, base, + *metabole*, change]. The amount of energy needed for maintenance of life when the subject is at physical, emotional, and digestive rest.

ba'sal metab'olic rate [G.]. The metabolic rate as measured under so-called *basal conditions*, expressed in large Calories/M² of body surface/hr. Abbr: **BMR.**

bāse [G. *basis*, base]. The lower or broader part of anything; an alkali, any substance that will neutralize an acid.

benign (bē-nīn') [L. *benignus*, mild]. Mild; not malignant.

beta- (bā'tā) [G. *β*, Second letter of alphabet]. Prefix: to note isomeric position or variety in compounds of substituted groups.

beta-oxidation (ōk'sĭ-dā'shŭn) [" + *oxys*, sour, pungent]. A metabolic scheme that catabolizes fatty acids by removal of two carbon units.

biceps (bī'sĕps) [L. *bis*, two, + *caput*, head]. A muscle with two heads.

bifid (bī'fid) [" + *findere*, to cleave]. Cleft or split into two parts.

bilirubin (bil-ī'ru-bin) [L. *bilis*, bile, + *ruber*, red]. The the orange or yellow pigment in bile; derived from breakdown of heme ($C_{33}H_{36}O_6N_4$).

biliverdin (bil-ĭ-ver′dĭn) [″ + G. *viridis,* green]. The green pigment of bile formed by oxidation of bilirubin ($C_{33}H_{34}O_6N_4$).

bio- [G. *bios,* life]. Prefix: life.

-blast [G. *blastos,* germ or bud]. Suffix: immature or primitive.

blastocele (blas′tō-sēl) [″ + *koilos,* hollow]. An embryonic stage of development, the cavity of the blastula.

blastocyst (blas′tō-sist) [″ + *kystis,* bag]. Stage in embryonic development that follows the morula; also called the *blastodermic vesicle,* it is a hollow mass of cells.

blastomere (blas′to-mēr) [″ + *meros,* a part]. One of the cells resulting from the segmentation or cleavage of a fertilized ovum; the cells composing the *morula.*

blastula (blas′tu-lă). A stage in the early development of an ovum that has a hollow sphere of cells enclosing the *blastocele.* Also, *blastocyst.*

blind spot [A. S. *blind,* unable to see]. Area in the retina in which there are no light receptor cells (no rods nor cones); point where optic nerve enters the eye (*optic disk*).

blood-brain bar′rier [A. S. *blŏd,* blood, + *braegen,* cranial portion CNS, + *barrier,* an obstacle or impediment]. Special mechanism that prevents the passage of certain substances, such as dyes or toxins, from the blood to the cerebrospinal fluid.

blood pressure. The pressures measured on a peripheral artery during contraction and relaxation of the heart.

    b. p., diastolic. The pressure remaining in the arteries during relaxation of the left ventricle.

    b. p., systolic. The peak pressure created by contraction of the left ventricle of the heart, and the resistance to flow offered by the blood vessels.

bo′lus [G. a mass]. A moss of food ready for swallowing.

bone marrow [A. S. *bŏn,* bone, + *mearh,* marrow]. Soft tissue within the hollow of certain bones of the skeleton; responsible for production of certain blood cells.

booster (boos′ter) [colloq. *bouse,* to haul up]. An additional dose of an immunizing agent given to maintain or increase the protection or immunity originally conferred.

-borne (born, [*bore,* to carry]. Refers to types of methods of transmission of communicable diseases, e.g., airborne, waterborne, foodborne, etc.

Bowman's capsule [Sir Wm. Bŏwman, Eng. phys., 1816-1892, + L. *capsula,* small box]. The capsule containing the glomeruli of the kidney; the cuplike depressions on the expanded ends of the renal tubules or nephrons that surround the capillary tufts (*glomeruli*).

brachium (brā′kĭ-um) [G. *brachiŏn,* arm]. The upper arm from shoulder to elbow.

brady- [G. *bradys,* slow]. Prefix: slow, as in *bradycardia* or slow heart rate.

breathe (brēth) [A. S. *braeth,* odor]. To take in and release air from the lungs and respiratory system; *respire.*

bronchi (bron′kī) (sing. bronchus) [G. *bronchos,* windpipe]. The first two divisions of the trachea that penetrate the lungs and terminate in the bronchioles; air tubes to and from the lungs.

bronchiectasis (bron-kĭ-ek′tă-sis) [″ + *ektasis,* dilatation]. Condition characterized by dilatation of the bronchus(i) with secretion of large amounts of foul substance.

bronchiole (bron'kĭ-ōl) [L. *bronchiolus,* air passage]. A smaller division of the bronchi that lead to the alveolar ducts to the alveoli.

bucca (buk'a) [L. mouth, cheek]. The mouth; hollow part of the cheek.

buffer (bŭf'ĕr) [Fr. *buffe,* blow]. A substance preserving pH upon addition of acids or bases; reduces the change in H$^+$ concentration. *b. pair, b. system,* the combination of a weak acid and its conjugate base, which work together to maintain pH nearly constant.

bul'b [G. *bolbos,* bulb-shaped root]. An expansion of an organ, canal, or vessel.

bulk [M. E. *bulke,* a heap]. Size, mass, or volume; the main body of something.

bun'dle of His [M. E. *bundel,* bind, + W. His, Gr. anat., 1863-1934]. A bundle of fibers of the impulse-conducting system that initiates and controls contraction of the heart muscle. It extends in the A-V node and becomes continuous with the Purkinje fibers of the ventricles.

bur'sa [G. a leather sac]. A sac or cavity filled with a fluid that reduces friction between tendon and ligament or bone, or other structures where friction may occur.

buttocks (but'uks) [M. E. *butte,* thick end]. The gluteal muscle prominence, the "rump" or "seat."

-calcemia (kal-sē'mĭ-ă) [L. *calcarius,* pert. to calcium or lime, + G. *amia,* blood]. The calcium level of the blood.

calcification (kal-sif'ĭk-a shun) [L. *calx,* lime, + *ferre,* to make]. Deposit of lime salts in bones and other tissues.

calcitonin (kal-sĭ-tōn'ĭn) [G. *calci,* pert. to calcium + *tonos,* tone]. A hormone secreted by the thyroid gland that aids in regulating calcium-phosphorus metabolism.

calculus (kal'kū-lus) (pl. calculi) [L. pebble]. A "stone" formed within the body usually composed of mineral salts.

cal'lus [L. hard]. Circumscribed area of the horny layer of the skin that has thickened; osseous material formed between the ends of a fractured bone.

calorie, small (kăl'ŏ-rē) [L. *calor,* heat]. A unit of heat; amount of heat required to raise temperature of 1 gram of water 1°C. Abbr: **cal.**

Calorie, large [L.]. A large calorie or kilocalorie; 1000 times as large as a calorie; amount of heat required to raise temperature of 1 kilogram of water 1°C. (distinguished from small calorie by a capital C). Abbr: **Cal.,** or **kcal.**

calyx (kā'lix) (pl. *calyces*) [G. *kalix,* cup]. Cuplike divisions of the kidney pelvis.

canal (ka-nal) [L. *canā'lis,* canal]. A narrow channel, tube, duct, or groove.

canaliculus (kan-ā-lĭk'ū-lus). A tiny canal or channel.

can'cellous [L. *cancellus,* a grating]. Having a latticework or reticular structure, as the spongy tissue of bone.

cancer (kan'ser) [L. a crab; ulcer]. A malignant tumor or neoplasm (new growth); a carcinoma or sarcoma.

capillary (kap'ĭ-lā-rē) (pl. *capillaries*) [L. *capillaris,* hairlike]. Minute blood vessels that connect the arterioles with the venules to carry blood; minute lymphatic ducts. Capillary *networks* allow passage of oxygen and nutrients from the blood to the tissues, and of wastes from the tissues into the blood.

capitulum (ka-pĭt'ū-lum) [L. *caput,* head]. A small rounded end of a bone that articulates with another bone.

ca'put [L.]. The head; upper part of an organ.

carbaminohemoglobin (karb'ă-mē-nō-hē'mō-glō-bin). The compound formed by the combination of carbon dioxide with the amine groups of the globin molecule of hemoglobin.

carbohyd'rates (kar-bō-hĭd'ratz) [L. *carbo*, carbon, + G. *ydōr'* water]. A class of organic compounds composed of C, H, and O, with H and O usually in the ratio of 2:1; sugars, starches and cellulose; the *monosaccharides, disaccharides,* and *polysaccharides.*

car'bon [L. *carbo*, carbon or coal]. Constituent of organic compounds, found in all living things; combines with H, O, N to make life possible.

carbon dioxide. A colorless gas, heavier than air; produced by complete combustion, fermentation, or decomposition of carbon (compounds) found in air, and exhaled by all animals. CO_2.

carbonic anhydrase. An enzyme catalyzing the reaction of CO_2 and water to form *carbonic acid* (H_2CO_3) and vice versa; present primarily in red blood cells.

carcinogens (kar'sĭ-nō-jens') [G. *karkinos*, cancer, + *genere*, to produce]. Substances known to cause cancer or neoplasms.

carcinoma (kar-sĭ-nō'mă) [" + *oma*, tumor]. A malignant cancer or neoplasm of epithelial tissues.

cardiac cycle (sī'kle) [G. *kardia*, heart, + Fr. *cycle*, circle]. The period from the beginning of one heart beat to the next beat, includes systole and diastole.

cardiac output. Amount of blood ejected from one ventricle per minute.

caries (ka'rēz) [L. rottenness]. Tooth or bone decay.

carotene (car'ō-ten) [G. *karóton*, carrot]. A yellow crystalline pigment obtained from yellow vegetables; stored and converted to *vitamin A* in the liver.

carpus (kar'pus) [G. *karpos*, wrist]. The eight bones of a wrist. Also, *carpals.*

carrier [Fr. *carier*, to bear]. A large molecule transporting substances across cell membranes in *active transport.*

carotid (kă-rot'ĭd) [G. *karótides*, from *karos*, heavy with sleep]. Arteries providing the main blood supply to the head and neck; pressure on them may produce unconsciousness.

cartilage (kar'til-ăj) [L. *cartilagō*, gristle]. A form of connective tissue usually with no blood nor nerve supply of its own; it is firm, eleastic, and a semiopaque bluish-white or gray; cells lie in cavities called *lacunae.*

catabolism (kă-tab'ō-lism) [G. *katabolē*, a casting down]. The destructive phase of metabolism; breaking down of complex chemical compounds into simpler ones, usually with the release of energy.

catalyst (kat'ă-list) [G. *katalysis*, dissolution]. A substance that alters or speeds up the rates of chemical reactions without being itself altered in the process.

cat'aract [G. *katarraktēs*, a rushing down]. Opacity or loss of transparency of the lens of eye, its capsule, or both.

cation (kat'ĭ-on) [G. *katiōn*, descending]. A positively charged ion or radical.

cauda equina (kaw'dă e-kwin'a) [L. *cauda*, tail, + *equus*, horse]. The terminal portion of the roots of the spinal nerves and spinal cord below the first lumbar nerve: resembles a horse's tail.

cavernous body (kăv'ĕr-nūs) [L. *caverna*, a hollow, + *corpus*, body]. One of several columns of tissue containing blood spaces in the dorsum of the clitoris or penis that aids in the erection of these organs.

cavity (kav'ĭ-tĭ) [L. *cavitas*, hollow]. A hollow space, such as in a body organ or a decayed tooth.

cecum (sē'kŭm) [L. *caecum*]. The blind pouch that forms the first portion of the large intestine.

-cele [G. *hernia*, tumor]. Suffix: a swelling.

celiac (se'lĭ-ak) [G. *koilia*, belly]. Pert. to the abdominal region.

celom (sē'lom) [G. *koilōma*, a hollow]. The embryonic cavity between the layers of mesoderm that develops into the peritoneal, pleural, and pericardial cavities. Also **coelom.**

center (sen'ter) [G. *kentron*, middle]. Midpoint of a body; usually refers to a group of nerve cells in the brain or spinal cord; controls some specific activity.

centimeter [L. *centum*, a hundred, + *metron*, measure]. Abbr. **cm.** 2.5 cm equals 1 inch; 1 cm equals $1/100$ meter, or about $2/5$ inch (0.3937 inch).

centriole (sĕn'trĭ-ōl). A cell organelle that is involved in cell division; forms a spindle.

centromere (sen'trō-mēr). The structure at the junction of the two arms (chromatids) of a chromosome.

centrosome (sen'trō-sōm) [G. *kentron*, center, + *soma*, body]. An area of the cell cytoplasm lying near the nucleus that contains one or two *centrioles;* active during cell division.

cephalic (sef-ăl-ic) [G. *kephalē*, head]. Cranial, pert. to the head; superior in position.

cerebellum (ser-ĕ-bel'um) [L.]. A dorsal portion of the brain that is involved in coordinating activity of skeletal muscles, and in the coordination of voluntary muscular movements.

cerebrum (ser'ĕ-brum) [L.]. The upper largest part of the brain; contains the motor, and sensory areas, and the association areas that are concerned with the higher mental faculties.

cerumen (se-rū'men) [L. *cera*, wax]. The waxlike secretion in the external ear canal.

cervical (ser'vĭ-kal) [L. *cervicalis*, pert. to neck]. The neck region, or neck of an organ, e.g., **cervix** or neck of the uterus. The first **8** spinal nerves; the first 4 cervical spinal nerves form the **c. plexus.** The first **7** vertebrae.

chambers (cham'ber) [G. *kamara*, vault]. A compartment or closed space, as the chambers of the heart.

chem'oreceptor. A sense organ or sensory nerve ending that is stimulated by a chemical substance or change, as in the taste buds.

chemotactic (kem-o-tak'tĭk) [G. *chēmeia*, chemistry, + *taxikos*, arranging]. Responding to a chemical stimulus by being attracted or repelled.

chiasm, chiasma (ki'azm) [G. *chiazen*, to mark with letter X]. A crossing or decussation; optic chiasm; the point of crossing of the optic nerve fibers.

chemotherapy (kem-o-ther'a-pi) [" + *therapeia*, treatment]. Treatment of disease by use of chemical agents that have a specific effect on the causative microorganism or agent, usually without producing toxic effects on the person.

chloride shift (klō'rĭd) [G. *chlōrus*, green, + A. S. *sciftan*, to divide]. The movement of chloride ions into red blood cells as bicarbonate ions are formed and leave the cell during the reaction of CO_2 and water in the cell.

choana (kō'ă-nă) (pl. choanae) [G. *choane*, funnel]. A funnel-shaped opening; the communicating passageways between the nasal cavities and the pharynx; the posterior nares or nostrils.

cholecystokinin (ko-le-sĭs-tō-kī'nin) [G. *cholē*, bile, + *kystis*, cyst, *kinein*, to move]. A hormonelike chemical secreted by the duodenal mucosa that stimulates contraction of the gallbladder.

cholesterol (ko-les'ter'ol) [" + *stereos*, solid]. A sterol (monohydric alcohol, $C_{27}H_{45}OH$) forming the basis of many lipid hormones in the body, e.g., sex hormones, adrenal corticoids; synthesized in the liver; a constituent in the bile.

cholinergic (kō'len-er-gik). Term applied to nerve endings that liberate acetylcholine.

chondroblast (kŏn' drō-blăst) [G. *chondros*, cartilage, + *blastos*, germ]. A primitive cell that forms cartilage.

chondrocyte (kŏn'drō-sīt) [G.]. A cartilage cell.

cholinesterase (kō-lĭn-ĕs'ter'ās). Any enzyme that catalyzes the hydrolysis of choline esters (organic acid + an alcohol), e.g., *acetylcholinesterase* catalyzes the breakdown of *acetylcholine*.

chorion (kō'rĭ-ŏn) [G.]. Outer embryonic membrane of the blastocyst from which *villi* connect with the endometrium to give rise to the placenta.

chordae tendinae (kor'da tĕn'dĭn-ē) [G. *chorde*, cord, + L. *tendinōsus*, like a tendon]. Small tendinous cords that connect the edge of an atrioventricular valve to a papillary muscle, and prevent valve reversal.

chord'ee (kor-dē') [Fr. corded]. Downward painful curvature of the penis on erection as in *gonorrhea*.

chorea (ko-re'ā) [G. *choreia*, dance]. Rapid, irregular, and wormlike movements of body parts; a nervous affliction.

chorioid (kō'rĭ-oid) [G. *chorioeidēs*, skinlike]. The middle vascular layer of the eyeball between the sclera and retina; a part of the *uvea*. Also, **choroid.**

choroid plexus (kō'roid plĕk'sŭs) [G. *chorioeidēs*, skinlike, + L. *plexus*, interwoven]. Vascular structures in the roofs of the four ventricles of the brain that produce *cerebrospinal fluid*.

chromatid (kro'mă-tĭd) [G. *chrōma*, color, + *tĭd*, time]. One of the two bodies resulting from longitudinal separation of duplicated chromosomes, each goes to a different pole of the dividing cell.

chromatin (krō'ma-tin) [" + *in*, within]. Darkly staining substance within the nucleus of most cells; consists mainly of *DNA*, and is considered to be the physical basis of *heredity*; forms *chromosomes* during mitosis and meiosis.

chromosome (kro'mō-sōm) [" + *soma*, body]. A microscopic J- or V-shaped body developing from chromatin during cell division. Contains the gene or hereditary determinants of body characteristics. In human, total **46** (23 pairs).

chronaxie (kro'nak-sĭ) [G. *chronos*, time, + *axia*, value]. A value expressing sensitivity of nerve fibers to stimulation; the time that a current which is twice threshold value must last to cause depolarization.

chronic [G. *chronos*, duration]. Long drawn out with slow progress; a disease that is not acute.

chyle (kīl) [G. *chylos*, milklike]. The contents of intestinal lymph vessels. Consists mainly of absorbed products of fat digestion.

chyme (kīm) [G. *chymos*, juice]. The mixture of partially digested foods and digestive juices found in the stomach and small intestine.

cilia (sil'ĭ-ā) [L. *ciliaris*, pert. to eyelashes]. Motile hairlike projections from epithelial cells; eyelashes.

circadian (sir-kā′dĭ-en) [L. *circa*, about, + *dies*, day]. Describes events that repeat in a length of time approximating a 24-hour cycle.

circumduction (sir-kum-duk′shun) [L. *circum*, around + *ducere*, to lead]. Movement of a part in a circular direction, involves flexion, extension, abduction, adduction, and rotation.

cirrhosis (sĭ-rō′sĭs) [G. *kirros*, yellow, + *osis*, infection]. A chronic liver disease characterized by degenerative changes in the liver structure.

cister′na [L. a vessel]. A cavity or enlargement in a tube or vessel.

cleavage (kle′vej) [A. S. *cleofian*, to adhere]. Mitotic cell division of a fertilized egg; splitting a complex molecule into two or more simpler ones.

cleft [M. E. *clyft*, crevice]. A fissure; divided or split.

cleido (kli′do) [G. *kleis*, clavicle]. Prefix: pert. to the clavicle (collarbone).

clitoris (klĭ′to-rĭs, klit′ō-ris) [G. *kleitoris*, clitoris]. Female erectile genital organ that is homologous to the penis of the male.

clone (klōn) [G. *klon*, a cutting used for propagation]. A group of cells descended from a single cell; something reproduced exactly like its predecessor.

clot (klōt) [A. S. *clott*]. The semisolid mass of fibrin threads plus trapped blood cells, which forms when the blood coagulates; to coagulate.

Co-A A dinucleotide that activates many substances in metabolic cycles.

coagulation [L. *coagulatiō*]. The process of clotting by which liquid blood is changed into a gel to aid in preventing blood loss through injured vessels.

coccygeal (kok-sij′ē-al) [G. *kokkyx*, coccyx]. Pert. to the coccyx, tailbone, last four fused spinal bones.

cochlea (kok′lē-ā) [G. *kochliās*, a spiral]. A cone-shaped winding tube that forms the portion of the inner ear that contains the organ of Corti, the receptor for hearing.

codon. The sequence of three nitrogenous bases of messenger RNA that specifies a given amino acid and its position in a protein.

coen′zyme [L. *co*, together, + G. *en*, with, + *zymē* leaven]. A nonprotein substance that activates enzymes or chemical compounds.

cofactor A substance essential to or necessary for the operation of another material such as an enzyme.

coitus (ko′i-tus) [L. a uniting]. Sexual intercourse between man and woman. Also, *coition, copulation.*

collagen (kol′ā-jen) [G. *kolla*, glue, + *gennan*, to produce]. A fibrous protein forming the main organic structures of connective tissues.

collateral (ko-lat′er-al) [L. *con*, together, + *lateralis*, pert. to a side]. Side branch of a nerve axon or blood vessel.

colloid (kol′oid) [G. *kollōdēs*, glutinous]. A system formed by large particles that remain suspended and do not settle. Size of particles lies between 1 and 100 millimicrons.

colon [G. *kōlon*]. The portion of the large intestine from the end of the ileum to the rectum. It consists of the ascending, transverse, descending, and sigmoid portions.

co′ma [G. *kōma*, a deep sleep]. An abnormal deep stupor from which the person cannot be aroused by external stimuli.

combust′ [L. *comburere*, to burn up]. To burn or to oxidize.

compact′ bone [L. *compactus*, joined together]. The hard or dense part of bones containing subunits of structure called osteons or Haversian systems.

compensate [L. *cum*, with + *pensdāre*, to weigh]. To make up for a defect or deficiency.

compensatory. Serving to balance or offset; returning to a normal state.

compliance In the lung, ease of expanding.

component A constituent part.

compound [L. *componere*, to place together]. A substance composed of two or more parts and having properties different from its parts.

commissure (kŏm′ĭ-shūr) [L. *commissura*, a joining together]. Band of nerve fibers passing transversly over the midline in the central nervous system.

concave (kon′kāv) [L. *con*, with, + *cavus*, hollow]. Having a rounded hollow or depressed surface.

concentration (kon-sen-tra′shun) [L. *contratiō*, in the center]. Amount of a solute per volume of solvent, increase in strength of a fluid by evaporation or removal of water by other means.

concep′tion [L. *conceptiō*, a conceiving]. The point in time when a sperm unites with an ovum to initiate formation of a new individual; *fertilization*.

concha (kong′kā) [G. *kogchē*, shell]. A shell-shaped structure, as the nasal turbinates.

conduc′tion [L. *conducere*, to lead]. The transmission of a nerve impulse by exciting progressive segments of a nerve fiber; the transfer of electrons, ions, heat, or sound waves through a conducting media.

congenital (kon-jen′ĭt-al) [L. *congenitus*, born together]. Present at birth.

cone (kōn) [G. *kōnos*, cone]. A receptor cell in the retina concerned with color vision.

conjugate (kon′jū-gāt) [L. *con*, with, + *jugum*, yoke]. To pair or to join something with something else.

conjunctiva (con-junk-tī′vă) [L. " + *jungere*, to join]. Mucous membrane lining of the eyelid that is reflected onto the eyeball.

conscious (kon′shus) [L. *conscius*, aware]. Awake, being aware and having perception.

consolidate (kon-sal-ĭ-dā′shun) [L. *consolidāre*, to make firm]. To become solid or firm.

constric′tion [" + *stringere*, to draw]. The narrowing or becoming smaller in diameter as of a pupil of the eye or blood vessel.

commu′nicable disease. A disease that may be transmitted directly or indirectly from one person to another; due to an infectious agent, or toxic products produced by it.

continuity (kon-tĭ-nū′ĭ-tĭ) [L. *continuus*, continued]. The state of being intimately united, continuous, or held together.

contraception (kon-tra-sep′shun) [L. *contra*, against, + *conceptiō*, a conceiving]. The prevention of conception, impregnation, or implantation.

contrac′tion [L. *contractio*, a drawing up]. Shortening or tightening as of a muscle, or a reduction in size.

convection (kon-vek′shun) [L. *convehere*, to convey]. Transference of heat in liquids or gases by means of currents.

convergent (kon-ver′jent) [" + *vergere*, to incline]. Tending toward a common point; coming together.

conver′sion [L. *convertere*, to turn around]. Change from one state to another.

con′vex [L. *convexus*, vaulted, arched]. Having an outward curvature or bulge.

convul'sion [L. *convulsio*, a pulling together]. Paroxysms or involuntary muscular contractions and relaxations.

coordination (ko-or-din-a'shun) [L. *co*, together + *ordināre*, to arrange]. The working together of different body systems in a given process; the working together of muscles to produce a certain movement.

cope (kōp). The ability to effectively deal with and handle the stresses to which one is subjected.

copulation (kop-u-la'shun) [L. *copulātiō*]. Sexual intercourse.

cor'nea [L. *corneus*, horny]. The transparent anterior portion of the fibrous coat of the eye, continuous with the sclera.

corium (ko'rĭ-um) [G. *chorion*, skin]. The dermis or true skin that lies immediately under the epidermis; contains nerve endings, capillaries, and lymphatics, and is composed of connective tissue.

coronary (kor'o-na-rĭ) [L. *coronarius*, pert. to a circle or crown]. Blood vessels of the heart that supply blood to its walls.

cor'pus (pl. *corpora*) [L. body]. Any mass or body; the principal part of any organ.

cor'puscle [L. *corpusculum*, little body]. An encapsulated sensory nerve ending; any small rounded body; old term for blood cell, erythrocyte or leukocyte.

cor'pus luteum (lū'tē-um) [" + *luteus*, yellow]. Yellow body of cells that fills the ovarian follicle in the stage following ovulation.

coro'na [G. *korōnē*, crown]. Any structure resembling a crown; **coronal** plane or section, same as *frontal*.

cor'tex (pl. *cortices*) [L. rind]. An outer portion or layer of an organ or structure.

cor'ticoid. One of many adrenal cortical steroid hormones. Also, *corticosteroid*.

corticospi'nal [" + *spīna*, thorn]. Pert. to cerebral cortex and spinal cord.

cos'tal [L. *costa*, rib]. Pert. to a rib.

countercurrent (kown'ter-ker'rent) [L. *contra*, against, + *currere*, to run]. When flow of gases or fluids is in opposite directions in two closely spaced limbs of a bent tube.

crā'nium [L. from G. *kranion*, skull]. That portion of the skull that encloses the brain; consists of eight bones.

creatine (krē'ă-tĭn) [G. *kreas*, flesh]. A crystalline substance found in organs and body fluids; combines with phosphate and serves as a source of high-energy phosphate released in the anaerobic phase of muscle contraction.

crest [L. *crista*, tuft]. A ridge or long prominence, as on bone.

crista galli [" + *galea*, helmet]. A crest or shelf on the ethmoid bone that attaches the *falx cerebri*.

criterion (pl. *criteria*) [G. *krites*, judge]. A standard or means of judging a condition or establishing a diagnosis.

crura (krōō'ra) (sing. *crus*) [L. legs]. The legs; a pair of long bands or masses resembling legs.

cross linkage. Chemical bonds linking fibrous molecules.

crypt (kript) [G. *kryptein*, to hide]. A small cavity or sac extending into an epithelial surface; a tubular gland.

crys'talloid [G. *krystallos*, clear ice, + *eidos*, form]. A substance that is capable of forming crystals, and that in solution diffuses rapidly through membranes; less than 1 millimicron in diameter; smaller than a *colloid*.

currettage (ku-ret'aj) [Fr. *curette*, a cleanser instrument]. Scraping of a cavity.

cuta'neous [L. *cutis*, skin]. Pert. to the skin.

cyanosis (sī-an-o'sis [G. *kyanos,* dark blue, + *-osis,* state of]. A condition of blue color being given to skin and mucous membranes by excessive amounts of reduced hemoglobin in the bloodstream.

cybernetics (sī'ber-net'iks) [G. *kybernetes,* helmsman]. The study of self-monitoring and regulating mechanisms for maintenance of near constant values of a function.

cycle (sī'kl) [G. *kyklos,* circle]. A series of events; something that has a predictable period of repetition; a sequence.

cystoscope (sīst'o-skōp) [G. *kystis,* bladder, sac, + *skopein,* to examine or view]. An instrument for examining or viewing the interior of the bladder.

-cyte (sīt) [G. *kytos,* cell]. Suffix denoting cell.

cyto- [G.]. Prefix indicating the cell.

cytochrome (sī'to-krōm) [" + *chrōma,* color]. A yellow, iron-containing protein pigment that functions to transport electrons and/or hydrogens in cellular respiration.

cytokine'sis [" + *kinēsis,* movement]. Division of cytoplasm in latter stages of mitosis.

cytoplasm (sī'to-plazm) [" + *plasma,* matter]. The cellular substance between the membrane and nucleus of a cell.

damping. Steady decrease in amplitude of successive vibrations, as in the cochlea.

de-. Prefix: down; from.

deamination (dē-am-ĭ-na'shun). Enzymatic removal of an amine group ($-NH_2$) from an amino acid to form ammonia.

death (deth) [A. S. *death*]. Permanent cessation of all vital functions. Absence of life. A flat EEG for 48 hours is taken as a sign of loss of spontaneous brain activity and death.

decompression [" + *compressio,* a squeezing together]. Release or removal of pressure from an organ or organism.

decussate (de-kus'āt) [L. *decussāre,* to cross, as an X]. To undergo crossing, or crossed.

defecation (def-ĕ-ka'shun) [L. *defaecāre,* to remove the dregs]. Evacuation or emptying fecal matter from the large bowel.

de'fect. A flaw or imperfection.

deficit (def'e-sit) [L. *deficere,* to be lacking]. Lack of or lacking in; to be less than.

degeneration (di-jen'er-ā-shun) [L. *degenerāre,* to degenerate]. To "go bad" or deteriorate; to lose quality.

degrade (di-grāde) [L. *degradare,* to reduce in rank]. To tear down or break down, as to degrade a chemical compound.

dehydration (de-hī-drā'shun) [" + G. *ydōr,* water]. Removal of water from something; the result of water removal.

den'drite [G. *dendritēs,* pert. to a tree]. A branched process of a neuron that conducts impulses to the cell body; form synapses.

dener'vate [L. *dē,* from, + G. *neuron,* nerve]. To block, remove, or cut the nerve supply to a structure.

de novo (dē-nō'vō) [L.]. Anew; again; once more.

dense [L. *densus,* compact]. Packed tightly together; in a bone, the outer layers of bony tissue that do not contain osteons.

deoxyribonucleic acid (dē-ok-sī-ri-bō-nu′klē-ik). Composed of nitrogenous bases, deoxyribose sugar, and a phosphate group. It is believed to be the site of determination of the body's hereditary characteristics. **DNA.**

depolarization (dē-pō-lar-ĭ-zā′shun) [" + *polus,* pole]. Loss of polarity or the polarized state.

depressed (de-prest′). To push down; flattened or hollow; lowered in intensity; low in spirits.

depression (de-presh′un) [L. *depressiō,* a pressing down]. A lowered or hollowed region; lowering of a vital function; a mental state of dejection, lack of hope or absence of cheerfulness.

derivative (dē-riv′a-tiv) [L. *derivare,* to turn a stream from its channel]. A substance or structure originating from another substance or structure and having different properties than its source.

-derm- (durm) [L. *derma,* skin or covering]. A word used as a prefix or suffix referring to the skin or a covering.

derma (dermis). Pert. to the skin; the connective tissue layer of the skin under the epidermis or covering layer.

desquamate (des-kwam-āte) [" + L. *squamāre,* to scale off]. Shedding of the epidermal surface cells.

detoxify (de-toks′ĭ-fī) [" + G. *toxikon,* poison, + L. *facere,* to make]. To remove the toxic or poisonous quality of a substance.

development (Fr. *de′velopper,* to unwrap). Progress from embryonic to adult status; change from a less specialized to a more specialized state.

dextrose (deks′trōs) [L. *dexter,* right + *ose,* sugar]. Another name for glucose, a simple sugar (monosaccharide). Also, *grape sugar.*

di- [G.]. Prefix: twice or double.

dialysis (di-al′ĭ-sis) [" + *lysis,* loosening]. Passage of a diffusible solute through a membrane that restricts passage of colloids; separation of solutes by their rates of diffusion.

diameter (di-am′ē-ter) [" + *metron,* a measure]. A distance from any point on the periphery of a surface to the opposite point.

diapedesis (di-ă-ped-e′sis) [" + *pēdan,* to leap]. Movement of white blood cells through undamaged capillary walls.

diaphragm (di′ă-fram) [" + *phragma,* wall]. Any thin membrane; a rubber or plastic cup placed over the cervix of the uterus for contraceptive purposes; the muscle and connective tissue partition between chest and abdomen (muscle of respiration).

diaphysis (di-af′i-sis) [" + *plassein,* to grow]. The shaft or cylindrical part of a long bone.

diarrhea (di-ă-re′ā) [G. *dia,* through, + *rein,* to flow]. Passage of frequent watery stools: a symptom of gastrointestinal disturbance.

diastasis (di-as′tă-sis) [G. a separation]. In the cardiac cycle, the time when there is little change in length of muscle fibers and filling of the chamber is very slow. It is followed by contraction of the muscle of the chamber.

diastole (di-as′tō-le) [G. *diastellein,* to expand]. The relaxation phase of cardiac activity, when fibers are elongating and filling of a chamber is rapid.

diencephalon (di-en-sef′ă-lon) [" + *egkephalos,* brain]. The second portion of the brain between telencephalon and mesencephalon; includes thalamus and hypothalamus.

differentiation (dif'er-en'shi-a'shun) [L. *dis*, apart, + *ferre*, to bear]. Acquiring a structure or function different from those originally present.

diffuse (dĭ-fūs') [" + *fundere*, to pour]. To spread, scatter, or move apart; usually caused by movement of molecules in a solution or suspension.

diffusion (dĭ-fu'zhun). Eventual even mixing of solutes and solvents as a result of motion of molecules.

digest [" + *gerere*, to carry]. To break down foods using chemical or mechanical means; to convert foods to an absorbable state.

digitalis (dij-ĭ-tal'is) [L. *digitus*, finger]. The dried leaves of the purple foxglove plant; used in powdered form as a heart stimulant.

dilate (di'lāt) [L. *dilatāre*, to expand]. To become larger in size or diameter.

dilute (di-loot') [L. *dilutus*, to wash away]. To thin down or weaken, by addition of water or other liquids.

dimension (dĭ-men'shun) [Fr. *dimensio*, a measuring]. A measurable extent, such as length, width; extent or size; importance.

diopter (dĭ-op'ter) [G. *dioptron*, something that can be seen through]. The refractive power of a lens with a focal distance of 1 meter.

dip'loe [G. a fold]. The spongy bone between the inner and outer tables (layers) of dense bony tissue in the skull bones.

diploid (dĭp'loyd). Cell having twice the number of chromosomes present in the egg or sperm of a given species.

disaccharide (dĭ-sak'i-rid) [G. *dis*, two, + *sakcharon*, sugar]. A sugar composed of two simple sugar molecules.

dissect (dis-sekt') [L. *dissecare*, to cut up]. To separate tissues or organs of a cadaver for study.

dissociate (dis-so'si-āt) [" + *sociatiō*, union]. To separate into components or parts.

dissolve (dĭ-zolv) [L. *dissolvere*, to dissolve]. To make liquid, liquefy or melt by placing in water or other liquids.

distal (dis'tal) [L. *distāre*, to be distant]. Fartherest from the center of the body or point of attachment; opposite of proximal.

diuresis (di-u-re'sis) [G. *dia*, through, + *ourein*, to urinate]. Secretion and passage of abnormally large amounts of urine.

diuretic (di-u-ret'ik). A substance causing diuresis.

diverticulum (di-ver-tik'u-lum) [L. *diverticulāre*, to turn aside]. An outpocketing, sac, or pouch in the wall of a canal or hollow organ.

dominance (dom'i-nense) [L. *dominare*, to rule]. In inheritance of characteristics, the effects of one allele overshadow the effects of the other and the character determined by the dominant (stronger) gene prevails.

donor (dō'ner) [Fr. *doneor*, to give]. One who gives or donates (such as blood or organs).

dor'sal [L. *dorsum*, back]. Pert. to the upper or back portion of the body.

drug [Fr. *drogue*]. A medicinal substance used in the treatment of disorder or disease.

duct [L. *ducere*, to lead]. A tubular vessel or channel carrying secretions from a gland onto an epithelial surface.

ductus. A channel from one organ, or part of an organ, to another.

duplication (doo'plĭ-cā-shun) [L. *duplicare*, to double]. To double something; to form a copy.

dura (du′ră) [L. *durus,* hard]. Hard or tough; *dura mater,* the tough outer membrane around the brain and spinal cord.

dwarf (M. E. *dwerf,* little). An abnormally small, short, or disproportioned person.

dysmenorrhea (dis-men-o-re′a) [L. *dys,* bad, difficult, or painful, + *men,* mouth + *rein,* to flow]. Painful or difficult menstruation.

dyspnea (disp-ne′a) [″ + *pnoē,* breathing]. Labored or difficult breathing.

dystrophy (dis′tro-fĭ) [″ + G. *trephein,* to nourish]. Degeneration of an organ resulting from poor nutrition, abnormal development, infection, or unknown causes.

ecchymosis (ek-ĭ-mo′sis) [G. *ek,* out, + *chymos,* juice, + *osis,* caused by, state of]. A blue-black, greenish-brown, or yellow area of hemorrhage in the skin.

ectoderm (ek′tō-derm) [G. *ekto,* outside, + *derma,* skin]. The outer layer of cells in the embryo, from which develops the nervous system, special senses, and certain endocrines.

-ectomy (ek′tō-my) [G. *ektomē,* a cutting out]. Suffix pert. to surgical removal of an organ or gland.

ectopic (ek-top′ik) [G. *ek,* out, + *topia,* place]. In an abnormal place or position.

eczema (ek′zĕ-mă) [G. *ekzein,* to boil out]. A chronic or acute inflammation of the skin characterized by the development of skin lesions.

edema (e-de′ma) [G. *oidema,* swelling]. Swelling due to increase of extracellular fluid volume; dropsy.

effect′or [L. *effectus,* accomplishing]. An organ that responds to stimulation; a gland or muscle cell.

ef′ferent [L. *ex,* out + *ferre,* to carry]. Carrying away from a center or specific point of reference.

egest (ē-jest′) [L. *egestus,* to discharge]. To rid the body of something. Also, *eliminate.*

ejaculation (ē-jak-u-la′shun) [L. *ejaculāri,* to throw out]. Ejection of semen from the male urethra during sexual excitement, or the discharge of secretions from vaginal glands.

elastance (ē-last′ance) [G. *elasticos,* elastic]. The power of elastic recoil after being stretched. Also, *elasticity.*

electro- (ē-lec′trō) [G. *elektron,* amber]. Prefix pert. to electrical phenomena of one sort or another.

electrolyte (e-lek′trō-līt) [″ + *lytos,* solution]. A substance that dissociates into electrically charged ions; a solution capable of conducting electricity because of the presence of ions.

elec′tron [G.]. A minute body or charge of negative electricity; a component of atoms.

el′ement [L. *elementum,* a rudiment]. A substance that cannot be separated into its components by conventional chemical means.

eliminate (e-lim′ĭ-nāte) [L. *ē,* out, + *limen,* threshold]. To rid the body of wastes by emptying of a hollow-organ; to expel.

em′bolus [G. plug]. A floating undissolved mass or bubble of gas brought into a blood vessel by the fluid flow.

embryo (em′brĭ-o) [G.]. The developing human from the *third* to *eighth weeks* after conception.

embryonic disc. The primitive two-layered structure formed about two weeks after fertilization, from which the embryo develops.

emeiocytosis (ēm-ē-ōsī-tō'sis) [G. *emein*, to vomit, + *cyte*, cell, + *osis*, state of]. Elimination of materials from a cellular vacuole by fusion of the vacuole with the cell membrane and release of contents to the outside of the cell.

emesis (em'ĕ-sis) [G. *emein*, to vomit]. Vomiting.

-emia (ē'mi-a) [G. *aima*, blood]. Suffix: blood.

emmetropic (em-me-trō'pik) [G. *emmetros*, in due measure, + *opsis*, sight]. Normal in vision.

emphysema (em-fi-se'mă) [G. *emphysan*, to inflate]. Condition resulting from rupture or expansion of alveoli of the lungs.

en- [G. *en*, in]. Prefix: in.

encephalo- (en-sef'a-lō) [G. *egkephalos*, brain]. Prefix pert. to brain or cerebrum.

en'dō- [G. *endon*, within]. Prefix: within; inner.

endocrine (ĕn'do-krĭn, -krĭn) [" + *krinein*, secrete]. Internally secreting into the bloodstream.

endoderm (en'do-derm) [" + derma, skin]. Inner layer of cells of the embryo; gives rise to *linings* of gut and all outpocketings of the gut (e.g., lungs, pancreas, liver).

endometrium (en-do-me'trĭ-um) [" + *mētra*, uterus]. The mucous membrane lining the inner surface of the uterus.

endoplasmic reticulum (en-dō-plas'mik re-tĭk'ū-lum). The series of tubular structures found in the cytoplasm of cells; transports substances through cells.

endothelium (en-dō-the'lĭ-um) [" + *thēlē*, nipple]. The simple squamous epithelium forming the inner lining of blood and lymphatic vessels, and the heart.

end product. The final product or waste of a digestive or metabolic process.

energy (en'er-ji) [" + *ergon*, work]. Capacity to do work; heat or chemical bond capacity for work. Energy is seen as motion, heat, sound, etc.

engorged (en-gorjd') [Fr. *engorger*, to obstruct or devour]. Swollen or distended as with blood.

engulf (in-gulf') [Fr. *engolfer*, to swallow up]. To surround, take in, or swallow.

en'tero- [G. *enteron*, intestine]. Prefix pert. to the intestines.

enteroceptive (en-ter-ō-sep'tĭv) [" + L. *capere*, to take]. Originating within body viscera.

environment (en-vī'rŏn-ment) [M. E. *environuen*, about, + ment]. or [L. *in*, in, + *virer*, to turn]. The surroundings of a cell, organ, or organism.

enzyme (en-zīm) [L. *en*, in + *zymē*, leaven]. An organic catalyst causing alterations in rates of chemical reactions without being consumed in the reaction.

eosin (ē'o-sin) [G. *ēōs*, dawn]. A rose-colored dye used for staining cells and tissues. Something exhibiting a preference for the dye is called *eosinophilic*.

epi- [G.]. Prefix: upon, at, outside of, in addition to.

epididymis (ep-i-did'ĭ-mis) [" + *didymos*, testes]. A long convoluted tubule or duct resting on the testis and conveying sperm to the vas deferens.

epiglottis (ep-ĭ-glot'is) [" + *glōttis*, glottis]. A leaf-shaped structure located over the superior end of the larynx. Aids in closing the larynx during swallowing.

epileptic (ep-ĭ-lep'tik) [G. *epilēptikos*, pert. to a seizure]. Pert. to a disturbance of consciousness occurring during epilepsy; an individual suffering from the disorder of epilepsy.

epiphysis (ē-pif'ĭ-sis) [G. *epiphysis,* a growing upon]. One of the two ends of a long bone, separated from the diaphysis or shaft by a growth line.

epistaxis (ē-pis-tax'is) [G. *epistaxein,* to bleed from the nose]. Hemorrhage from the nose.

epithelial (ep-ĭ-thē'lĭ-al). Pert. to epithelia, the covering or lining tissues of the body.

equilibrium (e-kwil-ib'rĭ-um) [L. *aequus,* equal, + *libra,* balance]. A state of balance or rest, in which opposing forces are equal.

eruption (e-rup'shun) [L. *eruptio,* a breaking out]. Becoming visible, such as a lesion or rash appearing on the skin.

erythropoietin. A substance produced by the kidney which stimulates production of red blood cells.

erythropoiesis (ē-rith-ro-poy-e'sis) [G. *erythema,* redness, + *poiesis,* making]. The formation of erythrocytes or red blood cells.

eschar (es'kar) [G. *eschara,* scab]. A slough (dead matter or tissue); *debris* developing following a burn.

esthesia (es-the'zĭ-ā) [G. *aisthēsis,* sensation or feeling]. Perception, sensation or feeling.

estrogen (es'trō-jen) [G. *oistros,* mad desire, + *gennan,* to produce]. A substance producing or stimulating female characteristics; the follicular hormone.

eu- (ū) [G. *eu,* well]. Prefix: normal, well, or good.

euphoria (u-fo'rĭ-ā) [" + *pherein,* to bear]. An exaggerated feeling of well-being.

evacuate (e-vak'ū-āte) [L. *evacuāre,* to empty]. To empty or discharge the contents of, especially the bowels.

evaporate (ē-va'por-āte) [L. *ē,* out, + *vaporāre,* to steam]. To change from liquid to gaseous form usually by addition of heat, as to evaporate water by boiling, or on the skin.

evoke (e-vōk) [Fr. *évoquer,* to call out]. To cause to happen, to call forth.

ex- [L.]. Prefix: out, away from, completely.

excitable (ēk-sĭ'tā-bl) [L. *excitāre,* to rouse]. The property of being capable of responding to stimuli by alterations in ion separation.

excitatory [L.]. Something that causes changes in state of excitability; something that stimulates a function.

excrete (eks-krēt') [L. *excernere,* to separate]. To separate and expel useless substances from the body.

exocrine (eks'ō-krēn) [" + *krinein,* to separate]. External secretion by a gland, through ducts, upon an epithelial surface.

exogenous (eks-oj'ē-nus) [" + *gennan,* to produce]. Originating outside the cell or organism.

expiration (eks-pi-rā'shun) [L. *ex,* out, + *spirare,* to breathe]. Expulsion of air from the lungs. Death.

exponential (ek-spō-nen'shul) [L. *exponere,* to put forth]. Change in a function according to multiples of powers (e.g., square or cube), rather than numerically (e.g., 1 time, 2 times, etc.).

external (ek-sturn'al) [L. *externus,* outside]. Outside or toward the surface.

exteroceptive (eks-ter-o-sep'tiv) [" + *ceptus,* to take]. Sensations or reflex acts originating from stimulation of receptors at or near a body surface.

extracellular [L. *extra,* outside of, + cell]. Fluids outside of the cells. Abbr. **ECF.**

extraneous (eks-tra'ne-us) [L. *extraneus,* external]. Outside of or unrelated to something.

extrinsic (eks-trĭn'sik) [L. *extrinsecus,* ouside]. From without or coming from without; separate or outside of an organ or cell.
exudate (eks'u-dāt) [" + *sudāre,* to sweat]. Accumulation of fluid in a body cavity or on a body surface. Pus, serum, or the passing of the same.

facial (fā'shul) [L. *facies,* the face]. Pert. to the face.
facilitation (fah-sĭl-ĭ-tā'shun) [L. *ficilis,* to make easier]. An increased ease of passage of a nerve impulse across a synapse.
FAD. Abbr. for *flavine adenine dinucleotide,* one of several hydrogen acceptors.
fascia (fash'ĭ-ă) [L. *fascia,* a band]. Fibrous membranes covering, supporting, and separating muscles. *Superficial f.* is the hypodermis; *deep f.* is around muscles.
fascicle (fas'ik-l) [L. *fasciculus,* a little bundle]. An arrangement like a bundle of rods. Applied to bundles of nerve and muscle fibers.
fasciculation (fas-ĭck-u-lā'shun). A localized contraction of one or a few motor units in a skeletal muscle.
fatigue (fă-tēg) [L. *fatigare,* to tire]. A feeling of tiredness; the state of an organ, synapse, or tissue in which it no longer responds to stimulation.
feces (fē'sēz) [L. *faeces,* dregs]. The waste matter expelled from the bowel through the anus.
feedback. Detection of the nature of an output and using that to control the process producing the output. May be negative (*inhibitory*) or positive (*stimulating*).
fe'male [L. *femella,* woman]. The designation given to the sex that produces ova and bears offspring; a girl-child or woman.
fertilization (fur-tĭl-ĭ-zā'shun) [L. *fertilis,* to bear]. The impregnation of an ovum by a sperm.
fetus (fē'tus) [L. *fetus,* progeny]. The name given to a developing human after it assumes *clearly human form;* the period from 6 to 8 weeks after conception to birth.
fiber (fī'ber) [L. *fibra,* thread]. A larger threadlike or ribbonlike structure; a muscle cell; usually composed of smaller units.
fibril (fī'brĭl) [L. *fibrilla,* a little fiber]. A very small, threadlike structure, often the component of a muscle or nerve cell, or a connective tissue fiber.
fibrillate (fī'bril-āte) [L. *fibrilla,* a little fiber]. Spontaneous uncoordinated quivering of a muscle fiber, as in cardiac muscle.
fibrin (fī-brĭn) [L. *fibra,* a fiber]. A white or yellowish insoluble fibrous protein formed when blood clots.
fibrinogen (fī-brĭn'ō-gen). A soluble plasma protein that is acted upon to produce fibrin as blood or lymph clots.
fibrosis (fi-brō'sis). Abnormal deposition or increase in fibrous tissue in a body part, organ, or tissue.
fil'iform [L. *filum,* thread, + *form,* form]. Having a threadlike or hairlike form.
fil'ter [L. *filtare,* to strain through]. To strain, or separate on the basis of size; a device to strain liquids.
filtrate (fĭl'trāt). The name given to the fluid that has passed through a filter.
filtration (fĭl-trā'shun). The process of forming a filtrate by passing a fluid, under pressure, through a filter or selective membrane.
flaccid (flak'sid) [L. *flaccidus,* flabby]. Relaxed, flabby, having no tone, as in a muscle.

flatus (fla'tus) [L. *flatulentiā*, a blowing]. The gas in the digestive tract.

flora (flō'ra) [L. *flos*, flower]. Plant life; the microorganisms of the bowels, adapted to live in that organ.

follicle (fŏl'ĭ-cul) [L. *folliculus*, a little bag]. A small hollow structure containing cells or secretion.

foodstuff (food'stŭf) [M. E. *fode*, to eat]. Any substance made into or used as a food; nutritive substances providing heat and energy to the body when metabolized.

fornix (for'nĭks) [L. *fornix*, arch]. Any structure having an arched or vaultlike shape; a band connecting lobes of a cerebral hemisphere; a cleftlike space of the vagina around the neck of the uterus.

free energy. Energy released from metabolic processes that is capable of doing work.

frequency (frē'qwĕn-sĭ) [L. *frequens*, often or constant]. Number of repetitions of something in a given time period (e.g., cycles per second, *cps*).

fulcrum (fŭl,krŭm) [L. *fulcire*, to prop or support]. The point about which a lever turns or pivots.

function (fŭnk'shŭn) [L. *functia*, to perform]. The action performed by living matter (cell, tissue, organ, system, or organism); to carry on an action.

fundus (fŭn'dŭs) [L. *fundus*, sling or base]. The larger, usually blind end of an organ or gland.

fu'siform [L. *fusus*, spindle, + *forma*, shape]. Having a shape tapered at both ends.

fusion (fū'shun) [L. *fusio*, to meet]. Coming together; meeting or joining.

gait (gāt) [M. E. *gaite*, street]. The manner of walking.

galactose (gă-lak'tōs) [G. *gala*, milk, + *ose*, sugar]. Milk sugar; a simple hexose sugar found in milk. $C_6H_{12}O_6$.

gamete (gam'ēt) [G. *gamēte*, a wife or spouse]. A male or female reproductive cell, that is, a sperm or egg (ovum).

gametogenesis (gam-ē-tō-jen'ĕ-sis) [" + *genesis*, birth or origin]. The formation of gametes; spermatogenesis or oögenesis.

gamma (găm'ŭh) [G. letter **g** of alphabet]. The third item of a series; a microgram (one millionth gram).

gamma globulins. A plasma protein carrying most of the blood-borne antibodies.

ganglion (gang'lĭ-ŏn) [G. *gagglion*, a tumor or swelling]. A mass of nerve cell bodies outside of the brain and spinal cord; a cystic tumor developing on a tendon.

gangrene (gan'grēn) [G. *gaggraina*, an eating sore]. Decay of tissue when blood supply is obstructed.

gas'tric [G. *gaster*, stomach]. Pert. to the stomach.

gene (gēn) [G. *gennan*, to produce]. The unit of heredity responsible for transmission of a characteristic to the offspring; believed to be a part of a nucleic acid chain.

genetic (jen-ĕt'ik) [G. *genesis*, origin]. Pert. to genesis or origin of something.

genital (jen'ĭ-tal) [L. *genitalis*, genital]. Pert. to the organs of reproduction; *external genitalia*, the external organs of reproduction.

genotype (jēn'ō-tĭp) [G. *gennan*, to produce, + *typos*, type or kind]. The hereditary makeup of an individual as determined by his genes.

germinal (jer'mĭn-āl) [L. *germen*, microbe]. Pert. to a reproductive cell, or a "germ" (microorganism).

germ layer. One of the three basic tissue layers (ectoderm, endoderm, mesoderm) formed in the embryo that give rise to all body tissues and organs.

gerontology (jer-on-tol'ō-ji) [G. *geron*, old man, + *logos*, study of]. The study of the phenomena of old age.

gestation (jes-ta'shun) [L. *gestāre*, to bear]. The period of intrauterine development of a new organism; the time of pregnancy.

gingiva (jin'jĭ-vă) [L. *gingiva*, gum]. The gums, or tissues surrounding.

girdle (gĭr'dŭl) [A. S. *gyrdel*, to encircle]. A belt, zone, or a structure resembling a circular belt or band.

gland [L. *glans*, a kernel]. A secretory organ or structure.

glia (glē'ā) [G. *glia*, glue]. The nonnervous tissues of the nervous system; supporting and nutritive cells of various types.

globular (glob'ū-lar) [L. *globus*, a globe]. Having a rounded or spherical shape.

glomerulus (glō-mer'ū-lus) [L. a little skein, a tangle of thread or yarn]. A small round mass of cells or blood vessels.

glottis (glŏt'ĭs) [G. *glottis*, back of tongue]. The opening between the vocal cords in the larynx (voice box).

-glyc- (glīk) [G. *glykus*, sweet]. Pert. to glucose.

glyceride (glĭs'ĕr-īd) [G. *glykus*, sweet]. Glycerin (an alcohol) together with one or more fatty acids.

glycogen (glī'ko-jen) [G. *glykos*, sweet, + "]. A complex polysaccharide; "animal starch"; stored in liver and muscle.

goblet cell. A one-celled mucus secreting gland found in epithelia of the respiratory and digestive systems.

goiter (goi'tĕr) [L. *guttur*, throat]. An enlargement of the thyroid gland, irrespective of cause.

gon'ads [G. *gonē*, a seed]. A general name for an organ producing sex cells; the name of the embryonic sex gland before differentiation into ovary or testis.

gradient (grā'dĭ-ent). A slope or grade; a curve representing increase or decrease of a function.

graft [L. *graphium*, grafting knife]. A tissue or organ inserted into a similar substance or area to correct loss or absence of that structure; a transplant; to carry out the procedure of grafting or transplantation.

-gram. Combining form pert. to the record drawn by a machine (e.g., cardiogram, telegram).

granulation (grăn-ū-lā'shun) [L. *granulum*, little grain]. To form granules; tissue appearing in the course of wound healing, characterized by its large numbers of blood vessels.

gran'ule [L.]. Any minute or tiny grain in a cell.

grav'ity [L. *gravitās*, weight]. The force of the earth's gravitation attraction that gives objects weight.

gray matter. The tissue of the central nervous system consisting mainly of nerve cell bodies.

gristle (grĭs'l) [A. S.]. Cartilage.

gross (grōs) [L. *grossus*, thick]. Large; anatomy seen with the naked eye.

ground substance. The fluid or semifluid material occupying the intercellular spaces in connective tissues.

growth [A. S. *grōwan*, to grow]. Progressive increase in size of a cell, tissue, or whole organism.

gut [A. S.]. The primitive or embryonic digestive tube (fore-, mid-, hindgut); Colloq. for intestine.

gyrus (jī'rus) (pl. *gyri*) [G. *gyros*, circle]. An upfold of the cerebrum.

[H⁺]. Symbol for hydrogen ion concentration.

hallucination (hă-lu-sī-nā'shun) [L. *alucinari*, to wander in mind]. A false perception having no basis in reality; may be auditory, visual, olfactory, etc.

hap'loid [G. *aploos*, simple]. Having one half the normal number of chromosomes characteristic of the species; sperm and ova are haploid cells.

hap'toglobin [G. *aptein*, to seize]. A protein in the plasma that combines specifically with hemoglobin released from red blood cells.

haustra (haws'tra) [L. *haurīre*, to draw water]. The sacculations or pouches of the colon.

heat (hēēt) [G. *heito*, fever]. High temperature; being hot; a form of energy; to make hot.

helix (pl. *helices*) (hē'liks) [G. *helix*, a coil]. A coil or spiral.

hematocrit (he-mat'ō-krit) [G. *aima*, blood, + *krinein*, to separate]. The volume of red blood cells in a tube after centrifugation.

heme (hēm) [G.]. The iron containing red pigment that, with globin, forms hemoglobin.

hemiplegia (hem-ĭ-plē'jĭ-ă) [G. *hemi*, half, + *plēgē*, a stroke]. Paralysis of one half of the body in a right-left direction.

hemoglobin (hē-mō-glō'bin) [G. *aima*, blood + *globus*, globe]. The respiratory pigment of red blood cells that combines with O_2, CO_2, and acts as a buffer.

hemolysis (he-mol'ĭ-sis) [" + *lysis*, dissolution]. Destruction of red blood cells with release of hemoglobin into the plasma.

hemorrhage (hem'o-rij) [" + *rēgnunai*, to burst forth]. Abnormal loss of blood from the blood vessels, either internally or externally.

hemorrhoids (hem'o-royds) [G. *aimorrois*, a vein liable to discharge blood]. Dilated veins in the rectal columns.

hemostasis (he-mō-stā'sis) [" + *stasis*, stopping]. Stoppage of blood loss through a wound due to vascular constriction and coagulation.

he'par [G. *ēpar*, liver]. The liver.

hepatic (he-pat'ik). Pert. to the liver.

hereditary (hĕ-red'ĭ-tar-ē) [L. *hereditas*, heir]. Passed or transmitted (as a genetic characteristic) from parent to offspring; handed down.

hernia (her'nĭ-ă) [G. *ernos*, a young shoot]. Protrusion or projection of an abdominal organ through the wall that normally contains it.

heterogeneous (het-er-ō-je'nē-us) [G. *eteros*, other, + *gennos*, type]. Composed of things different or contrasting in type or nature.

heterograft [" + *graphium*, grafting knife]. Grafting of tissues or organs between different species.

hex'ose [G. *hex*, six, + *ose*, sugar]. A six-carbon sugar (e.g., glucose, fructose).

hi'lus [L. *hilus*, trifle]. A depression or recess on the side of an organ; vessels and nerves usually enter and/or leave at this point.

histio-, histo- [G. *histos*, tissue]. Prefix: tissue.

holocrine (hōl-o-krĭn) [G. *olos*, whole + *krinein*, to secrete]. A manner of production of a secretion by a gland in which a cell is the product or in which the whole cell enters the secretion.

homeostasis (hō-mē-ō-stā'sis) [G. *omoios*, like, + *stasis*, a standing]. The state of near constancy of body composition and function.

homogenous (hō-moj'ĕn-ŭs) [G. *homo*, likeness, + *genos*, kind]. Like or uniform in structure or composition.

homograft [" + L. "]. Use of tissues or organs of the same species for grafting purposes. *Isograft.*

homologous (hō-mol'o-gus) [" + *logos*, relation]. Similar in origin and structure.

hormone (hor'mōn) [G. *ormanein*, to excite]. The chemical produced by an endocrine gland.

humor (hū'mŭr) [L. *humor*, fluid, moisture]. Any fluid or semifluid substance in the body.

hyaline (hī'ă-lĭn) [G. *hyalinos*, glassy]. Clear, translucent, or glassy.

hyaluronidase (hi-ă-lur-on'ĭ-dās). An enzyme that liquefies the intercellular cement holding cells together.

hydrate (hī'drāt) [G. *ydōr*, water]. To cause water to combine with a compound; the compound after water has been added.

hydrogen (hi'drō-jen) [" + *gennan*, to produce]. An inflammable gaseous element, the lightest of all known elements.

hydrostatic pressure [" + *statikos*, standing]. Pressure exerted by liquids.

hyperemia (hī-per-ē'mĭ-ă) [G. *yper*, above, + *aima*, blood]. An extra amount of blood in an area.

hyperesthesia (hī-per-es-the'zē-ā) [" + *aisthēsis*, sensation]. Excessive sensitivity to sensory stimulation.

hypermetropic (hi-per-mĕ-trō'pik) [" + *metron*, measure + *ōps*, eye]. Pert. to farsightedness.

hyperphagia (hī-per-fā-jē-ā) [" + G. *phagein*, to eat]. Excessive intake of food.

hyperplasia (hī-per-plā'zē-ā) [" + *plassein*, to form]. Increase in size due to increased numbers of cells derived by division.

hyperpnea (hī-perp'nē-ā) [" + *pnoe,* breath]. An increase in depth of breathing.

hypertension (hī-pĕr-tĕn'shun) [" + *tensio*, tension]. High blood pressure.

hypertonic (hi-per-tŏn'ik) [" + *tonos*, tension]. A solution containing a greater solute concentration and thus osmotic pressure than the blood, or having an osmotic pressure greater than another reference solution.

hypertrophy (hi-per'trō-fē) [" + *trophē*, nourishment]. Increase in size by adding substance and *not* by increasing numbers of cells.

hyperventila'tion [" + *ventiatio*, ventilate]. Increased exchange of air by increasing both rate and depth of breathing.

hypnotic (hip-nŏ'tik) [G. *ypnos*, sleep]. An agent producing sleep or depression of the senses.

hypokalemia (hī-pō-kăl-ē'mē-ā) [G. *ypo*, under, + L. *kali*, potash, potassium, + *aima*, blood]. Low blood potassium levels.

hypoten'sion [G.]. Low blood pressure.

hypothesis (hī-poth'ē-sis) [" + *thesis*, a placing]. An assumption made to explain something. It is unproved and is made to enable testing of its soundness.

hypotonic (hi-pō-tŏn′ik) [″]. A solution containing less solute, and thus having a lower osmotic pressure, than the blood, or a solution having a lower osmotic pressure than some reference solution.

hypoxemia (hī-poks-ē′mē-ă) [″ + *oxys*, acid, + ″]. Lowered blood oxygen levels.

hypoxia (hī-poks′ē-ă) [″]. Low oxygen levels in inspired air or reduced tension in the tissues.

icteric (ik-tĕr′ik) [G. *ikteros*, jaundice]. Pert. to jaundice; excessive accumulation of bile pigments in the body tissues.

idiopathic (id-ĭ-ō-path′ik) [G. *idio*, own, + *pathos*, disease]. Any condition arising without a clear-cut cause; of spontaneous origin.

immune (ĭm-ūn′) [L. *immunis*, safe]. Protected from getting a given disease.

impermeable (im-pĕrm′ē-ă-bl) [L. *in*, not, + *permeāre*, to pass through]. Not allowing passage.

implantation (im-plan-tā′shun) [L. *in*, into, + *plantāre*, to plant]. Embedding of the blastocyst in the uterine lining; inserting something into a body organ.

impulse (im′puls) [L. *impellere*, to drive out]. A physicochemical or electrical change transmitted along nerve fibers or membranes.

inactivate (ĭn-āk′tĭ-vāt) [L. *in*, not, + *activus*, acting]. To make inactive or inert.

inclusion (ĭn-klū′zhun) [L. *inclusus*, enclosed]. A lifeless, usually temporary constituent of a cell's cytoplasm.

independent irritability. Capable of reacting to external stimuli, or those delivered by other than the normal route.

indigestion (in-dī-jes′chūn) [L. *in*, not, + *digerere*, separated]. Inability to digest food; incomplete digestion; dyspepsia. Usually accompanied by pain, nausea, gas (belching), and heartburn.

induce (in-dūs) [L. *inducere*, to lead in]. To cause an effect; induction may be used to express the way that a gene causes an effect.

infarct (in′farkt) [L. *infacire*, to stuff into]. An area of an organ that dies following blockage of blood supply.

infection (in-fek′shun) [L. *inficere*, to taint]. The state when the body has been invaded by a pathogenic (disease-producing) agent.

infectious (in-fek′shus) [″]. Capable of being transmitted with or without contact. Usually used in connection with microorganism caused diseases.

inflame (in-flām) [L. *inflammare*, to set on fire]. To cause to become warm, swollen, red, sore, and feverish.

infundibulum (in-fun-dib′ū-lum) [L. *infundibulum*, funnel]. A funnel-shaped passage.

infusion (in-fū′zhun) [L. *infusiō*, from, + *fundere*, to pour]. Introduction of liquid into a vein.

ingest (in-jest′) [L. *ingerere*, to carry in]. To take foods into the body.

inguinal (in′gwi-nal) [L. *inguinalis*, the groin]. Pert. to the groin.

inherit (in-her′ĭt) [L. *inhereditare*, to appoint as heir]. To receive from one's ancestors.

inhibit (in-hĭ′bĭt) [L. *inhibere*, to restrain]. To repress or slow down.

innervate (ĭn-nŭr′vāt) [L. *in*, in, + *nervus*, nerve]. To supply with nerves; to stimulate to action.

inorgan′ic [L. *in*, not, + G. *organon*, an organ]. A chemical compound not containing both hydrogen and carbon; not associated with living things.

input (in'poot). That which is put into something, as in the passage of nerve impulses to the brain for processing.

insertion (in-sūr'shen) [L. *insertio,* to join into]. The addition of something by "setting it into" something else, as insertion of parts of a chromosome into another chromosome.

in situ [L.]. In position.

inspiration (in-spi-rā'shun) [L. *inspiratio,* to breath in]. Taking air into the lungs.

integrate (in'tē-grāt) [L. *intergratus,* to make whole]. To blend, put together or unify.

integument (in-teg'ū-ment) [L. *integumentum,* a covering]. The outer covering of the body; skin.

intensity (in-ten'sĭ-tĭ) [L. *intensus,* tight]. Degree of strength, force, loudness, or activity.

intermediary (in-ter-me'dĭ-a-rĭ) [L. *inter,* between, + *mediāre,* to divide]. Something occurring between two time periods. In metabolism, the compounds formed as foodstuffs are utilized.

internuncial (in-tĕr- nun'sē-al) [" + *nuncius,* messenger]. A connector; between two other items, as neurons.

interphase. The "resting state" of a cell, when it is not dividing.

interstitial (in-ter-stish'al) [L. *interstitium,* thing standing between]. Lying between, as, interstitial fluid lies between vessels and cells or between cells.

in'tima [L. innermost]. The inner coat of a blood vessel.

intracellular (in-tra-sel'ū-lar) [L. *intra,* within, + *cellula,* cell]. Within cells.

intravenous (in-tră-ve'nus) [" + *vena,* vein]. Within or into a vein.

intrin'sic [L. *intrinsicus,* on the inside]. Located or originating within a structure.

in utero [L.]. Within the uterus.

inversion (in-ver'shun) [L. *in,* into, + *versiō,* a turning]. Turning, as upside down, or inside out; reversal.

in vitro [L.]. In glass, as in a test tube; *not* within the body.

in vivo [L.]. Within the body.

involuntary (in-vol'un-tĕr-i) [L. *involuntarius,* without act of will]. Occurring without an act of will.

involution (in-vō-lū'shun) [L. *in,* into, + *volvere,* to roll]. Change in a backward or diminishing direction.

i'on. [G. *iōn,* going]. An atom or group of atoms (radical) carrying an electric charge.

ipsilateral (ip-sĭ-lăt'er-al) [L. *ipse,* same, + *latus,* side]. On the same side; opposite of contralateral (*contra,* opposite).

irritability (ĭ-rĭ-tā-bĭl'ĭ-tē) [L. *irritare,* to tease]. Excitability, or the ability to respond to a stimulus.

ischemia (ĭs-kē'mĭ-ă) [G. *ischein,* to hold back, + *aima,* blood].

i'so- [G. *isos,* equal]. Prefix: equal to.

i'sograft [G. *isos,* equal, + *graphium,* grafting knife]. A graft or transplant from another animal of the same species. Also, *homograft.*

isometric (ī-sō-mē'trik) [G. *isos,* equal, + *metron,* measure]. Refers to no change of length, as an isometric muscle contraction.

isoton'ic [" + *tonos,* tension]. A solution having the same osmotic pressure as the blood or a reference solution; a muscular contraction in which shortening is allowed.

isthmus (ĭs-mus) [G. *isthmos*, narrow passage]. A narrow structure connecting two other structures.

itch [A. S. *giccan*, to itch]. An irritation of the skin, causing a desire to scratch. Also, *pruritus*.

-itis (ī'tĭs) [G.]. Suffix: inflammation of.

jaundice (jawn'dis) [M. E. *jaundis*, yellow]. A condition in which the skin is yellowed due to excessive bile pigment (bilirubin) in the blood.

joint [L. *junctura*, a joining]. An articulation or junction between bones or bone and cartilage.

jug'ular [L. *jugulum*, neck]. Large veins in the neck; pert. to the neck.

juice (jūs) [L. *jus*, broth]. Liquid secreted or expressed from an organism or any of its parts; esp., one of the digestive secretions.

ju'venile [L. *juvenilis*, young]. Pert. to youth or childhood; an immature white cell (metamyelocyte).

juxta- [L. near to]. Close or near to; in close proximity, as in juxtaglomerular.

karyoplasm (kar'ĭ-ōplăs-m) [G. *karyon*, nucleus, + *plasma*, a thing formed].

karyotype (kar'ĭ-ō-tīp) [" + *typos*, form]. A grouping or arrangement of the 46 human chromosomes based on the size of the individual chromosomes.

keratin (ker'ă-tĭn) [G. *keras*, horn]. A protein found in the superficial epidermal cells.

keto (kē'tō). Prefix denoting the presence in an organic compound of the carbonyl or "keto" group.

ke'tone. Any substance (e.g., acetone) having the carbonyl group in its molecule. Also, *acetone* or *ketone bodies*.

ketosis (kē-tō'sĭs) [ketone + G. *-osis*, disease]. Accumulation of ketone bodies in blood or body.

kilocalorie (kĭl-ō-căl-ŏ-rē) [G. *chilioi*, one thousand, + *calor*, heat]. One thousand small calories, or one large Calorie.

kil'ogram [" + *gramma*, a weight]. A metric measurement of weight; equals 1000 gm or 2.2 lb.

kinesthesia (kin-es-thē'zĭ-ă) [G. *kinesis*, motion, + *aisthēsis*, sensation]. The sensation concerned with appreciation of movement and body position.

kinetic (kĭ-net'ĭk) [G. *kinēsis*, motion]. Pert. to movement or work.

-kinin. Suffix denoting causing motion or action.

kyphosis (ki-fō'sis) [G. *kyphosis*, humpback]. An exaggeration of the normally anteriorly concave thoracic spinal curvature.

labium (lā'bĭ-um) [L. *labium*, lip]. A lip or liplike structure.

lacrimal (lak'rĭm-al) [L. *lacrima*, tears]. Pert. to the tears or any structure involved in producing or carrying tears.

lact- [L. *lac*, milk]. Prefix: milk.

lactation (lak-tā'shun) [L. *lactatio*, a suckling]. The function of secreting milk.

lactose (lak'tōse) [" + ose, sugar]. Milk sugar, a double sugar (disaccharide).

lacuna (lā-kū'nă) [L. *lacuna*, a pit]. A hollow space in the matrix of cartilage and bone in which lie characteristic cells.

lamella (lam-el'ă) [L. *lamella*, a little plate or leaf]. Circularly arranged layers, as in the elastic layers in a large artery; a ring of bony tissue around a Haversian canal.

la'tent [L. *latēre,* to be hidden]. Not active, quiet.

leg [M. E.]. Specifically, the part of the lower limb between the knee and ankle; Colloq.: a lower appendage.

lemniscus (lĕm-nis'kŭs) [G. *lemniskōs,* a filet]. A bundle of sensory nerve fibers in the brain stem.

lesion (lē'zhun) [L. *laesio,* a wound]. A wound, injury, or infected area.

leuco-, leuko- (lu'ko) [G. *leukos,* white]. Prefix: white.

lever (lĕv'ĕr, lē'vĕr) [M. E. *lever,* to raise]. A rigid elongated structure used to change direction, force, or movement.

ligament (lĭg'ă-mĕnt) [L. *ligamentum,* a band]. A strong band or sheet of fibrous connective tissue commonly used to connect bones together; a structure formed by degeneration of a fetal blood vessel.

lim'bic system [L. *limbus,* border]. A group of nervous structures in the cerebrum and diencephalon serving emotional expression.

liminal (lĭm'ĭ-nal) [L. *līmen,* threshold]. Threshold; just perceptible.

linear (lĭñ'ē-ar) [L. *linea,* line]. Pert. to a line or lines; extended in a line.

linkage (link'ĭj) [M. E. *linke,* akin to]. In genetics, when two or more genes on a chromosome tend to remain together during sex cell formation.

lipid (lĭp'id) [G. *lipos,* fat]. Fats or fatlike substances that are not soluble in water.

lipoprotein (lī-pō-prō'tēen) [" + *proteios,* major, of first importance]. A combination between a lipid and a simple protein.

liter (lē'tĕr) [G. *litra,* a pound]. Metric measurement of fluid volume. Equals 1000 milliliters or 1.06 quarts or 61 cubic inches.

-lith- [G. *lithos,* stone]. Stone; presence of stones, *lithiasis.*

lobe (lōb) [G. *lobos,* part or section]. A section or part of an organ separated by clear boundaries from other sections or parts.

localize (lō'kăl-īz) [L. *locus,* place]. To restrict to a small area.

locomotion (lō-kō-mō'shun) [L. *locus,* place, + *motus,* moving]. To move from one place to another; movement.

locus (lō'kŭs) ["]. A place; in genetics the location or position of a gene in a chromosome; a place in the brain or heart where abnormal electric discharge may originate.

logarithmic (lŏg-ă-rĭth'mĭk) [G. *logos,* a proportion, + *arithmos,* number]. Progression of a function of powers (e.g., square or cube) and not by individual numbers.

loin (loyn) [Fr. *loigne,* long part]. The sides and back of the trunk between ribs and pelvis.

lozenge (loz-ĕnj) [Fr. *lozenge,* diamond-shaped]. A solid medicine to be held in the mouth until dissolved.

lucid (lū'sĭd) [L. *lucidus,* clear]. Clear, as of thought, mind, or speech.

lumen (lū'mĕn) (pl. *lumina*) [L. *lumen,* light]. The cavity of any hollow organ; a unit of light.

luteo- (lū'tē-ō) [L. *luteus,* yellow]. Prefix: yellow.

lympho- (lĭm'fō) [L. *lympha,* lymph]. Prefix: lymph or lymphatic system.

-lysis (lī'sĭs) [G. *lysis,* dissolution]. To destroy or break down.

lysosome (lī'sō-som) [" + *soma,* body]. Cell organelle concerned with digestion of large molecules.

L-tubule. Longitudinally arranged tubules of the sarcoplasmic reticulum in muscle cells.

macro- (mak′rō) [G.]. Prefix: large or long.

mal- (mahl) [L. *malum*, an evil]. Ill, bad, or poor.

male (māl) [L. *masculus*, man]. The designation given to the sex that produces sperm, fertilizes ova, and begets offspring; a boy-child or man.

malignancy (mă-lĭg′năn-sĭ) [L. *malignus*, of bad kind]. A severe form of something; tending to grow worse.

malnutrition (mal-nū-trĭ′shun). Literally, "poor nutrition"; most often used to refer to absence of essential foodstuffs in the diet.

maltose (mawl′tōs) [A. S. *mealt*, grain]. Malt sugar, a double sugar (disaccharide) found in malt and seeds. $C_{12}H_{22}O_{11}$.

mammary (mam′ā-rĭ) [L. *mamma*, breast]. Pert. to breast.

mammillary (mam′ĭl-lar-ĭ) [L. *mammilla*, nipple]. Resembling a nipple.

mastication (măs-tĭ-kā′shŭn) [L. *masticāre*, to chew]. Chewing of food in the mouth.

matrix (mā′trĭks) [L. *matrix*, mother or womb]. Intercellular substance of cartilage; formative portion of a tooth or nail; the uterus.

mature (ma-tūr) [L. *maturus*, ripe]. Fully developed or ripened.

maximal (maks′ĭ-mal) [L. *maximum*, greatest]. Highest; greatest possible.

mean (mēn) [L. *medius*, in the middle]. The average (sum of numbers divided by the number of numbers).

meatus (mē-ā′tūs) [L. *meatus*, opening]. A passage or opening.

mediastinum (mē-dĭ-ăs-tĭ′nŭm) [L. in the middle]. A septum or cavity between two parts of an organ or between two other cavities.

medulla (mē-dul′lă) [L. marrow]. Inner or central portion of an organ; the medulla oblongata of the brain stem.

medullated (med′ū-lāt-ĕd). ["]. Having a myelin sheath.

mega- [G]. Large; one million.

meiosis (mī-ō′sĭs) [L. *mĕiosis*, to make smaller]. A form of cell division that reduces chromosome number to haploid. Occurs in formation of sex cells.

melanophores (mel-ăn′ō-fōr) [G. *melas*, black, + *phoros*, a bearer]. A cell carrying dark pigment(s).

membrane (mĕm′brān) [L. *membrana*, membrane]. A thin soft layer of tissue lining a tube or cavity, or covering or separating one part from another.

meninges (mĕn-ĭn′jēz) [G. *mĕnigx*, membrane]. The membranes around the central nervous system; *dura mater, arachnoid, pia mater*.

meniscus (men-ĭs′kus) [G. *mĕniskos*, crescent]. The crescent-shaped cartilages of the knee joint.

menstrual (men′strū-al) [L. *menstruāre*, to discharge menses]. Pert. to menstruation or the sloughing of the uterine endometrium.

merocrine (mer′ō-krĭn) [G. *meros*, a part, + *krinein*, to secrete]. A method of production of a secretion in which no part of the secreting cell enters the secretion.

mesentery (mes′en-ter-ĭ) [G. *mesos*, middle, + *enteron*, intestine]. A double-layered fold of peritoneum suspending the intestine from the posterior abdominal wall.

mesoderm (mĕs′ō-derm [" + *derma*, skin]. The middle germ layer of the embryo; gives rise to connective tissue, muscle, blood, and the cellular part of many organs.

meta- Prefix: after or beyond, later or more developed.

metabolism (mĕ-tăb'ŏl-ĭzm) [G. *metabolē*, change, + *ismos*, state of]. The sum total of all chemical reactions in the body; any product of metabolism is a *metabolite*.

metaphase (mĕt'ă-fāz) [" + *phasis*, a shining out]. A stage of mitosis in which chromosomes line up on the equator of the dividing cell.

metastasis (mĕ-tas'tă-sis) [" + *stasis*, a standing]. Movement from one part of the body to another (e.g., cancer cells).

micelle (mī-sĕl') [L. a little crumb]. A small complete unit, usually of colloids.

micro- [G.]. Small; one millionth.

micron (mi'kron) [G. *mikros*, small]. Metric unit of length; equals one one-thousandth of a millimeter. *u.*

microvillus (mī-kro-vĭl'ūs) [" + *villus*, tuft of hair]. Very small fingerlike extensions of a cell surface.

micturition (mĭk-tū-rĭ'shŭn) [L. *micturīre*, to urinate]. Voiding of urine; urination.

milliequivalent (mil-e-e-kwĭv'a-lent) [L. *mille*, thousandth, + *aequis*, equal, + *valere*, to be worth]. One one-thousandth of an equivalent weight. (An *equivalent* weight is the quantity, by weight, of a substance that will combine with one gram of H, or 8 grams of oxygen.) Abbr: **meq.**

millimeter (mil'ĭ-mēt-er) [" + *metron*, measure]. One one-thousandth of a meter. Abbr: **mm.** One **meter** equals 39.36 inches.

millimicron. One-thousandth of a micron. Abbr: **mμ.**

milliosmol. One one-thousandth of an osmol. (An *osmol* is the molecular weight of a substance in grams divided by the number of particles each molecule releases in solution.) Abbr: **mos.**

mimetic (mi-met'ĭk) [G. *mimētikos*, to imitate]. Imitating or causing the same effects as something else.

miotic (mī-ŏ'tĭk) [G. *meiōn*, less]. An agent causing pupillary contraction.

mitosis (mī-tō'sis) [G. *mitos*, thread]. A type of cell division that results in production of daughter cells like the parent.

mitral (mi'tral) [L. *mitra*, a miter, a bishop's hat]. Pert. to the bicuspid (mitral) valve between the left artrium and ventricle of the heart.

modality (mō-dal-ĭ-tē) [L. *modus*, mode]. The nature or type of a stimulus; a property of a stimulus distinguishing it from all other stimuli.

molar solution. A solution in which one gram molecular weight of a substance is present in each liter of the solution.

molecular weight. Weight of a molecule obtained by adding the weights of its constituent atoms. It carries no units. **m. w., gram.** The molecular weight of something expressed in grams. Also, a **mol(e).**

molecule (mŏl'ē-kūl) [L. *molecula*, little mass]. The smallest unit into which a substance may be reduced without loss of its characteristics.

monitor (mon'ĭ-tŭr) [L. *monere*, to warn]. To watch, check, or keep track of; one or something that watches.

monosaccharide (mŏn-ō-sak'ar-id) [G. *mono*, one, single, + *sakcharon*, sugar]. A simple sugar that cannot be further decomposed by hydrolysis.

morphology (mor-fol'ō-ji) [G. *morphē*, form, + *logos*, study]. The structure or form of something; study of the same.

morula (mor'ū-lā) [L. *morus*, mulberry]. A solid mass of cells resulting from division of the cells of a fertilized ovum.

motor (mō'tor) [L. *motus*, moving]. Refers to movement or those structures (e.g., nerves and muscles) that cause movement.

motor end plate. The specialized ending of a motor nerve on a skeletal muscle fiber.

motor unit. One motor nerve fiber (axon) and the skeletal muscle fibers it supplies.

mucous (mū'kŭs) [L. *mucus*, mucus]. A mucus-secreting structure.

mucus (mū'kŭs) [L.]. The thick sticky fluid secreted by a mucous cell or gland.

multi- [L.]. Prefix: many.

mutation (mū-tā'shŭn) [L. *mutatio*, to change]. A change or transformation of a gene; the evidence of a gene alteration.

mydriatic (mid-rĭ-at'ik) [G. *midriasis*, dilation]. An agent causing pupillary dilation.

myelin (mī'ĕ-lĭn) [G. *myelos*, marrow]. A fatty substance forming a sheath or covering aroud many nerve fibers. Speeds impulse conduction.

myelo- [G.]. Prefix: pert. to the spinal cord or to bone marrow.

myeloid (mī'el-oid) [" + *eidos*, form]. Formed in bone marrow; resembling marrow.

myoglobin (mī-ō-glō'bĭn) [G. *myo*, muscle, + L. *globus*, globe]. A respiratory pigment of muscle that binds oxygen.

myopia (mī-ō'pē-ă) [G. *myein*, to shut, + *ōps*, eye]. Nearsightedness. Also, *hypometropia*.

myosin (mī'ō-syn) [G. *myo*, muscle]. A muscle protein acting as an enzyme to aid in initiating muscle contraction.

myotatic (mī-ō-tă'tik) [" + *tactus*, touch]. Refers to muscle or kinesthetic sense (*kinesthesia*).

nasal (nā'zl) [L. *nasus*, nose]. Pert. to the nose.

NAD. Abbr. for *nicotinamide adenine dinucleotide,* one of several hydrogen acceptors (NADP. NAD phosphate, another hydrogen acceptor).

narcotic (nar-kŏt'ik) [G. *narkōtikos*, benumbing]. A drug that depresses the central nervous system, relieving pain and producing sleep.

nares (nar'ēz) [L. *naris*, nostril]. Nostrils.

natremia (na-trē'mĭ-ă) [L. *natrium*, sodium, + *aima*, blood]. Blood sodium.

necrosis (nĕk-rō'sis) [G. *nekrōsis*, a killing]. Death of tissue or cells.

negative pressure. A pressure less than atmospheric pressure, that is, <760 mm Hg.

neoplasm (nē'ō-plăzm) [G. *neo-*, new, recent, + *plasma*, a thing formed]. A new and abnormal formation of tissue, as a cancer, tumo₁, or growth.

nephron (nĕf'ron) [G. *nephros*, kidney]. The functional unit of the kidney that forms urine and regulates blood composition.

nerve (nŭrv) [L. *nervus*, sinew]. A bundle of nerve fibers outside the central nervous system.

neurilemma (nŭr-ĭ-lĕm'mă) [G. *neuron*, sinew, + *lemma*, rind]. A thin living membrane around some nerve fibers; aids in myelin formation and fiber regeneration.

neuroglial (nū-rŏg'lĭ-ăl) [" + *glia*, glue]. Pert. to the glial or nonnervous cells of the nervous system.

neuron (nū'rŏn) [G.]. The cell serving as the unit of structure and function of the nervous system; is excitable and conductile.

neurosis (nu-rō'sis [" + *osis*, disease]. A disorder of the mind in which contact with the real world is maintained.

neutralize (nū'tral-īz) [L. *neuter*, not, + *uter*, either]. To counteract, make inert, or destroy the properties of something.

newborn. A child under 6 weeks of age.

nigra (nī'grā) [L. *nigra*, black]. Black or blackness.

nitrogenous (nī-troj'ĕn-ŭs) [G. *nitron*, soda, + *gennan*, to produce]. Pert. to or containing nitrogen.

node (nōd) [L. *nodus*, knot]. A constricted region; a knob, protuberance or swelling; a small rounded organ.

nostril [A. S. *nosu*, nose, + *thyrl*, a hole]. External opening of the nasal cavity.

nuchal (nū'kal) [L. *nucha*, the back of the neck].

nuclear (nū'klē-ăr) [L. *nucleus*, a kernel]. Pert. to a nucleus (cell) or a central point.

nucleic acids (nū-klay'ik). Large molecules formed by nucleotides. They form the basis of heredity and protein synthesis. **DNA; RNA.**

nucleolus (nū-klē'ō-lŭs) [L. little kernel]. A spherical mass of nucleic acid within the nucleus. Stores and may synthesize **t-RNA..**

nucleotide (nū'klē-ō-tīd) ["]. A unit or compound formed of a nitrogenous base, a five-carbon sugar, and a phosphoric acid radical. Unit of structure of *DNA* and *RNA*.

nucleus (nū'klē-ŭs) ["]. The controlling center of a cell; a group of nerve cells in the brain; the heavy central atomic region in which mass and positive charge are concentrated.

nutrient (nū'trĭ-ent [L. *nutriens*, to nourish]. A food substance necessary for normal body functioning.

nutrition (nūtrī'shun) [L. *nutritiō*, a feeding]. The total of all processes involved in intake, processing, and utilization of foods.

nystagmus (nīs-tag'mŭs) [G. *nystazein*, to nod]. Involuntary cyclical movements of the eyeball.

obese (ō-bēs') [G. *obesus*, fat]. Excessively fat; overweight.

obligatory (ob-lĭg'ă-tō-rē) [L. *obligatorius*, to bind to]. Carries the idea of not having a choice; bound to or fixed in function.

oblique (ŏb-līk) [L. *oblique*, slanting]. A slanting direction or position.

obliterate (o-blĭt'er-āte) [L. *obliterate*, to blot out]. To erase, leaving no traces; extinction; occlusion of a part by surgical, degenerative, or disease processes.

occlude (ŏ-klŭd) [L. *occludere*, to shut up]. To close, obstruct or block something, such as a blood vessel.

occlusion (ŏ-klū'zhŭn) [L. *occlusiō*, a closing up]. The state of being closed, blocked, or obstructed.

ocular (ok'ū-lăr) [L. *oculus*, eye]. Pert. to eye or vision.

-oid [G.]. Suffix: like, similar to, resembling.

olfactory (ŏl-fak'tō-rĭ) [L. *olfacere*, to smell]. Pert. to smell or the sense of smell.

oligo- (ŏl'ĭ-gō) [L. *oligos*, little]. Prefix: small, few, scanty or little.

oliguria (ol-ĭg-ū'rĭ-ă) [" + *ouron*, urine]. Secretion of small amounts of urine.

-ology [G.]. Suffix: science of, study of, knowledge of.

omentum (ō-mĕn'tūm) [L. *omentum*, a covering]. A four-layered fold of peritoneum lying between the stomach and other abdominal viscera; a site of fat storage.

oncotic (ŏng-kŏt'ĭk) [G. *ogkos,* tumor]. Pert. to the colloid osmotic pressure created by the plasma proteins.

opaque (ō-pāk') [L. *opacus,* dark]. Not transparent; not allowing light to pass; dark.

operon (ŏp'er-ŏn) [L. *operatiō,* a working]. The combination of an operator gene and a structural gene.

operon concept. The current theory of how genes operate to control body function.

ophthalmic (ŏf-thăl'mĭk) [G. *ophthalmos,* eye]. Pert. to the eye.

-opsin (ŏp'sĭn) [G. *opson,* food]. Suffix: pert. to the protein component of the visual pigments.

oral (ō'răl) [L. *os, or-,* mouth]. Pert. to the mouth or the mouth cavity.

orbicular (ŏr-bĭk'ū-lăr) [L. *orbiculus,* a small circle]. Circular in arrangement.

orbit (or'bĭt) [L. *orbita,* track]. The cavity holding the eyeball; to go around.

orchido- [G. *orchis,* testicle]. Combining form meaning testicle.

organ (or'găn) [G. *organon,* organ]. Two or more tissues organized to a particular job.

organelle (or-găn-ĕl') [G. *organelle,* a small organ]. Submicroscopic formed structures within the cytoplasm that carry out particular functions.

organic (or-găn'ĭk) [G. *organon,* organ]. Pert. to organs; compounds containing carbon and hydrogen.

organic acids. An acid containing one or more carboxyl groups (-COOH).

organism (or'găn-ĭzm) [G. *organon,* organ, + *ismos,* condition]. A living thing.

orifice (or'ĭ-fĭs) [L. *orificium,* outlet]. The entrance or outlet to any chamber or hollow.

origin (or'ĭ-jĭn) [L. *origo,* beginning]. A source; the beginning of a nerve; the more fixed attachment of a muscle.

os (ōs) [L.]. Bone; mouth; opening.

oscillate (ŏs'ĭll-āt) [L. *oscillāre,* to swing]. To move back and forth; to swing.

-ose (ōs). Suffix: pert. to sugar.

-osis [G.]. Suffix: caused by; state of; disease or intensive.

osmo- [G.]. Combining form, pert. to smell; pert. to osmosis.

osmol (os'mōl). The molecular weight of a substance, in grams, divided by the number of particles it releases in solution.

osmolarity (os-mō-lar'ĭ-tĭ). The number of osmols per liter of solution.

osmoreceptor (ŏz-mō-rē-cĕp'tor). A receptor sensitive to osmotic pressure of a fluid.

osmosis (ŏz-mō'sĭs) [G. *osmōs,* a thrusting, + *-osis,* intensive]. The passage of solvent through a selective membrane.

osmotic pressure. The pressure developing when two solutions of different concentrations are separated by a membrane permeable to the solvent.

osseous (ōs'ē-ūs) [L. *osseus,* bony]. Bonelike, bony, or pert. to bones.

osteo- [G.]. Prefix: bone(s).

osteoid (ŏs'tē-oyd) [" + *eidos,* resembling]. A substance resembling bone.

osteon (ŏs'tē-on) [G. *osteon,* bone]. The unit of structure of compact bone. Also, *Haversian system.*

oto- [G.]. Combining form: the ear.

-otomy (ŏt'ō-mē) [G. *tomē,* incision]. Suffix: pert. to opening or repair of an organ without its removal.

output. The result or product of the operation of an organ or a machine.

ova (ō'vă) (sing. ovum) [L. *ovum,* egg]. Female reproductive cells; eggs.

ovale (ō'val'ē) [L. *ovum,* egg]. Egg-shaped or oval.

ovulate [L. *ŏvulum,* little egg]. Release of an egg from the ovary.

-oxia (ŏks'ĭ-ă). Suffix: Pert. to oxygen or oxygen concentration.

oxidation (ŏk'sĭ-dā'shŭn) [G. *oxys,* sour]. The combining of something with oxygen; loss of electrons.

oxidative phosphorylation. A metabolic scheme that transfers H ions and electrons to produce ATP and water.

oxygen (ŏk'sĭ-jĕn) [G. *oxys,* sharp, + *gennan,* to produce]. A colorless, ordorless, tasteless gas; forms more than 75 percent of organisms; symbol: **O.**

oxygen debt. A temporary shortage of oxygen necessary to combust products (lactic acid, pyruvic acid) of glucose catabolism.

oxyhemoglobin (ŏk-sĭ-hē-mō-glō'bĭn) ["]. The combination of oxygen and hemoglobin.

pacemaker (pās'māk-ĕr) [L. *passus,* a step, + A. S. *macian,* to make]. The sinoatrial (SA) node that sets the basic rate of heartbeat; artificial pacemaker is an electrical device substituting for the SA node.

palate (păl'ăt) [L. *palatum,* palate]. The roof of the mouth; composed of hard and soft portions.

palpate (păl'pāt) [L. *palpāre,* to touch]. To examine by feeling or touching.

papilla (pă-pĭl'ă) [L. *papilla,* a nipple]. A small elevation or nipplelike protuberance.

para- [G.]. Combining form: near, past, beyond, the opposites, abnormal, or irregular.

paralysis (pă-ral'ĭ-sĭs) [G. *paralyein,* to disable at the side]. Temporary or permanent loss of the ability of voluntary movement; loss of sensation.

parasympathetic (păr-ă-sĭm-pă-thĕt'ĭk) [G. *para,* beside, + *sympathētikos,* suffering with]. Pert. to the portion of the autonomic nervous system that controls normal body functions; the craniosacral division.

paresthesia (păr-ĕs-thē'zĭ-ă) [G. " + *aisthēsis,* sensation]. Abnormal sensation without demonstrable cause.

parietal (pă-rī'ĕ-tăl) [L. *paries,* wall]. Pert. to outer lining or covering, or wall of a cavity; a cell of the stomach secreting HCl.

paroxysm (păr'ŏk-sĭzm) [G. *para,* beside, + *oxynein,* to sharpen]. A sudden, periodic attack or recurrence of symptoms of a disease.

pars (parz) [L. *pars,* a part]. Part or portion.

passive (păs'ĭv) [L. *passivus,* enduring]. Not active; in immunity, the acquisition of antibodies by administration from the outside rather than producing them oneself.

patent (păt'ĕnt, pā'tĕnt) [L. *patens,* to be open]. Wide open; not closed.

pathological (păth-ō-lŏj'ĭk-l) [G. *pathos,* disease, + *logos,* study]. Diseased or abnormal.

PCO₂. Symbol for partial pressure of carbon dioxide.

peduncle (pēdung'kl) [L. *pedunculus,* a little foot]. A band of nerve fibers connecting parts of the brain.

peptide (pĕp'tīd) [G. *peptein,* to digest]. A compound containing two or more amino acids formed by cleavage of proteins.

perforate (pŭr'fō-rāt) [L. *perforāre*, to pierce through]. To make a hole through; puncture.

perfuse (pŭr-fūz) [L. *perfundere*, to pour through]. To pass a fluid through something.

peri- [G.] Prefix: around, or about.

peripheral (pĕr-ĭf'ĕr-ăl) [" + *pherein*, to bear]. Located away from the center; to the outside.

peristalsis (pĕr-ĭs-tăl'sĭs) [G. *perissos*, odd, + *stalsis*, contraction]. A progressive wavelike contraction of visceral muscle that propels liquids through hollow tubelike organs.

peritoneum (pĕr-ĭ-tō-nē'ŭm) [G. *peritonaion*]. The serous membrane lining the abdominal cavity and covering the abdominal organs.

permeable (per'me-ă-bl) [L. *per*, through, + *meare*, to pass]. Allowing passage of solutes and solvents in solutions.

pH. A symbol used to express the acidity or alkalinity of a solution. Strictly, the logarithm of the reciprocal of H ion concentration.

-phag- [G. *phagein*, to eat]. Combining form: an eater, or pert. to engulfing or ingestion.

phagocytosis (făg-ō-sī-tō'sĭs) [" + *kytos*, cell]. Engulfing of particles by cells.

-phil (fĭl). Combining form: love for; having an affinity for.

phospholipid (fŏs-fō-lĭp'ĭd) [G. *phōs*, light, + *pherein*, to carry, + *lipos*, fat]. A fatty substance combined with some form of phosphorus.

pia (pī'ă) [L. tender]. Tender or delicate; the inner meninge carrying blood vessels to cord and brain. Also, *pia mater*.

pigment (pĭg'mĕnt) [L. *pigmentum*, paint]. Any coloring substance.

pinocytosis (pi-nōsī-tō'sĭs) [G. *pinein*, to drink, + *kytos*, cell]. Intake of solution by cells through "sinking in" of the cell membrane.

pitch. That quality of a sound making it high or low in a scale; depends on frequency of vibration.

placenta (plă-sĕn'ta) [L. a flat cake]. The structure attached to the inner uterine wall through which the fetus gets its nourishment and excretes its wastes.

plane (plān) [L. *planus*, flat]. A flat surface formed by making a real or imaginary cut through the body or a portion of it.

-plasia (pla'zi-a) [G. *plasis*, a molding]. Combining form: Pert. to change or development. Also, *-plastic*.

-plasm (plăs'm) [G. *plasm*, a thing formed]. Combining form: Pert. to the fluid substance of a cell.

plasma (plăs'mă). The liquid portion of the blood.

plasma cell. One capable of producing antibodies in response to antigenic challenge. Also, *plasmocyte*.

plasma membrane. The cell membrane.

pleural (plū'răl) [G. *pleura*, a side]. Pert. to the membrane(s) lining the two lateral cavities of the thorax, or covering the lung; or, the cavities themselves that house the lungs.

plexus (plĕk'sūs) [L. a braid]. A network of nerves or vessels (blood or lymphatic).

plica (plī'kă) [L. a fold]. A fold.

-pnea (nē'ă). Suffix: pert. to breathing. Also, *pneo*.

pneumo- [G.]. Combining form: pert. to lungs or air.

pneumonia (nū-mō′nĭ-ă). Inflammation of the lungs.

PO₂. Symbol for partial pressure of oxygen.

-pod, -poda, -podo- [G.]. Combining forms referring to foot or feet.

poikilocytosis (poy-kĭl-ō-sī-tō′sĭs) [G. *poikilos*, various, + *kytos*, cell, + *osis*, intensive]. Variation in shape of red blood cells.

polarize (pō′lar-īz) [Fr. *polaire*, polar]. To create the state in which ions are in unequal concentrations on two sides of a membrane, with production of an electric potential across the membrane.

poly- [G.]. Prefix: many, much, or great.

polymer (pŏl′ĭ-mer) [G. *poly*, many, + *meros*, a part]. A substance formed by combining two or more molecules of the same substance.

pore (pōr) [G. *pŏros*, a passage or hole]. An opening of small size.

positive pressure. A pressure greater than atmospheric, that is, >760 mm Hg.

post- [L.]. Prefix: after; in back of.

postganglionic (pōst-găng-lĭ-ŏn′ĭk). Beyond or past a ganglion or synapse.

postmortem (pōst-mor′tĕm) [L.]. After death.

potency (pō′tĕn-sī) [L. *potentia*, power]. Strength; power; force.

potential (pō-tĕn′shăl) ["]. An "electrical pressure." Implies a measurable electric current flow or state between two areas of different electrical strength.

pre- [L.]. Prefix: in front of; before.

precipitate (prē-sĭp′ĭt-āt) [L. *praecipitāre*, to cast down]. Something usually insoluble which forms in a solution; to cause precipitation or a casting down of an insoluble mass.

precocious (prē-kō′shŭs) [L. *praecoquere*, to mature before]. Matured or developed earlier than normal.

precursor (pri-kŭr′sŭr) [L. *praecurrere*, to run ahead]. Anything preceding something else or giving rise to.

preganglionic (prēgăng-lĭ-ŏn-ik). In front of, or before a ganglion or synapse.

pregnancy (prĕg′năn-sī) [L. *praegnans*, with child]. The condition of carrying a child *in utero.*

premature (prē-mă-tūr′). Before term or full development.

primitive (prĭm′ĭ-tĭv) [L. *primitivus*, first]. Original; early in development; embryonic.

primordial (prī-mor′dĭ-ăl) [L. *primordium*, the beginning]. Existing first or in undeveloped form.

pro- [L; G]. Prefix: for, from, in favor of.

process (prŏs′ĕs) [L. *processus*, a going before]. A method of action; a projection from something.

prognosis (prŏg-nō′sĭs) [G. foreknowledge]. Prediction of the course and outcome of a disease or abnormal process.

projection (prō-jĕk′shŭn) [L. *pro*, forward, + *jacere*, to throw]. Throwing forward; the efferent connection of a part of the brain.

proliferate (prō-lĭf′ĕr-āt) [L. *proles*, offspring, + *ferre*, to bear]. To increase by reproduction of similar forms or types.

propagation (prŏp-ă-gā′shŭn) [L. *propagāre*, to fasten forward]. Carrying forward; act of reproducing or giving birth.

prophase (prō′fāz) [G. *pro*, before, + *phasis*, an appearence]. A stage in mitosis characterized by nuclear disorganization and formation of visible chromosomes.

proprioceptive (prō-prĭ-ō-sĕp'tĭv) [L. *proprius,* one's own, + *cepius,* to take]. Pert. to awareness of posture, movements, equilibrium and body position.

protein (prō'tē-in, prō'tēn) [G. *prōtos,* first]. A large molecule composed of many amino acids.

pseudo- (sū'dō) [G. *pseudēs,* false]. Prefix: false.

psychosis (sī-kō'sĭs) [G. *psyche,* mind]. A disorder of the mind in which contact with reality is lost.

puberty (pū'bĕr-tĭ) [L. *pubertās,* puberty]. The time of life when both sexes become functionally capable of reproduction.

pulmo- [L.]. Combining form: lung.

pulse (pūls) [L. *pulsāre,* to beat]. A throbbing caused in an artery by the shock wave resulting from ventricular contraction.

pupil (pū'pĭl) [L. *pupilla,* pupil]. The opening in the center of the iris of the eye.

pyelo- [G.]. Combining form: the pelvis.

pyramidal (pĭ-răm'ĭd-āl) [G. *pyramis,* pyramid]. Having the shape of a pyramid.
  p. cell. Characteristic cell of cerebral cortex.
  p. tract. The bundle of nerve fibers from the primary motor area to the cerebrum.

pyrogen (pī'rō-jĕn) [G. *pyr,* fire, + *gennan,* to produce]. An agent causing a rise of body temperature.

quad- (kwŏd) [L. *quadri,* four]. Prefix: four.

quality (kwŏl'ĭ-tĭ) [L. *qualitās,* quality]. The nature or characteristic(s) of something.

quantity (kwŏn'tĭ-tĭ) [L. *quantitās,* quantity]. Amount.

radial (rā'dĭ-āl) [L. *radius,* a spoke]. To pass outward from a specific center, like spokes of a wheel.

radiation (rā-dĭ-ā'shŭn) [L. *radiāre,* to emit rays]. Sending out heat or other forms of energy as electromagnetic waves; emission of rays from a center; to treat a disease by using a radioactive substance.

radical (răd'ĭ-kāl) [L. *radix,* root]. A group of several atoms that act as a unit in a chemical reaction.

ramus (rā'mŭs) (pl. rami) [L. *ramus,* a branch]. A branch of a vessel, nerve, or bone.

re- (rē) [L.]. Prefix: again; back.

receptor (rē-sĕp'tor) [L. a receiver]. A sense organ responding to a particular type of stimulus.

recessive (rĭ-sĕs'ĭv) [L. *recessus,* to go back]. To go back; in genetics, a characteristic that does not usually express itself because of suppression by a dominant gene (allele).

recipient (rē-sĭp'ĭ-ĕnt) [L. *recipiens,* receiving]. One who receives, as blood or a graft.

reciprocal (rē-sĭp'rō-kāl) [L. *reciprocus,* turning back and forth]. The opposite; interchangeable in nature.

reduction (rē-dŭk'shŭn) [L. *reductio,* a leading back]. Uptake of H by a compound, gain of electrons; restoring a broken bone to normal relationships.

referred (rē-ferd') [L. *re,* back, + *ferre,* to bear]. Sent to another area or place, as, referred pain.

reflex (rē′flĕks) [L. *reflexus,* bent back]. An involuntary, stereotyped response to a stimulus.

reflex arc. A series of nervous structures serving a reflex (receptor, afferent nerve, center, efferent nerve, effector).

refract (rē-frăkt′) [L. *refractus,* to break or bend]. To bend or deflect a light ray.

refractory (rē-frăk′tō-rĭ) [″]. Resistant to stimulation.

regeneration (rē-jĕn-ĕr-ā′shŭn) [″ + *generāre,* to beget]. Regrowth or repair of tissues or restoring a body part by regrowth.

regulate (rĕg′ū-lāt) [L. *regulatus,* to direct or control]. To control or govern.

regurgitate (rē-gur′jĭ-tāt) [″ + *gurgitāre,* to flood]. To return to a place just passed through; backflow.

relative to (rĕl′e-tĭv) [L. *relativus,* to bring back]. Considered in comparison to something else.

remnant (rĕm′nent) [″ + *manere,* to stay]. Something left over or behind; a fragment.

renal (rē′năl) [L. *renalis,* kidney]. Pert. to the kidney.

replicate (rĕp′lĭ-cāt) [L. *replicatio,* a folding back]. To duplicate.

repression (rē′prĕsh-ŭn) [L. *repressus,* to check]. To put down or prevent expression of something; in genetics, the way a gene shuts off its effect.

reproduction (rē-prō-dūk′shŭn) [L. *re,* again, + *productio,* production]. Creation of a similar structure or individual.

respiration (rĕs-pĭr-ā′shŭn) [L. *respirāre,* to breathe]. Cellular metabolism; the act of exchanging gases between the body and its environment.

response (rĭ-spons′) [L. *respondum,* to answer]. A reaction to a stimulus.

resuscitation (rē-sūs-ĭ-tā′shŭn) [L. *resuscitātus,* to revive]. Bringing back to consciousness.

reticular (rĕ-tĭk′ū-lar) [L. *reticula,* net]. In the form of a network, interlacing.

reticuloendothelial (re-tĭk′ū-lō-ĕn-dō-thē′lĭ-al) [″] Pert. to the tissues of the reticuloendothelial system, that is, the fixed phagocytic cells of the body and the plasma cells.

retina (rĕt′ĭ-nă) [L. *rētē,* a net]. The third and innermost coat of the eye; contains visual receptors (rods and cones).

retro- [L.]. Prefix: behind or backward.

rheobase (rē′ō-bās) [G. *rheos,* current, + *basis,* step]. The minimum strength of a stimulus required to produce a response.

rhin-, rhino- [G.]. Combining form: nose.

rhodopsin (rō-dŏp′sĭn) [G. *rhodon,* a rose, + *opsis,* vision]. A visual pigment found in rod cells; "visual purple."

rhythm (rĭth′ŭm) [G. *rhythmos,* measured motion]. A regular recurrence of action or function.

ribo- (rī′bō). Combining form: pert. to ribose, a five-carbon sugar (*pentose*).

rickets (rĭk′ĕts) [G. *rachitis,* spine]. A bone disease resulting from deficient or defective deposition of inorganic salts in new bone; due primarily to vitamin **D** deficiency.

RNA. Ribonucleic acid, composed of nitrogenous bases, ribose sugar, and phosphate. Three types: messenger-RNA, transfer-RNA, and ribosomal-RNA.

sac (săk) [G. *sakkos,* a bag]. Any bag or sacklike part of an organ.

saccharide (săk'ă-rīd) [G. *sakcharon,* sugar]. A sugar; a carbohydrate containing two or more simple sugar units.

sagittal (săj'ĭ-tăl) [L. *sagitta,* arrow]. Like an arrow or arrowhead; a suture in the skull; a plane dividing the body into right and left portions.

salt (sawlt) [A. S. *sealt*]. NaCl (sodium chloride); a chemical compound that results from replacing a hydrogen in the carboxyl group of an acid with a metal or cation.

saltatory (sal'tă-tō-rĭ) [L. *saltātio,* a leaping]. Dancing, skipping, or leaping.
  s. conduction. Conduction of a nerve impulse down a myelinated nerve by leaping from node to node.

sarcolemma (sar-kō-lĕm-ă) [G. *sarco,* flesh, + *lemma,* rind]. Cell membrane of a skeletal muscle fiber.

sarcoma (sar-kō'mă) [" + *ōma,* tumor]. A neoplasm of cancer arising from muscle, bone, or connective tissue.

sarcomere (sar'kō-mēr) [" + *meros,* part]. The portion of a myofibril lying between two **Z** lines.

sarcoplasm (sar'kō'plăzm) [" + *plasma,* a thing formed]. Muscle protoplasm.

saturated (săt'ū-ra-tēd) [L. *saturāre,* saturate]. Holding all it can. A saturated fat has all the hydrogen it can hold on its chemical bonds.

scheme (skēm) [L. *schema,* a plan]. An orderly series of events or changes, as, a metabolic scheme.

sciatic (sī-ăt'ĭk) [G. *ischiadikos,* pert, to the ischium]. Pert. to the hip or ischium.
  s. nerve. Largest nerve in the body. Runs from pelvis to knee.

sclera (sklē'ră) [G. *sklēros,* hard]. The outer coat of the eyeball; the "white of the eye."

sclerosis (sklē-rō'sĭs) [G. *sklerosis,* a hardening]. Hardening or toughening of a tissue or organ, usually by increase in fibrous tissue.

sebaceous (sē-bā'shŭs) [L. *sebaceus,* fatty]. Pert. to sebum, a fatty secretion of the sebaceous (oil) glands.

secrete (sē-krēt') [L. *secretus,* separated]. To separate; to make a product different from that originally presented as a starting material; active movement of substances into a hollow organ.

secretion. The product of secretory activity.

segmental (sĕg-mĕn'tăl) [L. *segmentum,* a portion]. Composed of segments, that is, individual parts or portions.

seizure (sē'zhŭr) [M. E. *seizen,* to take possession of]. A sudden attack of a disease, or pain.
  s. convulsive. Epilepsy.

selective (sē-lēk'tĭv). Exhibiting choice, as a selective membrane passes some materials and not others.

semi- [L.]. Prefix: half.

seminiferous (sĕm-ĭn-ĭf'ĕr-ŭs) [L. *sēmen,* seed, + *ferre,* to produce]. Pert. to production and transport of sperm and/or semen.

senescence (sĕn-es'ĕns) [L. *senescere,* to grow old]. Growing old; the period of old age.

sensation (sĕn-sā'shŭn) [L. *sensatio,* a feeling]. An awareness of a change of condition(s) inside or outside the body because of stimulation of a receptor (requires the brain to interpret the change).

sensitivity (sĕn-sĭ-tĭv'ĭ-tē) [L. *sensitivus*, feeling]. The quality of being sensitive or able to receive and transmit sensory impressions; affected by external conditions.

sensitize (sĕn'sĭ-tīz). To make sensitive to an antigen by repeated exposure to that antigen.

sensory (sĕn'sō-rī) [L. *sensorius*]. Pert. to sensation or the afferent nerve fibers from the periphery to the central nervous system.

septum (sĕp'tŭm) [L. *saeptum*, a partition]. A membranous wall separating two cavities.

serological (sĕ-rō-lŏj'ĭk-ăl) [L. *serum*, whey, + G. *logos*, study]. Pert. to study of serum.

serous (sĕ-rŭs). Pert. to a cell producing a secretion having a watery nature; a watery product.

serum (sĕ'rum). [L. *serum*, whey]. The watery portion of the blood remaining after coagulation.

sesamoid (ses'am-oyd) [G. *sēsamon*, sesame, + *eidos* form]. Like a sesame seed; specifically refers to bones that develop in tendons.

sex (sĕks) [L. *sexus*]. The quality that distinguishes between male and female.
  s. chromosomes. Chromosomes determining sex (XX, female; XY, male).
  s., nuclear. The genetic sex as determined by the presence or absence of sex chromatin *(Barr bodies)* in the nuclei of body cells.

shaft. The central portion of a long bone. Also, *diaphysis*.

sheath (shēth) [A. S. *scēath*]. A covering surrounding something.

shock (shōk) [M. E. *schokke*]. A state resulting from circulatory collapse (low blood pressure and weak heart action).

shunt (shŭnt) [M. E. *shunten*, to avoid]. A "shortcut"; passage between arteries and veins which by-passes the capillaries; a scheme for metabolism of a foodstuff that eliminates certain steps found in a scheme that metabolizes the same substance.

sickle cell. An erythrocyte that is crescent-shaped because it contains an abnormal hemoglobin.

sinus (sī'nŭs) [L. a hollow or curve]. A cavity in a bone, or a large channel for venous blood to flow in.

sinusoid (sī'nŭs-oyd) [" + G. *eidos*, like]. A small irregular blood vessel found in the liver and spleen.

site (sīt). Position, place, or location.

solubility (sŏl-ū-bĭl'ĭ-tī) [L. *solubilis*, to dissolve]. Capable of being dissolved or going into solution.

solute (sŏl'ūt) [L. *solutus*, dissolved]. The dissolved, suspended, or solid component of a solution.

solution (sō-lū'shŭn) [L. *solutio*, a dissolving]. A liquid containing dissolved substance.

solvent (sŏl'vĕnt) [L. *solvens*, to dissolve]. A dissolving medium.

soma (sō'mă) [G. *sōma*, body]. Pert. to the body, as a cell body, or the body as a whole.

somatic (sō-măt'ĭk) ["]. Pert. to the body exclusive of reproductive cells; pert. to skeletal muscles and/or skin.

somesthesia (som-es-thē'sĭ-ă) [" + *aisthēsis*, sensation]. The awareness of body sensations.

somite (sō'mīt) [''']. An embryonic blocklike segment of mesoderm formed alongside the neural tube.

spasm (spăzm) [G. *spasmos,* convulsion]. A sudden, involuntary, often painful contraction of a muscle.

spastic (spăs'tĭk) [G. *spastikos,* convulsive]. Contracted or in a state of continuous contraction.

spatial (spā'shăl). Pert. to space.

species (spē'shēz) [L. a kind]. A specific type, kind or grouping of something having distinguishing and unique characteristics.

specific gravity (spē-sĭf'ĭk grăv'ĭ-tē). Weight of a substance compared to that of an equal volume of pure water, with water assigned a value of 1.

sperm (spŭrm) [G. *sperma,* seed]. The male sex cells produced in the testes. Also, *spermatozoan; spermatozoa.*

sphincter (sfĭngk'ter) [G. *sphigktēr,* a binder]. A band of circularly arranged muscle that narrows an opening when it contracts.

spirometer (spī-rŏm'ĕt-ĕr) [L. *spirāre,* to breathe, + G. *metron,* measure]. An apparatus that measures air volumes of the lung.

spontaneous (spŏn-tā'nē-ŭs) [L. *spontaneus,* voluntary]. Occurring without apparent causes; activity seeming to occur "on its own."

squamous (skwā'mŭs) [L. *squama,* scale]. Flattened; scalelike.

stage (stāj) [L. *stāre,* to stand]. A period, interval, or degree in a process of development, growth, or change.

stagnant (stăg'nănt) [L. *stagnare,* to stand]. Without motion, not flowing or moving.

stasis (stā'sĭs) [G. *stasis,* halt]. Stoppage of fluid flow.

stenosis (stĕn-ō'sĭs) [G. *stenōsis,* a narrowing]. Narrowing of an opening or passage.

stereotyped (stĕr'ē-ō-typ'd) [G. *stereos,* solid, + *typos,* type]. Repeated, predictable response to stimulation.

steroid (stĕr'oyd). A lipid substance having as its chemical skeleton the phenanthrene nucleus.

stimulate (stĭm'ū-lāt) [L. *stimulare,* to goad on]. To increase an activity of an organ or structure.

stimulus (stĭm'ū-lŭs). An agent acting to cause a response by a living system.

strabismus (strā-bĭz'mŭs). When the eyes do not both focus at the same point. Crossed eyes or "squint."

stratified (străt'ĭ-fīd) [L. *steatificāre,* to arrange in layers].

stratum (străt'ŭm) [L. *stratum,* layer]. A layer, as of the epidermis of the skin.

stressor. An agent tending to upset homeostasis and to produce strain or stress on the organism.

striated (strī'āt-ĕd) [L. *stria,* channel]. Striped or streaked.

stridor (strī'dor) [L. a harsh sound]. A harsh sound occurring during breathing; like the sound of wind "whining."

stroke (strōk) [A. S. *strāk,* a going]. The total set of symptoms resulting from a cerebral vascular disorder. Also, *apoplexy.*

structural gene. A gene that directs the synthesis of specific protein.

sub- [L.]. Combining form: under, beneath, in small quantity, less than normal.

subconscious (sŭb-kŏn'shŭs) ['' + *conscius,* aware]. Operating at a level of which one is not aware.

subcutaneous (sŭb-kū-tā′nē-ŭs) [″ + *cutis,* skin]. Beneath the skin, as an injection. Abbr. **Sq.**

subjective (sŭb-jĕk′tĭv). Arising as a result of activity by a person; a *personal* reaction or conclusion.

subliminal (sŭb-lĭm′ĭn-ăl) [″ + *limen,* threshold]. Less than that required to get a response; not perceptible.

substrate (sŭb′strāt) [L. *substratum,* a strewing under]. The substance an enzyme acts on.

sudoriferous (sū-dor-ĭf′ĕr-ŭs) [L. *sudor,* sweat, + *ferre,* to bear]. Pert. to production or transport of sweat.

s. gland. Sweat gland.

sulcus (sŭl′kĭs, sŭl′sus) [L. a groove]. A slight depression or groove, esp. in the brain or spinal cord.

super-, (supra) [L.]. Combining form: above, beyond, superior.

superficial (sū-pĕr-fĭsh′ăl) [″ + *facies,* shape]. At or to a surface.

surface tension (alveolus). The phenomenon whereby liquid droplets tend to assume the smallest area for their volume. The surface acts as though it had a ″skin″ on it as a result of cohesion among the surface water molecules.

surfactant (sŭrf-ăk′tănt). A lipid-protein substance that reduces surface tension in the lung alveoli and thus reduces chances of alveolar collapse. Also, *surface active agent.*

suspension (sŭs-pĕn′shŭn) [L. *suspensio,* a hanging]. A state in which solute molecules are mixed but are not dissolved in a solvent and do not settle out.

sym-, syn [G.]. Combining form: with, along, together with, beside.

sympathetic (sĭm-pă-thĕt′ĭk) [G. *syn,* with, + *pathos,* suffering]. Pert. to the portion of the autonomic nervous system that controls response to stressful situations.

symphysis (sĭm′fĭs-ĭs) [G. *symphysis,* a growth together]. A type of joint in which the two bones are held together by fibrocartilage.

synapse (sĭn′ăps [G. *synapsis,* to touch together]. A point of junction between two neurons (functional continuity).

synchronous (sĭn′krŏn-ŭs) [″ + *chronos,* time]. Occurring at the same time.

syncytium (sĭn-sĭsh-yŭm) [″ + *kytos,* cell]. Cells running together anatomically or functionally, and behaving as a single multinucleated mass.

syndrome (sĭn-drōm) [G. *syndromē,* a running together]. A complex or group of symptoms.

syneresis (sĭn-ĕr-ē′sĭs) [″ + *airesis,* a taking]. Shrinking of a gel, as a clot.

synergist (sĭn′ĕr-jĭst) [″ + *ergon,* work]. A muscle working with other muscles to ″firm″ an action.

synonym (sĭn′ō-nĭm) [″ + *onoma,* name]. Names used to refer to the same process, characteristic, or structure; an additional or substitute name.

synovial (sĭn-ō′vĭ-ăl) [″ + L. *ovum,* egg]. Pert. to the cavity or fluid in the space between bones of a freely movable joint.

synthesis (sĭn′thĕ-sĭs) [″ + *tithenai,* to place]. To form new or more complex substances from simple precursors.

syphilis (sĭf′ĭ-lĭs) [Origin uncertain; G. *syn,* with, + *philos,* love, or from *Syphilus,* shepherd who had the disease]. An infectious, chronic venereal disease transmitted by physical contact with an infected person.

system (sĭs'tĕm) [G. *systema*, an arrangement]. An organized group of related structures.

systemic (sĭs-tĕm'ĭk). Pert. to the whole or the greater part of the body.

systole (sĭs'tō-lē) [G. *systolē*, contraction]. Contraction of the muscle of a heart chamber.

tachy- [G.]. Combining form: rapid, fast.

tamponade (tăm-pōn-ād') [Fr. *tampon*, plug]. To plug up.

  t., cardiac. A condition resulting from accumulation of excess fluid in the pericardial sac.

taxis (tăk'sĭs) [G. arrangement]. Response to an environmental change; a turning toward (positive) or away from (negative) the change.

telodendria (tĕl-ō-dĕn'drĭ-ă) [G. *telos*, end, + *dendron*, tree]. The treelike branching ends of an axon or its branches.

telophase (tĕl'ō-fāz) [" + *phasis*, a phase]. The final stage of mitosis characterized by cytoplasmic division and reformation of nuclei.

temperature (tĕm'per-ă-tūr) [L. *temperatura*, proportion]. The degree(s) of heat of a living body.

temporal (tĕm'por-ăl) [L. *temporalis*, pert. to time]. Pert. to time.

tension (tĕn'shŭn) [L. *tensio*, a stretching]. The state of being stretched; a concentration of a gas in a fluid; the force developed by a muscle contraction as measured by a gauge.

terminal (ter'mĭn-al) [L. *terminus*, a boundary]. End or placed at the end; a disease that will end with death.

tetanus (tĕt'ă-nŭs) [G. *tetanos*, tetanus]. A sustained contraction of a muscle; a disease caused by a bacterium characterized by sustained contraction of jaw muscles (*"lockjaw"*).

tetany. Muscular spasm caused by deficient parathyroid hormone. Also, *hypoparathyroidism.*

therapy (thĕr'ă-pī) [G. *therapein*, treatment]. The treatment of a condition or disease.

thermal (ther'măl) [G. *thermē*, heat]. Pert. to heat.

thorax (thō'răks) [G.]. The chest.

threshold (thrĕsh'ōld) [A. S. *therscwold*]. Just perceptible; the lowest strength stimulus that results in a detectable response or reaction. Also, liminal.

thrombus (thrŏm'bŭs) [G.]. A blood clot obstructing a vessel.

thrombosis. The formation of the clot, or the result of blocking the vessel.

tissue (tĭsh'ū) [L. *texere*, to weave]. A group of cells that are similar in structure and function (e.g., epithelial, connective, muscular, and nervous tissues).

-tome (tōm) [G.]. Combining form: a cutting, or a cutting instrument.

tone (tōn) [L. *tonus*, a stretching]. A state of slight constant tension or contraction exhibited by muscular tissue.

tonic (tŏn'ĭk) [G. *tonikos*, pert. to tone]. Having tone; continual.

tonicity (tō-nĭs'ĭ-tĭ) [G. *tonos*, tone]. The property of having tone; a reference to the osmotic pressure of a solution.

tortuous (tŏr'tū-ŭs) [L. *tortuosus*, twisted]. Twisted and turned; full of windings.

toxic (tŏks'ĭk) [G. *toxikon*, poison]. Poisonous.

toxin (tŏks'ĭn) ["]. A poisonous substance derived from plant or animal sources.

  t., bacterial. Toxin produced by bacteria.

toxoid (tŏks'oyd) [" + *eidos,* form]. A toxin treated so as to lower or decrease its toxicity, but which will still cause antibody production.

trace (trās) [L. *tractus,* a drawing]. A very small quantity.

t. element. Metals or organic substances essential in minute amounts for normal body function.

tract (trăkt) [L. *tractus,* a track]. A bundle of nerve fibers in the central nervous system that carries particular kinds of motor or sensory impulses.

trans- (trăns) [L.]. Prefix: across, through, over, or beyond.

transamination (trăns-ăm-ĭ-nā'shŭn). Transfer of an amino (-NH₂) group from one compound to another without formation of ammonia.

transcellular (trăns-cĕl'ū-lăr). Literally, through cells. Used to describe those body fluids separated from other fluids by an epithelial membrane (e.g., fluids in stomach, intestine).

transcription (trăns-krĭp'shŭn). The process in which DNA gives rise to RNA.

transduce (trăns-dūs) [L. *trans,* across, + *ducere,* to lead]. To change one form of energy into another.

transferrin. The plasma protein (a beta-globulin) that binds iron and transports it.

transient (trăn'shĕnt) [L. *trans,* over, + *ire,* to go]. Temporary, not permanent.

transition (trăns-ĭ'shŭn) [L. *transitio,* a going across]. Changing from one state or position to another.

translation (trăns-lā'shŭn) [L. *translatio,* to translate]. The process by which the code in messenger RNA is utilized to synthesize a specific protein.

transparent (trăns-păr'ĕnt) [L. *trans,* across, + *parere,* to appear]. Allowing light through to permit a clear view through a substance.

transplantation (trăns-plăn-tā'shŭn) [" + *plantāre,* to plant]. To take living tissue from one person or species and place it elsewhere on that person's body, or in another person or species.

transport (trăns-pŏrt') [" + *pōrare,* to carry]. To transfer or carry across.

transverse (trăns-vĕrs) [L. *transversus,* turned across]. Crosswise.

trauma (traw'mă) [G. *trauma,* wound]. An injury or wound.

tremor (trĕm'or) [L. *tremor,* a shaking]. A quivering or involuntary rhythmical movement of a body part.

treppe (trĕp'eh) [Ger. *treppe,* staircase]. An increase in strength of muscular contraction when a muscle is stimulated maximally and repeatedly.

tri- [G.]. Combining form: three.

trophic (trō'fĭk) [G. *trophē,* nourishment]. Literally, nourishing. Used to refer to those hormones that control activity of other endocrine structures (e.g., ACTH, TSH, LH, FSH, ICSH).

-tropic (trō'pĭk) [G. *tropos,* turning]. Combining form: turning, changing, responding to stimulus.

T-tubule. Transversely arranged microscopic tubules that convey ECF to the interior of muscle cells.

tumor (tū'mor) [L. a swelling]. A swelling or enlargement. Also, neoplasm.

tunic (tū'nĭk) [L. *tunica,* a sheath]. A coat, covering, or layer.

turbinate (tŭr'bĭn-āt) [L. *turbo,* a whirl]. Nasal concha.

turbulent (tŭr'bū-lĕnt) [L. *turbulentus,* to trouble]. Disturbed or not "smooth." Used to describe flow of blood in the circulatory system.

twitch (twĭch) [M. E. *twicchen*]. A single muscular contraction in response to a single stimulus.

ulcer (ŭl'ser) [L. *ulcus*, ulcer]. An open sore or lesion in the skin or a mucous membrane lined organ (e.g., stomach, mouth, intestine, etc).

umbilical (ŭm-bĭl'ĭ-kăl) [L. *umbilicus*, navel]. Pert. to the navel.

   u. cord. The structure connecting the fetus to the placenta.

un- [A. S.]. Prefix: not, back, reversal.

unconscious (ŭn-kŏn'shŭs) [A. S. *un*, not, + L. *conscius*, conscious]. Insensible, not aware of surroundings.

uni- [L.]. Combining form: one.

unit (ū'nĭt) [L. *unus*, one]. One of anything; a single distinct object or part.

urea (ū-rē'ă) [G. *ouron*, urine]. A waste of metabolism formed in the liver from NH_3 and CO_2. $CO(NH_2)_2$.

uremia (ū-rē'mĭ-ă) [" + *aima*, blood]. A toxic disorder associated with elevated blood urea levels. A result of poor kidney function.

-uria (ūr'ĭ-ă) Suffix: pert. to urine.

uric acid (ū'rĭk) ["]. An end product of purine (a nitrogenous base) metabolism found in urine. $C_5H_4N_4O_3$.

urine (ū'rĭn) [L. *urina*, urine]. The fluid formed by the kidney and eliminated from the body via the urethra.

uro- [G.]. Combining form: pert. to urine.

vaccinate (văk'sĭn-āt) [L. *vaccinus*, pert. to a cow]. To inoculate with a vaccine to produce immunity against a disease.

vaccine (văk'sēn) [L. *vacca*, a cow]. Killed or attenuated bacteria or viruses prepared in a suspension for inoculation.

vacuole (văk'ū-ōl) [L. *vacuolum*, a tiny empty space]. A fluid- or air-filled space in a cell.

vagus (vā'gŭs) [L. wandering]. The tenth cranial nerve.

valve (vălv) [L. *valva*, a fold]. A structure for temporarily closing an opening so as to achieve one-way fluid flow.

varicose (văr'ĭ-kōs) [L. *varix*, a twisted vein]. Pert. to distended, knotted veins.

vascular (văs'kū-lăr) [L. *vasculum*, a small vessel]. Pert. to or having blood vessels.

vasomotor (văs-ō-mō'tor) [L. *vaso*, vessel, + *motor*, a mover]. Pert. to nerves controlling activity of smooth muscle in blood vessel walls. May cause vasodilation (enlargement) or vasoconstriction (narrowing) of the vessel.

vein (vān) [L. *vena*]. A vessel carrying blood toward the heart or away from the tissues.

venereal (vē-nē'rē-ăl) [L. *venerus*, from *Venus*, goddess of love]. Pert. to or resulting from sexual intercourse.

ventilation (vĕn-tĭl-ā'shŭn) [L. *ventilāre*, to air]. To circulate air, esp. into and out of the lungs.

   v., alveolar. Supplying air to alveoli.

   v., pulmonary. Supplying air to the conducting division of the respiratory system.

ventral (vĕn'trăl) [L. *venter*, the belly]. Pert. to belly or front side of the body.

ventricle (vĕn'trĭk-l) [L. *ventriculus*, a little belly]. One of the two lower chambers of the heart.

venule (vĕn'ūl) [L. *venula*, a little vein]. A small vein; gather blood from capillary beds.

vermis (vĕr′mĭs) [L. *vermis*, worm]. The middle lobe of the cerebellum. *Vermiform*, wormlike, as vermiform appendix.

vesicle (vĕs′ĭ-kl) [L. *vesicula*, a little bladder]. A small, fluid-filled sac.

vestibular (vĕs-tĭb′ū-lăr) [L. *vestibulum*, vestibule]. Used to refer to the equilibrium structures of the inner ear and their nerves; literally, pert. to a small space or cavity.

villus (vĭl′ŭs) [L. *villus*, tuft of hair]. Small fingerlike vascular projections from or of certain mucous membranes (e.g., intestine).

virus (vī′rŭs) [L. *virus*, poison]. An ultramicroscopic infectious organism that requires living tissue to survive and reproduce.

viscera (vĭs′ĕr-ă) [L. *viscus*, viscus]. The internal body organs, esp. those of the abdominal cavity.

visceral. Pert. to the viscera or an outer lining of an organ.

viscous (vĭs′kŭs) [L. *viscosus*, sticky]. Sticky or of thick consistency.

visual (vĭzh′ū-ăl) [L. *visio*, a seeing]. Pert. to vision or sight.

vital (vī′tăl) [L. *vitalis*, pert. to life]. Essential to or contributing to life.

vitamin (vī′tă-mĭn) [L. *vita*, life, + *amine*]. An essential organic substance not serving as a source of body energy, but working with enzymes to control body function.

vitreous (vĭt′rē-ŭs) [L. *vitreus*, glassy]. Glassy; the vitreous body (humor) of the eyeball found between the lens and retina.

voluntary (vŏl′ŭn-tā-rĭ) [L. *voluntas*, will]. Pert. to being under willful control.

Wallerian degeneration [A. V. *Waller*, Eng. phys.-physiologist, 1816-1870]. Describes the degenerative changes occurring in an axon that has been severed from its cell body.

waste (wāst) [L. *vastāre*, to devastate]. Useless end products of body activity; to shrink or become smaller in size and/or strength.

white matter (hwīt măt′tĕr). That part of the central nervous system composed primarily of myelinated nerve fibers.

Willis, circle of [Thos. *Willis*, Eng. phys., 1621-1675]. An arterial circle on the base of the brain composed of ant. cerebral, ant. communicating, internal carotid, post. communicating, and post. cerebral arteries.

xeno (zē′nō) [G. *xeno*, foreign]. Combining form: strange, foreign.

xenograft (ze′nō-grăft) [" + L. *graphium*, grafting knife]. A graft of tissues or organs between different species.

XX. Complement of sex chromosomes giving rise to a female.

XY. Complement of sex chromosomes giving rise to a male.

yolk sac (yōk săk). A membranous sac surrounding the yolk in an embryo that contains yolk; in the human, a transitory structure incorporated into the gut.

Z line. An intermediate line in the striation of skeletal muscle; lies within the **I** line, and serves as the limits of a sarcomere.

zygote (zī′gōt) [G. *zygotōs*, yoked]. A cell produced by the union of an egg and sperm; a fertilized ovum.

Index